....*with*   *of the*
  A   ny
  Cr   ica
  S   ca

American Society of Agronomy, Inc.
Crop Science Society of America, Inc.
Soil Science Society of America, Inc.
5585 Guilford Road, Madison, WI 53711-1086 USA

Agronomy Monograph Series
ISSN 0065-4663 (print)
ISSN 2156-3276 (online)
Available online at https://www.agronomy.org/publications/browse

ISBN: 978-0-89118-175-0 (print)
Library of Congress Control Number: 2010911235

Cover design: Patricia Scullion
Cover photo: Istock

Printed on recycled paper.
Printed in the United States of America.

Urban Ecosystem Ecology
Jacqueline Aitkenhead-Peterson and Astrid Volder, Editors

Book and Multimedia Publishing Committee
David Baltensperger, Chair
Warren Dick, ASA Editor-in-Chief
E. Charles Brummer, CSSA Editor-in-Chief
Sally Logsdon, SSSA Editor-in-Chief
Mary Savin, ASA Representative
Hari Krishnan, CSSA Representative
April Ulery, SSSA Representative

Managing Editor: Lisa Al-Amoodi

Agronomy Monograph 55

American Society of Agronomy

Crop Science Society of America

Soil Science Society of America

Contents 📊

Foreword

Humans significantly affect their ecosystem. Nowhere is this more evident than in our towns and cities around the world. We have options when it comes to the sustainability of our cities. We have to pay attention to meeting the needs of today without compromising the needs of the future. However, to truly understand those options, we need to have the best knowledge available so that we can make the best decisions possible. It has everything to do with the stewardship of our natural resources, while at the same time using those resources to meet the needs of today's urban population.

This volume deals with the impacts of urbanization on the environment—soils, air, and water quality, the animals and humans who live there—and offers some solutions to contemporary problems. Its coverage exemplifies the wide range of specialties in the agronomic sciences and shows how scientists define ecosystem functions and solve problems. To read this book will give instruction in how agronomy, an art and science, serves society in the urban landscape.

The authors of the publication have summarized hundreds of research articles dealing with urban ecosystems. Students, city planners, and scientists will find this book to be a useful reference. Appreciation is expressed to the editors, chapter authors, the book publishing committee, and reviewers for the production of a truly outstanding book.

F.J. Pierce
President American Society of Agronomy

J.G. Lauer
President Crop Science Society of America

N.B. Comerford
President Soil Science Society of America

Preface 📊

Living organisms interact with every other biotic and abiotic element in their local environment, and this is central to the concept of an ecosystem. The founder of the term *ecology*, Eugene Odum, stated: "Any unit that includes all of the organisms (i.e., the "community") in a given area interacting with the physical environment so that a flow of energy leads to clearly defined trophic structure, biotic diversity, and material cycles (i.e., exchange of materials between living and nonliving parts) within the system is an ecosystem." Urban ecosystems are modified from the initial ecosystem that was originally present at that specific spatial location. For example, under the Bailey Ecosystem Classification, New York was an Eastern Broadleaf Forest (Oceanic) Province, Las Vegas was an Intermountain Semi-desert and Desert Province, and Dallas was a Southwest Plateau and Plains Dry Steppe and Shrub Province. In an urban ecosystem the physical, chemical and biological components are altered relative to the original ecosystem present. The urban physical environment has an affect on clouds, precipitation and lightning (Shepherd et al., Chapter 1), on temperature (Heisler and Brazel, Chapter 2) and on air quality (Santosa, Chapter 3). Habitats and food sources differ for wildlife (Shochat et al., Chapter 4; McCleery, Chapter 5; Ehrenfeld and Stander, Chapter 6). Vegetation in the urban ecosystem is very different from the vegetation that was once present prior to urbanization and includes turfgrass species (Cook and Ervin, Chapter 8), horticultural and nonnative forest species (Volder, Chapter 9; Volder and Watson, Chapter 11), agriculture in the form of allotments and community gardens (Iaquinta and Drescher, Chapter 10), and novel and invasive species of plant life (Reichard, Chapter 12). The physical, chemical, and biological components of the soil on which this vegetation grows is also altered (Pouyat et al., Chapter 7), and the function of riparian areas may be drastically changed (Stander and Ehrenfeld, Chapter 13). The hydrology and surface water biogeochemistry are different from those found in natural ecosystems, primarily due to larger proportions of impervious surfaces and waste inputs into the watershed (Burian and Pomeroy, Chapter 14; Steele et al., Chapter 15). These changes in surface water volume and chemistry impact surface water ecology in watersheds that have a high proportion of urbanization (Roy et al., Chapter 16). Use of water in urban ecosystems is described by Jenerette and Alstad (Chapter 17) in context of its complexity, cost, and the service of urban ecohydrology. The final section of the book examines the role of the services an urban ecosystem provides (Aitkenhead-Peterson et al., Chapter 18) and sustainable development in urban ecosystems through roof gardens (Rowe and Getty, Chapter 19), bioretention and low impact development (Li et al., Chapter 20), and the provision of green spaces in planning (Beer, Chapter 21). Finally, as editors we would like to sincerely thank all of our contributors who made this book possible.

Jacqueline Aitkenhead-Peterson and Astrid Volder, Editors
Texas A&M University, College Station, TX

Contributors

Aitkenhead-Peterson, J.A.   Dep. of Soil and Crop Sciences, Texas A&M Univ., 2474 TAMU, College Station, TX 77843 (jpeterson@ag.tamu.edu)

Alstad, K.P.   Dep. of Botany and Plant Sciences, Univ. of California, Riverside, CA 92571 (karrin.alstad@ucr.edu)

Beer, A.R.   Professor Emeritus, Dep. of Landscape, Univ. of Sheffield, Sheffield, UK (anne.beer@btinternet.com)

Brazel, A.J.   School of Geographical Sciences and Urban Planning, Arizona State Univ., Coor Hall, 975 S. Myrtle, Tempe, AZ 85287-5302 (abrazel@asu.edu)

Burian, S.J.   Dep. of Civil and Environmental Engineering, The Univ. of Utah, 122 S. Central Campus Dr., Salt Lake City, UT 84112 (burian@eng.utah.edu)

Cook, T.W.   Dep. of Horticulture, Oregon State Univ., 4017 Ag. and Life Sciences Bldg., Corvallis, OR 97331-7304 (cookt@hort.oregonstate.edu)

Drescher, A.W.   Inst. of Physical Geography, Albert-Ludwigs-Univ., Werthmannstr. 4, D-79085 Freiburg, Germany (Axel.Drescher@sonne.uni-freiburg.de)

Dvorak, B.D.   Dep. of Landscape Architecture and Urban Planning, Texas A&M Univ., 3731 TAMU, College Station, TX 77843-3137 (bdvorak@tamu.edu)

Ehrenfeld, J.G.   Dep. of Ecology, Evolution, and Natural Resources, Rutgers Univ., 14 College Farm Rd., New Brunswick, NJ 08901 (ehrenfel@rci.rutgers.edu)

Ervin, E.H.   Dep. of Crop and Soil Environmental Sciences, Virginia Polytechnic Inst. and State Univ., 330 Smyth Hall, Blacksburg, VA 24061-0404 (eervin@vt.edu)

Fernández-Juricic, E.   Dep. of Biological Sciences, Purdue Univ., G-420 Lily Hall, 915 W. State St., West Lafayette, IN 47907 (efernan@purdue.edu)

Getter, K.L.   Dep. of Horticulture, Michigan State Univ., East Lansing, MI 48824 (smithkri@msu.edu)

Groffman, P.M.   Cary Inst. of Ecosystem Studies, Millbrook, NY 12545 (groffmanp@caryinstitute.org)

Heisler, G.M.   USDA Forest Service, Northern Research Station, SUNY ESF, 5 Moon Library, Syracuse, NY 13210 (gheisler@fs.fed.us)

Iaquinta, D.L.   Dep. of Sociology, Anthropology, and Social Work, Nebraska Wesleyan Univ., 5000 Saint Paul Ave., Lincoln, NE 68504-2794 (dli@NebrWesleyan.edu)

Jenerette, G.D.   Dep. of Botany and Plant Sciences, Univ. of California, Riverside, CA 92571 (Darrel.Jenerette@ucr.edu)

Jin, M.L.   Dep. of Meteorology, San Jose State Univ., Duncan Hall 620, San Jose, CA 95192 (jin@met.sjsu.edu)

Lerman, S.   Dep. of Organismic and Evolutionary Biology, Univ. of Massachusetts, 319 Morrill Science Ctr. South, 611 N. Pleasant Street, Amherst, MA 01003 (slerman@nsm.umass.edu)

Li, M.-H.   Dep. of Landscape Architecture, Texas A&M Univ., College Station, TX 77843-3137 (minghan@tamu.edu)

McCleery, R.   Western Illinois Univ.. Dep. of Biological Sciences. 1 Univ. Cir-
                       cle, Macomb, IL 61455 (bmcc@neo.tamu.edu)

McDowell, W.H.   Dep. of Natural Resources, James Hall, Univ. of New Hamp-
                       shire, Durham, NH 03824 (bill.mcdowell@unh.edu)

Mote, T.L.   Dep. of Geography, Univ. of Georgia, Rm. 213 Geog-Geol Bldg.,
                       204 Field St., Athens, GA 30602-2502 (tmote@uga.edu)

Paul, M.J.   Tetra Tech, Inc., Ctr. for Ecological Sci., 400 Red Brook Blvd., Ste.
                       200, Owings Mills, MD 21117 (Michael.Paul@tetratech.com)

Pomeroy, C.A.   Dep. of Civil and Environmental Engineering, The Univ. of
                       Utah, 122 S. Central Campus Dr., Salt Lake City, UT 84112
                       (christine.pomeroy@utah.edu)

Pouyat, R.V.   U.S. Forest Service, North Research Station, c/o Baltimore Eco-
                       system Study, 5200 Westland Blvd., Baltimore, MD 21227
                       (rpouyat@fs.fed.us)

Reichard, S.H.   School of Forest Resources, Univ. of Washington, Box 354115,
                       Seattle, WA 98195 (reichard@u.washington.edu)

Rowe, D.B.   Dep. of Horticulture, Michigan State Univ., East Lansing, MI
                       48824 (rowed@msu.edu)

Roy, A.H.   USEPA, National Risk Management Research Lab., 26 West
                       Martin Luther King Dr., Cincinnati, OH 45268; currently
                       at Dep. of Biology, Kutztown Univ., Kutztown, PA 19530
                       (roy@kutztown.edu)

Santosa, S.J.   Dep. of Chemistry, Universitas Gadjah Mada, Sekip Utara Kotak
                       Pos Bls. 21, Yogyakarta, Indonesia (sjuari@yahoo.com)

Schwarz, K.   Dep. of Ecology, Evolution and Natural Resources, Rutgers
                       Univ., New Brunswick, NJ 08901; currently at Cary Inst.
                       of Ecosystem Studies, Box AB, Millbrook, NY 12545-0129
                       (schwarzk@caryinstitute.org)

Shepherd, J.M.   Dep. of Geography, Univ. of Georgia, Athens, GA 30602
                       (marshgeo@uga.edu)

Shochat, E.   Global Inst. of Sustainability, Arizona State Univ., Box 875411,
                       Tempe, AZ 85287 (eyal.shochat@asu.edu)

Stallins, J.A.   Dep. of Geography, Florida State Univ., Rm. 323 Bellamy Bldg.,
                       Tallahassee, FL 32306-2190 (jastallins@fsu.edu)

Stander, E.K.   USEPA, Urban Watershed Management Branch, 2890 Woodbridge
                       Ave., MS-104, Edison, NJ 08837 (stander.emilie@epa.gov)

Steele, M.K.   Dep. of Soil and Crop Sciences, Texas A&M Univ., 2474 TAMU,
                       College Station, TX 77843 (mkbilek@gmail.com)

Sung, C.Y.   Texas Transportation Inst., Texas A&M Univ., College Station,
                       TX 77843-3135 (cysung@tamu.edu)

Szlavecz, K.   Dep. of Earth and Planetary Sciences, The Johns Hopkins Univ.,
                       3400 N. Charles St., Baltimore, MD 21218 (szlavecz@jhu.edu)

Volder, A.   Dep. of Horticultural Sciences, Texas A&M Univ., TAMU 2133,
                       College Station, TX 77843-2133 (a-volder@tamu.edu)

Watson, W.T.   Dep. of Ecosystem Science and Management, Texas A&M
                       Univ., TAMU 2138, College Station, TX 77843-2138 (t-wat-
                       son@tamu.edu)

Wenger, S.J.   Univ. of Georgia, River Basin Ctr., 100 Riverbend Rd., Athens,
                       GA 30602 (swenger@uga.edu)

Yesilonis, I.D.   U.S. Forest Service, North Research Station, c/o Baltimore Eco-
                       system Study, 5200 Westland Blvd., Baltimore, MD 21227
                       (iyesilonis@fs.fed.us)

Conversion Factors for SI and Non-SI Units

To convert Column 1 into Column 2 multiply by	Column 1 SI unit	Column 2 non-SI unit	To convert Column 2 into Column 1 multiply by
Length			
0.621	kilometer, km (10^3 m)	mile, mi	1.609
1.094	meter, m	yard, yd	0.914
3.28	meter, m	foot, ft	0.304
1.0	micrometer, μm (10^{-6} m)	micron, μ	1.0
3.94×10^{-2}	millimeter, mm (10^{-3} m)	inch, in	25.4
10	nanometer, nm (10^{-9} m)	Angstrom, Å	0.1
Area			
2.47	hectare, ha	acre	0.405
247	square kilometer, km^2 (10^3 m)^2	acre	4.05×10^{-3}
0.386	square kilometer, km^2 (10^3 m)^2	square mile, mi^2	2.590
2.47×10^{-4}	square meter, m^2	acre	4.05×10^3
10.76	square meter, m^2	square foot, ft^2	9.29×10^{-2}
1.55×10^{-3}	square millimeter, mm^2 (10^{-3} m)^2	square inch, in^2	645
Volume			
9.73×10^{-3}	cubic meter, m^3	acre-inch	102.8
35.3	cubic meter, m^3	cubic foot, ft^3	2.83×10^{-2}
6.10×10^4	cubic meter, m^3	cubic inch, in^3	1.64×10^{-5}
2.84×10^{-2}	liter, L (10^{-3} m^3)	bushel, bu	35.24
1.057	liter, L (10^{-3} m^3)	quart (liquid), qt	0.946
3.53×10^{-2}	liter, L (10^{-3} m^3)	cubic foot, ft^3	28.3
0.265	liter, L (10^{-3} m^3)	gallon	3.78
33.78	liter, L (10^{-3} m^3)	ounce (fluid), oz	2.96×10^{-2}
2.11	liter, L (10^{-3} m^3)	pint (fluid), pt	0.473
Mass			
2.20×10^{-3}	gram, g (10^{-3} kg)	pound, lb	454
3.52×10^{-2}	gram, g (10^{-3} kg)	ounce (avdp), oz	28.4
2.205	kilogram, kg	pound, lb	0.454
0.01	kilogram, kg	quintal (metric), q	100
1.10×10^{-3}	kilogram, kg	ton (2000 lb), ton	907
1.102	megagram, Mg (tonne)	ton (U.S.), ton	0.907
1.102	tonne, t	ton (U.S.), ton	0.907
Yield and Rate			
0.893	kilogram per hectare, kg ha^{-1}	pound per acre, lb acre^{-1}	1.12
7.77×10^{-2}	kilogram per cubic meter, kg m^{-3}	pound per bushel, lb bu^{-1}	12.87
1.49×10^{-2}	kilogram per hectare, kg ha^{-1}	bushel per acre, 60 lb	67.19
1.59×10^{-2}	kilogram per hectare, kg ha^{-1}	bushel per acre, 56 lb	62.71

Table continued.

To convert Column 1 into Column 2 multiply by	Column 1 SI unit	Column 2 non-SI unit	To convert Column 2 into Column 1 multiply by
1.86×10^{-2}	kilogram per hectare, kg ha^{-1}	bushel per acre, 48 lb	53.75
0.107	liter per hectare, L ha^{-1}	gallon per acre	9.35
893	tonne per hectare, t ha^{-1}	pound per acre, lb acre^{-1}	1.12×10^{-3}
893	megagram per hectare, Mg ha^{-1}	pound per acre, lb acre^{-1}	1.12×10^{-3}
0.446	megagram per hectare, Mg ha^{-1}	ton (2000 lb) per acre, ton acre^{-1}	2.24
2.24	meter per second, m s^{-1}	mile per hour	0.447
Specific Surface			
10	square meter per kilogram, m^2 kg^{-1}	square centimeter per gram, cm^2 g^{-1}	0.1
1000	square meter per kilogram, m^2 kg^{-1}	square millimeter per gram, mm^2 g^{-1}	0.001
Density			
1.00	megagram per cubic meter, Mg m^{-3}	gram per cubic centimeter, g cm^{-3}	1.00
Pressure			
9.90	megapascal, MPa (10^6 Pa)	atmosphere	0.101
10	megapascal, MPa (10^6 Pa)	bar	0.1
2.09×10^{-2}	pascal, Pa	pound per square foot, lb ft^{-2}	47.9
1.45×10^{-4}	pascal, Pa	pound per square inch, lb in^{-2}	6.90×10^3
Temperature			
$1.00 (K - 273)$	kelvin, K	Celsius, °C	$1.00 (°C + 273)$
$(9/5 °C) + 32$	Celsius, °C	Fahrenheit, °F	$5/9 (°F - 32)$
Energy, Work, Quantity of Heat			
9.52×10^{-4}	joule, J	British thermal unit, Btu	1.05×10^3
0.239	joule, J	calorie, cal	4.19
10^7	joule, J	erg	10^{-7}
0.735	joule, J	foot-pound	1.36
2.387×10^{-5}	joule per square meter, J m^{-2}	calorie per square centimeter (langley)	4.19×10^4
10^5	newton, N	dyne	10^{-5}
1.43×10^{-3}	watt per square meter, W m^{-2}	calorie per square centimeter minute (irradiance), cal cm^{-2} min^{-1}	698
Transpiration and Photosynthesis			
3.60×10^{-2}	milligram per square meter second, mg m^{-2} s^{-1}	gram per square decimeter hour, g dm^{-2} h^{-1}	27.8
5.56×10^{-3}	milligram (H_2O) per square meter second, mg m^{-2} s^{-1}	micromole (H_2O) per square centimeter second, μmol cm^{-2} s^{-1}	180
10^{-4}	milligram per square meter second, mg m^{-2} s^{-1}	milligram per square centimeter second, mg cm^{-2} s^{-1}	10^4
35.97	milligram per square meter second, mg m^{-2} s^{-1}	milligram per square decimeter hour, mg dm^{-2} h^{-1}	2.78×10^{-2}
Plane Angle			
57.3	radian, rad	degrees (angle), °	1.75×10^{-2}

Table continued.

To convert Column 1 into Column 2 multiply by	Column 1 SI unit	Column 2 non-SI unit	To convert Column 2 into Column 1 multiply by
Electrical Conductivity, Electricity, and Magnetism			
10	siemen per meter, S m^{-1}	millimho per centimeter, mmho cm^{-1}	0.1
10^4	tesla, T	gauss, G	10^{-4}
Water Measurement			
9.73 × 10^{-3}	cubic meter, m^3	acre-inch, acre-in	102.8
9.81 × 10^{-3}	cubic meter per hour, m^3 h^{-1}	cubic foot per second, ft^3 s^{-1}	101.9
4.40	cubic meter per hour, m^3 h^{-1}	U.S. gallon per minute, gal min^{-1}	0.227
8.11	hectare meter, ha m	acre-foot, acre-ft	0.123
97.28	hectare meter, ha m	acre-inch, acre-in	1.03 × 10^{-2}
8.1 × 10^{-2}	hectare centimeter, ha cm	acre-foot, acre-ft	12.33
Concentration			
1	centimole per kilogram, cmol kg^{-1}	milliequivalent per 100 grams, meq 100 g^{-1}	1
0.1	gram per kilogram, g kg^{-1}	percent, %	10
1	milligram per kilogram, mg kg^{-1}	parts per million, ppm	1
Radioactivity			
2.7 × 10^{-11}	becquerel, Bq	curie, Ci	3.7 × 10^{10}
2.7 × 10^{-2}	becquerel per kilogram, Bq kg^{-1}	picocurie per gram, pCi g^{-1}	37
100	gray, Gy (absorbed dose)	rad, rd	0.01
100	sievert, Sv (equivalent dose)	rem (roentgen equivalent man)	0.01
Plant Nutrient Conversion			
	Elemental	Oxide	
2.29	P	P_2O_5	0.437
1.20	K	K_2O	0.830
1.39	Ca	CaO	0.715
1.66	Mg	MgO	0.602

Urbanization: Impacts on Clouds, Precipitation, and Lightning

J.M. Shepherd
J.A. Stallins
M.L. Jin
T.L. Mote

Abstract

Precipitation variability and water cycle changes affect many components of the Earth's natural and human system. Further, it is clear that natural and anthropogenic activities can perturb key components of the hydrological cycle. Understanding and quantifying such changes is vital for a range of meteorological, hydrological, ecological, and climate problems. In an era of heightened sensitivity concerning climate change, the primary discussion of cloud-precipitation variability has been linked to greenhouse gas emissions. Scientific literature has presented theories and observational studies describing how urbanization influences convective processes and cumulative precipitation. The urban environment's (i.e., its land use, aerosols, thermal properties) impact on precipitation will be increasingly vital to climate diagnostics and prediction, global water and energy cycle assessment and prediction, weather forecasting, freshwater resource management, agriculture, and urban planning. This chapter presents a contemporary review of findings and methods related to urban effects on precipitation and related convective processes, particularly lightning. Herein, we present historical and current literature, prevailing hypotheses, critical analysis of challenges facing the research topic, and recommendations for the future.

> "...humans have become a geologic agent comparable to erosion and eruptions... it seems appropriate to emphasize the central role of mankind in geology and ecology by proposing to use the term 'anthropocene' for the current geological epoch."

This statement by noted scholar Paul Crutzen (Crutzen and Stoermer, 2000) clearly captures the notion that humans can alter the Earth system and related ecological processes. The built environment represents one mechanism by which anthropogenic (i.e., "human-induced") activities alter the Earth. A

J.M. Shepherd, Dep. of Geography, University of Georgia, Athens, GA 30602 (marshgeo@uga.edu); J.A. Stallins, Dep. of Geography, Florida State University, Rm. 323 Bellamy Bldg., Tallahassee, FL 32306-2190 (jastallins@fsu.edu); M.L. Jin, Dep. of Meteorology, San Jose State University, Duncan Hall 620, San Jose, CA 95192 (jin@met.sjsu.edu); T.L. Mote, Dep. of Geography, University of Georgia, Rm. 213 Geog-Geol Bldg., 204 Field St., Athens, GA 30602-2502 (tmote@uga.edu).

doi:10.2134/agronmonogr55.c1

well-established body of literature chronicles how urban environments modify weather and the hydroclimate. Scholars from antiquity up to the present day, through observation and speculation, have considered the nature of how human activity shapes Earth system properties (Ruddiman, 2003; Von Storch and Stehr, 2006; Yow, 2007). By constructing cities, humans initiate feedbacks between natural and anthropogenic processes. However, in the past few decades, interest in the feedbacks between human activity and climate has been focused on the global and hemispheric scales. Both natural and human-induced climate variations manifest themselves in the global water cycle (Chahine, 1992) and ecosystems. The Intergovernmental Panel on Climate Change (IPCC) (Trenberth et al., 2007) noted that Earth's mean temperature has increased in recent decades in response to primarily anthropogenic activities. In this context, higher evaporation and precipitation rates might occur, which could lead to an overall acceleration of the global water cycle and an increase in weather extremes and durations of major flood and drought episodes. While the focus has been on greenhouse gas forcing and its effects, the IPCC (Trenberth et al., 2007) noted a growing interest in understanding what role urban land cover, land use, and pollution have on climate change. Cities in themselves modify weather and climate, and global-scale forcings are superimposed on the effects of the built environment on local transfers of heat and moisture. Conversely, the impacts of cities on weather and climate may extend to regional scales.

The most comprehensively studied built environment–climate change manifestation is the urban heat island (UHI). Warmer skin, air, and canopy temperatures generally characterize the urban heat island (Chapter 2, Heisler and Brazel, 2010, this volume). Urban heat islands are evident in the thermal satellite image over the eastern United States (Fig. 1–1). The UHI can be understood by summarizing how the built environment affects the surface energy budget. The surface energy budget equation is:

$$(1 - a)\, S^- + LW^- - eT_{skin}^{\ 4} - SH - LH - G = 0 \tag{1}$$

where SH is sensible heat flux, LH is latent heat flux, and G is the ground heat flux.

This equation forms the basis for most land surface model calculations. SH, LW, and G compete for surface net radiation, which is the downward minus

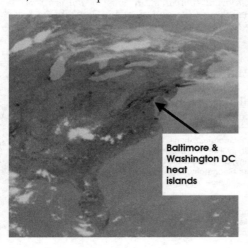

Fig. 1–1.Urban heat islands as indicated by weather satellite. Heat islands are relatively dark features in the image (from Shepherd, 2005).

Baltimore &
Washington DC
heat
islands

upward shortwave and longwave radiation (the first four terms in Eq. [1]). a is surface albedo, and S is downward solar radiation; thus, $(1 - a)\,S^-$ is reflected solar radiation. LW^- is longwave radiation. Emissivity (e) and surface skin temperature (T_{skin}), through the Stefan–Boltzmann Law, describe the upward longwave radiation, or surface emission. In an urban environment, an urban heat storage and anthropogenic term are also added to Eq. [1]. Arnfield (2003) is an appropriate review for readers interested in the energetics, thermodynamics, and turbulent flow associated with the UHI.

Urban heat island circulations are caused by differential heat capacity and thermal inertia between rural and urban regions. Urban heat island–related temperature gradients depend strongly on both land use and urban parameters (e.g., built-up ratio, green surface ratio, sky view factor) (Oke, 1987). A surplus of surface energy over urban regions can be traced to increased surface sensible heat flux, ground heat storage, and anthropogenic heating, as well as reduced evapotranspirational cooling. Urban regions typically have lower albedo values than rural areas and, therefore, more absorption of shortwave radiation energy at the surface. A reduction in sky view factor below roof level reduces radiative loss and turbulent heat transfer and contributes positively to the UHI anomaly (Unger, 2004). The coupling of the urban environment to that atmosphere at various scales is illustrated in Fig. 1–2 (Hidalgo et al., 2008).

The UHI is a firmly established signature of an anthropogenic perturbation to the weather–climate–ecological system. The effects of the built, urban environment on other aspects of the Earth's climate system are less understood (Mills, 2007; Shepherd and Jin, 2004; Seto and Shepherd, 2009). Herein, we discuss how the urban environment is modifying aspects of the Earth's hydroclimate, particularly precipitation and related convective processes. Although it is a hyperbole to suggest that "rain followed the plow" during the settling of the central U.S. plains (Worster, 1979), scientific studies have established the physical basis for how human activity in urban environments shapes spatiotemporal evolution and distribution of convection, precipitation, and lightning. When these changes are considered within the context of a rapidly urbanizing planet, it is apparent why

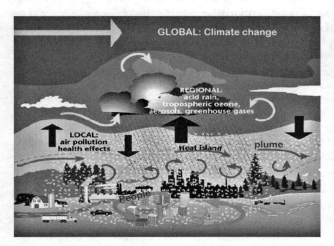

Fig. 1–2. The coupling of the urban environment to the atmosphere at various scales (following GURME WMO 1995–2008, as presented in Hidalgo et al., 2008).

Dabberdt et al. (2000) urged the scientific community to invest more time and resources in the study of urban weather and climate.

Hand and Shepherd (2009) noted that in 2008, more than one-half of the world's population lived in urban areas. The global view of the Earth at night vividly depicts the urban footprint (Plate 1—1; see the color images insert near the center of the book). This view of Earth is a landmark representation of the increasing role that human influence has on the natural system. Projections by the United Nations place this number at 81% by 2030 (UNFPA, 2007). Although urban areas are local to regional in scale, it is increasingly evident that they have impacts at local to global scales by altering atmospheric composition, perturbing components of the water cycle, and modifying cloud systems and related hazards. The 2007 IPCC report established that the notion of climate change must extend far beyond comprehension, analysis, and prediction of greenhouse gas forcing and its effects. In fact, the IPCC noted a growing interest in understanding what role urban land cover/land use (LCLU) and pollution has on climate change.

Convection and precipitation are vital nodes in the global water cycle and a likely proxy for changing climate. Proper assessment of the urban environment's impact on clouds, precipitation, and associated hazards (e.g., lightning, flash flooding) also have societal implications related to climate diagnostics and prediction, weather forecasting, freshwater resource management, urban planning and design, and agriculture.

Historical and Contemporary Perspectives on Urban Effects on Convection

As early as 1921, Horton (1921) observed a tendency for thunderstorm formation over large cities. Kratzer (1937, 1956) extended the notion that urban environments could alter precipitation. Helmut Landsberg, a pioneering urban climatologist, called attention to this process in the classic "The Climate of Towns" (Landsberg, 1956). He discussed possible impacts of large urban areas on rainfall patterns. In the 1960s, a group of researchers led by Stout (1962) and Changnon (1968) began to pose science questions related to the so-called "La Porte Anomaly." Hypotheses of the day argued that this industrialized urban region of Indiana (and southeast of Chicago, IL) received increased cumulative precipitation because of some type of urban or industrial aerosol effect. Studies on the La Porte Anomaly were generally inconclusive (Lowry, 1998), but the work stimulated an important era of increased research on urban—hydroclimate relationships.

The prevailing consensus in early studies was that increased rainfall downwind of urban areas represented the most common evidence of precipitation enhanced by urban effects (Landsberg, 1970; Huff and Changnon, 1972a,b). Though many of the early hypotheses concerning urban rainfall were born in Europe, it was a major North American field study, the Metropolitan Meteorological Experiment (METROMEX), that synthesized the research on urban climates. A series of publications from this multicity investigation provided the first formal hypotheses about drivers of urban weather and climate. Key findings from METROMEX were:

- Precipitation was enhanced by urban effects typically 25 to 75 km downwind of a city during summer months (Huff and Vogel, 1978; Changnon, 1979; Braham, 1981).

- Cumulative amounts were enhanced between 5 and 25% over background values (Changnon et al., 1981, 1991).
- The size of an urban area influenced the horizontal extent and magnitude of urban enhanced precipitation (Changnon, 1992).

Lowry (1998) questioned whether appropriate methodologies or data were utilized during METROMEX and some post-METROMEX studies. He also suggested that methodological deficiencies might explain so-called anomalies or enhancement regions. Other investigators have published findings counter to the results from the METROMEX era. Tayanc and Toros (1997) found no significant evidence that four large Turkish cities had altered rainfall in their analysis. Robaa (2003) even suggested that an inverse relationship existed between urbanization and precipitation around Cairo, Egypt. Kaufmann et al. (2007) used a statistical model and ground-based rainfall data to establish that urban land cover growth in the Pearly River Delta region of China was associated with decreased rainfall. Later in the chapter we will discuss factors related to the particulars of aerosols — their concentration, size, and to a lesser extent composition — that could diminish precipitation or delay its occurrence within these areas of observation.

Most post-METROMEX studies were consistent with the consensus METRO-MEX results. Balling and Brazel (1987) described more frequent late-afternoon storms in Phoenix during and after a rapid period of urbanization in an arid region of the southwest United States. Bornstein and colleagues (Bornstein and LeRoy, 1990) offered one of the first studies of a "megalopolis" city and its impact on rainfall processed. Bornstein's work established that New York City affects evolution, distribution, and movement of summer thunderstorms. Their radar analysis established a tendency for maximum convective rainfall to be produced on the lateral edges and downwind of the city. Jauregui and Romales (1996) observed that the daytime UHI was well correlated with intensification of rain showers during the rainy season (May–October) in Mexico City. They also presented an analysis of historical records showing that the frequency of intense rain showers had increased in recent decades in correlation with the growth of the urban area. Selover's (1997) results for moving summer convective storms over Phoenix were similar to the aforementioned studies. Changnon and Westcott (2002) noted a trend toward increasing heavy rainstorms in recent decades and suggested that the trend in intensity (and frequency) might continue in the future. Shepherd et al. (2002) applied spaceborne radar-derived rainfall data to the urban precipitation problem. Space-borne rainfall estimates enabled the study of multiple cities simultaneously, which mitigates one of the primary criticisms of urban precipitation studies (Lowry, 1998). Shepherd et al. (2002) revealed rainfall anomalies downwind of major U.S. cities. The conceptualization of the "urban rainfall effect" (URE) based on Shepherd et al. (2002) is illustrated in Fig. 1–3. Diem et al. (2004) and Shepherd (2004) debated some of the initial findings and methodologies presented in Shepherd et al. (2002). Although the scales of observation and analysis in Shepherd et al. (2002) limited fine-scale generalization, this work reinvigorated questions about the propensity for cities to modify the rainfall in their vicinity (Souch and Grimmond, 2006).

Dixon and Mote (2003) concluded that under a priori moist conditions some nocturnal convective events in Atlanta, GA might have been initiated by city. Around the same time, Fujibe (2003) attributed increased convection downwind

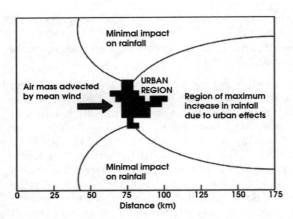

Fig. 1–3. Conceptualization of the downwind urban rainfall anomaly (following Shepherd et al., 2002, as presented in Simpson, 2006).

of large urban areas such as Tokyo to enhanced surface convergence over the urban region (Fujibe, 2003). Shepherd and Burian (2003) applied the satellite-based methodology to Houston, TX and found rainfall anomalies that were spatially consistent with lightning flash density anomalies in Orville et al. (2001). In 2005, Shepherd offered a review of the literature concerning urban effects on convection and rainfall (Shepherd, 2005). He also provided critical recommendations on how the scientific community needed to move forward on the topic.

Lowry (1998) offered several deficiencies in much of the historical literature. A review of recent studies (Shepherd, 2005) suggests that the recommendations of Lowry and issues raised during scholarly debates are now being addressed more carefully in contemporary studies. It is useful to review Lowry's key recommendations. He argued that urban rainfall studies required:

1.   Designed experiments—especially legitimate controls and, where appropriate, stratification schemes—in which explicitly stated hypotheses are tested by means of standard statistical methods.

2.   Replication of the experiments in several urban areas.

3.   Use of spatially small, and temporally short, experimental units reflecting the discontinuous nature of precipitating systems.

4.   Disaggregation of standard climatic data to increase sample size and avoid merging effects between dissimilar synoptic weather systems. (Lowry, 1998)

Post-2000 studies continue to highlight linkages between urbanization and precipitation throughout the world (Table 1–1). More importantly, contemporary studies are applying methodologies that consider recommendations offered by critical commentaries.

Burian and Shepherd (2005), using historical gauge data, concluded that the urban area and the downwind regions of Houston, TX had 59 and 30%, respectively, cumulative rainfall between noon to midnight in the warm season compared with the upwind region. Diem and Brown (2003) and Shepherd (2006) found evidence that urban attributes might alter precipitation during the monsoon season in Phoenix, AZ, particularly in the northeastern suburbs and exurbs. Shepherd (2006) reported statistically significant increases (12–14%) in mean precipitation from a pre-urban (1895–1949) to post-urban (1950–2003) period. Shepherd (2006) hypothesized that urban–topographical convective interactions

Table 1–1. Selected post 2000 urban rainfall studies.

City or Region	Study
Atlanta	Bornstein and Lin, 2000; Shepherd et al., 2002; Shepherd et al., 2004; Dixon and Mote, 2003; Diem and Mote, 2005; Mote et al., 2007; Rose et al., 2008; Diem, 2008; Shem and Shepherd, 2009; Bentley et al., 2009
Cairo (Egypt)	Robaa, 2003
China (Multiple Cities)	Meng et al., 2007; Kaufmann et al., 2007; Zhang et al., 2007; Guo et al., 2006; Jin and Shepherd, 2008; Zhang et al., 2008; Zhang et al., 2009
Europe	Trusilova et al., 2008; Freud et al., 2008
Fairbanks (USA)	Molders and Olson, 2004
Houston (USA)	Shepherd and Burian, 2003; Burian and Shepherd, 2005; Shepherd et al., 2009
Oklahoma City (USA)	Niyogi et al., 2006; Hand and Shepherd, 2009
Phoenix (AZ)	Diem and Brown, 2003; Shepherd, 2006
Paris (France)	Thielen et al., 2000
Riyadh (Saudi Arabia)	Shepherd, 2006
Sydney (Australia)	Gero and Pitman, 2006
St. Louis (USA)	Rozoff et al., 2003; van den Heever and Cotton, 2007
Taipei (Taiwan)	Chen et al., 2007
Japan (Multiple Cities)	Ohashi and Kida, 2002; Inoue and Kimura, 2004; Ikebuchi et al., 2007

were the cause of the anomaly. Chen et al. (2007) offered a similar hypothesis as the cause of the Taipei, Taiwan nocturnal rainfall anomaly. Mote et al. (2007), using radar analysis, showed that the eastern (e.g., downwind) suburbs of Atlanta, GA receive upward of 30% more warm season cumulative rainfall during the evening and early morning hours, possibly because of the URE (Plate 1–2). In their analysis, Mote et al. (2007) stratified case days to remove any large-scale forced weather days. This study was particularly compelling because the radar analysis clearly revealed other likely mesoscale or topographic rainfall features, which suggests that the urban feature is real. Diem (2008) extended the work of Diem and Mote (2005) using a long-term climatological analysis. His statistical analysis confirmed the existence of a northeast suburban Atlanta rainfall anomaly.

More recently, the recognition of the propensity for cities to modify precipitation has stimulated investigations that integrate rainfall observations with other convective phenomena. Rose et al. (2008) coupled cloud-to-ground lightning data with the North American Regional Reanalysis (NARR) rainfall dataset and found enhancement of lightning and rainfall around Atlanta as a function of prevailing wind. The radar-based analysis of Meng et al. (2007) also showed that Guangzhou City (China) strengthened thunderstorms associated with a tropical cyclone. Radar echoes were maximized directly over the urban area. Hand and Shepherd (2009) used TRMM Multisatellite Precipitation Analysis (MPA) and a mesoscale observational network (Fig. 1–2b) to reveal a statistically significant rainfall anomaly in the north-northeast suburbs of Oklahoma City, OK. This study was the first application of a merged satellite product that incorporated infrared, passive microwave, and rain gauge data ($\sim$25 km resolution). The previous satellite studies had only utilized a relatively coarse ($\sim$50 km) radar dataset. Like Rose et al. (2008), Hand and Shepherd (2009) found a strong relationship between prevailing wind and the downwind anomaly region. More importantly, Hand and Shepherd (2009) confirmed that new satellite-based estimates are relatively accurate compared to ground-based rain gauge networks, if applied properly. Therefore satellite data might be appropriate for urban rainfall studies in rapidly urbanizing regions around the globe, particularly if robust ground-based data are not available.

Although methods and data availability continue to improve, there is still uncertainty and scientific debate about whether urban environments increase rainfall, decrease rainfall, or have no effect on rainfall. Trusilova et al. (2008) used a coupled atmosphere—land surface model to simulate European climate sensitivity to urban land cover. They found statistically significant increases (decreases) in winter (summer) precipitation in their urban simulations as compared to the pre-urban simulations. Their study used relatively coarse (~10 km) model resolution and relied heavily on cumulus-parameterized rainfall that might misrepresent explicitly resolved urban convection. Further studies like Trusilova et al. (2008) should be repeated with more explicitly resolved microphysics.

Kaufmann et al. (2007), using econometric and statistical modeling, revealed that urban development in the Pearl River Delta of China has reduced local precipitation. They suggested that such reductions might be due to changes in surface hydrology. Along similar lines, Guo et al. (2006) also found decreased cumulative rainfall around Beijing. Zhang et al. (2009, 2007) argued that Beijing's urban land cover reduced ground evaporation and evapotranspiration, which was conducive to rainfall formation. However, the studies that found decreasing rainfall around Chinese cities did not consider the effects of pollution.

Urban aerosols (e.g., pollutants) have been shown to affect precipitation processes. Smaller cloud droplet size distributions and delayed or suppressed rainfall have been linked to increased aerosol concentrations from anthropogenic sources over and downwind of urban areas (Rosenfeld, 2000; Givati and Rosenfeld, 2004; Lensky and Drori, 2007). Very recent studies (Rosenfeld et al., 2008, 2007; van den Heever and Cotton, 2007) are beginning to shed light on the possible role of giant cloud condensation nuclei (CCN) (enhancement) and smaller CCN (suppression). Investigation of how polluted versus relatively unpolluted urban air modifies convective processes is in part a reflection of the interest in how aerosols impact global climate, particularly in the context of how convection differs over oceans and continental regions. This fact was corroborated by Jin and Shepherd (2008). They found that aerosol effects might be more dominant on maritime cloud droplet formation. They argued that aerosol effects are less dominant in convective rainfall processes over land.

Prevailing Hypothesis Concerning Urbanization and Convection

Overwhelmingly, the "urban rainfall effect" (URE) has been most associated with convective rainfall rather than the lighter, stratiform rainfall. Therefore, it is logical that researchers have expressed interest in whether an "urban lightning effect" (ULE) exists. Lightning is inherently a manifestation of convective processes. METROMEX studies observed that thunderstorms increased in the vicinity of St. Louis. With the advent of the ground-based lightning detection networks a spatially explicit means for assessing thunderstorm activity was available. Two prevailing findings for lightning have emerged (Stallins and Rose, 2008): (i) cloud-to-ground negative polarity flash densities increase, particularly downwind of the city center; (ii) the percentage of positive polarity cloud-to-ground flashes decrease in the vicinity of cities. A range of urban areas, across different continents and population sizes, confirm these trends (Table 1—2). Most of these studies

Table 1–2. Global urban lighting studies (adapted from Stallins and Rose, 2008).

Cities (reference)	Population	Urban–rural flash extent	Study area extent
	(Millions)	— km —	
Sao Paul, Brazil (Naccarato et al., 2003)	17.7	150 × 100	300 × 200
Houston, Texas (Orville et al., 2001)	4.7	75 × 75	250 × 250
Atlanta, Georgia (Stallins et al., 2006)	4.2	50 × 50	160 × 160
Bilbao, Spain (Areitio et al., 2001)	0.4	50 × 50	150 × 150
Belo Horizonte, Brazil (Pinto et al., 2004)	2.5	45 × 60	100 × 120

have found higher negative cloud-to-ground flash densities and decreased positive flash production within 100 km of the city center (Stallins and Rose, 2008). Lightning is inherently a convective process so it would be expected that URE and ULE signatures would be consistent. The spatial consistency between lightning and rainfall distributions around Atlanta, GA as presented by Stallins and Rose (2008) and Mote et al. (2007), respectively, is shown in Plate 1–2.

There is a general consensus that lightning originates from the noninductive charge separation process (Yair, 2008). In the noninductive charge separation process, collisions between graupel and ice crystals in the presence of super-cooled water within the updrafts and downdrafts of a thundercloud lead to the transfer and accumulation of positive and negative charges in different parts of the cloud. Mechanisms to explain how urban areas modify the noninductive charge separation process are extensions of better-developed and more-substantiated hypotheses to explain how urban areas modify convection and precipitation, the precursors of lightning formation.

Several plausible hypotheses for urban effects on convection have emerged within the literature (Shepherd, 2005). However, there is no conclusive theory on what mechanisms, independently or synergistically, affect urban convective processes. A large body of literature has been devoted to pattern description and visualization. These methods are vital for generating and testing these hypotheses, but emphasis has now shifted to documenting mechanism and elucidating relative contributions (Teller and Levin, 2008). Nevertheless, the trend toward documentation of mechanism and the inclusion of multiple phenomena has its weaknesses. Mechanism yields a high degree of precision for a single event, but low generality when a larger climatological perspective, spanning multiple years and mechanisms, is employed.

Enhanced Convergence, Surface Fluxes, and Urban Heat Island Destabilization

The literature has validated the propensity of land-use change and land cover to alter precipitation (Pielke et al., 2007), less so for lightning. Contemporary studies have relied on model simulations incorporating a range of observational data, theoretical frameworks, and land cover assumptions. Physically based coupled atmosphere–land surface (CALS) models provide a "laboratory" for experimentation. Since urban areas cannot be removed or altered easily, CALS models provide scientists with a tool for controlled experiments with the ability to replicate various meteorological and land cover scenarios. The atmospheric model simulates the dynamic, thermodynamic, and moisture content evolution of the

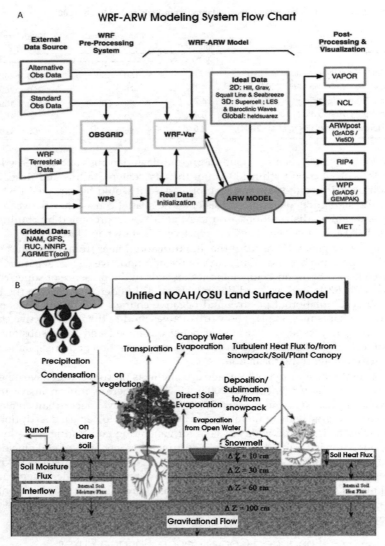

Fig. 1–4. (A) Typical schematic for a mesoscale weather model and (B) NOAH land surface model (courtesy of UCAR).

atmospheric fluid. The land surface model represents relevant land cover characteristics and exchanges with the lowest levels of the atmosphere. Figure 1–4 is a schematic describing the concept of numerical model simulations involving land–atmosphere interactions.

Early model experiments indicated that urban convection was likely a function of enhanced convergence and surface fluxes. Vukovich and Dunn (1978) used a three-dimensional model to show that intensity of the UHI–rural gradient and boundary layer stability are dominant in the development of UHI-induced

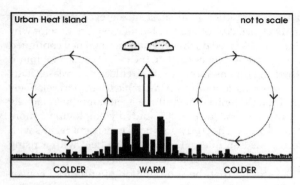

Fig. 1–5. A schematic of a typical urban mesoscale circulation (from Simpson, 2006).

mesocirculations (Fig. 1–5). Conceptually, an urban circulation is similar to a sea breeze circulation that forms due to a temperature—pressure gradient between the warmer land and water. Huff and Vogel (1978) further noted that the urban circulation was enhanced by the transfer of heat from the land surface to the atmosphere through sensible heat fluxes and surface roughness—induced transport. Hjelmfelt (1982) modeled the UHI of St. Louis, a key METROMEX city, and linked rising motion downwind of the city to surface roughness—induced (e.g., mechanical turbulence from buildings) convergence and the downwind translation of the urban circulation by the large-scale flow.

Two-dimensional model experiments by Thielen et al. (2000) further corroborated earlier findings. They found that when UHIs were weak, surface sensible heat fluxes, convergence, and buoyancy variations were most effective at a distance from the heat source. Craig and Bornstein's (2002) simulations described how UHI induces convergence and convection around Atlanta. Bornstein and Lin (2000) had found evidence of such urban-induced convection using a special observation network during the 1996 Atlanta Olympic Games. Rozoff et al. (2003) examined a 1999 storm case in St. Louis (from METROMEX studies) to ascertain the role of urban-generated surface convergence mechanisms on initiating deep, moist convection. They found that nonlinear interactions among the friction of urban surfaces, momentum drag, and urban heating could induce downwind convergence.

Niyogi et al. (2006) simulated a mesoscale convective system in Oklahoma City, OK, using a land surface—urban canopy model system. Urban canopy models enable representation of the three-dimensional urban morphology and its heterogeneity within the land surface model rather than using simplified slab-type representation of the urban surface. The complexity of the urban morphology directly affects temperature and wind flow in the urban environment. Like previous studies, Niyogi et al. (2006) found that the city concentrated the precipitation in the downwind region. Ikebuchi et al. (2007) found that urban land cover and anthropogenic heating altered the location and intensity of convective rainfall in the Tokyo, Japan metropolitan area.

Baik et al. (2007) and Han and Baik (2008) have provided some of the most compelling modeling studies to describe why there is a "downwind" tendency observed in urban convective studies. Using numerical and analytical models, Baik et al. (2007) found that as the boundary layer destabilizes, a wind region of rising motion, induced by the UHI, intensifies. Climatological studies (Arnfield, 2003) have clearly indicated that UHI magnitude is greatest in the evening and early morning hours. Therefore, Baik et al. (2007) offered a plausible explanation for why the urban convective effect is often observed during daytime hours. They continued to argue that as the boundary layer destabilizes further, the height of the maximum rising motion and the depth of the downwind updraft circulation increases. During daytime hours, stability conditions generally are favorable to support relatively strong UHI circulations even though the UHI magnitude itself might be relatively weak. Han and Baik's (2008) work extended the 2007 study by employing a more realistic three-dimensional framework. The study produced consistent results with the previous study but also revealed an internal gravity wave field with a rising branch downwind of the theoretical heating center (e.g., the city).

Shem and Shepherd (2009) found that the combination of enhanced convergence associated with the urban circulation on the perimeter of Atlanta's urban land cover footprint (Fig. 1–6) and increased sensible heat flux likely contributed to producing more rainfall around Atlanta, GA. Utilizing "no urban" and "urban" scenarios, they clearly showed that the storms were initiated at the same time by forcing other than the urban environment. However, subsequent cumulative rainfall may have been enhanced by the mesoscale dynamic interactions. Shepherd et al. (2009) applied an urban growth model to establish a projection of Houston's land cover in the year 2025. Future land cover realization was integrated into CLAS model and run under "current" meteorological conditions. The results indicated that a spatially larger UHI caused a dynamic–thermodynamic response that significantly altered the precipitation evolution and by inference, the regional hydroclimate. This result is significant because it suggests that

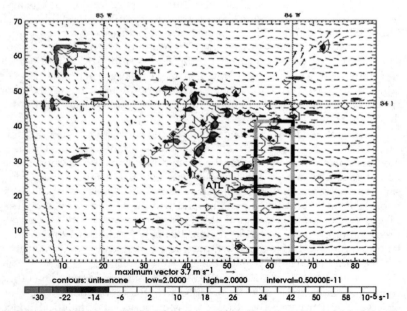

Fig. 1–6. Enhanced total accumulated rainfall zone-eastern side (selected rectangle) of the city superimposed on the difference plot for convergence (10^{-5} s^{-1}) between the URBAN and NOURBAN scenarios. The selected rectangle covers a strip approximately 20 to 50 km from the city center, marked ATL (from Shem and Shepherd, 2009).

future changes in precipitation may be driven by other anthropogenic factors in addition to greenhouse gas emissions.

Such studies have demonstrated the ability of CALS to contribute to describing mechanisms related to the urban convective effect. However, the studies referenced in aforementioned sections lack adequate representation of aerosols or pollution within the model cloud—precipitation processes. Van den Heever and Cotton (2007) considered the combined effects of urban land cover and aerosols on precipitation and offer a potential pathway forward. The following discussion highlights why the next generation of CALS model studies must integrate both aerosols and urban land cover.

Aerosol or Pollution Effects

Aerosols are particles in the atmosphere, ranging from 0.01 to 10 mm (Fig. 1–7). Urban regions usually have the highest concentration of aerosols, up to 10^8 to 10^9 particles per cc (Seinfeld and Pandis, 1998). Aerosols can be divided by size: fine mode and coarse mode particles (see Chapter 3, Santosa, 2010, this volume). Coarse particles are composed of mechanical materials such as tire dust, sea salt, and natural dust in urban areas. Fine particles tend to be produced locally by construction, air conditioning, transportation, and industry or by chemical interactions involving sulfates, nitrates, ammonium, and organics (Santosa, 2010, this volume).

Over urban regions, there are various types of aerosols, for example, sulfate aerosols and black carbon aerosols (Santosa, 2010, this volume). According to Seinfeld and Pandis (1998), urban aerosols are primarily sulfates, nitrates, ammonium, organics, crustal rock particulate matter, sea salt hydrogen ions, and water. Aerosols affect local to global climate through radiative forcing. The direct radiative effect of aerosols in the atmosphere is to scatter or reflect solar radiation and thus inhibit solar radiation from reaching the surface. Most aerosols promote a cooling effect on the atmosphere, as well as at the surface, since surface insolation is reduced by the reflection of incident solar radiation. Carbon aerosols absorb

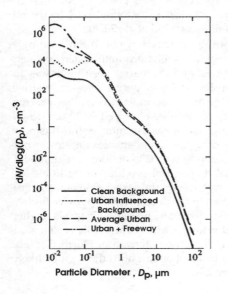

Fig. 1–7. Aerosol number distributions next to a source (freeway), for average urban, for urban influenced background, and for background conditions. (Reproduced from *Atmospheric chemistry and physics: From air pollution to climate change*, Seinfeld and Pandis, 1998, with permission.)

solar radiation in the atmosphere, thus warming the atmosphere and the surface. The indirect effect means that aerosols interact with clouds by serving as cloud condensation nuclei (CCN).

Urban regions typically have high aerosol content, and these aerosols serve as CCN on which cloud droplets form. Twomey (1977) proposed the theory that under the condition of finite water vapor content in the air, more CCN induce more competition for water vapor to form cloud droplets. Thus, the size of cloud droplets is reduced. Polluted clouds generally lead to increasing cloud nucleus concentrations, and hence increasing cloud optical thickness, finite cloud thicknesses, and cloud albedo. Jin and Shepherd (2008) used satellite-derived aerosol, cloud, and precipitation properties to validate Twomey's theory. Plate 1–3 illustrates that water cloud droplet size (effective radius) decrease with increasing aerosols in air, for all kinds of rainfall conditions

By serving as CCN, aerosols indirectly influence precipitation and lightning formation. A rich body of literature has considered how maritime and continental aerosol populations affect cloud and precipitation development (Rosenfeld and Lensky, 1998; Rosenfeld, 2000; Givati and Rosenfeld, 2004; Kaufman et al., 2005; Jin et al., 2005; Lensky and Drori, 2007; Rosenfeld et al., 2008; Jin and Shepherd, 2008). Large maritime aerosols tend to enhance the collision–coalescence process, stimulate larger droplet sizes, and cause earlier rainout. Smaller aerosols over continental land masses impede the collision–coalescence process. This leads to delayed rainout and subsequently, deeper convection. The maritime–continental difference in convective processes has been investigated in an analogous manner for the urban–rural gradient of air quality (Freud et al., 2008; Givati and Rosenfeld, 2004). For example, fine urban aerosols behave similarly to continental aerosols by slowing rainout, delaying the onset of the downdrafts, and producing strong, deep updrafts (Rosenfeld, 2000; Rosenfeld et al., 2008). Such updrafts are more efficient at transporting precipitable water to the mixed-phase region (i.e., super-cooled water and ice crystals coexist) and where the noninductive charge separation process of lightning production takes place. Under this scenario, suppressed precipitation by urban aerosols may be followed by increased electrification and intensification of rainfall (Rosenfeld et al., 2008).

Linkages between aerosols and lightning activity have been most strongly established through studies that found spatial correlations among aerosols-generating infrastructure, higher flash densities, and decreased positive flashes in nonurban locations. Stallins et al. (2006) found lower percentages of positive flashes downwind of large stretches of major transportation corridors passing through Atlanta, GA. This aerosol effect may also be observed ~100 km downwind from the congested traffic emerging out of the Atlanta central business district. Steiger and Orville (2003) detected higher flash densities and lower percentages of flashes in the Lake Charles–Baton Rouge, LA corridor, a region with significant aerosol-producing refineries and chemical production facilities. For locales devoid of significant urban land cover, the spatial association among aerosols, increased flash densities, and decreases in the percentage of positive flashes support a mechanistic role for atmospheric particulate matter. This fact is further corroborated by reviewing Plate 1–2. In this image, Mote et al.'s (2007) radar analysis indicates significant rainfall in the mountainous terrain of North Georgia. However, the lighting anomalies seem to be most apparent in the regions likely associated with higher anthropogenic aerosols.

Bifurcation and Other Dynamic Effects

The vertical structure of building and thermodynamic signature of urban land cover and their effects on local mesoscale circulation are hypothesized to bifurcate or redirect existing thunderstorms around the periphery as they approach urban areas. Casual observers have raised the question of whether approaching storms "avoid" cities. Bornstein and LeRoy (1990) hypothesized that thunderstorms bifurcated around New York City due to a building "barrier" or "dome" effect. Stallins and Bentley (2006) noted that frontal thunderstorms had peaks in flash density on the perimeter of the city, while high flash densities in air mass thunderstorms were more aligned with the urban core. Ntelekos et al. (2007) observed that the urban environment in the Baltimore—Washington, DC corridor strongly shaped the movement of a thunderstorm and associate outflow boundaries. Shearing instability and daytime heating can also create horizontal convective rolls in the vicinity of urban areas (Miao and Chen, 2008). Further research should examine the bifurcation effect, with the goal of determining whether a building barrier, thermodynamic-boundary layer barrier, or some combination of both causes such dynamic responses. It is even interesting to speculate on whether such effects reduce severe weather (e.g., tornadoes) in central business districts.

Toward Reconciling Mechanisms and Methodology

In time, it is likely that studies will conclusively establish that a combination of mechanisms and cycles most appropriately describes the urban convection effect. Stallins and Bentley (2006) hypothesized that aerosol mechanisms contribute to flash count variability over an urban region, while the number of thunderstorm events is a better indicator of the influence of mechanisms anchored to urban land cover. Aerosols vary on time scales ranging from daily to seasonally and in relation to human decisions. For example, Bell et al. (2008) identified day of the week and traffic volume as indicators of aerosol loads to document midweek precipitation enhancement over the southeastern USA. Bell et al. (2009) suggested that the weekly cycle is also apparent in lighting data. Sato and Takahashi (2000) also detected a weekly cycle in precipitation. However, studies by DeLisi et al. (2001), Schultz et al. (2007), and Lacke et al. (2009) found no statistically significant evidence for a weekly cycle.

Van Den Heever and Cotton (2007), in an important synergistic modeling study, established a plausible role for urban aerosols and land cover. They noted that urban-induced wind convergence was the primary factor determining whether thunderstorms developed in their simulations. Aerosols influenced the amount of cloud liquid water and ice present in the atmosphere. This aerosol effect also affects the surface precipitation totals, the strength and timing of updrafts and downdrafts, and the longevity of updrafts. The suggestion is that aerosols affect the efficiency of the rain cloud while land cover affected convective forcing. Shepherd et al. (2001) studied rainfall efficiency in Florida sea-breeze thunderstorms and found an analogous relation between integrated moisture convergence (initiating the storms) and mid-level moisture (affecting the rainfall efficiency).

As a strategy to further mechanistic understanding, scholars have also shifted to the coordinated study of multiple weather phenomena. These integrated approaches leverage fundamental knowledge that clouds, rainfall, and lightning develop from the same physical processes and that their coordinated examination may reveal subtleties of mechanism. Atlanta, for example, has been the subject of several independent studies of urban precipitation and lightning. Herein, we have discussed studies on Atlanta that qualitatively or quantitatively described rainfall patterns and trends (Bornstein and Lin, 2000; Shepherd et al., 2002; Diem and Mote, 2005; Mote et al., 2007; Diem, 2008), lightning patterns (Stallins and Rose, 2008), lightning and rainfall patterns (Rose et al., 2008; Stallins and Bentley, 2006; Stallins et al., 2006), and mechanisms (Shem and Shepherd 2009). Nevertheless, there remains a lack of urban studies that synergistically document the variability in the timing of precipitation and flash production outside of the context of urban flash flood prediction. Pertinent research questions include:

- What is the geographic correlation between peak lightning and precipitation?
- Is there a resolvable lightning signature associated with delayed rainfall or dry urban thunderstorms?
- What are the relationships between atmospheric stability, aerosols, and lightning (Rosenfeld et al., 2008)?

For example, Boussaton et al. (2007) found variability in the time sequencing of radar reflectivity intensity with lightning production over Paris. In some thunderstorm events, high reflectivities indicative of heavy precipitation lagging peaks in lightning production. In other storms, reflectivity peaks led the maxima in flash production. These contrasts may provide insight into how urban aerosols influence updrafts, rainfall evolution, and vertical transport of water to the levels where the noninductive charge separation process of cloud electrification occurs.

Mitigating Factors

Several mitigating factors constrain the expression of aerosol, thermodynamic, and surface roughness mechanisms. Their shared quality is that they are context specific (Lowry, 1998). Topographic influences complicate any simple attribution of mechanism. Topography implies the local configuration of physiography that can enhance or dampen convection more than any discrete orographic effect. Westcott (1995) hypothesized that river valleys may initiate convection in the downwind direction. River valleys in Taipei, Taiwan interact with sea breezes to augment thunderstorms and rainfall (Chen et al., 2007). Urban-influenced weather along the U.S. east coast also reflects the influence of the Appalachians to the west and land—ocean boundaries to the east. Sea breezes can also be are collinear influences on the character of urban thunderstorms (Gero and Pitman, 2006; Chen et al., 2007; Shepherd et al., 2009). All of these local influences complicate any simple assumption of causality as well as a limiting of urban weather to the predominant downwind direction (Hand and Shepherd, 2009; Rose et al., 2008). Furthermore, aerosol effects may decrease with their increasing concentration, suggesting that more heavily polluted areas may not necessarily have greater aerosol-driven influences on precipitation and cloud electrification (Van Den Heever and Cotton, 2007). Aerosol composition and secondary chemical reactions are also relevant (Blanchard et al., 2008), although composition may not ultimately be as important as the size distribution of the aerosols (Dusek et al., 2006). Ultrafine particulate matter from forest fires has been linked to increased positive flash activity (Lyons et al., 1998) and decreased precipitation (Rosenfeld et al., 2007), suggesting that in developing regions of the world where vegetation burning contributes a large component of the atmosphere, different urban weather drivers will unfold. Sao Paulo, for example, experiences high particulate matter from vehicles, but during the dry season biomass burning in the Amazon also influences the city's atmosphere (Landulfo et al., 2005). Industrial processes release aerosols that may undergo secondary alteration. Even biogenic particles such as pollen could conceivably influence cloud formation, precipitation, and the noninductive charge separation (Diehl et al., 2001).

Many cities may not exhibit flash and precipitation anomalies, or do so only under particular conditions. Westcott (1995) noted that Detroit, MI and Cincinnati and Toledo, OH did not exhibit significant differences in flash patterns. Lake effects may have limited development of urban flashes in Detroit. Cincinnati's complex terrain was invoked to explain its lack of flash augmentation. Studies have found significantly positive correlations (Naccarato et al., 2003) between aerosols and lightning, but some have found variable (Westcott, 1995) and weak correlations (Soriano and de Pablo, 2002; Kar et al., 2007). In part, this may reflect the resolution of point-specific air pollution measurements (typically only a few stations per city) relative to the wide areal distribution of flashes. As a precautionary principle, investigations should rely on a suite of conditions—air mass vs. frontal flow regime (Lowry, 1998), severe vs. nonsevere weather (Niyogi et al., 2006)—to characterize the propensity for urban modifications. Urban effects must account for variability in synoptic regime.

Another mitigating factor is size and shape of cities. METROMEX indicated that larger urban areas are more likely to exhibit urban rainfall anomalies. However, very little observational or theoretical work has considered what the threshold size is to trigger urban-induced convection. Further, are more circular, oval, linear, or irregular shaped urban land areas more likely to alter or initiate convective processes?

Human–Natural System Implications

Given the burgeoning migration of human populations to urban habitats, the facets of public health and safety are immense. Heat-related mortality has received the greatest share of attention (Zhou and Shepherd, 2009), particularly after the European heat wave of 2003 (e.g., Beniston, 2004). Point-specific aerosol loads have been correlated with increased hospital room visits for asthma and heart disease by many scholars in a range of cities (Peters et al., 2001; Sarnat and Holguin, 2007). However, how urban heating and aerosols influence human health via weather modification has received less attention. Thunderstorms have been associated with increased hospital admissions for respiratory distress and asthma for the region around Atlanta, GA (Grundstein et al., 2008). Enhanced suspension and aerosolization of pollen has been hypothesized to account for this small but significant increase. Speculatively, delayed precipitation and enhanced updraft formation may lead to considerable concentrations of aerosols in developing thunderstorms, which are then redistributed over populated areas as the thunderstorms dissipate.

In the developing world urban flooding is a reality for many (Chapter 14, Burian and Pomeroy, 2010, this volume), and its effects on the public safety and health of city dwellers is considerable. The National Weather Service (NWS) office in Charlotte, NC has become very concerned that urban-enhanced precipitation events in conjunction with increased impervious surface extent within the

Charlotte metro area has resulted in an increase in heavy runoff/urban flooding events (Quattrochi, NASA Geographer at Marshall Space Flight Center, personal communication, 2006). Changnon and Westcott (2002) found that impacts from the record high number of eight storms in 2001 revealed that efforts to control flooding, including the Deep Tunnel system, had reduced street and basement flooding in the moderate intensity storms. The two most intense storms, each with 100-yr rainfall intensities, led to excessive flooding (physical process) and forced managers to release flood waters into Lake Michigan (decision). Burian et al. (2004) discussed the implications of urban-induced precipitation on the design of urban drainage systems.

Although socioeconomics and public health practices constrain disease transmission rates, given two areas of similar land cover type and demographics, it is not unrealistic to hypothesize that areas receiving more urban-enhanced rainfall might also be susceptible to more water-borne diseases, such as dysentery or cholera. Mosquitoes, under conditions of greater urban heating and precipitation, might also export malaria and dengue fever to larger numbers of residents. Lightning is a widely acknowledged public safety risk addressed by many governmental entities. In most cases risk is assumed to be a function of time (morning versus afternoon) or the vertical characteristics of a specific location (water or open area versus indoors). Less attention is given to geographic location relative to documented areas of flash enhancement.

Urban weather and climate modification is woven into the economics of urban areas. To some extent, the costs have been normalized and are considered a requisite for the development of dense population and commercial centers. Although figures are difficult to obtain and integrate, the costs of airport closures, electrical outages, and fires in flash-prone urban and surrounding suburban areas may be immense. Many of the datasets for these losses are partial or lacking full geographic information. This arises because insurers often compete with each other and do not wish to divulge their financial risk strategies. In addition, geographically detailed loss information raises issues of privacy and liability when the full extent of weather losses is more visible.

Urban areas and their weather and climate can be thought of as a model system (Seto and Shepherd, 2009) to better understand global climate change. Cities have higher carbon dioxide emissions. Urban weather and climate, as with the global system, are influenced by the radiative and CCN effects of aerosols. Land cover and albedos in urban areas may set up feedbacks that heighten and perpetuate atmospheric contrasts with surrounding rural areas. Enhanced convection and thunderstorm activity has also been predicted under global warming (Del Genio et al., 2007; Trapp et al., 2007). Greater dialog between the global and urban scales of climate change study may help refine mutual understanding of human-modified atmospheric convection and aerosol-driven mechanisms, and it may tighten linkages between socioeconomic drivers and atmospheric responses. Urban weather and climate change present an opportunity for urban planning and design. Can inadvertent weather modification ever be used strategically? For example, Chen et al. (2007) noted that increased rainfall and thunderstorm activity in the Taipei basin in Taiwan worsens traffic hazards but eases the stress of water supply, reduces ground subsidence, and cleans up pollution. Shepherd and Mote (2009) argued that the state of Georgia could consider urban rainfall anomalies when planning future water supply reservoir locations. Can this information be employed for future urban planning and environmental management? When selecting the locations for sensitive electrical infrastructure, dangerous industrial facilities, water impoundments or water treatment plants, or commerce that produces large quantities of atmospheric aerosols, should knowledge of local precipitation or lightning anomalies be consulted? Although the history of weather modification is fairly criticized for its incomplete understanding of mechanism and political wrangling, urban areas and their study are providing insights into the mechanisms of precipitation enhancement. Weather modification may ultimately involve a degree of planned human inputs and a coincidental history of land use change.

In closing, urban weather and climate studies should deploy a plurality of methods, especially those that balance description and mechanism (Schröder and Seppelt, 2006; Grimaldi and Engel, 2007). Geographic studies coupling nomothetic (generalizing) and ideographic (context specific) perspectives (Phillips 2001) will stimulate more understanding of mechanism and lead to improvements in prediction. A synoptic typology for cities, deployed across a large range of latitudes and maritime versus continental locations will help disentangle the aerosol, thermodynamic, and surface roughness-based mechanisms and local mitigating influences.

Recommendations for Studying Urban Convective Effects

As scientific investigation on urban-convective processes move forward, we consider and extend recommendations put forth by Shepherd (2005):

Observing Systems

First, new spaceborne observing systems are required to monitor and track spatiotemporal trends in anthropogenic and natural aerosols, land cover and land use changes, cloud microphysics, lightning (e.g., GOES-R Lightning Mapper), and precipitation processes. Second, high spatiotemporal mesoscale and microscale in situ networks (e.g., existing mesonets and Long-Term Ecological Research [LTER] sites and emerging micronet and ecological observing networks [Basara et al., 2009, http://www.neoninc.org/documents/publications, verified 2 Feb. 2010]) are needed for model verification, process studies, and trend detection. Finally, exploitation of emerging capabilities like dual-polarimetric systems and lidar are needed.

Urban Canopy Representation

Improved observations and representation of urban canopies are needed because currently only a few cities have robust urban canopy assessments. For example, at the time of writing (2009), the lead author of this chapter had the technological capacity to conduct CALS simulations for Atlanta, GA using an urban canopy model. However, there was no available dataset of urban morphological parameters for Atlanta. Burian et al. (2004) discussed recent applications of lidar and synthetic aperture radar for assessing urban canopies. The role that urban canopy-scale dynamics (UCD) play in larger scale precipitation process is uncertain. Recent papers by Dupont et al. (2004) and Kusaka and Kimura (2004a,b) identified the importance of complex surface characterization and urban canopies on predicting several meteorological fields, but precipitation was not included. Urban canopy datasets will increasingly be needed as CALS models begin to routinely employ urban canopy models.

Urban Land Cover–Aerosol Process in Models

Modeling systems must explicitly resolve aerosols, cloud microphysics, complex land surfaces, and precipitation evolution so that a more complete understanding of the feedbacks and interactions can be gained. Recent work by Khain et al. (2004) has advanced the use of spectral (bin) microphysics to simulate aerosols in cloud–mesoscale modeling systems. Van den Heever and Cotton (2007) and Molders and Olson (2004) also demonstrated methodologies for introducing aerosols into mesoscale modeling systems. Additionally, high performance land surface model and data assimilation systems (e.g., NASA Land Information System, http://lis.gsfc.nasa.gov, verified 2 Feb. 2010) can improve representation of complex and heterogeneous land surfaces. Diem and Brown (2003) hinted at possible influences of irrigation-related moisture in urban convection in an arid city so soil moisture and vegetation representation is not trivial (Niyogi et al., 2006). The logical next step is to couple these land surface, atmosphere, and aerosol-microphysics schemes.

Field Campaigns

Emerging technologies have broadened the data available for studying the urban climate problem, but field studies enable detailed interrogation of the underlying physical processes for specific cases and candidate case days for modeling studies. As noted earlier, no focused field study has comprehensively leveraged new knowledge, instrumentation, and models to adequately focus on the urban convective problem since METROMEX, although the Houston Environmental Aerosol Thunderstorm Project (HEAT) was an unsuccessful attempt (due to lack of funding) in the early 2000s.

Climate Model–Urban Convergence

Urban footprints are converging with the increasingly high spatial resolution of regional and global climate models. As the evidence that urban areas impact precipitation mounts, the climate modeling community must begin to represent urban land surface and aerosol processes to better understand the aggregate influences of built-up land and urban aerosols on short- and long-term climate change. Recent articles (Jin et al., 2007; Jin and Shepherd, 2005) discuss potential pathways to achieving adequate representation of complex urban–atmosphere interactions in climate models.

Acknowledgments

Dr. Shepherd acknowledges support from NASA grant NNX07AF39G, Precipitation Measurement Missions Program. Dr. Stallins acknowledges support from NSF BCS (Award Number 0241062). Dr. Jin acknowledges support from NSF ATM (Award Number 0701440). Dr. Mote and Dr. Shepherd acknowledge support from the Southern High Resolution Modeling Consortium and USDA Forest Service contract AG-4568-C-08-0063.

References

Areitio, J., A. Ezcurra, and I. Herrero. 2001. Cloud-to-ground lightning characteristics in the Spanish Basque Country area during the period 1992–1996. J. Atmos. Sol. Terr. Phys. 63:1005–1015.

Arnfield, A.J. 2003. Two decades of urban climate research: A review of turbulence, exchanges of energy and water, and the urban heat island. Int. J. Climatol. 23:1–26.

Baik, J.J., Y.H. Kim, J.J. Kim, and J.Y. Han. 2007. Effects of boundary-layer stability on urban heat island-induced circulation. Theor. Appl. Climatol. 89:73–81.

Balling, R.C., and S.W. Brazel. 1987. Diurnal variations in Arizona monsoon precipitation frequencies. Mon. Weather Rev. 115:342–346.

Basara, J.B., B. Illston, T.E. Winning, and C.A. Fiebrich. 2009. Evaluation of rainfall measurements from the WXT510 Sensor for use in the Oklahoma City Micronet. Open Atmos. Sci. J. 3:39–45.

Bell, T.L., D. Rosenfeld, and K.M. Kim. 2009. Weekly cycle of lightning: Evidence of storm invigoration by pollution. Geophys. Res. Lett. 36, L23805, doi:10.1029/2009GL040915.

Bell, T.L., D. Rosenfeld, K.M. Kim, J.M. Yoo, M.I. Lee, and M. Hahnenberger. 2008. Midweek increase in US summer rain and storm heights suggests air pollution invigorates rainstorms. J. Geophys. Res. Atmos. 113:D02209, doi:10.1029/2007JD008623.

Beniston, M. 2004. The 2003 heat wave in Europe: A shape of things to come? An analysis based on Swiss climatological data and model simulations. Geophys. Res. Lett. 31:2022–2026, doi:10.1029/2003GL018857.

Bentley, M., W. Ashley, and J.A. Stallins. 2009. Climatological radar delineation of urban convection for Altlanta, Georgia. Int. J. Climatol. doi:10.1002/joc.2020.

Blanchard, C.L., G.M. Hidy, S. Tanenbaum, E. Edgerton, B. Hartsell, and J. Jansen. 2008. Carbon in southeastern U.S. aerosol particles: Empirical estimates of secondary organic aerosol formation. Atmos. Environ. 42:6710–6720.

Bornstein, R., and M. LeRoy. 1990. Urban barrier effects on convective and frontal thunderstorms. p. 120–121. In Extended Abstracts, Fourth Conf. on Mesoscale Processes, Boulder, CO. 25–29 Jan. 1990. Am. Meteorol. Soc., Boston, MA.

Bornstein, R., and Q.L. Lin. 2000. Urban heat islands and summertime convective thunderstorms in Atlanta: Three case studies. Atmos. Environ. 34:507–516.

Boussaton, M.P., S. Soula, and S. Coquillat. 2007. Total lightning activity in thunderstorms over Paris. Atmos. Res. 84:221–232.

Braham, R.R. 1981. Urban precipitation processes. p. 75–116. In METROMEX: A review and summary. Meteor. Monogr. No. 40. Am. Meteorol. Soc., Boston, MA.

Burian, S., W. Stetson, W.S. Han, J.K.S. Ching, and D.W. Byun. 2004. High-resolution dataset of urban canopy parameters for Houston, Texas. Proceedings of the AMS Fifth Conference on the Urban Environment, Vancouver.

Burian, S.J., and C.A. Pomeroy. 2010. Urban impacts on the water cycle and potential green infrastructure implications. p. 277–296. In J. Aitkenhead-Peterson and A. Volder (ed.) Urban ecosystem ecology. Agron. Monogr. 55. ASA, CSSA, and SSSA, Madison, WI.

Burian, S.J., and J.M. Shepherd. 2005. Effect of urbanization on the diurnal rainfall pattern in Houston. Hydrol. Processes 19:1089–1103.

Chahine, M.T. 1992. The hydrological cycle and its influence on climate. Nature 359:373–380.

Changnon, S.A. 1968. The La Port weather anomaly—Fact or fiction? Bull. Am. Meteorol. Soc. 49:4–11.

Changnon, S.A. 1979. Rainfall changes in summer caused by St. Louis. Science 205:402–404.

Changnon, S.A. 1992. Inadvertent weather modification in urban areas: Lessons for global climate change. Bull. Am. Meteorol. Soc. 73:619–627.

Changnon, S.A., R.G. Semonin, A.H. Auer, R.R. Braham, and J. Hales. 1981. METROMEX: A review and summary. Meteorol. Monogr. 18. Am. Meteorol. Soc., Boston, MA.

Changnon, S.A., R.T. Shealy, and R.W. Scott. 1991. Precipitation changes in fall, winter and spring caused by St. Louis. J. Appl. Meteorol. 30:126–134.

Changnon, S.A., and N.E. Westcott. 2002. Heavy rainstorms in Chicago: Increasing frequency, altered impacts, and future implications. J. Am. Water Resour. Assoc. 38:1467–1475.

Chen, T.-C., S.-Y. Wang, and M.-C. Yen. 2007. Enhancement of Afternoon Thunderstorm Activity by Urbanization in a Valley: Taipei. J. Appl. Meteorol. Climatol. 46:1324–1340.

Craig, K., and R. Bornstein. 2002. MM5 simulation of urban induced convective precipitation over Atlanta. p. 5–6. In Preprint Vol., Proc. of the Fourth AMS symposium on the Urban Environment, Norfolk, VA.

Crutzen, P.J., and E.F. Stoermer. 2000. The Anthropocene. IGBP Newsl. 41:17–18.

Dabberdt, W.F., J. Hales, S. Zubrick, A. Crook, W. Krajewski, J.C. Doran, C. Mueller, C. King, R.N. Keener, R. Bornstein, D. Rodenhuis, P. Kocin, M.A. Rossetti, F. Sharrocks, and E.M. Stanley. 2000. Forecast issues in the urban zone: Report of the 10th Prospectus Development Team of the U.S. Weather Research Program. Bull. Am. Meteorol. Soc. 81:2047–2064.

Del Genio, A.D., M.S. Yao, and J. Jonas. 2007. Will moist convection be stronger in a warmer climate? Geophys. Res. Lett. 34(16), doi:10.1029/2007GL030525.

DeLisi, M.P., A.M. Cope, and J.K. Franklin. 2001. Weekly precipitation cycles along the northeast corridor? Weather Forecast. 16:343–353.

Diehl, K., C. Quick, S. Matthias-Maser, S.K. Mitra, and R. Jaenicke. 2001. The ice-nucleating ability of pollen, part I: Laboratory studies in deposition and condensation freezing modes. Atmos. Res. 58:75–87.

Diem, J.E. 2008. Detecting summer rainfall enhancement within metropolitan Atlanta, Georgia USA. Int. J. Climatol. 28:129–133.

Diem, J.E., and D.P. Brown. 2003. Anthropogenic impacts on summer precipitation in central Arizona, USA. Prof. Geogr. 55:343–355.

Diem, J.E., L.B. Coleman, P.A. Digirolamo, C.W. Gowens, N.R. Hayden, E.E. Unger, G.B. Wetta, and H.A. Williams. 2004. Comments on "Rainfall modification by major urban areas: Observations from spaceborne rain radar on the TRMM satellite". J. Appl. Meteorol. 43:941–950.

Diem, J.E., and T.L. Mote. 2005. Interepochal changes in summer precipitation in the southeastern United States: Evidence of possible urban effects near Atlanta, Georgia. J. Appl. Meteorol. 44:717–730.

Dixon, P.G., and T.L. Mote. 2003. Patterns and causes of Atlanta's urban heat island-initiated precipitation. J. Appl. Meteorol. 42:1273–1284.

Dupont, S., T. Otte, and J. Ching. 2004. Simulation of meteorological fields within and above urban and rural canopies with a mesoscale model. Boundary-Layer Meteorol. 113:111–158.

Dusek, U., G.P. Frank, L. Hildebrandt, J. Curtius, J. Schneider, S. Walter, D. Chand, F. Drewnick, S. Hings, D. Jung, S. Borrmann, and M.O. Andreae. 2006. Size matters more than chemistry for cloud-nucleating ability of aerosol particles. Science 312:1375–1378.

Freud, E., J. Strom, D. Rosenfeld, P. Tunved, and E. Swietlicki. 2008. Anthropogenic aerosol effects on convective cloud microphysical properties in southern Sweden. Tellus Ser. B Chem. Phys. Meteorol. 60:286–297.

Fujibe, F. 2003. Long-term surface wind changes in the Tokyo metropolitan area in the afternoon of sunny days in the warm season. J. Meteorol. Soc. Jpn. 81:141–149.

Gero, A.F., and A.J. Pitman. 2006. The impact of land cover change on a simulated storm event in the Sydney basin. J. Appl. Meteorol. Climatol. 45:283–300.

Givati, A., and D. Rosenfeld. 2004. Quantifying precipitation suppression due to air pollution. J. Appl. Meteorol. 43:1038–1056.

Grimaldi, D.A., and M.S. Engel. 2007. Why descriptive science still matters. Bioscience 57:646–647.

Grundstein, A., S.E. Sarnat, M. Klein, J.M. Shepherd, L. Naeher, T.L. Mote, and P. Tolbert. 2008. Thunderstorm-associated asthma in Atlanta, Georgia. Thorax 63:659–660 10.1136/thx.2007.092882.

Guo, X., D. Fu, and J. Wang. 2006. Mesoscale convective precipitation system modified by urbanization in Beijing City. Atmos. Res. 82:112–126.

GURME WMO. 1995–2008. GURME—The WMO. Gaw Urban Research Meteorology and Environmental Project. Available at http://www.cgrer.uiowa.edu/people/carmichael/GURME/GURME.html (verified 2 Feb. 2010).

Han, J.Y., and J.J. Baik. 2008. A theoretical and numerical study of urban heat island-induced circulation and convection. J. Atmos. Sci. 65:1859–1877.

Hand, L., and J.M. Shepherd. 2009. An investigation of warm season spatial rainfall variability in Oklahoma City: Possible linkages to urbanization and prevailing wind. J. Appl. Meteor. Climatol. 48:251–269.

Heisler, G.M., and A.J. Brazel. 2010. The urban physical environment: Temperature and urban heat islands. p. 29–56. *In* J. Aitkenhead-Peterson and A. Volder (ed.) Urban ecosystem ecology. Agron. Monogr. 55. ASA, CSSA, and SSSA, Madison, WI.

Hidalgo, J., V. Masson, A. Baklanov, G. Pigeon, and L. Gimenoa. 2008. Advances in urban climate modeling. Trends and directions in climate research. Ann. N. Y. Acad. Sci. 1146:354–374.

Hjelmfelt, M.R. 1982. Numerical simulation of the effects of St. Louis on mesoscale boundary-layer airflow and vertical air motion: Simulations of rrban vs. non-urban effects. J. Appl. Meteorol. 21:1239–1257.

Horton, R.E. 1921. Thunderstorm breeding spots. Mon. Weather Rev. 49:193.

Huff, F., and S.A. Changnon. 1972a. Climatological assessment of urban effects on precipitation at St. Louis. J. Appl. Meteorol. 11:823–842.

Huff, F.A., and S.A. Changnon. 1972b. Climatological assessment of urban effects on precipitation St. Louis: Part II. Final Report. NSF Grant GA-18781. Illinois State Water Survey, Champaign.

Huff, F.A., and J.L. Vogel. 1978. Urban, topographic and diurnal effects on rainfall in the St. Louis region. J. Appl. Meteorol. 17:565–577.

Ikebuchi, S., K. Tanaka, Y. Ito, Q. Moteki, K. Souma, and K. Yorozu. 2007. Investigation of the effects of urban heating on the heavy rainfall event by a cloud resolving model CReSiBUC. Ann. Disas. Prev. Res. Inst. Kyoto Univ. No. 50C:105–111.

Inoue, T., and F. Kimura. 2004. Urban effects on low-level clouds around the Tokyo metropolitan area on clear summer days. Geophys. Res. Lett. 31:L05103, doi:10.1029/2003GL018908.

Jauregui, E., and E. Romales. 1996. Urban effects on convective precipitation in Mexico city. Atmos. Environ. 30:3383–3389.

Jin, M.L., and J.M. Shepherd. 2005. Inclusion of urban landscape in a climate model- How can satellite data help? Bull. Am. Meteorol. Soc. 86:681–689.

Jin, M., and J.M. Shepherd. 2008. Aerosol relationships to warm season clouds and rainfall at monthly scales over east China: Urban land versus ocean. J. Geophys. Res. 113:D24S90 doi:10.1029/2008JD010276.

Jin, M.L., J.M. Shepherd, and M.D. King. 2005. Urban aerosols and their variations with clouds and rainfall: A case study for New York and Houston. J. Geophys. Res. Atmos. 110:D10S20, doi:10.1029/2004JD005081.

Jin, M.L., J.M. Shepherd, and C. Peters-Lidard. 2007. Development of a parameterization for simulating the urban temperature hazard using satellite observations in climate model. Nat. Hazards 43:257–271.

Kar, S.K., Y.A. Liou, and K.J. Ha. 2007. Characteristics of cloud-to-ground lightning activity over Seoul, South Korea in relation to an urban effect. Ann. Geophys. 25:2113–2118.

Kaufman, Y.J., I. Koren, L.A. Remer, D. Rosenfeld, and Y. Rudich. 2005. The effect of smoke, dust and pollution aerosol on shallow cloud development over the Atlantic Ocean. Proc. Natl. Acad. Sci. USA 102:11207–11212.

Kaufmann, R.K., K.C. Seto, A. Schneider, Z. Liu, L. Zhou, and W. Wang. 2007. Climate response to rapid urban growth: Evidence of a human-induced precipitation deficit. J. Clim. 20:2299–2306.

Khain, A., A. Pokrovsky, M. Pinsky, A. Seifert, and V. Phillips. 2004. Simulation of effects of atmospheric aerosols on deep turbulent convective clouds using a spectral microphysics mixed-phase cumulus cloud model. Part I: Model description and possible applications. J. Atmos. Sci. 61:2963–2982.

Kratzer, P.A. 1937. Das stadtklima. Friedr. Vieweg, Braunschweig.

Kratzer, P.A. 1956. Das stadtklima. 2nd ed. Friedr. Vieweg, Braunschweig. Transl. by the U.S. Air Force, Cambridge Research Laboratories, Bedford, MA.

Kusaka, H., and F. Kimura. 2004a. Thermal effects of urban canyon structure on the nocturnal heat island: Numerical experiment using a mesoscale model coupled with an urban canopy model. J. Appl. Meteorol. 43:1899–1910.

Kusaka, H., and F. Kimura. 2004b. Coupling a single-layer urban canopy model with a simple atmospheric model: Impact on urban heat island simulation for an idealized case. J. Meteorol. Soc. Jpn. 82:67–80.

Lacke, M., T.L. Mote, and J.M. Shepherd. 2009. Aerosols and associated precipitation patterns in Atlanta. Atmos. Environ. doi:10.1016/j.atmosenv.2009.04.022.

Landsberg, H. 1956. The climate of towns. p. 584–603. In W.L. Thomas Jr. (ed.) Man's role in changing the face of the Earth. University of Chicago Press, Chicago, IL.

Landsberg, H.E. 1970. Man-made climate changes. Science 170:1265–1274.

Landulfo, E., A. Papayannis, A.Z. De Freitas, N.D. Vieira Jr., R.F. Souza, A. Gonçalves, A.D.A. Castanho, P. Artaxo, O.R. Sánchez-Ccoyllo, D.S. Moreira, and M.P.M.P. Jorge. 2005. Tropospheric aerosol observations in Sao Paulo, Brazil using a compact lidar system. Int. J. Remote Sens. 26:2797–2816.

Lensky, I.M., and R. Drori. 2007. The satellite-based parameter to monitor the aerosol impact on convective clouds. J. Appl. Meteorol. Climatol. 46:660–666.

Lowry, W.P. 1998. Urban effects on precipitation amount. Prog. Phys. Geogr. 22:477–520.

Lyons, W.A., T.E. Nelson, E.R. Williams, J.A. Cramer, and T.R. Turner. 1998. Enhanced positive cloud-to-ground lightning in thunderstorms ingesting smoke from fires. Science 282:77–80.

Meng, W., J. Yen, and H. Hu. 2007. Urban effects and summer thunderstorms in a tropical cyclone affected situation over Guangzhou city. Sci. China Ser. Earth Sci. (Paris) 50:1867–1876.

Miao, S.G., and F. Chen. 2008. Formation of horizontal convective rolls in urban areas. Atmos. Res. 89:298–304.

Mills, G. 2007. Cities as agents of global change. Int. J. Climatol. 27:1849–1857.

Molders, N., and M.A. Olson. 2004. Impact of urban effects on precipitation in high latitudes. J. Hydrometeorol. 5:409–429.

Mote, T.L., M.C. Lacke, and J.M. Shepherd. 2007. Radar signatures of the urban effect on precipitation distribution: A case study for Atlanta, Georgia. Geophys. Res. Lett. 34:L20710, doi:10.1029/2007GL031903.

Naccarato, K.P., O. Pinto, and I. Pinto. 2003. Evidence of thermal and aerosol effects on the cloud-to-ground lightning density and polarity over large urban areas of Southeastern Brazil. Geophys. Res. Lett. 30:1674–1677.

Niyogi, D., T. Holt, S. Zhong, P.C. Pyle, and J. Basara. 2006. Urban and land surface effects on the 30 July 2003 mesoscale convective system event observed in the southern Great Plains. J. Geophys. Res. Atmos. 111(D19):D19107, doi:10.1029/2005JD006746.

Ntelekos, A.A., J.A. Smith, and W.F. Krajewski. 2007. Climatological analyses of thunderstorms and flash floods in the Baltimore metropolitan region. J. Hydrometeorol. 8:88–101.

Ohashi, Y., and H. Kida. 2002. Local circulations developed in the vicinity of both coastal and inland urban areas: A numerical study with a mesoscale atmospheric model. J. Appl. Meteorol. 41:30–45.

Oke, T.R. 1987. Boundary layer climates. 2nd ed. Methuen Co., London.

Orville, R.E., G. Huffines, J. Nielsen-Gammon, R. Zhang, B. Ely, S. Steiger, S. Phillips, S. Allen, and W. Read. 2001. Enhancement of cloud-to-ground lightning over Houston, Texas. Geophys. Res. Lett. 28:2597–2600, doi:10.1029/2001GL012990.

Peters, A., D.W. Dockery, J.E. Muller, and M.A. Mittleman. 2001. Increased particulate air pollution and the triggering of myocardial infarction. Circulation 103:2810–2815.

Phillips, J.D. 2001. Human impacts on the environment: Unpredictability and the primacy of place. Phys. Geogr. 22:321–332.

Pielke, R.A., J. Adegoke, A. Beltrán-Przekurat, C.A. Hiemstra, J. Lin, U.S. Nair, D. Niyogi, and T.E. Nobis. 2007. An overview of regional land-use and land-cover impacts on rainfall. Tellus Ser. B. Chem. Phys. Meteorol. 59:587–601.

Pinto, I. R.C.A., O. Pinto Jr., M.A.S.S. Gomes, and N.J. Ferreira. 2004. Urban effect on the characteristics of cloud-to-ground lightning over Belo Horizonte-Brazil. Ann. Geophys. 22:697–700.

Robaa, S.M. 2003. Urban-suburban/rural differences over greater Cairo, Egypt. Atmosfera 16:157–171.

Rose, L.S., J.A. Stallins, and M.L. Bentley. 2008. Concurrent cloud-to-ground lightning and precipitation enhancement in the Atlanta, Georgia (United States), Urban Region. Earth Interact. 12, doi:10.1175/2008EI265.1.

Rosenfeld, D. 2000. Suppression of rain and snow by urban and industrial air pollution. Science 287:1793–1796.

Rosenfeld, D., M. Fromm, J. Trentmann, G. Luderer, M.O. Andreae, and R. Servranckx. 2007. The Chisholm firestorm: Observed microstructure, precipitation and lightning activity of a pyro-cumulonimbus. Atmos. Chem. Phys. 7:645–659.

Rosenfeld, D., and I.M. Lensky. 1998. Satellite-based insights into precipitation formation processes in continental and maritime convective clouds. Bull. Am. Meteorol. Soc. 79:2457–2476.

Rosenfeld, D., U. Lohmann, G.B. Raga, C.D. O'Dowd, M. Kulmala, S. Fuzzi, A. Reissell, and M.O. Andreae. 2008. Flood or drought: How do aerosols affect precipitation? Science 321:1309–1313.

Rozoff, C.M., W.R. Cotton, and J.O. Adegoke. 2003. Simulation of St. Louis, Missouri, land use impacts on thunderstorms. J. Appl. Meteorol. 42:716–738.

Ruddiman, W.F. 2003. The anthropogenic greenhouse era began thousands of years ago. Clim. Change 61:261–293.

Santosa, S.J. 2010. Urban air quality. p. 57–74. In J. Aitkenhead-Peterson and A. Volder (ed.) Urban ecosystem ecology. Agron. Monogr. 55. ASA, CSSA, and SSSA, Madison, WI.

Sarnat, J.A., and F. Holguin. 2007. Asthma and air quality. Curr. Opin. Pulm. Med. 13:63–66.

Sato, N., and M. Takahashi. 2000. A weekly cycle of summer heavy rainfall events in Tokyo. Tenki. 47:643–648.

Schröder, B., and R. Seppelt. 2006. Analysis of pattern-process interactions based on landscape models—Overview, general concepts, and methodological issues. Ecol. Modell. 199:505–516.

Schultz, D.M., S. Mikkonen, A. Laaksonen, and M.B. Richman. 2007. Weekly precipitation cycles? Lack of evidence from United States surface stations. Geophys. Res. Lett. 34:L22815.

Seinfeld, J.H., and S.N. Pandis. 1998. Atmospheric chemistry and physics: From air pollution to climate change. Wiley-Interscience, New York.

Selover, N. 1997. Precipitation patterns around an urban desert environment topographic or urban influences? Assoc. Am. Geogr. Convention, 25–29 May 1997. Assoc. Am. Geogr., Fort Worth, TX.

Seto, K., and J.M. Shepherd. 2009. Global urban land use trends and climate impacts. Curr. Opin. Environ. Sustain. 1:89–95.

Shem, W., and J.M. Shepherd. 2009. On the impact of urbanization on summertime thunderstorms in Atlanta: Two numerical model case studies. Atmos. Res. 92:172–189.

Shepherd, J.M. 2004. Comments on "Rainfall modification by major urban areas: Observations from spaceborne rain radar on the TRMM satellite"—Reply. J. Appl. Meteorol. 43:951–957.

Shepherd, J.M. 2005. A review of current investigations of urban-induced rainfall and recommendations for the future. Earth Interact. 9. Paper 12.

Shepherd, J.M. 2006. Evidence of urban-induced precipitation variability in arid climate regimes. J. Arid Environ. 67:607–628.

Shepherd, J.M., and S.J. Burian. 2003. Detection of urban-induced rainfall anomalies in a major coastal city. Earth Interact. 7:1–14.

Shepherd, J.M., M. Carter, M. Manyin, D. Messen, and S. Burian. 2009. The impact of urbanization on current and future coastal convection: A case study for Houston. Environ. Plan. B doi:10.1068/b34102t.

Shepherd, J.M., B.S. Ferrier, and P.S. Ray. 2001. Rainfall morphology in Florida convergence zones: A numerical study. Mon. Weather Rev. 129:177.

Shepherd, J.M., and M. Jin. 2004. Linkages between the urban environment and earth's climate system. EOS 85:227–228.

Shepherd, J.M., and T.L. Mote. 2009. Urban effects on rainfall variability: Potential implications for Georgia's water supply. *In* Proc. of the 2009 Georgia Water Resources Conference, Athens, GA.

Shepherd, J.M., H. Pierce, and A.J. Negri. 2002. Rainfall modification by major urban areas: Observations from spaceborne rain radar on the TRMM satellite. J. Appl. Meteorol. 41:689–701.

Shepherd, J.M., L. Taylor, and C. Garza. 2004. A dynamic multi-criteria technique for siting NASA-Clark Atlanta rain gauge network. J. Atmos. Ocean. Technol. 21:1346–1363.

Simpson, M.D. 2006. Role of urban land use on mesoscale circulations and precipitation. Ph.D. diss. North Carolina State Univ., Raleigh.

Soriano, L.R., and F. de Pablo. 2002. Effect of small urban areas in central Spain on the enhancement of cloud-to-ground lightning activity. Atmos. Environ. 36:2809–2816.

Souch, C., and S. Grimmond. 2006. Applied climatology: Urban climatology. Prog. Phys. Geogr. 30:270–279.

Stallins, J.A., and M.L. Bentley. 2006. Urban lightning climatology and GIS: An analytical framework from the case study of Atlanta, Georgia. Appl. Geogr. 26:242–259.

Stallins, J.A., M.L. Bentley, and L.S. Rose. 2006. Cloud-to-ground flash patterns for Atlanta, Georgia (USA) from 1992 to 2003. Clim. Res. 30:99–112.

Stallins, J.A., and L.S. Rose. 2008. Urban lightning: Current research, methods, and the geographical perspective. Geogr. Compass. doi: 10.1111/j.1749–8198.2008.00110.x.

Steiger, S.M., and R.E. Orville. 2003. Cloud-to-ground lightning enhancement over southern Louisiana. Geophys. Res. Lett. 30(19):1975, doi:10.1029/2003GL017923.

Stout, G.E. 1962. Some observations of cloud initiation in industrial areas. *In* Air over cities. Technical Report A62–5. U.S. Public Health Service, Washington, DC.

Tayanc, M., and H. Toros. 1997. Urbanization effects on regional climate change in the case of four large cities of Turkey. Clim. Change 35:501–524.

Teller, A., and Z. Levin. 2008. Factorial method as a tool for estimating the relative contribution to precipitation of cloud microphysical processes and environmental conditions: Method and application. J. Geophys. Res. Atmos. 113(D2):D02202, doi:10.1029/2007JD008960.

Thielen, J.W., A. Wobrock, A. Gadian, P.G. Mestayer, and J.-D. Creutin. 2000. The possible influence of urban surfaces on rainfall development: A sensitivity study in 2D in the meso-gamma scale. Atmos. Res. 54:15–39.

Trapp, R.J., N.S. Diffenbaugh, H.E. Brooks, M.E. Baldwin, E.D. Robinson, et al. 2007. Changes in severe thunderstorm environment frequency during the 21st century caused by anthropogenically enhanced global radiative forcing. Proc. Natl. Acad. Sci. USA 104:19719–19723.

Trenberth, K.E., P.D. Jones, P. Ambenje, R. Bojariu, D. Easterling, A. Klein Tank, D. Parker, F. Rahimzadeh, J.A. Renwick, M. Rusticucci, B. Soden, and P. Zhai. 2007. Observations: Surface and atmospheric climate change. *In* S. Solomon et al. (ed.) Climate Change 2007: The physical science basis. Contribution of Working Group I to the Fourth Assessment Report of the Intergovernmental Panel on Climate Change. Cambridge Univ. Press, Cambridge, UK.

Trusilova, K., M. Jung, G. Churkina, U. Karstens, M. Heimann, and M. Claussen. 2008. Urbanization impacts on the climate in Europe: Numerical experiments by the PSU-NCAR Mesoscale Model (MM5). J. Appl. Meteorol. Climatol. 47:1442–1455.

Twomey, S. 1977. Minimum size of particle for nucleation in clouds. J. Atmos. Sci. 34:1832–1835.

Unger, J. 2004. Intra-urban relationship between surface geometry and urban heat island: Review and new approach. Clim. Res. 27:253–264.

UNFPA. 2007. The state of world population 2007. United Nations Population Fund, United Nations Publications.

van den Heever, S.C., and W.R. Cotton. 2007. Urban aerosol impacts on downwind convective storms. J. Appl. Meteorol. Climatol. 46:828–850.

von Storch, H., and N. Stehr. 2006. Anthropogenic climate change: A reason for concern since the 18th century and earlier. Geograf. Annal. Ser. A. Phys. Geogr. 88A:107–113.

Vukovich, F.M., and J.W. Dunn. 1978. Theoretical study of St. Louis heat island—Some parameter variations. J. Appl. Meteorol. 17:1585–1594.

Westcott, N.E. 1995. Summertime cloud-to-ground lightning activity around major midwestern urban areas. J. Appl. Meteorol. 34:1633–1642.

Worster, D. 1979. Dust Bowl: The southern plains in the 1930s. Oxford Univ. Press, New York.

Yair, Y. 2008. Charge generation and separation processes. Space Sci. Rev. 137:119–131.

Yow, D.M. 2007. Urban heat islands: Observations, impacts, and adaptation. Geogr. Compass. doi:10.1111/j.1749–8198.2007.00063.x.

Zhang, C., F. Chen, S. Miao, Q. Li, X. Xia, and C.Y. Xuan. 2009. Impacts of urban expansion and future green planting on summer precipitation in the Beijing metropolitan area. J. Geophys. Res. 114:D02116, doi:10.1029/2008JD010328.

Zhang, C., S. Miao, Q. Li, and F. Chen. 2007. Incorporation of offline-resolution land use information of Beijing into numerical weather model and its assessing experiments on a summer severe rainfall. Chin. J. Geophys.

Zhang, H., N. Sato, T. Izumi, K. Hanaki, and T. Aramaki. 2008. Modified RAMS-Urban Canopy Model for Heat Island Simulation in Chongqing, China. J. Appl. Meteor. Climatol. 47:509–524.

Zhou, Y., and J.M. Shepherd. 2009. Atlanta's urban heat island under extreme heatconditions and potential mitigation strategies. Nat. Hazards 10.1007/s11069–009–9406-z.

2

The Urban Physical Environment: Temperature and Urban Heat Islands

Gordon M. Heisler
Anthony J. Brazel

Abstract

The term *urban heat island* (UHI) describes the phenomenon in which cities are gener-ally warmer than adjacent rural areas. The UHI effect is strongest with skies free of clouds and with low wind speeds. In moist temperate climates, the UHI effect causes cities to be slightly warmer in midday than rural areas, whereas in dry climates, irriga-tion of vegetation in cites may cause slight midday cooling compared to rural areas. In most climates, maximum UHIs occur a few hours after sunset; maximum intensities increase with city size and may commonly reach 10°C, depending on the nature of the rural reference. Since the recognition of London's UHI by Luke Howard in the early 1800s, UHIs of cities around the world have been studied to quantify the intensity of UHIs, to understand the physical processes that cause UHIs, to estimate the impacts of UHIs, to moderate UHI effects, and to separate UHI effects from general warming of Earth caused by accumulation of greenhouse gases in the upper atmosphere. This chapter reviews a portion of the literature on UHIs and their effects, literature that has expanded greatly in the last two decades spurred on by a series of successful interna-tional conferences. Despite considerable research, many questions about UHI effects remain unanswered. For example, it is still not clear what portion of the long-term trends of increasing temperatures at standard weather stations is caused by UHI effects and how much is contributed by greenhouse gas effects. Also not well quantified is the effect of increasing tree cover in residential areas on temperatures

The process of urbanization alters natural surface and atmospheric conditions so as to create generally warmer temperatures (Landsberg, 1981). Oke (1997) suggested that urban atmospheres provide the strongest evidence we have of the potential for human activities to change climate. In the 20th century, rapid urbanization occurred worldwide, and today the majority of the world's popu-lation lives in cities. Increased temperature in cities, termed the *urban heat island* (UHI) *effect*, is present all around the world and both contributes to global climate change and, in turn, is exacerbated by global climate change (Mills, 2007; San-chez-Rodriguez et al., 2005). With increasing energy shortages, the importance of

G.M. Heisler, USDA Forest Service, Northern Research Station, SUNY ESF, 5 Moon Library, Syracuse, NY 13210 (gheisler@fs.fed.us); A.J. Brazel, School of Geographical Sciences and Urban Planning, Arizona State University, Coor Hall, 975 S. Myrtle, Tempe, AZ 85287-5302 (abrazel@asu.edu).

doi:10.2134/agronmonogr55.c2

urban temperatures will increase, especially in climates in which passive cooling by opening windows can reduce reliance on air conditioning (Mills, 2006). The UHI effect creates one of the key challenges to evaluating the influence of greenhouse gases on global climate change because urban influences are present in archived historical weather data that are used to determine long-term climate trends (Karl and Jones, 1989).

Publication of observations of the different climate of cities began with the now-classic work of Howard (1833), who described the climate of London as being warmer than surrounding areas. Real growth of urban climatology dates from the 1920s, followed by increases in interest in urban climates between the 1930s and 1960s (especially in Germany, Austria, France, and North America). After World War II and into the environmental era of the 1960s and 1970s and beyond, there was an exponential increase in urban climatic investigations, and the investigations have simultaneously become less descriptive, more oriented to quantitative and theoretical modeling, and more integrative and interdisciplinary (Brazel and Quatrocchi, 2005).

Types of Urban Heat Islands

It is important to distinguish between the different types of UHIs and how they relate to urban built and vegetative structure (Table 2–1). For decades, urban climatologists have used an analogy with rural forests to describe urban climate in terms of the urban canopy layer (UCL), the space generally below the tops of trees and buildings. In humid climate forests, the active surface, where most of the exchange of radiant energy and turbulent transport of water vapor and heat takes place, is usually a layer from the tops of trees down to the point where tree crowns meet. Foresters think of the forest canopy layer as the space between the tops of tallest trees and the bottom of tree crowns that bear living foliage. The active surface in urban areas is more variable than in closed natural forests, and the urban canopy layer is usually considered to be the entire space from the tops of trees or buildings, depending on which dominates, down to ground level. In UHI studies, canopy-layer air temperatures are usually measured at about the height of people or the lower stories of buildings, between 1.5 and 3 m above ground. If that temperature is warmer than the temperature at the same height in nearby rural areas, then this is termed a UCL heat island (Oke, 1976, 1995). This chapter focuses on urban canopy layer heat islands.

The heat island that forms in the atmospheric boundary layer above the city is the urban boundary layer (UBL) heat island (Oke, 1987, 1995). The UBL varies greatly in thickness and turbulence over the course of a clear day (Stull, 2000), and thus the UHI in the UBL also varies. During the night, if the sky is

Table 2–1. Simple classification scheme of urban heat island types (after Oke, 2006a).

UHI Type	Location
Air temperature UHI:	
Urban canopy layer heat island	Found beneath roof or tree-top level
Urban boundary layer	Found above roof level; can be advected downwind with the urban plume
Surface temperature UHI	Different heat islands according to the definition of surface used (e.g. bird's eye view 2D vs. true 3D surface vs. ground)
Sub surface UHI	Found in the ground beneath the surface

not heavily overcast, radiative cooling lowers the temperature of surfaces at the bottom of the boundary layer, and the air just above these surfaces tends toward slow laminar flow horizontal to the Earth's surface and remains in a shallow layer, 20 to 300 m thick, even as air in the free atmospheric above the boundary layer may be moving at a much higher speed. During the day, the air at the bottom of the boundary layer becomes turbulent because of surface heating and it mixes with air throughout the boundary layer to form a "mixed layer" that expands vertically. This process increases UBL thickness to 1 km or more. Stull (2000) provided a good description of UBL dynamics, and Oke (1995) summarized UBL heat islands.

Urban heat islands may also be described by the temperatures of the upper surfaces of buildings, trees, streets, lawns, and so forth, as seen from above. This is sometimes called the urban "skin" temperature. This type of heat island should not be confused with "surface temperatures" as used in some climatology reports to refer to air temperatures near the ground, usually at a height of 1.5 m. The 1.5-m height is essentially *at* the surface of Earth compared to the elevations at which temperatures are measured in atmospheric soundings (balloon measurements through the atmosphere), which may go to 30 km above the Earth. During the day, temperatures *of* the surfaces ("skin" temperatures) of nonliving solid material can be much warmer than air temperatures (Hartz et al., 2006b). Temperatures of entire urban surfaces are generally measured by satellite (e.g., Gallo et al., 1993). With clear skies, upper surface heat islands are small at night and large during the day, the opposite of UCL heat islands (Voogt and Oke, 2003).

Subsurface or soil heat islands have received much less attention than air temperature or skin temperature heat islands, primarily because very small scale effects of surface cover or shading may affect near-surface soil temperatures much more than the general large scale UHI. Most studies of urban soil temperatures have concentrated on the effects of asphalt cover on temperatures of adjacent soil or of soil below the asphalt (e.g., Celestian and Martin, 2004; e.g., Halverson and Heisler, 1981). Urban soil temperatures are described in Chapter 7 (Pouyat et al., 2010, this volume).

Heat Island Impacts

The influences of UHIs on human society include effects on human health and comfort, energy use, air pollution, water use, biological activity, ice and snow, flooding, and even environmental justice (Harlan et al., 2006; Roth, 2002; Voogt, 2002). Urban heat island effects may also lead to modifications to precipitation and lightning (see Chapter 1, Shepherd et al., 2010, this volume).

Not all UHI effects are viewed as negative. In cold climates, UHI impacts may help reduce hazards of ice and snow in the city (Voogt, 2002), and winter comfort may be enhanced.

Both direct and indirect effects of temperature changes influence human health and comfort. Several studies have demonstrated that temperature threshold exceedance and air pollution in cities exacerbate human discomfort, heat-related health incidences, and mortality (Baker et al., 2002; Grass and Crane, 2008; Harlan et al., 2006; Kalkstein and Smoyer, 1993). Ozone concentrations, which influence human health, are amplified by the effects of higher daily maximum temperatures (Oke, 1997). For tourist information and promotion, city climate is generally reported by data from the main weather station, usually an airport, which may have significantly more or less disagreeable climate than most of the city (Hartz et al., 2006a). Given an assumed temperature change, the effects of that change on human comfort can be quantitatively modeled (Hartz et al., 2006a; Heisler and Wang, 2002; Matzarakis et al., 2007), but because of both physiological and psychological adaptation (Nikolopoulou and Steemers, 2003), the perception of climate among residents of a particular city may not change in proportion to the magnitude of temperature changes.

Coastal, tropical, more arid, and more rapidly growing cities are especially vulnerable to global climate change and higher temperatures, as well as impacts of urbanization. For example, sea level rise portends major impacts on coastal city infrastructures such as in New York (Rosenzweig et al., 2007). In arid-land cities, excessive heat waves and the UHI, together with rapid population growth, present challenges to city officials in coping with impacts on health, water, and energy (Baker et al., 2002). Most population growth in the world will take place in urban areas, and rapid growth in moderate-sized tropical cities is expected. In addition, growth will be a major issue in less developed countries with low adaptive capacity, and the impact of urban climate may be accentuated in these environments (Dabberdt, 2007).

Energy use and peak electricity loads are impacted by higher temperatures in cities (e.g., Akbari et al., 1989), and the UHI may increase other resource use. For example, in Phoenix, AZ, a 0.55°C (1°F) increase in minimum daily temperature was associated with a 1098-L (290-gallon) increase in monthly water use in a typical single-family dwelling (Guhathakurta and Gober, 2007). Of course, urban areas may also adversely impact water resources by increasing peak flooding in streams through the city (Brazel and Quatrocchi, 2005).

Heat waves may impact parts of a city differently as a function of exposed landscapes, lack of air conditioning, and citizen inability to adapt to intense heat. In some cities or at least parts thereof, including Phoenix, AZ, there is a positive correlation between residents' income and vegetation cover, which suggested in initial studies at the census tract level that lower income residents could be negatively impacted by the UHI effect (Jenerette et al., 2007). Subsequent detailed neighborhood-scale level research on this issue has substantiated the patterns of sparse vegetation cover, landscapes with exposed and barren soil, and lack of proper cooling associated with low income levels (Harlan et al., 2006; Ruddell et al., 2010).

There are important and complex interactions between biological components of urban ecosystems and the UHI effect. Pouyat et al. (1995) noted UHI effects on carbon and nitrogen dynamics, especially N mineralization, in forest remnants within urban areas. Carreiro and Tripler (2005) and Ziska et al. (2003, 2004) developed the case for using UHI effects on biotic components of ecosystems as surrogates for global climate change effects. The papers by Ziska also show that the UHI effect can interact with higher CO_2 in urban areas to increase ragweed (*Ambrosia artemisiifolia* L.) production and therefore have an important influence on public health.

Sources of Information about Heat Islands
Past Reviews
Many scholarly reviews of urban climatology and accompanying bibliographies have illustrated how cities alter their climatic environment (e.g., Beryland and

Kondratyev, 1972; Brazel, 1987; Chandler, 1976; Landsberg, 1981; Lee, 1984; Oke, 1974, 1979, 1980, 1987). Today, urban climatology has achieved its own recognition as a subdiscipline in climatology and among allied disciplines, such as planning, ecology, environmental science, and meteorology. Roth (2007) reviewed urban climates in the tropics and listed some earlier reviews, including one by Givoni (1991) that evaluated the effects of vegetation on urban climate. Arnfield (2003) described results of 20 years of urban climate research beginning about 1980, a period in which new electronics with fast-response sensors, advanced remote sensing tools, and computer modeling capability played increasingly important roles in enhanced understanding of urban climate dynamics. Grimmond (2006) reviewed methods of observing meteorological variables in urban atmospheres, and Souch and Grimmond (2006) presented a concise but well-referenced review that covered UHI development and other aspects of urban climate. A book by Gartland (2008) summarized urban heat island physical science, impacts, and mitigation for general audiences including activist nonprofit groups.

Conference Proceedings and Organizations

The increasing interest in urban climate including UHIs is evidenced in the increasing size and frequency of conferences on urban climate and environment. An early event devoted to urban environment was the Conference on Urban Environment in Philadelphia in 1972. The Metropolitan Physical Environment conference in 1975 (Heisler and Herrington, 1977) included a variety of papers on the urban environment, including urban temperatures. The American Meteorological Society continued the urban environment theme with a series of Urban Environment Symposia that were held beginning in 1998; the eighth occurred in January 2009. Beginning with the Third Urban Environment Symposium in 2000, the proceedings are freely available online (American Meteorological Society, 2009). The distribution of knowledge about urban climate is the sole purpose of the International Association for Urban Climate (http://www.urban-climate.org/), which has sponsored International Conferences on Urban Climate (ICUC) since 1989. The seventh ICUC conference was held in Yokohoma in 2009, and the published proceedings are available online for the last four conferences.

Special Journal Issues

Urban climate is also the subject of a number of special issues of journals, including 17 papers in Volume 84 of *Theoretical and Applied Climatology* in February 2006. A series of refereed articles from the sixth ICUC (ICUC6) in 2006 forms a special issue in *International Journal of Climatology* in 2007 (Grimmond et al., 2007).

Scope of this Chapter

In this chapter we examine the influence of urbanization on temperature, concentrating on the UHI effect in the canopy layer. Our approach is to illustrate temperature patterns and the physical processes that govern temperatures by reviewing examples of the various methods used in urban climate research. We use examples from our past research on urban climate, which includes studies in Phoenix and nearby locations in Arizona and in and near Baltimore, MD. In the past 10 years, research on the urban ecosystems of Phoenix and Baltimore has expanded with the selection of these locations by the National Science Foundation for inclusion as sites in the Long Term Ecological Research program. Baltimore and Phoenix provide contrasts of very different general climates—warm and moist versus hot and dry. Studies in San Juan, Puerto Rico and its vicinity illustrate urban influences in a tropical coastal city (Murphy et al., 2007, 2010). Our discussion focuses on factors that can be altered to modify UHIs, especially management of vegetation.

Energy Exchanges and Urban Heat Island Dynamics

Brazel and Quatrochi (2005) provided an overview of energy exchanges in urban environments. Grimmond (2007) presented a concise summary of urban heat island dynamics. Shepherd et al. (2010, Chapter 1, this volume) briefly framed energy exchanges in the city.

The Urban Energy Balance

The energy balance of a city can be expressed as follows with symbols similar to Eq. 1 in Shepherd et al. (2010, Chapter 1, this volume):

$$\text{NR [i.e., } (1 - \alpha)S\downarrow + LW\downarrow - \varepsilon T_{skin}^{4}] - SH - LH - G + A = 0 \qquad [1]$$

where NR is surface net radiation, SH is sensible heat flux, LH is latent heat flux, and G is the heat flux to storage in the ground or to buildings and vegetation aboveground. The storage in the ground is often separated from storage in the aboveground volume of buildings and vegetation. The SH, LW, and G terms compete for surface net radiation, which is the downward minus upward shortwave and longwave radiation. In the component of NR, α is surface albedo, and S is downward solar radiation, thus, $(1 - \alpha)S\downarrow$ is reflected solar radiation. $LW\downarrow$ is longwave radiation. Emissivity (ε) and surface skin temperature (T_{skin}), through the Stefan–Boltzmann Law, describe the upward longwave radiation, or surface emission. A is the anthropogenic emitted heat.

Shortwave Radiation

An urban area affects the exchanges of shortwave and longwave radiation by air pollution and complex changes of surface radiative characteristics. The atmospheric attenuation of incoming shortwave radiation has been analyzed in numerous urban climatic environments. It is thought that the attenuation in the atmospheric over cities is typically 2 to 10% more than in the surrounding rural areas. Generally, the very shortest wavelengths (<0.4 μm) of the electromagnetic spectrum to reach the surface of Earth, the ultraviolet (UV) portion, are commonly depleted by 50% or more (Heisler and Grant, 2000). However, total depletion across all solar wavelengths (0.15–4.0 μm) is <10%. The processes of scattering and absorption are greatly modified by the urban aerosol characteristics and concentrations (Gomes et al., 2008).

Albedo

The second major effect of urbanization is the change in the ratio of outgoing shortwave radiation to that of incident shortwave radiation in a three-dimensional environment. This ratio, expressed as a percentage, is the *albedo* and is typically less in urban areas than in the surrounding landscape. Lower albedo is due in part to darker surface materials making up the urban mosaic and also to the effects of trapping shortwave radiation by the vertical walls and the urban, canyon-like morphology. There is considerable variation of albedo within the city depending on the vegetative cover, building materials, roof composition, and land-use characteristics. The difference in albedo between a city and its surrounding environment also depends on the surrounding terrain. A city and a dense forest may differ little in albedo; both may range from 10 to 20%. In winter, a mid-latitude to high-latitude city with surrounding snow cover may display a much lower albedo than its surroundings. Thus, since cities receive 2 to 10% less shortwave radiation than their surroundings, yet have slightly lower albedos (by <10%), most cities experience very small overall differences in absorbed shortwave radiation relative to rural surroundings (Brazel and Quatrocchi, 2005).

Longwave Radiation

Longwave radiation is affected by city pollution and the warmer urban surfaces. Warmer surfaces promote greater thermal emission of energy vertically upward from the city surface compared to rural areas, especially at night. Some longwave radiation is reradiated by urban aerosols back to the surface and also from the warmer urban air layer (see Chapter 3, Santosa, this volume). Thus, increases in incoming longwave radiation and outgoing longwave radiation are usually experienced in urban areas. Outgoing longwave radiation increases are slightly greater than the incoming increases in the city, again especially on clear, calm nights. During daytime there is little difference between the city and its surroundings. However, surface emissivity (i.e., the amount emitted relative to black-body amounts for a given temperature) can be quite different between country and city areas, and can account for considerable longwave radiation differences between urban and rural (Yap, 1975). A major consideration is that in the city a three-dimensional surface temperature must be characterized to accurately estimate the flux values of radiation and the energy budget (Voogt and Oke, 1997, 1998).

Longwave emission from soils and soil heat capacity is determined by soil moisture and hence by antecedent precipitation. Therefore, temperatures depend on precipitation (Heisler et al., 2007; Kaye et al., 2003).

Anthropogenic Heat Sources

The A term (from Eq. [1]) ranges from 0 to 300% of net radiation, depending on the extent of industrialization. Generally A is higher in more industrialized cities, in high latitude cities, and in winter. It is composed of heat produced by combustion of vehicle fuels, stationary source releases such as from buildings, and heat released by human metabolism (Qfm) (Sailor and Lu, 2004). Combustion heat is a function of type and amount of gasoline used, number of vehicles, distance traveled, and fuel efficiency. It requires an analysis of consumer usage of fuel such as gas and electricity. Qfm can be evaluated by active and sleep rates, but it is generally less than 5% of total A. Methodology is described in detail by Sailor et al. (2003) and Sailor and Fan (2004) to evaluate the total A term in the energy budget.

Fan and Sailor (2005) estimated fluxes for summer and winter in Philadelphia, PA. The anthropogenic heating ranges from about 10 W m^{-2} at night to around 40 W m^{-2} during the day. This compares with typical peak daytime insolation levels of around 850 W m^{-2}. In winter the anthropogenic heating ranges from about 20 W m^{-2} at night to around 60 W m^{-2} during the day. Nocturnal anthropogenic heating in winter is about double that in summer. The daytime anthropogenic heating is also larger than that for summer (about 1.5 times). At the same time, the peak daytime insolation levels in winter are typically in the range of 400 W m^{-2} (about one-half of the summer magnitude). Inclusion of anthropogenic heating in heat island simulations (either in the air layer or ground surface) increased estimates of the UHI by 0.5 and 2°C during the day and night, respectively. For a winter simulation, the results suggested that about 30 to 50% of the error may be due to the original model not accounting for anthropogenic heating.

Heat Storage, Evapotranspiration, Heating the Air

The partitioning of energy in urban areas among sensible (SH), latent (LH), and storage heat (G) depends primarily on the variety of land uses in the city compared with rural areas. Generally, the drier urban building and road materials induce higher SH, less LH, and higher G in urban areas. Significant LH does, however, occur in some cities. It is theorized that this is due to urban irrigation effects and vegetation in the city (Kalanda et al., 1980). Marotz and Coiner (1973) indicated that vegetation in urban areas is not as limited as supposed, and Oke (1987) showed that G/NR ratios for rural areas vary by only about 0.10 from those for suburban and urban areas. Table 2–2 shows some recent results for selected U.S. cities (Grimmond and Oke, 1995). Note the substantive fluxes of SH and G, also significant LH, especially for the more moist city of Chicago, IL. In an evapotranspiration study of nine places, Grimmond and Oke (1999a) showed that the ratio of LH/NR ranged from 0.09 (Mexico City, Mexico, with very little external water use) to Chicago's ratio of 0.46 (Table 2–2). Goward (1981) listed thermal properties of typical interface materials, noting that most urban area materials (except for wood) have similar thermal properties and that urban thermal inertias (thermal admittance, μ) are higher than dry soils but lower than wet soils (Table 2–3).

Table 2–2. Summary of daytime mean summer energy balance fluxes for selected cities derived from tall tower observations of Grimmond and Oke (1995).†

Location	NR	SH	LH	S	G
			MJ m^{-2} d^{-1}		
Tucson, AZ	16.27	7.54	4.11	4.62	na
Sacramento, CA	12.65	5.19	3.79	3.67	12.73
Chicago, IL	17.20	5.58	7.11	4.51	2.65
Los Angeles, CA	16.40	5.74	4.12	6.54	1.37

† Symbols for fluxes are: NR, net radiation; SH, sensible heat; LH, latent heat; S, total storage both above and below ground estimated as a residual from the energy budget equation; G, storage in ground measured with a soil heat flux plate.

Table 2–3. Thermal properties of typical urban interface materials.

Material	Thermal conductivity	Specific heat	Density	Thermal admittance
	W m^{-1} C^{-1}	J kg^{-1} C^{-1} 10^3	kg m^{-3} × 10^3	J m^{-2} C^{-1} s$^{-1/2}$
Asphalt	0.7454	0.92	2.114	1204
Brick	0.6910	0.84	1.970	1067
Concrete	0.9338	0.67	2.307	1185
Glass	0.8794	0.67	2.600	1213
Granite	2.7219	0.67	2.600	2176
Limestone	0.9338	0.92	1.650	1182
Sand (dry)	0.3308	0.80	1.515	633
Wood	0.2094	1.38	0.500	377
Soil (wet)	2.4288	1.48	2.000	2681
Soil (dry)	0.2513	0.80	1.600	567
Water (20°C)	0.5988	4.15	0.998	1579
Air (20°C)	0.0251	1.01	1.001	56

The ubiquitous asphalt covering in urban areas strongly affects temperatures of soil below the asphalt. In 2.5- by 2.5-m tree planter boxes cut into the asphalt of a parking lot in New Brunswick, NJ, maximum summer temperature exceeded temperature in control tree planting spaces off the parking lot (Halverson and Heisler, 1981). Near the center of the planter spaces, 85 cm from the edge of the asphalt and at a depth of 15 cm, maximum temperature exceeded controls by up to 3°C. At the same depth but below the asphalt, maximum temperatures exceeded controls by up to 10°C. Asphalt covering the soil not only increased maximum temperatures through a 60-cm profile, but increased the rate of heat exchange since temperatures in the covered soil rose and fell more rapidly than control temperatures. Temperatures below the asphalt ranged from 0.5 to 34.2°C, which was well within the toleration of tree roots. In contrast, temperatures below the asphalt of a parking lot in the warmer climate of Phoenix reached the likely plant-damaging temperature of 40°C at a depth of 30 cm (Celestian and Martin, 2004).

Differences in the UCL structure and composition are also important to explaining heat excesses in cities, rather than just the thermal properties of city materials per se. Although much attention has been given to internal variability of climate conditions within the urban environment and to the importance of the UCL (Arnfield, 1982; Goldreich, 1985; Goward, 1981; Grimmond and Oke, 1995, 1999b; Grimmond, 2006; Johnson and Watson, 1984; Lowry, 1977; Oke et al., 1981; Oke, 1982; Terjung and O'Rourke, 1980), many questions remain and the effect of urban form and structure on energy budgets and air temperatures has become a focal point in the field of urban climatology.

Quantifying Urban Canopy Layer Temperature Regimes

Many methods are used to determine how much a city affects climate. Early methodologies were capable of studying urban temperature patterns within the urban canopy layer. These included sampling the differences between urban and rural environments, upwind minus downwind portions of the urban area, urban minus regional ratios of various climatic variables, time trends of differences and ratios, time segment differences such as weekday versus weekend, and point sampling in mobile surveys throughout the urban environment (Lowry 1977).

The intensity of an UHI depends in part on the rural basis for comparison (Hawkins et al., 2004). Though the rural reference is generally an agricultural landscape, the true impact of human development would use as a reference a vegetation cover that represented a natural climax vegetation community for the region.

Urban Structural Classification

In nearly all studies of urban climate, some means must be used to categorize the structure of the urban area. For many urban–rural comparisons, the city character may be described simply by population (Karl et al., 1988; Oke, 1973). A more precise characterization is the use of satellite-derived night light (Hansen et al., 2001).

In recent decades, the intensity of urban heat islands often has been related to urban structure characterized by remote sensing. The scale of the analysis may be large, such as 1 km as used by the Advanced Very High Resolution Radiometer (AVHRR) on a NOAA satellite, which has been used to derive the Normalized Difference Vegetation Index (NDVI), for example, by Gallo et al.

(1993). Another commonly used product is the National Land Cover Database (NLCD) (Homer et al., 2004) that covers all of the United States at a 30-m resolution of land-use categories (Plate 2–1, see the color insert near the center of the book) and tree canopy cover, impervious surface cover, and water cover. Recent tests of the NLCD indicated that it generally underestimates urban canopy and impervious cover (Greenfield et al., 2010). Other 30-m spatial resolution satellite products with higher spectral resolution have been tested in categorizing impervious cover (Weng et al., 2008). These are all two-dimensional products that do not implicitly consider the vertical dimension. The heights of trees and buildings were considered in developing an urban character database from samples of 1-ha (100-m^2) grid squares from a variety of data sources for Sacramento, CA; Uppsala, Sweden; El Paso, TX–Juarez, Mexico; and St. Louis, MO (e.g., Cionco and Ellefsen, 1998). Some current research is making use of the vertical dimension with high-resolution digital aerial imagery and light detection and ranging (LIDAR) data to resolve landscape objects, such as houses, trees, and pavements down to single-car driveways (Su et al., 2008; Zhou and Troy, 2008). The high resolution may make it possible to improve models of urban temperature patterns to estimate the influence of added tree cover in suburban neighborhoods.

In examining UHI effects on minimum temperature at the local scale in Phoenix, AZ, Brazel et al. (2007) found significant correlations of temperature increases with type of development zone (DZ) and the number of home completions within 1 km during the 14-yr study. The DZ types were urban core, infill, agricultural fringe, desert fringe, and exurban. The DZ concept originated with Oke (2006a), who proposed seven DZ types.

Short-Time Observations with Fixed Sensors

Influences on Temperature

The simplest observations of urban heat islands are by comparison of temperatures in the urban area and in the adjacent rural area. The length of the observations needs to be only sufficiently long to capture representative times conducive to large heat island formation, usually with clear skies and low wind speeds a few hours after sunset. More will be learned if the observations are continuous over at least a year, which will provide samples over a wider range of synoptic weather conditions. This method is most telling for cities with little topographic relief that are not near large bodies of water.

Measurements near Baltimore, MD (Heisler et al., 2006a) illustrated the influence of land cover and land use on temperature differences (Plate 2–1). Temperatures were measured at six suburban sites: a grassy area near a large apartment complex (Apartments, Plate 2–1a,c), a residential area with heavy tree cover but few buildings (Residential under Trees), a residential area with some trees and large lawn areas (Residential Open), a woodlot (Woods), a large open pasture (Rural Open), and at the Baltimore/Washington International Airport (Airport). The urban reference site was in downtown Baltimore (R in Plate 2–1a). None of the suburban sites were far from some developed land uses (Plate 2–1b). From May through September in the Baltimore study, average hourly temperature differences, ΔT, downtown site minus each of the other sites, were positive for all hours of the day. For most sites, ΔT through the day followed the usual UHI pattern of moist temperate climates—urban areas slightly warmer in mid-day, more rapid cooling of more rural areas after sunset leading to a maximum heat island in a few hours, and the cooler suburban areas heating more quickly after sunrise to approach the temperatures of urban areas that are heating more slowly. The Woods site was coolest both day and night, and the other site with many trees, Residential under Trees, was similarly cool during the day. However, the Residential under Trees site was unusual in not cooling as much as other suburban sites at night, in part perhaps because of cold air drainage away from the site into nearby valleys (Plate 2–1a,c).

Interactions with Terrain Effects

Studies in other cities have described interactions between topographic influences and land cover; two of note are descriptions of the effects of complex topography in Phoenix (Brazel et al., 2005) and Tucson (Comrie, 2000). For Phoenix, Brazel et al. (2005) reported local thermal winds (i.e., daytime upslope and evening downslope winds) that extended 50 km across the Phoenix area when synoptic winds were low. These conditions occur with a frequency between 13% of days in July to 70% of days in June. The topographic influences were noted for slopes as small as 0.5°. The topographic influence was analyzed to be about equal to the convective circulation effect of the UHI. Time of onset of the downslope flows could be hours after sunset so that in one day of observations, near center city the UHI peaked at about 5.5°C at 2200 h, then decreased to 0.5°C at 0300 h because of downslope flow, before increasing to a second peak of about 3°C at 0500 h after slope flow subsided. Comrie (2000) found that in Tucson, cool downslope flow extended at least 11 km from low mountains, and these flows could obscure urban warming influences.

Mobile Sampling

Methods

Mobile transects, often in combination with fixed-station observations and remote sensing, have frequently been used to measure UHI patterns. (e.g., Hart and Sailor, 2008; Hedquist and Brazel, 2006; Martin et al., 2000; Stabler et al., 2005; Sun et al., 2009). Unless judicious and rigid criteria are employed (e.g., Oke, 2006a), it is unlikely that any method will yield an adequate sampling of the effects of land-cover type and morphological zones on urban climate. Fixed stations in government or special networks suffer problems of representation (i.e., point-to-area extrapolation is inadequate); thus, mobile sampling offers a better method to sample across urban-to-rural gradients. In using this approach, there is typically a lack of thorough temporal sampling (e.g., season, diurnal), and also data are not instantaneously sampled across the gradients chosen. This is usually addressed by sampling "across the transect and back" and taking an average or time-correcting the transect via comparison with fixed points that are being sampled through time along the transect route (e.g., a standard continuously recording weather site).

Summer and Winter Differences

Martin et al. (2000) used automobile transects to evaluate temperature and humidity differences along roads through commercial, industrial, residential, agricultural, and greenbelt land-use classes during clear-sky early mornings (beginning at 0500 h) and afternoons (beginning at 1500 h) in Phoenix, AZ. In summer mornings, industrial areas, which had the lowest NDVI, were warmest; commercial areas were just 1°C cooler; residential and greenbelt 3°C cooler; and agricultural areas, which were irrigated, 6°C cooler. In the afternoons, all land uses averaged within 2°C of each other, with industrial being warmest and agriculture coolest. In winter the pattern was similar, with smaller temperature ranges: only 2°C in the morning and only 1°C in the afternoon. The smaller UHI in winter is consistent with results from some other climates, for example, Vancouver, BC (Fig. 2 in Oke, 1976). However, others (Sailor, 2006; Souch and Grimmond, 2006) have reported that most often winter UHIs are greater than summer UHIs. It seems that the difference in magnitude of UHIs between summer and winter is sufficiently small that careful analysis is needed to assess which season has the most intense UHI. This was the case for Melbourne, Australia in a study by Morris and Simmonds (2000).

Modeling

Mesoscale Meteorology Models

Mesoscale meteorology models carry out numerical simulations of atmospheric conditions over three-dimensional atmospheric space with horizontal extent of up to thousands of kilometers and vertical extent of the entire troposphere. Their development has been underway for more than three decades and has progressed as computer capabilities have progressed to be able to carry out the solutions of huge numbers of primitive (based on first principles) equations that begin with those describing the conservation of mass, heat, and motion (Pielke, 2002). Varying horizontal scales may be used. For modeling city-scale processes, the grid spacing is less than with synoptic-scale models, but still large, for example, 5 km in some examples (Sailor, 1995; Taha et al., 1997). Mesoscale models couple the ground surface to the atmosphere. Thus, in the terminology of Shepherd et al. (2010, Chapter 1, this volume), they are Coupled Atmosphere–Land Surface (CALS) models, and they require ground cover conditions as input for atmospheric predictions.

The ground cover input to mesoscale models can include varying albedo and amount of vegetation. Taha et al. (1997) found that increasing the albedo of streets and of residential, commercial, and industrial areas in the Los Angeles basin reduced predicted 1500-h air temperatures by 2°C, which caused a significant reduction in predicted ozone concentrations. In this case, the average albedo over the basin was increased from 0.139 to 0.155, which was deemed to be reasonable and "doable."

Estimating the effects of increased tree cover on UHIs may be a greater challenge than estimating effects of changed albedo, because trees exert a greater variety of physical influences. In their mesoscale modeling for Los Angeles, CA, Taha et al. (1997) simulated the effects of increased trees by proportional increases in evaporation and increased roughness at the lower boundary of the modeled atmospheric domain. The effect of trees in shading high thermal admittance building and paving surfaces was not implicitly included in the model. From a detailed analysis of urban structure (Horie et al., 1990), Taha et al. (1997) found that in 394 of the 2158 5-km cells, tree cover could be added. For a simulation of a "moderate" tree cover increase, they added tree canopy up to 0.15 of cell areas. They estimated that this increase would require the planting of 10 million trees. The mesoscale model predicted reduced temperatures of 2°C in the central Los

Angeles basin, and 1°C in surrounding areas. Similar results for the Los Angeles basin, using slightly different inputs, were reported by Sailor (1995).

Coutts et al. (2008) described options for modeling urban structure effects on air temperature. They considered the potential operation of climate models directly by urban planners, but concluded that this goal is probably unachievable currently, and therefore climate impact studies of urban development scenarios are best outsourced to urban climatologists. While continued model improvements and validation are needed and anticipated, urban climate models will still need to be run by those who know how to use them. An interdisciplinary and team-based approach is imperative in order for this to be effective (Oke, 2006a).

Empirical Modeling

To evaluate the influence of urban cover on below-canopy air temperatures, especially the influence of urban trees on temperature, regression analysis was used with hourly weather data to develop relationships for predicting temperature differences (ΔT) between the city's center and six weather stations in different land uses around Baltimore, MD (Heisler et al., 2006a,b; 2007). One predictor of ΔT was the difference in upwind land cover between stations as determined from the 2001 National Land Cover Database (Homer et al., 2004). Land cover had an influence on air temperature, but there were strong interactions between land cover and other predictors of ΔT, particularly atmospheric stability and topography. Land-cover differences out to 5 km in the upwind direction were significantly related to ΔT under stable atmospheric conditions.

The relatively simple Turner Class index of thermal stability (Panofsky and Dutton, 1984) was a useful indicator of urban heat island intensity in the Baltimore study. Thermal stability depends on the vertical profile of temperature in the atmosphere. A layer of the atmosphere is unstable when temperature decreases with height, such as occurs just above the ground when wind is light and the sun is strong during midday. The air in contact with the warm ground is heated; it expands and is thus lighter than cooler air above. Air near the ground tends to rise and be turbulent. We say the atmosphere is unstable. On clear, calm nights, outgoing longwave radiation cools the ground, which then cools air above it. The cool air tends to sink or, on level ground, stay put. The atmosphere is stable. When clouds obscure the sky and wind is strong, the layers of air are mixed and the air temperature is uniform with height. Thermal stability is neutral.

The Turner Index scale ranges from 1 for extremely unstable (little cloud cover, low wind speed near midday), to 4 for neutral (overcast sky or high wind speed or both), to 7 for very stable (clear sky, light wind at night). The conditions for a very stable atmosphere are also conditions that promote large UHI intensity. Thus, as anticipated, in the Baltimore study, the ΔTs, which are essentially indicators of the UHI intensity, were usually larger with Turner Classes 6 and 7, which indicate strong stability and occur at night. In this use of Turner Class, it was a predictor of stability in rural rather than urban areas because urban surfaces remain warmer after sunset, and the air usually does not reach the very stable condition of Turner Class 7 (Panofsky and Dutton, 1984).

The regression equations combined with recent geographic information systems (GIS) tools permitted mapping ΔT across a mesoscale-sized area of Baltimore and surroundings (Plate 2–2). The GIS methods have the potential for

testing the effects on temperature of changed land cover, for example, by input-
ting and mapping different scenarios of altered tree or impervious cover.

Microscale Energy Budget Models

Voogt and Oke (2000) and Szpirglas and Voogt (2003) used the zero-dimensional
surface heat island model (SHIM) to understand the roles of thermal admittance, μ,
and the sky view factor. It uses a so called "force-restore" equation that can derive
the nocturnal cooling of a homogeneous substrate. The model calculates the
change of surface temperature with time as a function of radiative loss from the
surface and a restoring of heat from the subsurface to the surface. The model is
able to simulate the cooling of all canyon facets, which include the canyon floor
and both canyon walls.

Brazel and Crewe (2002) evaluated the rates of nighttime cooling at four different sites containing different surface materials and building configurations
and employed SHIM in a discussion of values of inputs (building density and μ)
across a range of conditions. The rate of cooling simulated by SHIM depended
strongly on μ (units of J m^{-2} s$^{-0.5}$ K^{-1}), the property that controls rate of surface
temperature change for a given heat input or removal from the material. Typical
values of μ are 600 J m^{-2} s$^{-0.5}$ K^{-1} for sand and 1100 to 1200 J m^{-2} s$^{-0.5}$ K^{-1} for asphalt
(Table 2–3). The modeling results showed that the impact of varying μ across a
range from 500 to 2500 J m^{-2} s$^{-0.5}$ K^{-1} for an unobstructed sky horizon yields a nonlinear response of the total cooling amount at night (Fig. 2–1). For values lower
than the range of 1000 to 1500 J m^{-2} s$^{-0.5}$ K^{-1}, there is an increasing rate of cooling,
whereas for values greater than this range there is little rate-of-cooling response.
Across a range of sky view factor from 0.2 to 1.0, and at the same time μ of 600 to
3000 J m^{-2} s$^{-0.5}$ K^{-1}, it appears that changes in μ from 300 to 3000 J m^{-2} s$^{-0.5}$ K^{-1} caused
more cooling than the sky view factor impact across the range 0.2 to 1.0. Impli-
cations for UHI mitigation relate to simultaneously accounting for the delicate
balance between building density and thermal property effects on the nighttime
cooling rate of the materials and urban canyon environments.

Analysis of Long-Term Records

Analysis of long-term temperature records can yield indications of the influence
of urbanization, especially where temperature records are available from the

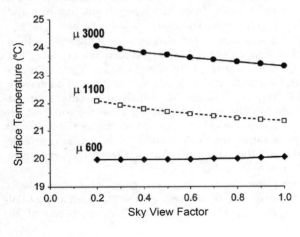

Fig. 2–1. Simulated surface
temperatures after 10 h of
cooling beginning at 1800 h
with 30°C as the starting point
for surfaces with different sky
view and thermal admittance,
μ (from Brazel and Crewe,
2002). Sky view factor 0.0 is
for complete obstruction by
buildings, etc.; 1.0 is for sky
completely unobstructed horizon to horizon. Units for μ are
J m^{-2} C^{-1} s$^{-1/2}$.

start of development. This was the case for a study in Columbia, MD (Landsberg, 1981), where in 1968, at the start of the development of the planned community, a heat island effect of 1°C was observed in a small residential area, and a 3°C heat island was found in a large parking lot. Six years later, the population had reached 20,000, and the maximum UHI increased to 7°C.

Brazel et al. (2000) analyzed long-term urban-minus-rural temperature trends using the Global Historical Climate Network (GHCN) database for several weather stations in and near Baltimore, MD and Phoenix, AZ. For the Baltimore area, the stations included a downtown Baltimore station, the Baltimore/Washington International Airport (BWI), a rural station near Woodstock, MD, about 9 km (15 miles) west of Baltimore, and two airports near Washington, DC. For Phoenix, the analysis included data from the Sky Harbor airport; downtown Phoenix; Mesa, AZ; and a rural location near Sacaton, AZ. The useable climate records began as early as 1908 and extended to 1997 for some stations. For the Baltimore region, the analysis used average daily maximum and minimum temperatures for July. For Phoenix, data were from May. Time series of the urban-minus-rural temperatures ($\Delta Tmax_{u-r}$) at the time of the daily maximum temperature showed a difference between the humid, forested East compared to the arid desert regions. In Baltimore, urban maximums are usually warmer than rural, whereas in the Phoenix area, urban maximum temperatures tend to be cooler than rural maximums. That is, values of $\Delta Tmax_{u-r}$ tend to be negative in Phoenix, an urban cool island (Plate 2–3). This results largely from extensive watering of plants in urban areas in this arid climate. There are only slight long-term trends of changing $\Delta Tmax_{u-r}$. The small or negative daytime heat island in Phoenix has consequences for urban convection effects on precipitation; the convection probably is greater just outside the urban core than within it (Shepherd, 2006).

In downtown Baltimore, $\Delta Tmax_{u-r}$ averaged about 1.5°C toward the end of the period, up about 1°C since 1950 (Plate 2–3). Generally, downtown Baltimore was warmer than BWI. Maximum temperatures at BWI were close to maximums at Woodstock. National Weather Service studies suggest that the temperatures measured at the Customs House in downtown Baltimore may be especially high because the station was located on a building roof (personal communication, Robert Leffler of NWS, 1999). The station was moved to a downtown ground-level location over grass but near water in May 1999 (see next section).

Differences in urban-minus-rural temperature ($\Delta Tmin_{u-r}$) at the time of the daily minimum temperature are greater than differences in maximums and tend to reflect population trends. The long-term average $\Delta Tmin_{u-r}$ for Baltimore peaked at 4.5°C about 1970 and decreased slightly since then (Plate 2–3), apparently because of development encroaching on Woodstock, rather than because population decreased in Baltimore. A similar trend appears for the BWI-Woodstock $\Delta Tmin_{u-r}$ since the BWI record keeping began in 1951. In Phoenix, long-term average $\Delta Tmin_{u-r}$ increased substantially from about 2.5°C in 1908 to 6.5°C in 1995. The rural comparison site for Phoenix, Sacaton, has developed little since the beginning of the century. Thus, as has been found in many other cities, the UHI in both Baltimore and Phoenix is primarily manifested in increased nighttime temperatures rather than in greatly increased temperatures during the warmest part of the day.

Historical Climatology Networks

The long-term records in the Global Historical Climate Network (GHCN) or the United States Historical Climate Network (USHCN) may be useful in evaluating UHI effects (Brazel et al., 2000), although these datasets have been undergoing revisions that should be considered (personal communication, Russell Vose, National Climatic Data Center, 2008).

Caution must be used in the interpretation of long-term temperature trends from standard weather observations. These can be influenced by change of instrument types, station location, or change of surrounding cover, or nearby cover may be unrepresentative of the general area (Brazel and Heisler, 2000; Davey and Pielke, 2005; Oke, 2006b; Vose et al., 2005). Stations are sometimes discontinued just when their records are becoming most valuable. This is the case for downtown Baltimore and rural Woodstock records used by Brazel et al. (2000) for the analysis in Plate 2–3. These were the only two stations in or near Baltimore in the USHCN. The downtown station that was on the roof of a four-story building is now at ground level over grass and only 40 m from a significant body of water (the Baltimore Inner Harbor). Runnalls and Oke (2006) suggested means of checking for discontinuities in station records.

Summary of Warming in Different Cities

When average temperatures over many years are examined, many cities show a warming trend. This warming can be attributed to both the UHI effect and global climate change. Over the 20th century, average annual temperatures have increased 1.72°C across Maricopa County, Arizona, which includes the city of Phoenix (Brazel, 2003). In urban areas of the county, however, temperatures rose by 4.22°C, or three times the 1.28°C increase in rural areas. In the last quarter of the century, Phoenix warmed at about 0.8°C per decade. This warming rate for Phoenix is one of the largest urban-warming rates in the world for its population (Hansen et al., 1999). Other rates of warming per decade for other cities were as follows: Los Angeles, 0.44°C; San Francisco, 0.11°C; Tucson, 0.33°C (Comrie, 2000); Baltimore, 0.11°C; Washington, DC, 0.28°C; Shanghai, 0.11°C; and Tokyo, 0.33°C.

A measure to separate UHI from global influences is the average heat island intensity, $\Delta T_{u\text{-}r}$. This measure varies over the course of a year with the cycles of wet and dry seasons. Roth (2007) graphed the monthly precipitation and nocturnal $\Delta T_{u\text{-}r}$ for eight tropical or subtropical cities. Although in all of the cities there was a definite relationship between monthly precipitation and average $\Delta T_{u\text{-}r}$, with drier months having larger $\Delta T_{u\text{-}r}$, precipitation was not a good predictor of $\Delta T_{u\text{-}r}$ relative to other cities. For example, the highest monthly precipitation was about 400 mm in July in Veracruz, Mexico, but average $\Delta T_{u\text{-}r}$ was still about 2.5°C, whereas in Bogotá, Colombia, July had a $\Delta T_{u\text{-}r}$ of about 2.5°C, but only 70 mm of precipitation. The largest average nocturnal $\Delta T_{u\text{-}r}$ among the eight cities was about 5.6°C in Singapore in July, when precipitation totaled about 150 mm.

Another pertinent comparison of the heat island in different cities is the maximum intensity of the urban heat island, $\Delta T_{u\text{-}r(max)}$. Oke (1973) compared $\Delta T_{u\text{-}r(max)}$ with population for cities in Europe and found that the relationship differed from that in the United States (Fig. 2–2). Roth (2007) compared the Oke (1973) relationships to $\Delta T_{u\text{-}r(max)}$ in tropical and subtropical cities and found generally smaller $\Delta T_{u\text{-}r(max)}$ values in tropical cities and generally lower $\Delta T_{u\text{-}r(max)}$ in wet than dry climate tropical and subtropical cities (Fig. 2–2). In San Juan, an urban area with a

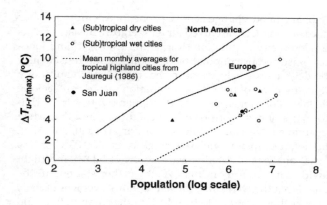

Fig. 2–2. Maximum night-time heat island intensity versus city population from Roth (2007). Solid lines are regressions through temperate cities of North America and Europe (Oke, 1973). The dashed line is a fit through maximum mean monthly urban heat island data from cities located primarily in tropical highland climates from Jauregui (1986). The point for San Juan, a subtropical wet location, is from Murphy et al. (2010).

population of about 2 million, the observed $\Delta T_{u-r(max)}$ of about 4.7°C was typical of wet climate tropical cities (Murphy et al., 2010).

The heat island, ΔT_{u-r}, depends in part on the nearby surroundings of measurement points. For San Juan, the rural location was in old-growth forest, and cooling patterns, which determine $\Delta T_{u-r(max)}$, differed from typical observations in temperate climates (Oke, 1987). The usually temperate climate pattern includes a $\Delta T_{u-r(max)}$ within a few hours after sunset because, beginning in midafternoon, both the rural and urban areas begin to cool, but the rural area cooling rate is greater than the urban rate until that time, a few hours after sunset, when the rural cooling rate decreases and $\Delta T_{u-r(max)}$ occurs. In San Juan, the forest continued cooling more rapidly than the urban area throughout the night, and indeed did not begin warming until an hour or two after sunrise, by which time the urban area had clearly begun warming. Thus, the $\Delta T_{u-r(max)}$ occurred shortly after sunrise, rather than within a few hours after sunset.

Relationship of Urban Canopy Layer Heat Islands to Global Climate

The UHI effect in even modest-sized cities is at times, much larger than the 100-yr trend (1906–2005) of 0.74°C average global temperature warming reported by the IPCC (2007). This is true especially on clear nights with low wind speeds. Global warming is caused by accumulation of "greenhouse" gases (GHG) in the stratosphere, a completely different phenomenon than the processes that cause UHIs. However, global warming and UHI effects are inextricably linked because a large portion of the GHGs are produced in urban areas, and the UHI effect modifies, either positively or negatively, the urban emissions of GHGs (Mills, 2007). Perhaps more importantly, the UHI effect makes terrestrial air temperature monitoring of the global effect uncertain because for many weather stations it is difficult to separate UHI influences from the global influences (Christy and Goodridge, 1995; Kalnay and Cai, 2003).

Confounding of Global Climate Temperature Analysis

The question of whether archived weather data is representative of global temperature trends caused by the greenhouse effect is perhaps the major bone of contention among those who urge major efforts to reduce GHG emissions and those who believe that global climate change is not a problem. Attempts at factoring out the urban influence on long-term archived air temperature measurements have sometimes been based on city population (Karl et al., 1988; Karl and Jones, 1989), at best an inexact exercise because population of a political subdivision may not correspond with the degree of development in the immediate vicinity of a weather station. Also, in many parts of the globe, population records are lacking or imprecise (Gallo and Hale, 2008). Jones et al. (1990) pointed out that large UHI intensities occur only for parts of days with favorable conditions and that when many stations are averaged in a large global-climate-analysis grid cell, the urban influences would be small—an estimated global average value of 0.05°C or less. Epperson et al. (1995) used analysis of satellite estimations of NDVI and night light brightness along with data from more than 2000 weather stations in the United States and estimated that the UHI effect caused monthly averages of daily minimum temperatures over these stations to be 0.40°C higher than they would be without the urban influence; monthly averages of daily mean temperatures were 0.25°C high, and monthly averages of maximum temperatures were 0.1°C high. They concluded that given NDVI and night light data, the urban bias could be eliminated satisfactorily. Hansen et al. (2001) found evidence of urban warming even in suburban and small-town surface air temperature records. They also found inherent uncertainties in the long-term temperature change at least of the order of 0.1°C for both the U.S. mean and the global mean. However, they judged that the urban effect "is modest in magnitude and conceivably could be an artifact of inhomogeneities in the station records" (Hansen et al., 2001). To clarify the "potential urban effect" they suggested further studies, including additional satellite night light analyses, which are used to define populated areas.

Another approach to sort out urbanization influences from GHG influences on temperature is to use the National Center for Environmental Prediction and National Center for Atmospheric Research (NCEP-NCAR) 50-yr Reanalysis (NNR) by the method of Kalnay and Cai (2003). The NNR, described by Kalnay et al. (1996), used a combination of 6-h forecasts and data from soundings of the atmospheric conditions to produce a very large set of predicted output variables through the atmosphere. Observations of temperature, moisture, and wind at the surface of land are not used in creating the NNR data; however, surface temperatures are estimated from the atmospheric values. Kalnay and Cai (2003) concluded that the NNR should not be sensitive to urbanization or land-use effects, although it will show climate changes to the extent that they affect the measurements above the surface on which the NNR is based. For the period 1960 through 1999, average daily maximum land-based temperature observations across the United States showed a small −0.017°C change per decade, while the NNR showed an average of +0.008°C per decade. Minimum daily land-based temperatures had a stronger positive trend in most of the United States, with an average of +0.193°C per decade. In the NNR, the minimum temperature had an average increase of only +0.113°C per decade. Thus, the difference in minimum temperature trends between observed and NNR values (Observed Minus Reanalysis, OMR) was positive in most of the United States, with an average of 0.080°C per decade, suggesting that 40% of the observed trend was urban related.

The results of Kalnay and Cai (2003) were challenged by other researchers. For example, Parker (2004) found long-term warming of minimum daily temperatures at 264 stations worldwide even for windy days when the urban effect should have been small. However, it must be noted that Parker assumed wind speed to be given by the daily average speed, which is usually greater than the wind speed at the time of minimum temperature; that is, winds are generally much higher during the day than at night when minimum temperatures and the maximum UHI generally occur. Another concern with the OMR method is the assumption that urban influences are restricted to the air at the bottom of the atmosphere. Urban influences do create vertical motion in the atmosphere (e.g., Baldi et al., 2008) that may significantly affect the NNR.

Gallo and Hale (2008) provided a concise summary of recent research on methods to factor out the land-use effects on global temperatures. They concluded that there is still the need for additional studies of urban influences on long-term trends in air temperature.

Amplification of Global Warming Effects

Another view of the global versus urban temperature change is that large urbanized areas are amplifying background rates of warming attributed to global-scale climate change (Stone, 2008). In an interesting analysis of data from 50 cities and associated rural areas between 1951 and 2000, Stone (2008) found both urban and rural warming and cooling trends. Of the 50 locations, 12 had cooling rural areas and 12 had cooling urban areas, but only 5 of the 12 were the same locations. As in any study of climate, the data collection and source of samples must be considered. In this case the study took urban temperatures to be the record from the primary airport for each city. This is problematic because airport temperatures are not necessarily representative of their urban areas. Two examples are Hartsfield Airport for Atlanta, GA, which is relatively warm compared to stations in the city (Heisler and Wang, 1998) and the Baltimore/Washington Airport near Baltimore, which is somewhat cooler than the city center (Heisler et al., 2006a). Of the 50 cities in Stone's (2008) analysis, the overall average for increasing UHI was a mean decadal increase in heat island intensity of 0.05°C. For the 29 cities experiencing an increasing trend in urban warming between 1951 and 2000, the mean decadal rate of increase in heat island intensity was 0.19°C. Stone (2008) concluded that planners and public health officials in large cities should be prepared to manage changes in temperature potentially in excess of those forecast by the Intergovernmental Panel on Climate Change (IPCC).

Similarly, Oleson et al. (2009) reported that efforts are underway at the U.S. National Center for Atmospheric Research to model UHI intensity given future global warming. Initial results suggest slightly smaller urban-minus-rural temperatures in temperate-climate winters because warmer global temperatures will reduce the anthropogenic input for space heating of buildings.

Mitigation

When considering the literature on the mitigation of urban heat islands, special attention should be paid to experimental design, assumptions of the study, and the language. Often experiments or data analyses are undertaken, either consciously or unconsciously, to prove the point of view that heat islands are universally detrimental. Possible winter benefits are often not considered. Rather than the scientifically sufficient "warmer" to describe urban temperatures, they are the value-laden "hotter". While these comments should not be taken as detracting from the overwhelmingly solid science in the field of urban climate, we remind the reader that some reports were planned and performed with the goal of proving the effectiveness of certain strategies.

The Heat Island Project at the Lawrence Berkley Laboratory in Berkley, CA was an early promoter of increasing urban albedo by use of white roofs and light-colored paving (Rosenfeld et al., 1995). Though much of this group's research results apply to effects of tree shade, light roofs, and insulation on energy use at the scale of individual buildings, there are implications for larger scale UHIs. For example, Rosenfeld et al. (1995) pointed out that in Los Angeles, the maximum air temperatures decreased during the city's early development, as dry arid regions were replaced with irrigated orchards and farmland. This is similar to Phoenix, where a small cool island exists during the day, apparently because of irrigation of vegetation in the city (Brazel et al., 2000) (Plate 2–3). In an experiment to test the effect of surface albedo changes on air temperature for comparison with results of mesoscale meteorological modeling, Rosenfeld et al. (1995) found that the experimental results showed less cooling than the model, but concluded that the experiment did not include all the factors in a full urban area (Rosenfeld et al., 1995). A 1998 report (Rosenfeld et al., 1998) predicted that the Los Angeles, CA heat island could be reduced by as much as 3°C by "cooler" (i.e., lighter) roof and paving surfaces and 11 million more shade trees.

Sailor (2006) reviewed then current and possible future UHI mitigation strategies, including albedo modifications, tree planting, and "ecoroofs". As in other reports on eco or green roofs, this one described the benefits for urban hydrology and energy use in the building with the roof without being able to say much about the effect of green roofs on general urban climate. Sailor also described the functions of U.S. national governmental and nonprofit organizations, as well as activities on mitigating UHIs in other countries.

The USEPA supported the development of a web-based computer program called MIST, the heat island Mitigation Impact Screening Tool (Sailor and Dietsch, 2007). The program (available at http://www.heatislandmitigationtool.com/, verified 9 Feb. 2010) provides estimates designed to assist urban planners and air quality management officials in assessing the potential of UHI mitigation strategies. The program can estimate effects of UHI modifications on city-wide urban climate, air quality, and energy consumption for more than 170 U.S. cities. To its credit, MIST provides estimates of both summer benefits and winter detriments of actions that affect UHIs. The estimates provided by MIST are table look-up values that come from a series of mesoscale modeling runs for 20 cities (Sailor and Dietsch, 2007). The authors warn that "results presented by MIST include a high degree of uncertainty and are intended only as a first-order estimate that urban planners can use to assess the viability of heat island mitigation strategies for their cities" (Sailor and Dietsch, 2007).

The New York State Energy Research and Development Authority sponsored a New York City Regional Heat Island Initiative to research effects of tree planting, white pavements and roofs, and green (living) roofs on summer "near-surface" air temperatures (Rosenzweig et al., 2006). Again, a mesoscale meteorological model was used to estimate the effects. The study concluded that all of the strategies could reduce summer UHIs, but the best was a combination of tree planting and living roofs. Any possible negative influences by reductions of winter air temperature or increases in heating costs for buildings by tree shade were not considered, and of course possible net benefits of trees and living roofs in winter were also not considered. A similar project was sponsored by the UK Engineering and Physical Sciences Research Council (EPSRC) and the UK Climate Impacts Program (UKCIP) for the region of Manchester, Great Britain (Gill et al., 2007). Their emphasis was on the role that the green infrastructure of a city can play in adapting for climate change, by which is meant both global and urban. Quantitative results from these and similar studies cannot be presented in brief because the results depend very much on the methods and assumptions of the studies.

Urban and global warming can also be mitigated at smaller scales by building and landscape architectural techniques such as by providing local appropriate shading and designing to permit natural ventilation. There are two recent reports from Great Britain that emphasized using these methods to foster human comfort (Smith and Levermore, 2008; Watkins et al., 2007).

The USEPA tried for many years to produce for planners and administrators a set of scientific explanations for UHI effects and guidelines for mitigation of UHIs and UHI effects on which most researchers in the field could generally agree. This effort was in part to update a previous guidebook on tree planting and light-colored surfacing (USEPA, 1992). The current online version (USEPA, 2009) includes separate documents that cover UHI basics, mitigation by trees and other vegetation, green (living) roofs, cool (light-colored) roofs, cool pavements, and activism in the cause of UHI reduction including tree planting programs, ordinances, and building codes and zoning.

Conclusions

- Developed areas in moist climates usually have warmer air temperatures than more rural areas, both day and night, creating an urban heat island effect. The UHI is usually not more than 3 or 4°C during midday. Depending on the rural reference site and synoptic weather conditions, the UHI effect in large cities may range up to about 11°C after sunset. Dry, desert climates have maximum UHIs of similar magnitude to moist climates, but during the daytime, the temperature island often turns out to be a small-magnitude cool island because of evaporative cooling of irrigated vegetation within the city.

- Urban heat islands are caused directly by differences in urban structure and materials from rural areas and indirectly by urban influences on hydroclimate and atmospheric pollutants. However, the primary cause is probably the high thermal admittance (high thermal entropy) of urban building and infrastructure materials that leads to slower rates of heating and cooling of surfaces in urban than rural areas.

- The UHI effect is generally considered to be detrimental. Warmer temperatures increase ozone production in urban atmospheres; increase use of energy for air conditioning, thereby increasing emissions of CO_2; and increase adverse effects on human health and mortality in heat waves. In temperate climates, UHIs are usually greater in summer than winter because of the greater amount of solar insolation in summer. However, substantial UHIs can also form in winter, with the benefits of reducing costs for heating buildings and less snow and ice hazard. The winter benefits of UHIs have seldom been quantified and compared with the detriments of summer.

Acknowledgments

The first author thanks the Baltimore Ecosystem Study Long-Term Ecological Research (LTER) Program, operating with contributions from NSF grant DEB 0423476, for providing instrumentation and technical support staff. GIS maps were prepared by Alexis Ellis.

References

Akbari, H., A. Rosenfeld, and H. Taha. 1989. Cooling urban heat islands. p. 50–57. *In* Proc. of Fourth Urban Forestry Conf., St. Louis, MO. 16–18 Oct. 1989. American Forestry Assoc., Washington, DC.

American Meteorological Society. 2009. 8th Symp. on the Urban Environment. American Meteorological Society Search Form. Available at http://ams.confex.com/ams/htsearch.cgi (verified 9 Feb. 2009). Am. Meteorol. Soc., Boston, MA.

Arnfield, A.J. 1982. An approach to the estimation of the surface radiative properties and radiation budgets of cities. Phys. Geogr. 3:97–122.

Arnfield, A.J. 2003. Two decades of urban climate research: A review of turbulence, exchanges of energy and water, and the urban heat island. Int. J. Climatol. 23:1–26.

Baker, L.A., A.J. Brazel, N. Selover, C. Martin, N. McIntyre, F.R. Steiner, A. Nelson, and L. Musacchio. 2002. Urbanization and warming of Phoenix (Arizona, USA): Impacts, feedbacks and mitigation. Urban Ecosyst. 6:183–203.

Baldi, M., G.A. Dalu, and R.A. Pielke, Sr. 2008. Vertical velocities and available potential energy generated by landscape variability—theory. J. Appl. Meteorol. Climatol. 47:397–410.

Beryland, M.E., and K.Y. Kondratyev. 1972. Cities and the global climate. A. Nurklik (transl.) Atmospheric Environment Service, Downsview, ON, Canada.

Brazel, A.J. 1987. Urban climatology. p. 889–901. *In* J. Oliver and R.W. Fairbridge (ed.) Encyclopedia of earth sciences. Vol. XI. Encyclopedia of climatology. Van Nostrand Reinhold, New York.

Brazel, A.J. 2003. Future climate in central Arizona: Heat and the role of urbanization. Available at http://caplter.asu.edu/docs/smartWebArticles/clim_futureclim.pdf. Consortium for the Study of Rapidly Urbanizing Regions, Arizona State Univ., Phoenix.

Brazel, A.J., and K. Crewe. 2002. Preliminary test of a surface heat island model (SHIM) and implications for a desert urban environment, Phoenix, Arizona. J. Arizona–Nevada Acad. Sci. 34:98–105.

Brazel, A.J., H.J.S. Fernando, J.C.R. Hunt, N. Selover, B.C. Hedquist, and E. Pardyjak. 2005. Evening transition observations in Phoenix, Arizona. J. Appl. Meteorol. 44:99–112.

Brazel, A., P. Gober, S.-J. Lee, S. Grossman-Clarke, J. Zehnder, B. Hedquist, and E. Comparri. 2007. Determinants of changes in the regional urban heat island (1990–2004). Clim. Res. 33:171–182.

Brazel, A.J., and G.M. Heisler. 2000. Some considerations in using climate data from existing weather stations or installing stations for research in Baltimore and Phoenix urban LTER sites. p. 187–188. *In* Third Urban Environment Symp., Davis, CA. 14–19 Aug. 2000. Am. Meteorol. Soc., Boston, MA.

Brazel, A.J., and D. Quatrocchi. 2005. Urban climatology. p. 766–779. *In* Encyclopedia of world climatology. Springer, New York.

Brazel, A., N. Selover, R. Vose, and G. Heisler. 2000. The tale of two climates—Baltimore and Phoenix urban LTER sites. Clim. Res. 15:123–135.

Carreiro, M.M., and C.E. Tripler. 2005. Forest remnants along urban-rural gradients: Examining their potential for global change research. Ecosystems 8:568–582.

Celestian, S.B., and C.A. Martin. 2004. Rhizophere, surface, and air temperature patterns at parking lots in Phoenix, Arizona, U.S. J. Arboricult. 30:245–252.

Chandler, T.J. 1976. Urban climatology and its relevance to urban design. Technical Note 149. World Meteorological Organization, Geneva.

Christy, J.R., and J.D. Goodridge. 1995. Precision global temperatures from satellites and urban warming effects of non-satellite data. Atmos. Environ. 29:1957–1961.

Cionco, R.M., and R. Ellefsen. 1998. High resolution urban morphology data for urban wind flow modeling. Atmos. Environ. 32:7–17.

Comrie, A. 2000. Mapping a wind-modified urban heat island in Tucson, Arizona (with comments on integrating research and undergraduate learning). Bull. Am. Meteorol. Soc. 81:2417–2431.

Coutts, A.M., J. Beringer, and N.J. Tapper. 2008. Investigating the climatic impact of urban planning strategies through the use of regional climate modelling: A case study for Melbourne, Australia. Int. J. Biometeorol. 28:1943–1957.

Dabberdt, W.F. 2007. A perfect storm?—Population growth, climate change and the urban environment. 7th Symp. on the Urban Environment, San Diego, CA. September 2007. Available at http://ams.confex.com/ams/htsearch.cgi (verified 9 Feb. 2010). Am. Meteorol. Soc., Boston.

Davey, C.A., and R.A. Pielke, Sr. 2005. Microclimate exposures of surface-based weather stations: Implication for the assessment of long-term temperature trends. Bull. Am. Meteorol. Soc. 86:497–504.

Epperson, D.L., J.M. Davis, P. Bloomfield, T.R. Karl, A.L. McNab, and K.P. Gallo. 1995. Estimating the urban bias of surface shelter temperatures using upper-air and satellite data. Part II: Estimation of urban bias. J. Appl. Meteorol. 34:358–370.

Fan, H., and D.J. Sailor. 2005. Modeling the impacts of anthropogenic heating on the urban climate of Philadelphia: A comparison of implementations in two PBL schemes. Atmos. Environ. 39:73–84.

Gallo, K., and R. Hale. 2008. Recognition of the influence of the urban climate in assessment of large-scale temperature trends. Available at http://www.urban-climate.org/ (verified 9 Feb. 2010). Urban Clim. News 29:5–7.

Gallo, K.P., A.L. McNab, T.R. Karl, J.F. Brown, J.J. Hood, and J.D. Tarpley. 1993. The use of a vegetation index for assessment of the urban heat island effect. Int. J. Remote Sens. 14:2223–2230.

Gartland, L. 2008. Heat islands, understanding and mitigating heat in urban areas. Earthscan, London.

Gill, S.E., J.F. Handley, A.R. Ennos, and S. Pauleit. 2007. Adapting cities for climate change: The role of the green infrastructure. Built Environ. 33:115–133.

Givoni, B. 1991. Impact of planted areas on urban environmental quality: A review. Atmos. Environ. 25B:289–299.

Goldreich, Y. 1985. The structure of the ground-level heat island in a central business district. J. Clim. Appl. Meteorol. 24:1237–1244.

Gomes, L., J.C. Roger, and P. Dubuisson. 2008. Effects of the physical and optical properties of urban aerosols measured during the CAPITOUL summer campaign on the local direct radiative forcing. Meteorol. Atmos. Phys. 102:289–306.

Goward, S.N. 1981. Thermal behavior of urban landscapes and the urban heat island. Phys. Geogr. 2:19–33.

Grass, D., and M. Crane. 2008. The effects of weather and air pollution on cardiovascular and respiratory mortality in Santiago, Chile, during the winters of 1988–1996. Int. J. Climatol. 28:1113–1126.

Greenfield, E., D.J. Nowak, and J.T. Walton. 2010. Accuracy assessment of 2001 NLCD tree canopy and impervious surface cover. Photogramm. Eng. Remote Sens. 75:1279–1286.

Grimmond, C.S.B. 2006. Progress in measuring and observing the urban atmosphere. Theor. Appl. Climatol. 84:3–22.

Grimmond, S. 2007. Urbanization and global environmental change: Local effects of urban warming. Geogr. J. 173:83–88.

Grimmond, C.S.B., W. Kuttler, S. Lindqvist, and M. Roth. 2007. Editorial: Urban climatology ICUC6. Int. J. Climatol. 27:1847–1848.

Grimmond, C.S.B., and T.R. Oke. 1995. Comparison of heat fluxes from summertime observations in the suburbs of four North American cities. J. Appl. Meteorol. 34:873–889.

Grimmond, C.S.B., and T.R. Oke. 1999a. Evapotranspiration rates in urban areas. p. 235–243. *In* Impacts of Urban Growth on Surface Groundwater Quality, Symp. HS5. International Union of Geodesy and Geophysics (IUGG) 99.

Grimmond, C.S.B., and T.R. Oke. 1999b. Heat storage in urban areas: Local-scale observations and evaluation of a simple model. J. Appl. Meteorol. 38:922–940.

Guhathakurta, S., and P. Gober. 2007. The impact of the Phoenix urban heat island on residential water use. J. Am. Plann. Assoc. 73:317–329.

Halverson, H.G., and G.M. Heisler. 1981. Soil temperatures under urban trees and asphalt. Research Paper NE-481. USDA, Northeastern Forest Exp. Stn., Broomall, PA.

Hansen, J., R. Ruedy, J. Glascoe, and M. Sato. 1999. GISS analysis of surface temperature change. J. Geophys. Res. 104:30997–31022.

Hansen, J., R. Ruedy, M. Sato, M. Imhoff, W. Lawrence, D. Easterling, T. Peterson, and T. Karl. 2001. A closer look at United States and global surface temperature change. J. Geophys. Res. 106(D20):947–963.

Harlan, S.H., A.J. Brazel, L. Prashad, W.L. Stefanov, and L. Larsen. 2006. Neighborhood microclimates and vulnerability to heat stress. Soc. Sci. Med. 63:2847–2863.

Hart, M.A., and D.J. Sailor. 2008. Quantifying the influence of land-use and surface characteristics on spatial variability in the urban heat island. Theor. Appl. Climatol. 95:397–406.

Hartz, D.A., A.J. Brazel, and G.M. Heisler. 2006a. A case study in resort climatology of Phoenix, Arizona, USA. Int. J. Biometeorol. 51:73–83.

Hartz, D.A., L. Prashad, B.C. Hedquist, J. Golden, and A.J. Brazel. 2006b. Linking satellite images and hand-held infrared thermography to observed neighborhood climate conditions. Remote Sens. Environ. 104:190–200.

Hawkins, T.W., A.J. Brazel, W.L. Stefanov, W. Bigler, and E.M. Saffell. 2004. The role of rural variability in urban heat island determination for Phoenix, Arizona. J. Appl. Meteorol. 43:476–486.

Hedquist, B., and A.J. Brazel. 2006. Urban, residential, and rural climate comparisons from mobile transects and fixed stations: Phoenix, Arizona. J. Arizona–Nevada Acad. Sci. 38:77–87.

Heisler, G., B. Tao, J. Walton, R. Grant, R. Pouyat, I. Yesilonis, D. Nowak, and K. Belt. 2006a. Land-cover influences on below-canopy temperatures in and near Baltimore, MD. In 6th Symp. on the Urban Environment, Atlanta, GA. 28 Jan.–2 Feb. 2006. Available at http://ams.confex.com/ams/pdfpapers/101404.pdf (verified 9 Feb. 2010).

Heisler, G., J. Walton, S. Grimmond, R. Pouyat, K. Belt, D. Nowak, I. Yesilonis, and J. Hom. 2006b. Land-cover influences on air temperatures in and near Baltimore, MD. p. 392–395. *In* 6th International Conf. on Urban Climate, Gothenburg, Sweden. Available at http://www.gvc2.gu.se/icuc6//index.htm (verified 9 Feb. 2010). International Association for Urban Climate.

Heisler, G., J. Walton, I. Yesilonis, D. Nowak, R. Pouyat, R. Grant, S. Grimmond, K. Hyde, and G. Bacon. 2007. Empirical modeling and mapping of below-canopy air temperatures in Baltimore, MD and vicinity. *In* 7th Urban Environment Symp., San Diego, CA. 7–13 Sept. 2007. Available at http://ams.confex.com/ams/pdfpapers/126981.pdf (verified 9 Feb. 2010). Am. Meteorol. Soc., Boston, MA.

Heisler, G.M., and R.H. Grant. 2000. Ultraviolet radiation in urban ecosystems with consideration of effects on human health. Urban Ecosyst. 4:193–229.

Heisler, G.M., and L.P. Herrington (ed.) 1977. Proc. of the Conf. on Metropolitan Physical Environment, Syracuse, NY. 25–29 Aug. 1975. Gen. Tech. Rep. NE-25. USDA Forest Service, Northeastern Forest Exp. Stn., Upper Darby, PA.

Heisler, G.M., and Y. Wang. 1998. Semi-empirical modeling of spatial differences in below-canopy urban air temperature using GIS analysis of satellite images, on-site photography, and meteorological measurements. p. 206–209. *In* 2nd Urban Environment Symp., Albuquerque, NM. 2–6 Nov. 1998. Am. Meteorol. Soc., Boston, MA.

Heisler, G.M., and Y. Wang. 2002. Applications of a human thermal comfort model. p. 70–71. *In* 4th Symp. on the Urban Environment, Norfolk, VA. 20–24 May 2002. Am. Meteorol. Soc., Boston, MA.

Homer, C., C. Huang, L. Yang, B. Wylie, and M. Coan. 2004. Development of a 2001 National Landcover Database for the United States. Photogramm. Eng. Remote Sens. 70:829–840.

Horie, Y., S. Sidawi, and R. Ellefsen. 1990. Inventory of leaf biomass and emission factors for vegetation in California's South Coast Air Basin. Final Rep. 90163. South Coast Air Quality Management District, Diamond Bar, CA.

Howard, L. 1833. Climate of London deduced from meteorological observations. 3rd ed. Harvery and Darton, London.

Intergovernmental Panel on Climate Change. 2007. Climate Change 2007 Synthesis Report. IPCC, Valencia, Spain.

Jauregui, E. 1986. Tropical urban climates: Review and assessment. p. 26–45. *In* T.R. Oke (ed.) Urban climatology and its applications with special regard to tropical areas. Publ. 652. World Meteorological Organisation, Geneva.

Jenerette, G.D., S.L. Harlan, A. Brazel, N. Jones, L. Larsen, and W.L. Stefanov. 2007. Regional relationships between surface temperature, vegetation, and human settlement in a rapidly urbanizing ecosystem. Landscape Ecol. 22:353–365.

Johnson, G.T., and I.D. Watson. 1984. The determination of view-factors in urban canyons. J. Clim. Appl. Meteorol. 23:329–335.

Jones, P.D., P.Y. Groisman, M. Coughlan, N. Plummer, W.C. Wang, and T.R. Karl. 1990. Assessment of urbanization effects in time series of surface air temperature over land. Nature 347:169–172.

Kalanda, B.D., T.R. Oke, and D.L. Spittlehouse. 1980. Suburban energy balance estimates for Vancouver, B.C. using the Bowen ratio-energy balance approach. J. Appl. Meteorol. 19:791–802.

Kalkstein, L.S., and K.E. Smoyer. 1993. The impact of climate change on human health: Some international implications. Experientia 49(49):969–979.

Kalnay, E., and M. Cai. 2003. Impact of urbanization and land-use change on climate. Nature 423:528–531.

Kalnay, E., M. Kanamitsu, R. Kistler, W. Collins, D. Deaven, L. Gandin, M. Iredell, S. Saha, G. White, J. Woollen, Y. Zhu, M. Chelliah, W. Ebisuzaki, W. Higgins, J. Janowiak, K.C. Mo, C. Ropelewski, J. Wang, A. Leetmaa, R. Reynolds, R. Jenne, and D. Joseph. 1996. The NCEP/NCAR 40-year reanalysis project. Bull. Am. Meteorol. Soc. 77:437–472.

Karl, T.R., H.F. Diaz, and G. Kukla. 1988. Urbanization: Its detection and effect in the United States climate record. J. Clim. 1:1099–1123.

Karl, T.R., and P.D. Jones. 1989. Urban bias in area-averaged surface air temperature trends. Bull. Am. Meteorol. Soc. 70:265–270.

Kaye, M.W., A. Brazel, M. Netzband, and M. Katti. 2003. Perspectives on a decade of climate in the CAP LTER Region. Available at http://caplter.asu.edu/docs/symposia/symp2003/Kaye_et_al.pdf. Central Arizona–Phoenix Long-Term Ecological Research (CAP LTER).

Landsberg, H.E. 1981. The urban climate. Academic Press, New York.

Lee, D.O. 1984. Urban climates. Prog. Phys. Geogr. 8:1–31.

Lowry, W.P. 1977. Empirical estimation of urban effects on climate: A problem analysis. J. Appl. Meteorol. 16:129–135.

Marotz, G.A., and J.C. Coiner. 1973. Acquisition and characterization of surface material data for urban climatological studies. J. Appl. Meteorol. 12:919–923.

Martin, C.A., L.B. Stabler, and A.J. Brazel. 2000. Summer and winter patterns of air temperature and humidity under calm conditions in relation to urban land use. p. 197–198. *In* 3rd Symp. on the Urban Environment, Davis, CA. 14–18 Aug. 2000. Am. Meteorol. Soc., Boston, MA.

Matzarakis, A., F. Rutz, and H. Mayer. 2007. Modeling radiation fluxes in simple and complex environments–application of the Rayman model. Int. J. Biometeorol. 51:323–334.

Mills, G. 2006. Progress toward sustainable settlements: A role for urban climatology. Theor. Appl. Climatol. 84:69–76.

Mills, G. 2007. Cities as agents of global change. Int. J. Climatol. 27:1849–1857.

Morris, C.J.G., and I. Simmonds. 2000. Associations between varying magnitudes of the urban heat island and the synoptic climatology in Melbourne, Australia. Int. J. Climatol. 20:1931–1954.

Murphy, D.J.R., M. Hall, C. Hall, G. Heisler, and S. Stehman. 2007. The relationship between land-cover and the urban heat island in Northeastern Puerto Rico. In 7th Urban Environment Symp., San Diego, CA. 10–13 Sept. 2007. Available at http://ams.confex.com/ams/pdfpapers/126931.pdf (verified 9 Feb. 2010). Am. Meteorol. Soc., Boston, MA.

Murphy, D.J.R., M. Hall, C. Hall, G. Heisler, S. Stehman, and A.-M. Carlos. 2010. The relation between land cover and the urban heat island in northeastern Puerto Rico. Int. J. Climatol. doi:10.1002/joc.2145.

Nikolopoulou, M., and K. Steemers. 2003. Thermal comfort and psychological adaptation as a guide for designing urban spaces. Energy Build. 35:95–101.

Oke, T.R. 1973. City size and the urban heat island. Atmos. Environ. 7:769–779.

Oke, T.R. 1974. Review of urban climatology 1968–1973. Tech. Note 134. World Meteorological Organization, Geneva.

Oke, T.R. 1976. The distinction between canopy and boundary-layer urban heat islands. Atmosphere 14:268–277.

Oke, T.R. 1979. Review of urban climatology 1973–1976. Technical Note 169. World Meteorological Organization, Geneva.

Oke, T.R. 1980. Climatic impacts of urbanization. p. 339–356. In W. Bach et al. (ed.) Interactions of energy and climate. Reidel, Boston, MA.

Oke, T.R. 1982. The energetic basis of the urban heat island. Q. J. R. Meteorol. Soc. 108:1–24.

Oke, T.R. 1987. Boundary layer climates. Methuen, London.

Oke, T.R. 1995. The heat island of the urban boundary layer: Characteristics, causes and effects. p. 81–107. In J. E. Cermak et al. (ed.) Wind climate in cities. Kluwer Academic Publishers, Dordrecht, The Netherlands.

Oke, T.R. 1997. Urban climates and global environmental change. p. 273–287. In R.D. Thompson and A. Perry (ed.) Applied climatology: Principles and practice. Routledge, London.

Oke, T.R. 2006a. Towards better communication in urban climate. Theor. Appl. Climatol. 84:179–189.

Oke, T.R. 2006b. Initial guidance to obtain representative meteorological observations at urban sites. Instruments and observing methods 81. WMO/TD-No. 1250. World Meteorological Organization, Geneva, Switzerland.

Oke, T.R., B.D. Kalanda, and D.G. Steyn. 1981. Parameterization of heat storage in urban areas. Urban Ecol. 5:45–54.

Oleson, K., G. Bonan, J. Feddema, and T. Jackson. 2009. Progress toward modeling global climate change in urban areas. Available at http://www.urban-climate.org/ (verified 9 Feb. 2010). Urban Clim. News 31:8–12.

Panofsky, H.A., and J.A. Dutton. 1984. Atmospheric turbulence. John Wiley and Sons, New York.

Parker, D.E. 2004. Large-scale warming is not urban. Nature 432:290.

Pielke, R.A., Sr. 2002. Mesoscale meteorological modeling. Academic Press, San Diego.

Pouyat, R.V., M.J. McDonnell, S.T.A. Pickett, P.M. Groffman, M.M. Carreiro, R.W. Parmelee, K.E. Medley, and W.C. Zipperer. 1995. Carbon and nitrogen dynamics in oak stands along an urban-rural gradient. p. 569–587. In J.M. Kelly and W.W. McFee (ed.) Carbon forms and functions in forest soils. SSSA, Madison, WI.

Pouyat, R.V., K. Szlavecz, I.D. Yesilonis, P.M. Groffman, and K. Schwarz. 2010. Chemical, physical, and biological characteristics of urban soils. p. 119–152. In J. Aitkenhead-Peterson and A. Volder (ed.) Urban ecosystem ecology. Agron. Monogr. 55. ASA, CSSA, and SSSA, Madison, WI.

Rosenfeld, A.H., H. Akbari, S. Bretz, B.L. Fishman, D.M. Kurn, D. Sailor, and H. Taha. 1995. Mitigation of urban heat islands: Materials, utility programs, updates. Energy Build. 22:255–265.

Rosenfeld, A.H., H. Akbari, J.J. Romm, and M. Pomerantz. 1998. Cool communities: Strategies for heat island mitigation and smog reduction. Energy Build. 28:51–62.

Rosenzweig, C., D.C. Major, K. Demong, C. Stanton, R. Horton, and M. Stults. 2007. Managing climate change risks in New York City's water system: Assessment and adaptation planning. Mitig. Adapt. Strategies Glob. Change 12:1391–1409.

Rosenzweig, C., W.D. Solecki, and R.B. Slosberg. 2006. Mitigating New York City's heat island with urban forestry, living roofs, and light surfaces. NYSERDA Rep. 06-06. Columbia Univ. Ctr. for Climate Systems Research and NASA/Goddard Inst. for Space Studies, New York.

Roth, M. 2002. Effects of cities on local climates. p. 1–13. *In* Proc. of Workshop of IGES/APN Mega-City Project. Kitakyushu, Japan. 23–25 Jan. 2002. Inst. for Global Environmental Strategies, Arlington, VA.

Roth, M. 2007. Review of urban climate research in (sub)tropical regions. Int. J. Climatol. 27:1859–1873.

Ruddell, D.M., S.L. Harlan, S. Grossman-Clarke, and A. Buyantuyev. 2010. Risk and exposure to extreme heat in microclimates of Phoenix, AZ. p. 179–202. *In* P.S. Showalter and Y. Lu (ed.) Geospatial techniques in urban hazard and disaster analysis. Geotechnologies and the Environment 2. Springer, New York.

Runnalls, K.E., and T.R. Oke. 2006. A technique to detect microclimatic inhomogeneities in historical records of screen-level air temperature. J. Clim. 19:959–978.

Sailor, D. 1995. Simulated urban climate responses to modifications in surface albedo and vegetative cover. J. Appl. Meteorol. 34:1694–1704.

Sailor, D.J. 2006. Mitigation of urban heat islands—Recent progress and future prospects. *In* Proc. of 6th Urban Environment Symp., Atlanta, GA. 28 Jan.–2 Feb. 2006. Available at http://ams.confex.com/ams/pdfpapers/105264.pdf (verified 9 Feb. 2010). Am. Meteorol. Soc., Boston, MA.

Sailor, D.J., and N. Dietsch. 2007. The urban heat island Mitigation Impact Screening Tool (MIST). Environ. Model. Software 22:1529–1541.

Sailor, D.J., and H. Fan. 2004. Mesoscale modeling of the impact of anthropogenic heating on the urban climate of Houston—The role of spatial and temporal resolution. *In* 5th Symp. on the Urban Environment, Vancouver, BC. 23–26 Aug. 2004. Available at http://ams.confex.com/ams/htsearch.cgi (verified 9 Feb. 2010). Am. Meteorol. Soc., Boston.

Sailor, D.J., and L. Lu. 2004. A top–down methodology for developing diurnal and seasonal anthropogenic heating profiles for urban areas. Atmos. Environ. 38:2737–2748.

Sailor, D.J., L. Lu, and H. Fan. 2003. Estimating urban anthropogenic heating profiles and their implications for heat island development. *In* Proc. of the 5th International Conf. on Urban Climate (ICUC-5), Lodz, Poland. 1–5 Sept. 2003. Available at http://nargeo.geo.uni.lodz.pl/~icuc5/ (verified 9 Feb. 2010). International Association for Urban Climate.

Sanchez-Rodriguez, R., K.C. Seto, D. Simon, W.D. Solecki, F. Kraas, and G. Laumann. 2005. Science plan: Urbanization and global environmental change. IHDP Rep. 15. International human dimensions programme on global environmental change, Bonn, Germany.

Santosa, S.J. 2010. Urban air quality. p. 57–74. *In* J. Aitkenhead-Peterson and A. Volder (ed.) Urban ecosystem ecology. Agron. Monogr. 55. ASA, CSSA, and SSSA, Madison, WI.

Shepherd, J.M. 2006. Evidence of urban-induced precipitation variability in arid climate regimes. J. Arid Environ. 67:607–628.

Shepherd, J.M., J.A. Stallins, M.L. Jin, and T.L. Mote. 2010. Urbanization: Impacts on clouds, precipitation, and lightning. p. 1–28. *In* J. Aitkenhead-Peterson and A. Volder (ed.) Urban ecosystem ecology. Agron. Monogr. 55. ASA, CSSA, and SSSA, Madison, WI.

Smith, C., and G. Levermore. 2008. Designing urban spaces and buildings to improve sustainability and quality of life in a warmer world. Energy Policy 36:4558–4562.

Souch, C., and S. Grimmond. 2006. Applied climatology: Urban climate. Prog. Phys. Geogr. 30:270–279.

Stabler, L., C.A. Martin, and A. Brazel. 2005. Microclimates in a desert city were related to land use and vegetation index. Urban For. Urban Green. 3:137–147.

Stone, B., Jr. 2008. Urban and rural temperature trends in proximity to large U.S. cities: 1951–2000. Available online http://www.urban-climate.org/ (verified 9 Feb. 2010). Urban Clim. News 30:7–10.

Stull, R.B. 2000. Meteorology for scientists and engineers. 2nd ed. Brooks/Cole, Pacific Grove, CA.

Su, W., J. Li, Y. Chen, Z. Lius, J. Zhang, T. Miin, I. Suppiah, and S.A.M. Hashim. 2008. Textural and local spatial statistics for the object-oriented classification of urban areas using high resolution imagery. Int. J. Remote Sens. 29:3105–3117.

Sun, C.-Y., A. Brazel, W.T.L. Chow, B.C. Hedquist, and L. Prashad. 2009. Desert heat island study in winter by mobile transect and remote sensing techniques. Theor. Appl. Climatol. 98:323–335.

Szpirglas, J., and J.A. Voogt. 2003. A validation and performance assessment of the surface heat island model. *In* International Conf. on Urban Climate, ICUC-5, Lodz, Poland. 1–5 Sept. 2003. Available online http://nargeo.geo.uni.lodz.pl/~icuc5/text/O_8_1.pdf (verified 9 Feb. 2010). International Association for Urban Climate.

Taha, H., S. Douglas, and J. Haney. 1997. Mesoscale meteorological and air quality impacts of increased urban albedo and vegetation. Energy Build. 25:169–177.

Terjung, W.H., and P.A. O'Rourke. 1980. Simulating the casual elements of urban heat islands. Boundary-Layer Meteorol. 19:93–118.

USEPA. 1992. Cooling our communities: A guidebook on tree planting and light-colored surfacing. U.S. Gov. Print. Office, Pittsburgh, PA.

USEPA. 2009. Reducing urban heat islands: Compendium of strategies. Available at http://www.epa.gov/hiri/resources/compendium.htm (verified 9 Feb. 2010).

Voogt, J.A. 2002. Urban heat island. p. 660–666. *In* I. Douglas (ed.) Causes and consequences of global environmental change. Vol. 3. John Wiley and Sons, Chichester, UK.

Voogt, J.A., and T.R. Oke. 1997. Complete urban surface temperatures. J. Appl. Meteorol. 36:1117–1132.

Voogt, J.A., and T.R. Oke. 1998. Radiometric temperatures of urban canyon walls obtained from vehicle traverses. Theor. Appl. Climatol. 60:199–217.

Voogt, J.A., and T.R. Oke. 2000. Multi-temporal remote sensing of an urban heat island. p. 505–510. *In* R. deDear et al. (ed.) Biometeorology and urban climatology at the turn of the millennium, from the Conferences ICB-ICUC'99, Sydney, Australia. 8–12 Nov. 1999. World Meteorological Organization, Geneva, Switzerland.

Voogt, J.A., and T.R. Oke. 2003. Thermal remote sensing of urban climates. Remote Sens. Environ. 86:370–384.

Vose, R.S., D.R. Easterling, T.R. Karl, and M. Helfert. 2005. Comments on "Microclimate exposures of surface-based weather stations." Bull. Am. Meteorol. Soc. 86:504–506.

Watkins, R., J. Palmer, and M. Kolokotroni. 2007. Increased temperature and intensification of the urban heat island: Implications for human comfort and urban design. Built Environ. 33:85–96.

Weng, Q., X. Hu, and D. Lu. 2008. Extracting impervious surfaces from medium spatial resolution multispectral and hyperspectral imagery: A comparison. Int. J. Remote Sens. 29:3209–3232.

Yap, D. 1975. Seasonal excess urban energy and the nocturnal heat island—Toronto. Arch. Meteorol. Geophys. Bioklim. Ser. B B23:69–80.

Zhou, W., and A. Troy. 2008. An object-oriented approach for analysing and characterizing urban landscape at the parcel level. Int. J. Remote Sens. 29:3119–3135.

Ziska, L.H., J.S. Bunce, and E.W. Goins. 2004. Characterization of an urban-rural CO_2/temperature gradient and associated changes in initial plant productivity during secondary succession. Oecologia 139:454–458.

Ziska, L.H., D.E. Gebhard, M. Frenz, A. David, S. Faulkner, B.D. Singer, and J.G. Straka. 2003. Cities as harbingers of climate change: Common ragweed, urbanization, and public health. J. Allergy Clin. Immunol. 111:290–295.

3

Urban Air Quality

Sri Juari Santosa

Abstract

The air in the troposphere surrounding our planet is a mixture of gases, vapors, aerosols, and suspended particulate matter (SPM). The gas component may be viewed as a solution of various gases in nitrogen solvent that is chemically far from equilibrium. The composition of the air constituents is originally governed by natural processes, mainly from geochemical and biological activities, and then modified substantially by human activity. In modern life, anthropogenic air constituents arise from a wide variety of activities, although they are mainly the result of combustion processes in industrialization and transportation, as well as rapid urbanization. Concentrations of sulfur dioxide (SO_2), SPM, polycyclic aromatic hydrocarbons (PAHs), and greenhouse gases such as CH_4, N_2O, CO_2, and chlorofluorocarbons are increased by human activities and are an increasing component of the atmosphere on which all life depends.

The air in the troposphere is perhaps one of the most vital natural resource for all creatures living on the planet Earth. It can be understood as a solution of various gases in dinitrogen (N_2), a relatively inert solvent. Oxygen (O_2) is the most abundance solute, followed by argon (Ar) and carbon dioxide (CO_2). Nitrogen and all these solutes are frequently identified as the major constituent of the composition of air, while other solutes with smaller abundance, such as neon (Ne), helium (He), methane (CH_4), hydrogen (H_2), nitrous oxide (N_2O), and carbon monoxide (CO), are the minor constituent. At sea level (total pressure is 1 atm), bulk dry air has the composition of the major and minor constituents, as shown in Table 3–1. In addition to major and minor constituents as listed in Table 3–1, the air still contains other components with minute abundance, such as ammonia (NH_3), krypton (Kr), Xenon (Xe), and ozone (O_3), and even particulate matter (PM). Water (H_2O) is also an important vapor component of the air, but its abundance varies greatly with temperature. The theoretical maximum attainable value of the pressure of H_2O in the air at various temperatures is given in Table 3–2. Mostly, the air is undersaturated with respect to H_2O. Relative humidity is used to show a percentage of the real pressure of the water vapor toward its maximum pressure at a given temperature. Thus a relative humidity of 70%, for example, corresponds to $p(H_2O)$ 611 Pa (6.03×10^{-3} atm) at 5°C or to 1636 Pa (16.15×10^{-3} atm) at 20°C.

S.J. Santosa, Dep. of Chemistry, Universitas Gadjah Mada, Sekip Utara Kotak Pos Bls. 21, Yogyakarta, Indonesia (sjuari@yahoo.com).

doi:10.2134/agronmonogr55.c3

Table 3–1. Natural composition of dry air.

No.			Pressure	Concentration†
	Constituent		Abundance	
	Name	Chemical symbol	atm	
	Major			
1	Nitrogen	N_2	0.78	78% (v/v)
2	Oxygen	O_2	0.21	21% (v/v)
3	Argon	Ar	0.93×10^{-2}	0.93% (v/v)
4	Carbon dioxide	CO_2	0.35×10^{-3}	0.035% (350 ppmv)
	Minor			
1	Neon	Ne	0.18×10^{-4}	18 ppmv
2	Helium	He	0.52×10^{-5}	5.2 ppmv
3	Methane	CH_4	0.15×10^{-5}	1.5 ppmv
4	Hydrogen	H_2	0.50×10^{-6}	0.5 ppmv
5	Nitrous oxide	N_2O	0.30×10^{-6}	0.3 ppmv
6	Carbon monoxide	CO	$\approx 0.10 \times 10^{-6}$	≈ 0.1 ppmv

† ppmv: ppm volume.

Table 3–2. Theoretical maximum pressure of H_2O attainable in the air at various temperatures.†

Temperature	$p(H_2O)$†	Temperature	$p(H_2O)$†
°C	10^{-3} atm	°C	10^{-3} atm
−10	2.57	15	16.83
−5	3.96	20	23.07
0	6.03	25	31.26
5	8.61	30	41.87
10	12.12	35	55.49

† Maximum pressure of H_2O vapor.

The bulk composition of air is quite similar all over the globe because of the high degree of mixing within air itself. This mixing is composed of horizontal and vertical mixing. Horizontal mixing is driven by the rotation of the globe, and vertical mixing is mostly the product of surface heating by incoming solar radiation. Gases in the air are not necessarily in chemical equilibrium, although this does not mean that the air is especially unstable. Many minor and minute, or trace, gases in the air are in steady state instead of equilibrium. Steady state simply describes the delicate balance between the input and output of the gas to the air. To be in steady state, the flux of gas entering the air must be equal to that leaving the air. This situation is written in terms of the equation:

$$F_{in} = F_{out} \qquad\qquad [1]$$

where F_{in} and F_{out} are fluxes in and out of the air, respectively.

The concept of steady state has no relation to the "fate" of a constituent or substance during its presence in the air. In the concept of steady state, the importance of fate is simply replaced by the term of residence time (τ), which is the measure of the amount of substance (A) in the air relative to its flux of inflow or outflow. Thus

$$\tau = \frac{A}{F_{in} \text{ or } F_{out}} \qquad\qquad [2]$$

Table 3–3. Estimated relative global warming potentials (GWP) of greenhouse gases (weight basis).†

Time horizon yr	CO_2	CH_4	N_2O	CCl_3F	CCl_2F_2	$CHClF_2$
20	1	72	289	6730	11000	5160
100	1	25	298	4750	10900	1810
500	1	7.6	153	1620	5200	549

† Source: IPCC, 2007.

Residence time is a very important parameter in determining whether a substance is widely distributed in the air. Substances with long residence times can accumulate and have the opportunity to become well-mixed throughout the air compared with substances with shorter residence times. Consequently, substances with greater residence times would be expected to have greater constancy in concentration all around the planet. However, it is important to note that even though gases with short residence times are removed more quickly, their high reactivity can yield reaction products that may be harmful. As occurs for highly reactive volatile organic compounds (HRVOCs), such as ethane, propene, butane, and 1,3-butediene, several hours after their injection from industrial emissions, they are immediately converted to hydroxyl radicals (OH·), which are responsible to the formation of harmful tropospheric ozone (Vizuete et al., 2008).

From the steady state concept premise, the air can be viewed as having source and removal processes that are all in balance. The balance is often fragile, especially for trace constituents. To maintain the balance, the sources of the air constituents should be stable in the long term. If they are not, the balance is disturbed, and there may be a shift to a new balance. As has been observed for CO_2, such a shift is also occurring for CH_4, which is a much more potent global warming gas than CO_2 (Table 3–3). The concentration of CH_4 is currently estimated to increase annually at 1 to 2% due to human agriculture activity (Khalil and Rasmussen, 1990).

Sources of Air Constituents

There are many sources of air constituents. All of them originally belong to either natural or anthropogenic sources. Geochemical and biological activities are the main natural sources air constituents, while industry and transportation are the main anthropogenic sources.

Natural Sources

Geochemical Activity

Volcanic activity is one of most important geochemical sources of trace and minor constituents of air (e.g., Ward, 2009). In addition to releasing fine particulates, volcanoes inject gases, such as sulfur dioxide (SO_2), CO_2, hydrogen fluoride (HF), hydrogen chloride (HCl), and many others (Orstom, 1990). These gases can react in the air with sulfuric acid (H_2SO_4) to provide further sources of particulates produced indirectly by volcanic activity. The contribution of volcanoes in maintaining the concentration of minor and trace gases in the air is significant. As an example, the most active volcano in the world, Merapi, in Central Java, Indonesia, has released 30 to 200 t of SO_2 per day for centuries (Orstom, 1990). This continuous injection from Merapi is further enhanced by relatively big eruptions that commonly take place every 3 to 4 yr. Injections of SO_2 from volcanic eruptions were also observed in the Tokyo area in 2000. Air masses arriving in the Tokyo area on 17 Sept. 2000, which had passed the recent volcanic eruption region of Miyake Island (150 km south of Tokyo), contained 200 ppb by volume of SO_2, which was approximately 20 times higher than the ambient SO_2 content of Tokyo air (Okuda et al., 2005).

In terms of particulates, wind-blown fine dusts and sea sprays are perhaps the largest natural source. Terrestrial fine particulates are easily blown and travel long distances. It has been well documented that in early spring fine particles from arid regions in Central Asia such as the Gobi and Takla Makan Deserts can reach eastern China, Korea, Japan, and eventually the central North Pacific Ocean (e.g., Hsu et al., 2008; Shaw, 1980; Fig. 3–1). This phenomenon is called a Kosa episode in Japan. Fine dust from the Sahara travels over the Atlantic Ocean and even to the Amazon region in Brazil (Swap et al., 1992; Fig. 3–1), enriching the soils in the Amazon area.

Compared with terrestrial fine particulates, wind-blown sea spray particulates are more reactive. They are sea salt particles predominantly in the form of sodium chloride (NaCl). These particles are hygroscopic and therefore are deliquescent and grow in size to form aqueous droplets. These droplets can act as sites for chemical reactions in the air. Strong acids such as nitric acid (HNO_3) and H_2SO_4 in the air dissolve in the droplets, and hydrogen chloride (HCl) gas is formed (Eq. [3]) The reaction in Eq. [3] is thought to be an important source of HCl in the air.

$$HNO_{3(g)} + NaCl_{(aq)} \rightarrow HCl_{(g)} + NaNO_{3(aq)}$$
$$H_2SO_{4(g)} + NaCl_{(aq)} \rightarrow HCl_{(g)} + NaHSO_{4(aq)} \quad\quad [3]$$

Fine particles in the submicrometer size class are able to scatter visible light. Light scattering is responsible for the opacity of clouds and therefore affects the heat balance of the air. Fine particles are also harmful to humans. The smaller the particles, the deeper they penetrate into the lungs as humans respire. The deep areas in the lungs have no cilia to move the particles upward, so there is no mechanism to remove these fine particles from the lung tissue, causing respiratory diseases (Dockery et al., 1993).

Minerals containing radioactive elements such as potassium (K) and heavy metals of uranium (U), radium (Ra), indium (In), and thorium (Th) are minor

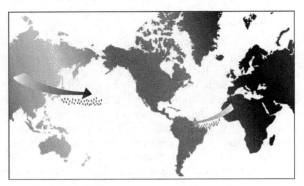

Fig. 3–1. Long range transport of wind-blown terrestrial fine particulate to the north Pacific Ocean, the Atlantic Ocean, and Amazon area.

constituents of air. A radioactive element is an element with an unstable nucleus that spontaneously decays, transforming into a nucleus of a different element. For example, the decay of K yields Ar (Eq. [4]). The series of radium–radon (Eq. [5]), thorium–radium (Eq. [6a] and [6b]), and uranium–thorium (Eq. [7] and [8]) decays produce α particles, which are helium nuclei with two protons and two neutrons. After capturing electrons, α particles change to helium (He) and are added to the air. Helium has never accumulated in large quantities in the Earth's atmosphere because it is light enough not to be gravitationally bound to the Earth and is constantly lost to space. In addition to α, β particles are often emitted as part of radioactive decay. The emitted β is actually an electron ejected from a neutron in the parent nucleus. The emission of α and β particles are accompanied by γ radiation as a form of excess energy to be released from an unstable nucleus. Gamma radiation may also occur from capturing one of the atom's orbital electrons by its unstable nucleus. This kind of electron capture is very rare.

$$^{40}K \rightarrow {}^{40}Ar + \gamma \tag{4}$$

$$^{226}Ra \rightarrow {}^{222}Rn + \alpha \rightarrow ... \rightarrow {}^{210}Pb + \beta \tag{5}$$

$$^{232}Th \rightarrow {}^{228}Ra + \alpha \rightarrow ... {}^{212}Bi + \beta \rightarrow {}^{212}Po + \beta \rightarrow {}^{208}Pb + \alpha \tag{6a}$$

$$^{232}Th \rightarrow {}^{228}Ra + \alpha \rightarrow ... {}^{212}Bi + \beta \rightarrow {}^{208}Tl + \alpha \rightarrow {}^{208}Pb + \beta \tag{6b}$$

$$^{238}U \rightarrow {}^{234}Th + \alpha \rightarrow ... {}^{210}Po + \beta \rightarrow {}^{206}Pb + \alpha \tag{7}$$

$$^{235}U \rightarrow {}^{231}Th + \alpha \rightarrow ... {}^{207}Tl + \alpha \rightarrow {}^{207}Pb + \beta \tag{8}$$

Lead-206 (^{206}Pb) is the final stable element yielded from the decay of ^{232}Th, but this final product can be yielded two different ways, as shown in Eq. [6a] and [6b]. Uranium in nature is consisted of uranium-235 (^{235}U) and uranium-238 (^{238}U). Uranium-238 composes more than 99%, while ^{235}U only contributes less than 1% of the total uranium. These two uranium(s) decay in different ways to produce ^{206}Pb for ^{238}U (Eq. [7]) and ^{207}Pb for ^{235}U (Eq. [8]).

Although small, incoming meteors may also contribute to both gas and particle constituents of the air. Their contribution is, of course, mainly to the upper part of the atmosphere where gases are at a low density.

Biological Activity

Photosynthetic organisms play a central role in exchanging gases in the air. Oxygen and CO_2 are directly involved in respiration and photosynthesis (Eq. [9]). Marine photosynthetic organisms contribute approximately 70% of O_2 in the air, while terrestrial photosynthetic organisms contribute less than one-half compared to their counterparts in the ocean. Photosynthetic microorganisms such as *Azotobacter* are able to bind N_2 from the air and exchange it with CO_2 (Eq. [10]). The bound N_2 is used as nitrogen source for the growth of vegetation and other microorganisms.

$$nH_2O_{(l)} + nCO_{2(g)} \leftrightarrows \{C(H_2O)\}_{n(s)} + nO_{2(g)} \tag{9}$$

$$3C(H_2O)_{(s)} + 2N_{2(g)} + 3H_2O_{(l)} + 4H^+_{(aq)} \rightarrow 3CO_{2(g)} + 4NH_4^+_{(aq)} \tag{10}$$

As opposed to nitrogen fixation, denitrification returns the inorganic nitrogen to the air. Denitrification is a special nitrate reduction reaction producing a nitrogen-containing gas, usually N_2 (Eq. [11]).

$$4NO_{3\ (aq)}^- + 5C(H_2O)_{(s)} + 4H^+_{(aq)} \rightarrow 2N_{2(g)} + 5CO_{2(g)} + 7H_2O_{(l)} \hspace{2cm} [11]$$

Other types of microorganisms, such as methane-producing bacteria, contribute large quantities of CH_4 to the air. Under anaerobic conditions, methane-producing bacteria accelerate the reduction of organic matter to CH_4 according to a simplified reaction shown in Eq. [12].

$$2C(H_2O)_{(s)} \rightarrow CH_{4(g)} + CO_{2(g)} \hspace{4cm} [12]$$

Many organic gases are emitted to the air by trees. Large amounts of volatile organic compounds, which belong to terpenes such as pinene and limonene, are emitted by trees, giving forests their characteristic aroma (e.g., Blanch et al., 2009), the emission rate can be enhanced by high CO_2 and O_3 concentrations typically found in urban areas (Li et al., 2009). Trees also act as a bridge between terrestrial systems and the air. Trees take up NH_3, mostly in the form of NH_4^+ from the soil, and, with the involvement of *Nitrosomonas* microorganisms, the NH_3 is oxidized to N_2O and H_2O as a side product according to the reaction (Jarvis et al., 2009):

$$2NH_{3(g)} + 2O_{2(g)} \rightarrow N_2O_{(g)} + 3H_2O_{(g)} \hspace{3cm} [13]$$

In addition to trees, animals also act as a potential source of trace gases. Animal urine, for example, contains urea [$(NH_2)_2CO$], which is easily hydrolyzed to NH_3 and CO_2 gases:

$$(NH_2)_2CO(aq) + H_2O_{(l)} \rightarrow 2NH_{3(g)} + CO_{2(g)} \hspace{2.5cm} [14]$$

As described, terrestrial microorganisms mediate the formation of various nitrogen-containing gases; however, the same situation is not true in the ocean. The ocean is depleted in nitrogen-containing compounds, and consequently it is not a significant source of nitrogen-containing gases. Instead of nitrogenous species, seawater is rich in dissolved chloride (Cl^-), and to a lesser extent other halogens like fluoride (F^-), bromide (Br^-), and iodide (I^-). Seawater also contains a large amount of sulfate (SO_4^{2-}). These ions are metabolized by marine organisms to form other halogen- and sulfur-containing compounds, and some of them are volatile enough to be emitted to the air.

Certain bacteria such as *Desulvovibrio* use SO_4^{2-} as an electron acceptor in the oxidation of organic matter and generate hydrogen sulfide (H_2S) gas to the air (Eq. [15]). This bacterial mediated formation of H_2S is the main source of sulfur in the air. The eruption from the ocean floor is due to anaerobic conditions formed by the decomposition of phytoplankton. In terrestrial ecosystems it occurs most often in anaerobic soils such as swamps or estuarine muds. To a lesser extent, H_2S is also yielded from the degradation of sulfur-containing organic compounds, such as amino acids, by the action of a number of different kinds of microorganisms. Because of instability toward oxidation, H_2S in the air is rapidly oxidized to SO_2.

$$SO_{4\ (aq)}^{2-} + 2C(H_2O)_{(s)} + 2H^+_{(aq)} \rightarrow H_2S_{(g)} + 2CO_{2(g)} + 2H_2O_{(l)} \hspace{1cm} [15]$$

Other important sulfur compounds released from seawater are dimethyl sulfide [$(CH_3)_2S$] and carbonyl sulfide (OCS). The flux of OCS is smaller than that of $(CH_3)_2S$, but because of its significantly higher residence time, OCS has a greater concentration than $(CH_3)_2S$. Dimethyl sulfide and OCS are oxidized to SO_2 and eventually to SO_4^{2-}. The oxidation is initiated by OH·.

Marine microorganisms also facilitate the formation of volatile halogenated organic compounds. Among them, methyl chloride (CH_3Cl), methyl bromide (CH_3Br), and methyl iodide (CH_3I) are the most important. It is believed that among these methyl halides, CH_3I is the most ubiquitously produced in surface and subsurface water of the ocean (Moore and Groszko, 1999; Yokouchi et al., 2001, 2008), and it can reach regions far from the ocean and become an important source of essential compounds for mammals (Yokouchi et al., 2008).

Anthropogenic Sources of Air Constituents

Anthropogenic sources of air constituents began when humans first used fire. The anthropogenic sources increased substantially with the advancement of human civilization. As human civilizations became increasingly advanced and increased in size, anthropogenic air constituents increased due to a wide a range of activities, mostly related to combustion processes. Industrialization and transportation, along with rapid urbanization, are the major activities contributing to the anthropogenic air constituents today.

Industrialization and Urbanization

Industrial activity is an important factor for fast economic growth. Establishment of a large number of industries in an area to form an industrial area is a common feature of industrial development in many countries. This is in turn catalyzes rapid local urbanization and the introduction of anthropogenic air constituents.

In most countries, urbanization is a natural consequence of economic development based on industrialization. Thus, the level of urbanization, as measured by the country's urban population relative to its total population, is highest in the most developed, high-income countries and lowest in the least developed, low-income countries. For example, in the high-income countries in Western Europe in the early 1980s, about 43% of the population already lived in urban areas (Turok and Mykhnenko, 2007), and at the same time, South Asia, East Asia, and Sub-Saharan Africa remained predominantly rural, with only 25% of the population living in urban centers (Jack, 2006). With increasing industrialization, urbanization increases as well. Industrialization attracts people to move to areas close to the industrial area. Urbanization presents a serious problem in many parts of the world. In Asia, urbanization continues to grow at an average rate of 2.2% per year, with higher rates being experienced in Southeast Asia (Jack, 2006). Taking Indonesia as an example, urbanization increased the percentage of the population living in urban areas from 35% in 1985 to 50% in 2005 (Prabowo, 2006). Indonesia's level of urbanization is higher than that in China, the most populous country in the world. In China, urbanization levels of 35% had not been realized until 2000, and the urbanization level in 2005 was approximately 43%.

Like the urbanization rate, injection of anthropogenic air constituents is also accelerated in developing countries. Because of generally tighter environmental regulations, higher labor cost, and more awareness related to environmental degradation in developed countries, many industries that pollute the environment such as fertilizer and pesticide industries, have moved to developing countries. Moreover, some developing countries are experiencing their own industrial revolution. China, India, and Vietnam had growth rates of more than 10% in the last 10 yr, followed by some countries in Southeast Asia, like Thailand and Indonesia, with growth rates of slightly more than 5%. On the other hand, industrial growth in developed countries in Western Europe, North America, and Japan grew by 3% or less (Central Intelligence Agency, 2007).

Sulfur Dioxide Concentrations

The industrial revolution in India and especially China is heavily fueled by coal. Burning of coal emits large quantities of anthropogenic SO_2 and suspended particulate matter (SPM). Some cities in China, such as Beijing and Shanghai, have a dangerously high concentration of SO_2 in the air (Chan and Yao, 2008; Plate 3–1, see color insert section). The ambient concentration of SO_2 in Beijing is in excess of the World Health Organization (WHO) guidelines (50 μg m^{-3}), while in Shanghai concentrations are close to and sometimes in excess of the WHO guidelines (Ha, 2005). The same situation is observed in India. Although the concentrations are within the WHO guidelines, there are days, or months, when the levels are higher in some cities. For instance, the concentrations of SO_2 in Kolkata are regularly exceeded during the non-monsoon months (Gupta et al., 2008). With the switch to oil and gas, ambient concentrations of SO_2 have decreased in some cities, such as Delhi and Mumbai (e.g., Chelani and Devotta, 2007), but this is an exception. Rapid industrialization has caused SO_2 emissions in Asia to increase steadily since the mid 1970s (Chan and Yao, 2008). Since the mid 1990s, Asia has replaced Eastern Europe as the biggest SO_2 emitter in the world. In other regions, including North America and Western Europe, SO_2 emissions have continuously decreased since the mid 1970s and SO_2 emissions in Eastern Europe started to decline in the early 1990s (e.g., UN DESA, 2005). The reduction of SO_2 emissions in developed countries is mainly a result of efforts to reduce emissions from power plants and industrial boilers, as well as the reduction of sulfur content in fuels.

While the concentration of SO_2 in urban air in developed countries has decreased substantially from year to year, that in some Asian countries is still relatively unchanged, such as that for Beijing and Shanghai in China (Fig. 3–2; Ha, 2005), or has even increased sharply, such as that for Jakarta and Surabaya in Indonesia (Fig. 3–2).

Suspended Particulate Matter

Suspended particulate matter, also commonly termed *total suspended particles* (TSP), is not a single pollutant, but rather a mixture of many subclasses of pollutants that occur in both solid and liquid forms. Each subclass contains many

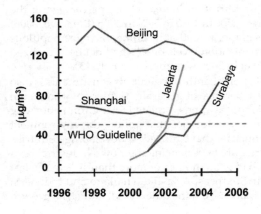

Fig. 3–2. Average annual concentrations of SO_2 (μg m^{-3}) in the two most important cities in China and Indonesia.

different chemical species. Particulate matter may be classified as primary or secondary. Primary particles are emitted directly by emission sources, whereas secondary particles are formed through the atmospheric reaction of gases, such as the reactions between NH_3 and NO_x or sulfur that lead to the formation of particles. Total suspended particles have historically been monitored and continue to be measured in developing countries. However, the WHO places special emphasis on suspended particles smaller than 10 μm in diameter (PM_{10}), also called *inhalable* PM, and those smaller than 2.5 μm ($PM_{2.5}$), called *fine* or *respirable* PM. Emerging scientific evidence points to increasing damage with decreasing particle diameter (e.g., Brook et al., 2004; Pope and Dockery, 2006; Zanobetti and Schwartz, 2009). Particles larger than about 10 μm are deposited almost exclusively in the nose and throat, whereas particles smaller than 1 μm are able to reach the lower regions of the lungs. The intermediate size range gets deposited in between these two extremes of the respiratory tract. A statistically significant association has been found between adverse health effects and ambient PM_{10} concentrations (e.g., Pope, 1991), and recent studies using $PM_{2.5}$ data have shown an even stronger association between health outcomes and particles in this size range (e.g., Jerrett et al., 2009). In response, industrial countries have switched from monitoring total suspended particulates (TSP), which is not directly correlated with health effects, to PM_{10}, and increasingly to $PM_{2.5}$.

Particularly high concentrations of suspended particulates are found in countries relying on coal for energy, notably in China and India (e.g., Chen et al., 2007; Rengarajan et al., 2007). Among main urban areas in the world, nearly all areas with the highest levels of suspended particulates are in Asia (Fig. 3–3). Indeed, the main urban areas in China, such as Beijing and Shanghai, and in India, like New Delhi and Kolkata, as well as in countries close to India like Kathmandu in Nepal and Dhaka in Bangladesh, are among the urban areas having the highest content of PM_{10} in the air. A special case is observed in Indonesia.

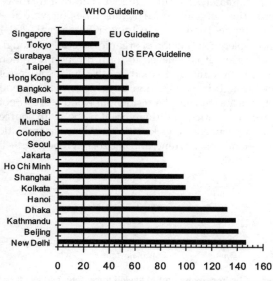

Fig. 3–3. Average annual concentrations of PM_{10} (μg m^{-3}) in the air of 20 urban areas in Asia (modified from Smith et al., 2006).

PM_{10} average annual concentration (μg/m³)

Instead of coal, this country uses fossil oil as the main energy source, but the level of PM_{10} in the air of main urban areas close to industrial areas, i.e., Jakarta, is also extremely high. This may be caused by the fact that in Indonesia nearly all power plants and industry boilers are not equipped with particulate control systems, vehicles are not set up with catalytic converters for better emission control, and pollution control measures are not effectively implemented. The opposite condition is found in developed counties, where since about 1970, particulate air pollution has been reduced by particulate control systems on power plants and industrial facilities, use of cleaner fuels such as natural gas, and requirements for catalytic converters on vehicles. Average annual concentrations of PM_{10} in the air for all 20 urban areas given in Fig. 3–3 exceeded the values set in the WHO guidelines (20 μg m^{-3}). Only two of them, Singapore and Tokyo, passed the requirement of EU guidelines (40 μg m^{-3}). Only four fulfilled the requirement of USEPA guidelines (50 μg m^{-3})—Singapore, Tokyo, Surabaya, and Taipei. The concentrations of PM_{10} in the air of the most polluted urban areas, New Delhi and Beijing, are more than seven times higher than the WHO guideline for PM_{10}.

The concentration of PM_{10} in the air of Beijing is seasonally independent. Recent measurements showed that the average concentration of PM_{10} during 5 months (November–March) in winter 2003 and 2004 (i.e., 150.0 ± 96.8 μg m^{-3}; n = 195) was similar to that observed during other seasons in the same years (i.e., 156.8 ± 90.6 μg m^{-3}; n = 323) (Okuda et al., 2006).

As observed for natural fine particulate (Fig. 3–1), its anthropogenic counterpart is also easily blown and travels long distances in the air. For instance, anthropogenic fine PM in the Canadian Arctic in late winter contains various combustion products of coal and gasoline, such as sulfate, soot, vanadium (V), and lead (Pb), originating from Russia and Europe (Bunce, 1994). It is well documented that the combustion of fossil fuels like coal and gasoline releases significant amounts of trace metals into the air. As can be seen in Fig. 1.3–3, the level of PM_{10} in Beijing was approximately five times higher than that in Tokyo (Okuda et al., 2004), and concentrations of many trace metals in Beijing were also approximately five or more times higher than those in Tokyo (Okuda et al., 2004).

Soot is a general term for incomplete combustion of organic materials although a more recent term used is *black carbon*. It is believed to form through accretion of graphite-like precursors. Coal, petroleum, and wood are the main sources of soot. Petroleum generates soot through its combustion in vehicles, while wood gives its portion through forest fires and burning in residential wood stoves. Soot is composed of various condensed aromatics, and therefore it is the main source of a group of substances containing benzene rings known as polycyclic aromatic hydrocarbons (PAHs), such as phenanthrene, pyrene, and benzo[*a*]pyrene (Fig. 3–4)

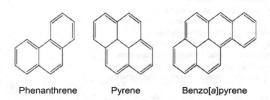

Phenanthrene   Pyrene   Benzo[a]pyrene

Fig. 3–4. Examples of polycyclic aromatic hydrocarbons (PAHs) having three, four, and five fused benzene rings.

with benzene ring numbers of 3, 4, and 5, respectively. Polycyclic aromatic hydrocarbons with small molar masses (i.e., PAHs with 3, 4, and 5 fused benzene rings)
will be in gas phase, while those with more than five fused rings are almost completely associated with solid particles.

The level of PAHs in the air is closely related to the combustion of coal. This
is why the extremely high contents of PAHs are found mostly in the air of countries that use coal as the fuel for their industrial development, including China
and India (Fig. 3–5). The air of Jakarta also contains a high concentration of PAHs.
This may be due to the same reasons as deduced for PM_{10} above. On the other hand,
the concentration of PAHs in the air of Hong Kong is comparatively low. The low
concentration of PAHs in Hong Kong is possibly associated with the implementation of stringent industrial and vehicular emission control measures since the
early 1990s. These include implementing natural gas for new or upgraded power
stations, enforcement of a smoky vehicle control program, reduction of diesel
vehicles in the fleet through using LPG taxis and public light buses, reduction of
sulfur in diesel to 0.05%, and phasing out of leaded petrol (Sin et al., 2003). Monitoring for 1.5 yr in Beijing, from October 2003 to April 2005, showed that the level
of PAHs during the period when coal combustion for residential heating is permitted (i.e., from 15 November to 15 March) was more than seven times higher
than when coal combustion is not permitted (Okuda et al., 2006). The average concentration of total PAHs during the residential coal combustion period was 305.1
± 279.0 ng m^{-3} (n = 33), while that in the noncombustion period was 42.3 + 32.0 ng
m^{-3} (n = 31). The enhancement was even greater for PAHs containing four benzene
rings (i.e., pyrene and benzo[a]anthracene); concentrations were approximately
ten times higher when coal combustion for residential heating was allowed. Similar result was obtained for the urban area of Hong Kong. The average monthly
PAHs concentration in 2000 ranged from 12.2 to15.8 ng m^{-3} and depicted obvious seasonal variations with maxima in winter (October–March) and minima in

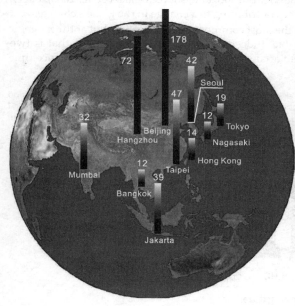

Fig. 3–5. Average concen-
trations (ng m^{-3}) of total
polycyclic aromatic hydrocarbons (PAHs) in various urban
areas in the Asia Pacific
region, with sampling dates
of Aug. 2000–March 2001 for
Bangkok, Oct. 2003–April
2005 and July 1999 for Hangzhou, Jan. 2000–Dec. 2001 for
Hong Kong, Dec. 1992–Dec.
2003 for Jakarta, Nov.–Dec.
1996 for Mumbai, July 1997–
June 1998 for Nagasaki,
Oct.–Nov. 1998, Feb.– Mar.
1999, May–June 1999, and
Sept.–Dec. 1999 for Seoul,
Aug. 1995–Jan. 1996 for Taipei, and Aug. 2000–March
2001 for Tokyo.

summer (April–September) months (Sin et al., 2003). The concentration of total PAHs in winter was 1.4 to 2.4 times higher than the concentration of PAHs in the summer. The climatic conditions in Hong Kong with a hot and rainy summer provided all the necessary criteria for effective removal of atmospheric PAHs through "rainout" (Panther et al., 1999).

Transportation

The growing population in urban areas triggers the need for transportation. Population growth in urban areas is commonly in imbalance with the development of transportation infrastructure, especially in developing countries. In the absence of an adequate and comfortable public transport system to provide the facilities for mobility, commuters are increasingly relying on private vehicles. The growing use of motor vehicles enhances the emission of anthropogenic constituents in urban areas. Particularly in developing countries, urban areas themselves have an insufficient road network, full with a mixture of inefficient vehicles that lack equipment for cleaner emission and are powered with poor quality fuels.

Transport contributes to anthropogenic SO_2 and SPM emissions, but contributes mostly to anthropogenic CO, hydrocarbon (HC), and NO_x, while industrialization contributes more SO_2 than transportation (Fig. 3–6; Santosa et al., 2008). The tailpipe emissions of CO, HC, NO_x, and PM vary with the air/fuel ratio, injection timing, and other settings. In general, with increasing vehicle speed or engine temperature, NO_x emissions increase, while CO, HC, and particulate emissions decrease (Gwilliam et al., 2004).

Greenhouse Gases

A greenhouse gas is a gas in the air that absorbs the infrared radiation emitted by the Earth's surface. Methane as the simplest HC, CO_2 as the product fossil fuel combustion, N_2O, chlorofluorocarbons (CFCs), and ozone are known as greenhouse gases that are increasing due to anthropogenic sources (Bates et al., 2008). Greenhouse gases are believed to be contributing to a change in the globe's climate.

Water vapor is actually the most common greenhouse gas, but anthropogenic water emissions are negligible. Indeed, on a planet with a surface that is two-thirds water, and taking into account that water has a very short residence time

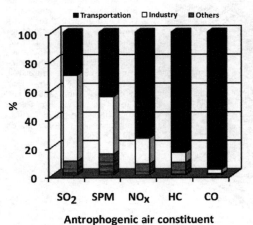

Fig. 3–6. Contribution of transportation and industrial sectors to the emission of SO_2, suspended particulate matter (SPM), NO_x, HC, and CO in Jakarta in 2001.

in the air, direct human emissions of water vapor do not have a significant impact on the global water cycle. Humans can definitely cause major perturbations to the water cycle on a local scale through, for example, irrigation, deforestation, creation of dams (Bates et al., 2008), but that doesn't have significant repercussions on the average proportion of water vapor in the air on a global scale, and therefore on the global greenhouse effect coming from water vapor. This is why water vapor is not taken into account when measuring the greenhouse gas emissions caused by human activities, except in some very particular cases.

As with CO_2 and CH_4, the concentration of N_2O in the air is also increasing at a rate of approximately 0.2% per year. Nitrous oxide absorbs infrared radiation very effectively. Nitrous oxide has no sink in the lower part of the atmosphere and eventually diffuses to higher layers, where it either decomposes photochemically or reacts with ozone, reducing ozone concentrations in the stratosphere. Although CO_2 has gained the most attention from the news media in terms of its increased concentration and the implications for global warming, recent estimations suggest that the other trace greenhouse gases (CH_4, N_2O, and CFCs) are likely have a combined effect similar to that of CO_2. The global warming potential (GWP) of a gas is defined as the cumulative radiative forcing "rate of energy change per unit area of the globe as measured at the top of the atmosphere" in watts per square meter of a given gas over a given period (20, 500, and more generally 100 yr) (IPCC, 2007; Le Treut et al., 2007). The global warming potential is actually never given as an absolute value, but relative to CO_2, which has a defined GWP of 1 (Table 3–3). For example, CH_4 (0.48 W m^{-2}) contributes almost one-third as much to greenhouse warming as CO_2 (1.66 W m^{-2}), although it is several orders of magnitude less abundant (Table 3–1). This is because its global warming potential is 25 times that of CO_2 (Table 3–3). Based on the atmospheric concentrations, efficiency of radiation trapping, and residence time of each gas, the relative radiative forcing since 1750 by CO_2, CH_4, CFCs, and N_2O are 63, 18, 13, and 6%, respectively (Forster et al., 2007).

Increased levels of greenhouse gases in the air could cause a variety of climatic changes. The global mean temperatures will rise, which will cause melting of glaciers and ice sheets, and in turn will cause a rise in sea level. Climatic change may also cause changes in rainfall patterns, the frequency and severity of storms, hydrological cycles, and cloud patterns (Bates et al., 2008). Current predictions showed that in the next 50 to 100 yr, the temperature is likely to rise in the range of 1 to 3°C (Meehl et al., 2007). Current greenhouse effect modeling suggests that the regional impacts from climate change may be quite disparate. No other area of the world may suffer more adverse effects than Asia. One meter rise in sea level would put 126,000 km^2 of land in China under water and would threaten coastal environments around the Pacific Rim. Southeast Asia and islands in the Pacific Ocean appear especially vulnerable to sea level rise since most their populations live in low-lying areas that are susceptible to flooding, and inhabitants of these regions depend significantly on fishing and agriculture, which may be disrupted by climate change.

Stratospheric Ozone and Air Pollution

Chlorofluorocarbons are a group of compounds having a general formula CF_nCl_{4-n}. Other compounds with similar function, such as hydrofluorocarbons

($C_xH_yF_z$, with $x + y + z = 4$), for technical purposes are incorporated also into the group, and they are often collectively referred to as CFCs. Chlorofluorocarbons are well known as fluids for filling refrigerators and any other devices that generate cold (e.g., air conditioners, freezers) and as propellants in sprays. They also involve in a number of industrial processes, such as in manufacturing of plastic foams. CFCs are a potent greenhouse gas (Table 3–3) and are also very effective at destroying the stratospheric ozone layer at approximately 12 to 50 km above the globe's surface.

Chlorine can be released from the photolysis of CFCs (Eq. [16]). The released chlorine is then involved in the O_3 destruction (Eq. [17]) and can be regenerated according to Eq. [18], allowing this cycle to be repeated over and over. This is why that one CFC-derived chlorine atom can destroy up to 100,000 O_3 before leaving the place of O_3 layer.

$$CF_2Cl_2 \xrightarrow{h(<250 \text{ nm})} CF_2Cl + Cl \tag{16}$$

$$Cl + O_3 \rightarrow ClO + O_2 \tag{17}$$

$$ClO + O \rightarrow Cl + O_2 \tag{18}$$

The discovery of a large O_3 hole the size of the continental United States over Antarctica in the mid 1980s catalyzed public and political concern over CFCs and the depletion of the O_3 layer. Negotiations to limit the release CFCs and other O_3 layer–depleting chemicals commenced and resulted in the 1985 Vienna Convention on Ozone-Depleting Substances, the 1987 Montreal Protocol on Substances that Deplete Ozone Layer, and several later amendments to the protocol. These agreements mandated a phase out of ozone layer–destructing products that were once widely used in refrigeration, plastic foam production, electronics, aerosol spray, and a number of other applications by 1996 for developed countries and by 2010 for developing countries.

By 2003, developed countries had reduced consumption of CFCs by more than 99% and developing countries by more than 50%. However, some challenges remain. Some of the chemicals replacing CFCs, such as methyl bromide (CH_3Br), are also ozone-depleting substances, although they are less damaging than CFCs. Currently, there are still difficulties in replacing CH_3Br.

The Importance of Trees for Air Quality in Urban Areas

Trees are planted for so many reasons, from aesthetic to economic to environmental. In the context of air quality, trees have beneficial and harmful effects. Trees can remove air pollutants (Nowak et al., 2006; Jim and Chen, 2008), reduce air temperatures (Huang et al., 2009), humidify the air through transpiration (e.g., Jarvis and McNaughton, 1986), and reduce building energy use (Heisler, 1986; Akbari et al., 1997, 2001). Trees can remove pollutants, especially O_3, NO_2, and particles from the air (Nowak et al., 2006; Jim and Chen, 2008). Trees also remove CO_2 from the air primarily through photosynthesis (Eq. [9]). Gaseous air pollutants are primarily removed via uptake into leaf stomata. Once inside the leaf, gases diffuse into intercellular spaces and may be absorbed by water films to form acids or react with inner-leaf surfaces. Trees also remove pollution by intercepting airborne particles (Nowak et al., 2006). Although most particles are retained on the plant surface, some particles may be absorbed into the tree. The retained particle is often resuspended to the air, wet deposited by rain, dry deposited, or dropped to the ground together with leaf and twig fall.

Tree canopies affect radiation absorption and heat storage, wind speed, turbulence, surface albedo, surface roughness, and consequently the evolution of the mixing-layer height, which directly governs local air temperature and relative humidity (e.g., Heisler et al., 1995). These changes in local meteorology can alter pollution concentrations in urban areas. Trees usually contribute to cooler summer air temperature, and on the other hand, may be able to warm winter temperature. Reduced air temperatures due to trees in the summer season can improve air quality because the emission of many pollutants and/or ozone-forming chemicals is temperature dependent.

Trees reduce building energy use by lowering temperatures and shading buildings during the summer, and blocking winds in winter. However, they can also increase energy use by shading buildings in winter, and may increase energy use by blocking summer breezes. Thus, proper tree placement near buildings is critical to achieve maximum building energy conservation benefits (see Chapters 11 [Volder and Watson, 2010] and 18 [Aitkenhead-Peterson et al., 2010], this volume). When building energy use is lowered, pollutant emissions from power plants are also lowered. While lower pollutant emissions generally improve air quality, lower NO_x emissions, particularly ground-level emissions, may lead to a local increase in O_3 concentrations under certain conditions, due to NO_x scavenging of O_3 (Rao and Sistla, 1993).

In fact, not all trees have beneficial effects on air quality. Some trees can have a negative effect and actually help to form pollutants in the atmosphere. Trees can emit gases known as volatile organic compounds (VOCs). Volatile organic compounds in combination with anthropogenic NO_x can contribute to the production of other pollutants, especially O_3 and particles. However, in atmospheres with low NO_x concentrations (e.g., some rural environments), VOCs may actually remove O_3 (Crutzen et al., 1985; Jacob and Wofsy, 1988). Because VOC emissions are temperature dependent and trees generally help lower air temperatures, increased tree cover can lower overall VOC emissions and, consequently, O_3 levels in urban areas (Cardelino and Chameides, 1990). Volatile organic compound emission rates also vary by tree species. The following trees—beefwood (*Casuarina* spp.), *Eucalyptus* spp., sweetgum (*Liquidambar* spp.), black gum (*Nyssa* spp.), sycamore (*Platanus* spp.), poplar (*Populus* spp.), oak (*Quercus* spp.), black locust (*Robinia* spp.), and willow (*Salix* spp.)—have very high isoprene emission rates and therefore have the greatest relative effect among genera on increasing O_3 formation (Geron et al., 1994).

Since trees can have positive and negative effects on air quality and the magnitude of those effects are dependent on the tree species, the selection of trees to be included for air quality improvement in urban areas should be carefully conducted. Trees that do not emit reactive VOCs, but do have large leaf surface areas for maximum cooling effect have the best effect on air quality. It was also documented that trees remove airborne pollutants at three times the rate of grassland (Hewitt, 2002) and that coniferous trees capture larger quantities than deciduous broadleaf trees (Beckett et al., 2000). Trees at the edge of woodland are more effective at removing air pollutants than trees in the center of woodland (Hewitt, 2002). This is due to both larger leaf areas and greater exposure to the wind.

References

Aitkenhead-Peterson, J.A., M.K. Steele, and A. Volder. 2010. Services in natural and human dominated ecosystems. p. 373–390. *In* J. Aitkenhead-Peterson and A. Volder (ed.) Urban ecosystem ecology. Agron. Monogr. 55. ASA, CSSA, and SSSA, Madison, WI.

Akbari, H., D.M. Kurn, S.E. Bretz, and J.W. Hanford. 1997. Peak power and cooling energy savings of shade trees. Energy Build. 25:139–148.

Akbari, H., M. Pomerantz, and H. Taha. 2001. Cool surfaces and shade trees to reduce energy use and improve air quality in urban areas. Sol. Energy 70:295–310.

Bates, B.C., Z.W. Kundzewicz, S. Wu, and J.P. Palutikof (ed.) 2008. Climate change and water. Technical Paper of the Intergovernmental Panel on Climate Change. IPCC, Geneva.

Beckett, K.P., P. Freer-Smith, and G. Taylor. 2000. Effective tree species for local air-quality management. J. Arboric. 26:12–18.

Blanch, J.S., J. Penuelas, J. Sardans, and J. Llusia. 2009. Drought, warming and soil fertilization effects on leaf volatile terpene concentrations in *Pinus halepensis* and *Quercus ilex*. Acta Physiol. Plant. 31:207–218.

Brook, R.D., B. Franklin, W. Cascio, Y.L. Hong, G. Howard, M. Lipsett, R. Luepker, M. Mittleman, J. Samet, S.C. Smith Jr., and I. Tager. 2004. Air pollution and cardiovascular disease—A statement for healthcare professionals from the expert panel on population and prevention science of the American Heart Association. Circulation 109:2655–2671.

Bunce, N. 1994. Environmental chemistry. 2nd ed. Wuerz Publishing Ltd., Winnipeg, Canada.

Cardelino, C.A., and W.L. Chameides. 1990. Natural hydrocarbons, urbanization, and urban ozone. J. Geophys. Res. 95(D9):13971–13979.

Central Intelligence Agency. 2007. CIA world factbook. Skyhorse Publishing, New York.

Chan, C.K., and X. Yao. 2008. Air pollution in mega cities in China. Atmos. Environ. 42:1–42.

Chelani, A.B., and S. Devotta. 2007. Air quality assessment in Delhi: Before and after CNG as fuel. Environ. Monit. Assess. 125:257–263.

Chen, Y.Y., Y.J. Siang, H. Wang, and D.L. Li. 2007. Assessment of ambient air quality in coal mine waste areas—A case study in Fuxin, China. N.Z. J. Agric. Res. 50:1187–1194.

Crutzen, P.J., A.C. Delany, J. Greenberg, P. Haagenson, L. Heidt, R. Lueb, W. Pollock, W. Seiler, A. Wartburg, and P. Zimmerman. 1985. Tropospheric chemical composition measurements in Brazil during the dry season. J. Atmos. Chem. 2:233–256.

Dockery, D.W., C.A. Pope, X.P. Xu, J.D. Spengler, J.H. Ware, M.E. Fay, B.G. Ferris, and F.E. Speizer. 1993. An association between air-pollution and mortality in 6 United-States cities. N. Engl. J. Med. 329:1753–1759.

Forster, P., V. Ramaswamy, P. Artaxo, T. Berntsen, R. Betts, D.W. Fahey, J. Haywood, J. Lean, D.C. Lowe, G. Myhre, J. Nganga, R. Prinn, G. Raga, M. Schulz, and R. Van Dorland. 2007. Changes in atmospheric constituents and in radiative forcing. *In* S. Solomon et al. (ed.) Climate Change 2007: The physical science basis. Contribution of Working Group I to the Fourth Assessment Report of the IPCC. Cambridge Univ. Press, Cambridge, UK.

Geron, C.D., A.B. Guenther, and T.E. Pierce. 1994. An improved model for estimating emissions of volatile organic compounds from forests in the eastern United States. J. Geophys. Res. 99(D6):12,773–12,791.

Gwilliam, K., M. Kojima, and T. Johnson. 2004. Reducing air pollution from urban transport. World Bank, Washington, DC.

Gupta, A.K., K. Karar, S. Avoob, and K. John. 2008. Spatio-temporal characteristics of gaseous and particulate pollutants in an urban region of Kolkata, India. Atmos. Res. 87:103–115.

Ha, K. 2005. Overview of international AQM practices: Options for china? Clean Air Initiative for Asian Cities, Beijing.

Heisler, G.M. 1986. Effects of individual trees on the solar radiation climate of small buildings. Urban Ecol. 9:337–359.

Heisler, G.M., R.H. Grant, S. Grimmond, and C. Souch. 1995. Urban forests: Cooling our communities? p. 31–34. *In* Inside Urban Ecosystems, Proc. 7th National Urban Forest Conference. American Forests, Washington, DC.

Hewitt, N. 2002. Trees and sustainable urban air quality—Using trees to improve air qualities in cities. Centre for Ecology and Hydrology, Lancaster Univ., UK.

Hsu, S.C., S.C. Lui, Y.T. Huang, S.C.C. Lung, F.J. Tsai, J.Y. Tu, and S.J. Kao. 2008. A criterion for identifying Asian dust events based on Al concentration data collected from northern Taiwan between 2002 and early 2007. J. Geophys. Res. Atmos. 113:D18306.

Huang, J.L., R.S. Wang, F. Li, W.R. Yang, C.B. Zhou, J.S. Jin, and Y. Shi. 2009. Simulation of thermal effects due to different amounts of urban vegetation within the built-up area of Beijing, China. Int. J. Sustain. Dev. World Ecol. 16:67–76.

IPCC. 2007. Climate Change 2007: Impacts, adaptation and vulnerability. In M.L. Parry et al. (ed.) Contribution of Working Group II to the Fourth Assessment Report of the IPCC. Cambridge Univ. Press, Cambridge, UK.

Jack, M. 2006. Urbanisation, sustainable growth and poverty reduction in Asia. IDS Bull. Inst. Dev. Stud. 37:101–114.

Jacob, D.J., and S.C. Wofsy. 1988. Photochemistry of biogenic emissions over the Amazon forest. J. Geophys. Res. 93(D2):1477–1486.

Jarvis, P.G., and K.G. McNaughton. 1986. Stomatal control of transpiration: Scaling up from leaf to region. Adv. Ecol. Res. 15:1–49.

Jarvis, A., C. Sundberg, S. Milenkovski, M. Pell, S. Smars, P.-E. Linden, and S. Halin. 2009. Activity and composition of ammonia oxidizing bacterial communities and emission dynamics of NH_3 and N_2O in a compost reactor treating organic household waste. J. Appl. Microbiol. 106:1502–1511.

Jerrett, M., M.M. Finklestein, J.R. Brook, M.A. Arain, P. Kanaroglou, D.M. Stieb, N.L. Gilbert, D. Verma, N. Finkelstein, K.R. Chapman, and M.R. Sears. 2009. A cohort study of traffic-related air pollution and mortality in Toronto, Ontario, Canada. Environ. Health Perspect. 117:772–777.

Jim, C.Y., and W.Y. Chen. 2008. Assessing the ecosystem service of air pollutant removal by urban trees in Guangzhou (China). J. Environ. Manage. 88:665–676.

Khalil, M.A.K., and R.A. Rasmussen. 1990. Atmospheric methane: Recent global trends. Environ. Sci. Technol. 24:549–553.

Le Treut, H., R. Somerville, U. Cubasch, Y. Ding, C. Mauritzen, A. Mokssit, T. Peterson, and M. Prather. 2007. Historical overview of climate change. In S. Solomon et al. (ed.) Climate Change 2007: The physical science basis. Contribution of Working Group I to the Fourth Assessment Report of the IPCC. Cambridge Univ. Press, Cambridge, UK.

Li, D.W., Y. Chen, Y. Shi, X.Y. He, and X. Chen. 2009. Impact of elevated CO_2 and O_3 concentrations on biogenic volatile organic compounds emissions from Ginkgo biloba. Bull. Environ. Contam. Toxicol. 82:473–477.

Meehl, G.A., T.F. Stocker, W.D. Collins, P. Friedlingstein, A.T. Gaye, J.M. Gregory, A. Kitoh, R. Knutti, J.M. Murphy, A. Noda, S.C.B. Raper, I.G. Watterson, A.J. Weaver, and Z.-C. Zhao. 2007. Global climate projections. In S. Solomon et al. (ed.) Climate Change 2007: The physical science basis. Contribution of Working Group I to the Fourth Assessment Report of the IPCC. Cambridge Univ. Press, Cambridge, UK.

Moore, R.M., and W. Groszko. 1999. Methyl iodide distribution in the ocean and fluxes to the atmosphere. J. Geophys. Res. C. Oceans 104(C5):11163–11171.

Nowak, D.J., D.E. Crane, and J.C. Stevens. 2006. Air pollution removal by urban trees and shrubs in the United States. Urban For. Urban Green. 4:115–123.

Okuda, T., T. Iwase, H. Ueda, Y. Suda, S. Tanaka, Y. Dokiya, K. Fushimi, and M. Hosoe. The impact of volcanic gases from Miyake island on the chemical constituents in precipitation in the Tokyo metropolitan area 2005. Sci. Total Environ. 341:185–197.

Okuda, T., J. Kato, J. Mori, M. Tenmoku, Y. Suda, S. Tanaka, K. He, Y. Ma, F. Yang, X. Yu, F. Duang, and Y. Lei. 2004. Daily concentrations of trace metals in aerosol in Beijing, China, determined by using inductively coupled plasma mass spectrometry equipped with laser ablation analysis, and source identification of aerosols. Sci. Total Environ. 330:145–158.

Okuda, T., D. Naoi, M. Tenmoku, S. Tanaka, K. He, Y. Ma, F. Yang, Y. Lei, Y. Jia, and D. Zhang. 2006. Polycyclic aromatic hydrocarbons (PAHs) in the aerosol in Beijing, China, measured by aminopropylsilane chemically-bonded stationary-phase column chromatography and HPLC/fluorescence detection. Chemosphere 65:427–435.

Orstom, R.G.S. 1990. An ecosystem under acid rain at Merapi volcano central Java, Indonesia. 5th Intl. Congr. on Ecology. Yokohama, Japan.

Panther, B.C., M.A. Hooper, and N.J. Tapper. 1999. A comparison of air particulate matter and associated polycyclic aromatic hydrocarbons in some tropical and temperate urban environments. Atmos. Environ. 33:4087–4099.

Pope, C.A. 1991. Respiratory hospital admissions associated with PM-10 pollution in Utah, Salt Lake, and Cache Valleys. Arch. Environ. Health 46:90–97.

Pope, C.A., and D.W. Dockery. 2006. Health effects of fine particulate air pollution: Lines that connect. J. Air Waste Manage. Assoc. 56:709–742.

Prabowo, A. 2006. Urban air quality improvement (UAQi) program development. Better Air Quality Conference. Yogyakarta, Indonesia.

Rao, S.T., and G. Sistla. 1993. Efficacy of nitrogen oxides and hydrocarbons emissions control in ozone attainment strategies as predicted by the Urban Airshed Model. Water Air Soil Pollut. 67:95–116.

Rengarajan, R., M.M. Sarin, and A.K. Sudheer. 2007. Carbonaceous and inorganic species in atmospheric aerosols during wintertime over urban and high-altitude sites in North India. J. Geophys. Res. Atmos. 112:D21307.

Santosa, S.J., T. Okuda, and S. Tanaka. 2008. Air pollution and urban air quality management in Indonesia. Clean Soil Air Water 36:466–475.

Shaw, G.E. 1980. Transport of Asian desert dust aerosol to the Hawaiian Islands. J. Appl. Meteorol. 19:1254–1259.

Sin, D.W.M., Y.C. Wong, Y.Y. Choi, C.H. Lam, and P.K.K. Louie. 2003. Distribution of polycyclic aromatic hydrocarbons in the atmosphere of Hong Kong. J. Environ. Monit. 5:989–996.

Smith, K.R. et al. 2006. Energy and air pollution. Geo Yearbook 2006. Available at http://www.unep.org/geo/yearbook/yb2006/054.asp (verified 9 Feb. 2010). United Nations Environment Prog., Nairobi.

Swap, R., M. Garstang, S. Grecos, R. Talbot, and P. Kallberg. 1992. Saharan dust in the Amazon Basin. Tellus Ser. B 44:133–149.

Turok, I., and V. Mykhnenko. 2007. The trajectories of European cities, 1960–2005. Cities 24:165–182.

UN DESA. 2005. Atmosphere and Air pollution, World Development Indicators 2005, World Bank. Available at http://www.un.org/esa/sustdev/publications/trends2006/atmosphere.pdf (verified 11 Feb. 2010).

Vizuete, W., B.-U. Kim, H. Jeffries, Y. Kimura, D.T. Allen, M.-A. Kioumourtzoglou, L. Biton, and B. Henderson. 2008. Modeling ozone formation from industrial emission events in Houston, Texas. Atmos. Environ. 42:7641–7650.

Volder, A., and W.T. Watson. 2010. Urban forestry. p. 227–240. In J. Aitkenhead-Peterson and A. Volder (ed.) Urban ecosystem ecology. Agron. Monogr. 55. ASA, CSSA, and SSSA, Madison, WI.

Ward, P.L. 2009. Sulfur dioxide initiates global climate change in four ways. Thin Solid Films 517:3188–3203.

Yokouchi, Y., Y. Nojiri, L.A. Barrie, D. Toom-Sauntry, and Y. Fujinuma. 2001. Atmospheric methyl iodide: High correlation with surface seawater temperature and its implication on the sea-to-air flux. J. Geophys. Res. D Atmos. 106(D12):12661–12668.

Yokouchi, Y., K. Osada, M. Wada, F. Hasebe, M. Agama, R. Murakami, H. Mukai, Y. Nojiri, Y. Inuzuka, D. Toom-Sauntry, and P. Fraser. 2008. Global distribution and seasonal concentration change of methyl iodide in the atmosphere. J. Geophys. Res. D Atmos. 113:18179–18188.

Zanobetti, A., and J. Schwartz. 2009. The effect of fine and coarse particulate air pollution on mortality: A national analysis. Environ. Health Perspect. 117:898–903.

4

Birds in Urban Ecosystems: Population Dynamics, Community Structure, Biodiversity, and Conservation

Eyal Shochat
Susannah Lerman
Esteban Fernández-Juricic

Abstract

With the global high rate of urbanization and the rapid loss of wild habitat land, cities are now viewed as challenging ecosystems for sustaining biotic communities and rich diversity. During the 2000s research on urban bird populations and communities focused on global patterns, as well as processes and mechanisms that lead to the two globally recognized patterns: increased overall population densities and decrease in species diversity compared with wildlands. Birds adapt to the urban ecosystem both physiologically (changes in stress hormones), and behaviorally (e.g., changes in foraging behavior, extending the breeding season). The increase in population density is related to the increase in food abundance, and probably to the reduction in predation pressure. The loss of diversity is related to loss of habitat, the high human density, and negative interactions with synanthropic species. Recognizing that the urban habitat will continue to grow, efforts to turn the city into a more friendly habitat for a variety of bird species should focus not only on habitat and vegetation structure, but also on niche opening for subordinate species, by excluding locally aggressive, synanthropic species.

Although the study of urban birds has a fairly long history, urban ecosystems have been largely ignored throughout many decades of ecological research (Miller and Hobbs, 2002; Collins et al., 2000). Since the early 1990s, a different view emerged, accepting urban settings as ecosystems that are structured and function like other natural ecosystems (McDonnell and Pickett, 1990; Rebele, 1994; Grimm et al., 2000; McKinney, 2002; Miller and Hobbs, 2002). This theoretical view represents an emerging realization that by now, most of the world's land is managed and dominated by humans (approximately 50% of the human population lives in cities). Wildlands are continuously converted to agricultural fields

E. Shochat, Global Inst. of Sustainability, Arizona State University, Box 875411, Tempe, AZ 85287 (eyal. shochat@asu.edu); S. Lerman, Dep. of Organismic and Evolutionary Biology, University of Massachusetts, 319 Morrill Science Ctr. South, 611 N. Pleasant St., Amherst, MA 01003 (slerman@nsm.umass. edu); E. Fernández-Juricic, Dep. of Biological Sciences, Purdue University, G-420 Lily Hall, 915 W. State St., West Lafayette, IN 47907 (efernan@purdue.edu).

doi:10.2134/agronmonogr55.c4

and urban areas. Consequently, urban environments can no longer be viewed as lost habitat for wildlife, but rather as new habitat that, with proper management, has the potential to support diverse bird communities. During the last decade urban ecosystems have therefore become ecological challenges in conservation, restoration, and reconciliation ecology (Miller and Hobbs, 2002; Rosenzweig, 2003). Designing sustainable urban ecosystems that support species-rich bird communities also includes maintaining key ecosystem services, such as clean air and water, waste decomposition, and pest control.

Knowledge of the patterns of urban bird populations and communities started emerging in the 1970s (e.g., Emlen, 1974). Compared with adjacent, more natural ecosystems, urban settings normally have higher bird abundances (Beissinger and Osborne, 1982; Marzluff, 2001a; Chace and Walsh, 2006). For example, in Tucson, overall bird density increased 26-fold from the Sonoran desert to urban habitats (Emlen, 1974). Patterns of high bird densities have been observed in tropical systems (Sodhi et al., 1999), grasslands (Sodhi, 1992; Bock et al., 2001), temperate forests (Beissinger and Osborne, 1982), deserts (Emlen, 1974), bushland (Sewell and Catterall, 1998), and oak woodlands (Blair, 1996). Increase in food abundance is the most common mechanism described in the literature for the increase in bird densities (Emlen, 1974; Bolger, 2001; Marzluff, 2001; Mennechez and Cleurgeau, 2001). This increase may reflect the combined effect of an increase in exotic vegetation, refuse, and, in many cases, the use of feeders. While all bird guilds increased their densities, the response of seed-eaters to urbanization has been the highest. Emlen (1974) associated this increase with both high supply of seeds in feeders and higher productivity in the urban environment due to urban lawns and weeds. While this "bottom-up control" of population size is accepted as the major cause of population growth in urbanized environments, the contribution of the "top-down control" (i.e., reduction in predation pressure) is still unclear. Cities have high abundance of birds despite the high densities of domestic and feral predators, creating a paradox (Shochat, 2004). We discuss possible solutions to this paradox later in this chapter.

In most cases, diversity in urban habitats decreases or remains similar to wildlands (Marzluff, 2001; Chace and Walsh, 2006). Although urbanization increases total bird densities, it appears that only a few species contribute to this increase. Cities consist of mixtures of built habitats and green patches. Only a few species can exist and thrive in the most built parts of the city where vegetation is almost absent, such as business districts and industrial zones. Thus, urbanization increases the abundance of feral pigeons, swallows, swifts, and a few other species that breed in walls. As vegetation cover increases toward the rural parts of the city, species diversity increases (Emlen, 1974; Mills et al., 1989; Chace and Walsh, 2006; Sandstrom et al., 2005). In areas of intermediate disturbances (i.e., suburban development) diversity also increases (Blair, 1999). Because the vegetation is usually exotic, the increase in diversity is attributed to many human-commensal or alien species, whereas native vegetation allows, in some cases, a higher proportion of native species (Emlen, 1974; Mills et al., 1989; Chace and Walsh, 2006; Daniels and Kirkpatrick, 2006). Since the majority of urban bird research addresses basic patterns of abundance and distribution, the generality and consequences of these patterns in urban bird ecology remain unclear. *Evenness*, a term denoting the similarity in relative abundance of all species in the community, appears to decrease with urbanization, although only few studies address this issue (Edgar and Kershaw, 1994; Marzluff, 2001). The reduction in evenness is the result of a few species becoming highly dominant in urban environments (Shochat et al., 2010). Exotic and synanthropic species generally thrive in the novel urban ecosystem, while many native species avoid it. This might lead to a loss of diversity if the dominant species "monopolize" resources. While these patterns have been described in many studies (reviewed by Chace and Walsh, 2006), the mechanisms underlying this community pattern have not been addressed (Shochat et al., 2006).

Until the early 2000s, an experimental and mechanistic approach had rarely been taken in urban bird research (Shochat et al., 2006). Currently, ecologists are trying to better understand the drivers of urban bird population dynamics and community structure, the role of habitat and vegetation profile vs. predator–prey interactions, and interspecific competition for food and other resources. In this chapter, we describe recent findings from a more process-oriented type of research on urban bird communities.

Physiological and Behavioral Adaptations to Urbanization

Birds respond to urban ecosystems by either avoiding cities or by adapting or exploiting the urban landscape. Many of the species able to adapt to or exploit the urban landscape undergo behavioral and/or physiological adaptations to survive and sometimes thrive in urbanized areas. If urbanization induces stress, one should expect to find differences in stress hormone levels and blood parasites between urban and wildland birds. Yet, it is not intuitive how these variables should change along the wildland–urban gradient due to the scant research on urban bird physiology. In Germany, urban blackbirds (*Turdus merula*) showed lower levels of corticosterone stress levels than forest birds. This may suggest that individuals that modify their stress response can adapt to the high stress level in cities (Partecke et al., 2006b). Yet, other studies show somewhat opposite trends. Urban male white-crowned sparrows (*Zonotrichia leucophrys*) sampled in Washington and California had higher corticosterone levels than rural ones, while no differences were found in females (Bonier et al., 2007). Furthermore, a comparison in several stress-associated variables between urban and desert birds in central Arizona indicated opposite trends for different species. At least two species, northern mockingbird (*Mimus polyglottos*) and curve-billed thrasher (*Toxostoma curvirostre*), appeared to be more stressed in the urban habitat (Fokidis et al., 2008).

Similarly to physiological adaptations, behavioral adaptations to urbanization do not show a general pattern. In some cases, urban birds appear more adapted to the presence of humans than rural birds. Such is the case of the black-billed magpie (*Pica pica*) in Colorado, where flushing response and flight distance were lower in urban than in rural habitats (Kenney and Knight, 1992). In contrast, the same species showed avoidance behavior in China, where birds built nests higher in urban than in rural and wildland habitats (Wang et al., 2008). Avoidance behavior can also occur temporarily. For example, in Madrid, Spain, the abundance of foraging individuals of several species decreased with an increase in the number of pedestrians in urban parks (Fernández-Juricic, 2000), suggesting that birds avoid foraging patches when their perceived risk increases. Species able to exist in the urban habitat enjoy higher resource abundances than species in wildlands. For example, in arid environments, birds have higher water availability (Shochat et al., 2006). Furthermore, fluctuations in these resources are minor compared with wildlands (Shochat, 2004). This allows some species to extend their breeding season, as seasonality does not restrict resource availability as in wildlands. For example, relying on food abundance as a cue, suburban Florida scrub jays (*Aphelocoma coerulescens*) start breeding earlier than wildland birds (Schoech and Bowman, 2001). Similarly, urban magpies begin breeding earlier than wildland magpies in Poland, taking advantage of higher food abundance and more advantageous microclimate conditions in cities. In addition, urban magpies also re-nested more often than wild magpies (Jerzak, 2001). In Germany, urban blackbirds extend their breeding season by developing their gonads 3 wk before forest individuals (Partecke et al., 2006a), again, due to the increase in food subsidies.

While food density is normally high in urban settings, the main source is low quality, anthropogenic refuse. Sauter et al. (2006) showed that although adult Florida scrub jays prefer to feed their nestlings with natural food (i.e., arthropods), they are forced to feed their nestlings with low quality food in suburban areas because of a low density of arthropods. This may negatively affect nestling growth and development. Thus, although the high density of resources may support high densities of birds, the low quality of these food resources may have costs in terms of bird health and growth. Pierotti and Annett (2001) showed how "urban diet" can lead to lower fitness in western gulls (*Larus occidentalis*). Gulls that nested close to urban areas relied mostly on refuse and had relatively low nesting success, whereas birds that nested far from urban areas relied on scavenging in marine habitats and had higher nesting success (Pierotti and Annett, 2001).

Another stressor birds have to cope with in urban settings is noise. The urban ecosystem is characterized by elevated noise levels, which can interfere with vocal communication (Warren et al., 2006). Birds use vocalizations to warn of danger, defend a territory, and attract mates. The most prominent noise source in urban ecosystems is traffic, and consequentially, the majority of urban acoustic studies concentrate around roads. The noise within urban ecosystems is at low frequencies, usually below 2000 Hz (Patricelli and Blickley, 2006; Warren et al., 2006); therefore, birds with higher frequencies or those with the ability to shift their frequency (Slabbekoorn and Peet, 2003) will have an advantage to communicate amid the urban noise. However, the interaction between responses to noise and resultant fitness is not well understood. A recent study showed that house finches (*Carpodacus mexicanus*) adjusted their songs in response to noisy areas within a city by raising the low frequency of their songs and decreasing the number of notes per song (Fernández-Juricic et al., 2005), which could potentially decrease their mating opportunities because females are generally more attracted to males with longer songs (Nolan and Hill, 2004). When communicating with their young, adults use low-frequency contact calls near the nest. Forman et al. (2002) suggested fledglings and nestlings cannot hear warnings from their parents because of the traffic noise. This interference may negatively impact reproductive success. In addition, urban noise might influence bird distribution. Rheindt (2003) demonstrated a strong correlation with song frequency and distance to roads, where birds with higher frequencies had a greater abundance closer to roads than birds with lower frequencies.

Top-Down and Bottom-Up Control of Urban Bird Populations

Urban bird densities are normally extremely high (Emlen, 1974; Marzluff, 2001; Chace and Walsh, 2006; Rodewald and Shustack, 2008). Increase in bird densities may be the result of high food density (bottom-up control), low predation pressure (top-down control), or the combination of both (Shochat, 2004). Although food abundance is normally difficult to quantify, the bottom-up concept has been accepted as the major driver of urban bird densities (Marzluff, 2001). Exotic vegetation, refuse, and bird feeders all provide food sources for urban birds. In a study on northern cardinals (*Cardinalis cardinalis*) in Ohio, food abundance was found to be 2.6 times higher in urban than in rural habitats, although based on bird densities total food abundance was expected to be four times higher in the urban habitat (Rodewald and Shustack, 2008). Fuller et al. (2008) also found a positive correlation between urban bird feeding stations and bird abundance in Sheffield, UK.

The role of top-down control, however, is more complex. Predators are known to affect prey on three different temporal scales: in the short term, prey may change its behavior; in the long term, prey population size may decrease; and in an evolutionary time scale, prey may show morphological adaptations to the presence of predators. Of these three possible responses, the second—the population level—is the most straightforward to address logistically, and thus, the most studied of the three levels. While many natural predators avoid urban areas, at least during daytime, when birds are active (Tigas et al., 2002), other feral or domestic predators that inhabit cities in high densities, especially cats (Haskell et al., 2001), can affect bird population regulation. In Britain, cats have been shown to hunt millions of birds per year (Woods et al., 2003). Baker et al. (2005) studied the impacts of cats on urban animals, including birds in Bristol, UK, and noted a prey preference for juvenile birds. In particular, house sparrows (*Passer domesticus*), dunnock (*Prunella modularis*), and robin (*Erithacus rubecula*) had a higher predation rate compared to their relative productivity. Thus, cat predation could negatively impact dispersal and recruitment in urban areas particularly. Other researchers (e.g., Sorace, 2002) suggested that due to the high densities of cats in urban ecosystems, predation pressure should be higher than or equal to that in wildlands. Yet, observations on bird behavior and population size do not appear to concur with the idea of strong top-down controls on urban bird populations (Shochat et al., 2006). For example, while direct top-down control predicts a negative correlation between predator and prey density, studies in urban settings consistently indicate that despite high cat densities, urban bird populations are denser than wildland populations. Thus, when correlated versus each other, cat and bird densities are positively correlated in urban settings (Sims et al., 2008). Shochat (2004) suggested that urban bird community composition may therefore represent the "ghost of predation past"; urban environments may have selected a small group of cat-resistant species. Having available water, high density of food, and a lack of native predators allows these species to flourish in the city.

The most effective way to test behavioral or survival responses of prey to predators is to manipulate predator abundance. Removal of black-billed magpies from city parks in Paris, France demonstrated that these nest predators have a minor effect on the abundance of 14 bird species or their reproduction (Chiron and Julliard, 2007). The only obvious effect of magpies appeared to be the shift in foraging niches by some of these species. Another study also showed that magpies in urban parks in Madrid influenced the antipredator behavioral responses of some bird species by reducing the number of species (and their neighbor distance) within a patch when magpies were present (Fernández-Juricic et al., 2004). This was attributed to the fact that magpies were opportunistic predators of adult birds; however, their capture success was relatively low (5%). Interestingly, differences in prey vigilance behavior were related to the probability of being attacked rather than mortality rate by magpies. A preferentially attacked prey species in relation to its abundance (e.g., blackbirds) enhanced their vigilance effort (i.e., increase in scanning time and scanning rate with magpies present) relative to a species (e.g., house sparrow) attacked infrequently in relation to its abundance, which showed no vigilance responses when magpies were present (Fernández-Juricic et al., 2004).

At the behavioral and morphological levels, the effect of predation on birds has not been studied thoroughly, and the few studies conducted showed mixed results regarding the low predation pressure hypothesis. Studies on foraging behavior of birds and squirrels in urban and wildland habitats suggest that the urban habitat is probably less risky than the wild habitat, whether it is forest or desert (Bowers and Breland, 1996; Shochat et al., 2004). In both cases, individuals quit foraging on artificial food patches earlier in wildland or rural habitats than in the urban habitat, as the costs of predation are apparently lower in the urban habitat. Furthermore, contrary to desert habitat, urban birds in Phoenix, AZ showed no differences in foraging behavior between food patches that are close to shelters (under bushes), and patches that are out in the open (Shochat et al., 2004). These findings suggest that animals view the urban habitat as safe from predators. However, other studies found that urban habitats are not as safe for some bird species as previously thought. Valcarcel and Fernández-Juricic (2009) found that house finches in more urbanized areas formed larger flocks, had less tolerance to human approaches, and increased their pecking rates to compensate for the lower amount of foraging time than those in less urbanized areas in Southern California. Interestingly, avian predator richness and abundance was lower in urban areas. These results suggest that urban house finch's perceived risk of predation is regulated by human activities, which could increase risk or decrease the ability to detect predators.

From a population perspective, the evidence supports a reduction in predation pressure in urban areas. The behavioral evidence is less conclusive and points to different mechanisms (predation risk, human disturbance, intra-guild predation) acting simultaneously, which can complicate interpretations. Furthermore, we should bear in mind that cats still cause damage to bird populations and that many cities are inhabited with natural predators (Chace and Walsh, 2006). Altogether, additional research is required to establish the role of top-down effects on urban bird populations.

Urban Bird Community Structure and Composition

Species interactions and the mechanisms underlying community structure in urban settings are the least studied issues in urban ecology. The presence of a few abundant synanthropic or alien species in cities may affect native species at different levels—behavioral, population size, and species diversity. Hints for such effects can be drawn from evenness, which is normally low in urban environments (Marzluff, 2001). The low evenness may be the result of a few synanthropic species that thrive and account for a high proportion of the whole community (Shochat et al., 2010).

To understand community structure, Shochat et al. (2004) studied foraging efficiency in urban and desert birds in Phoenix, AZ. Foraging efficiency is defined in terms of the ability to deplete food patches, especially in low resource density habitats. Inefficient species need to dominate over more efficient species to coexist. Using artificial food patches, Shochat et al. (2004) found that urban birds were more efficient foragers than desert birds. From a series of field experiments they concluded that the combination of high food and water resource density combined with low predation pressure allowed urban birds to increase their food intake.

In many cases, temporal partitioning is required for species coexistence. In this mechanism of coexistence, subordinate species are more efficient foragers than dominant species (Ziv et al., 1993). Thus, once dominant species quit food patches, they leave behind enough food for subordinate species in subsequent patch visits. However, it appears that in urban habitats, this situation changes. Shochat et al. (2010) demonstrated that the most efficient foragers in urban settings are probably the more dominant species in the community. Such a situation may constrain subordinate species, limit their population sizes, and in extreme cases, lead to local extinction.

Species Diversity

According to the random sampling hypothesis (Connor and McCoy, 1979), urban environments should have higher species diversity because cities attract more individuals from the regional species pool. However, most studies on urban bird species diversity detect a low diversity for the number of individuals "sampled" (Emlen, 1974; Mills et al., 1989; Sewell and Catterall, 1998; Marzluff, 2001; Chace and Walsh, 2006). These findings indicate that urban ecosystems do not draw a random set of species from the regional pool, but rather favor a small group of birds that appear to adapt well to this novel ecosystem. Indeed, cities are normally inhabited with high densities of human commensal or synanthropic species, many of which are invasive or alien (e.g., house sparrow, feral pigeon, Eurasian starling). Because such species were introduced by humans to many parts of the world, it has been argued that humans create a homogeneous avifauna in cities (Blair, 2001; McKinney, 2006).

Relatively few studies to date have tried to thoroughly investigate the mechanisms for the loss of species in urbanized areas. Whereas habitat fragmentation or destruction leads to extinction of many native species, urbanization also creates new habitat for other species. Yet, regardless of large-scale landscape composition or geography, urbanization results in low species diversity on a global scale. Two different mechanisms have been suggested for these phenomena. Using island biogeography as a framework, Marzluff (2005) suggested that the overall sum of bird species should be lower in the most urbanized parts of the landscape. In Seattle, bird diversity peaked at intermediate levels of urbanization, where the proportion of forest was still relatively high. The high number of species in these habitats was mostly the result of an increase in richness of early successional birds—species that are found in a variety of habitats around the area.

Extinction rates of native forest species and immigration rates of synanthropic species played a minor role in influencing species diversity.

Shochat et al. (2010) suggested a different scenario, involving competitive exclusion. They used data from Phoenix and Baltimore to link this phenomenon to community structure and species interactions. In their scenario, the increase in resource abundance, combined with the decrease in predation pressure, results in a "winner take all" situation. Not only do cities offer high amounts of resources to birds, resource input into the ecosystem is highly predictable, owing to human activity routines (Shochat, 2004). Bird species that cannot exist in less predictable or resource-poor environments may be able to flourish in cities, where these hurdles associated with food and water resources are removed. Growing in numbers, they dominate resource patches and out-compete many native species or cause a significant reduction in others. Such changes may be the reason for the low evenness pattern of urban bird communities.

Conservation

The changes in bird community structure and composition discussed here are summarized in Fig. 4–1. The role of species interactions in urban bird population dynamics and community structure may suggest that solutions for the loss of diversity cannot be based on habitat alteration per se. Creating proper habitat for a given species may not be sufficient to attract it into the city if it suffers from aggressive interactions from local urban species or human disturbance. Urban conservation ecology should therefore seek creative solutions based on the evolutionary differences between dominant and subordinate species, creating special breeding or feeding niches for the latter. Such solutions already exist and are widely used—e.g., squirrel-proof feeders. Their basic concept is that whereas dominant, aggressive species are impossible to control, we can still open a niche

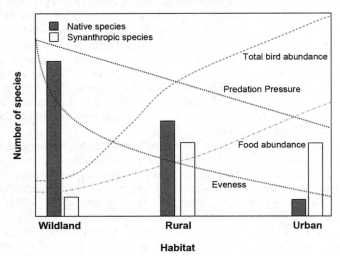

Fig. 4–1. Changes in bird community structure and composition along a wildland–urban gradient. As food abundance increases and predation pressure decreases, total bird abundance increases. This increase results from a few thriving synanthropic species, while many native species are lost due to habitat changes and competitive exclusion. These changes in community composition lead to lower evenness.

for subordinate species with simple manipulations that exclude the dominant species from a small but important part of the urban habitat. Squirrels are aggressive toward birds, but-squirrel proof feeders based on body mass turn a squirrel's advantage in adapting to the urban ecosystem into a disadvantage. Feeders allowing seed access only for animals below a given threshold body mass, such as small birds, allow for species coexistence by opening a niche to granivorous birds, which in turn increase local diversity. The same concept is used with nest boxes, where hole size prevents dominant species like starlings from occupying them, creating more available nesting sites for smaller, native cavity nesters. The foraging "niche opening" principle has been taken one step further with sparrow-proof feeders that allow small finches to persist in the urban habitat. This suggests that the principle can be applied for other species or on larger scales and presents an alternative to expensive programs to eradicate abundant aggressive synanthropic species (Shochat et al., 2010).

Whether one considers loss of habitat or negative interactions, addressing the mechanisms leading to the loss of diversity in urban settings is one of the fundamental challenges in conservation biology because of the widespread distribution of urbanized habitats on the planet. The above-described frameworks of Marzluff (2005) and Shochat et al. (2010) concerning the role of landscape structure, top-down, and bottom-up effects should be treated as starting points for future research on this topic, with the ultimate conservation goal of transforming urban environments into species-rich ecosystems.

References

Baker, P.J., A.J. Bentley, R.J. Ansell, and S. Harris. 2005. Impact of predation by domestic cats *Felis catus* in an urban area. Mammal Rev. 35:302–312.

Beissinger, S.R., and D.R. Osborne. 1982. Effects of urbanization on avian community organization. Condor 84:75–83.

Blair, R.B. 1996. Land use and avian species diversity along an urban gradient. Ecol. Appl. 6:506–519.

Blair, R.B. 1999. Birds and butterflies along an urban gradient: Surrogate taxa for assessing biodiversity? Ecol. Appl. 9:164–170.

Blair, R.B. 2001. Creating a homogeneous avifauna. p. 459–486. *In* J.M. Marzluff et al. (ed.) Avian ecology and conservation in an urbanizing world. Kluwer Academic Publ., Boston, MA.

Bock, C.E., J.H. Bock, and B.C. Bennett. 2001. Songbird abundance in grasslands at a suburban interface on the Colorado high plains. Stud. Avian Biol. 19:131–136.

Bolger, D.T. 2001. Urban birds: Population, community, and landscape approaches. p. 155–177. *In* J.M. Marzluff et al. (ed.) Avian ecology and conservation in an urbanizing world. Kluwer Academic Publ., Boston, MA.

Bonier, F., P.R. Martin, S.K. Sheldon, P.J. Jensen, S.L. Foltz, and J.C. Wingfield. 2007. Sex-specific consequences of life in the city. Behav. Ecol. 18:121–129.

Bowers, M.A., and B. Breland. 1996. Foraging of gray squirrels on an urban–rural gradient: Use of the GUD to assess anthropogenic impact. Ecol. Appl. 6:1135–1142.

Chace, J.F., and J.J. Walsh. 2006. Urban effects on native avifauna: A review. Landscape Urban Plan. 74:46–69.

Chiron, F., and R. Julliard. 2007. Responses of songbirds to magpie reduction in an urban habitat. J. Wildl. Manage. 71:2624–2631.

Collins, J.P., A. Kinzig, N.B. Grimm, W.F. Fagan, D. Hope, J.G. Wu, and E.T. Borer. 2000. A new urban ecology. Am. Sci. 88:416–425.

Connor, E.F., and E.D. McCoy. 1979. The statistics and biology of the species–area relationship. Am. Nat. 113: 791–833.

Daniels, G.D., and J.B. Kirkpatrick. 2006. Does variation in garden characteristics influence the conservation of birds in suburbia? Biol. Conserv. 133:326–335.

Edgar, D.R., and G.P. Kershaw. 1994. The density and diversity of the bird populations in three residential communities in Edmonton, Alberta. Can. Field Nat. 108:151–161.

Emlen, J.T. 1974. An urban bird community in Tucson, Arizona: Derivation, structure, regulation. Condor 76:184–197.

Fernández-Juricic, E. 2000. Local and regional effects of pedestrians on forest birds in a fragmented landscape. Condor 102:247–255.

Fernández-Juricic, E., J. Jokimaki, J.C. McDonald, F. Melado, A. Toledano, C. Mayo, B. Martin, I. Fresneda, and V. Martin. 2004. Effects of opportunistic predation on antipredator behavioural responses in a guild of ground foragers. Oecologia 140:183–190.

Fernández-Juricic, E., R. Poston, K. De Collibus, T. Morgan, B. Bastain, C. Martin, K. Jones, and R. Treminio. 2005. Microhabitat selection and singing behavior patterns of male house finches (*Carpodacus mexicanus*) in urban parks in a heavily urbanized landscape in the western U.S. Urban Habitats 3. Available at http://www.urbanhabitats.org/v03n01/finch_full.html (verified 11 Feb. 2010).

Fokidis, H.B., E.C. Greiner, and P. Deviche. 2008. Interspecific variation in avian blood parasites and haematology associated with urbanization in a desert habitat. J. Avian Biol. 39:300–310.

Forman, R.T.T., B. Reineking, and A.M. Hersperger. 2002. Road traffic and nearby grassland bird patterns in a suburbanizing landscape. Environ. Manage. 29:782–800.

Fuller, R.A., P.H. Warren, P.R. Armsworth, O. Barbosa, and K.J. Gaston. 2008. Garden bird feeding predicts the structure of urban avian assemblages. Divers. Distrib. 14:131–137.

Grimm, N.B., J.M. Grove, C.L. Redman, and S.T.A. Pickett. 2000. Integrated approaches to long-term studies of urban ecological systems. Bioscience 50:571–584.

Haskell, D.G., A.M. Knupp, and M.C. Schneider. 2001. Nest predator abundance and urbanization. p. 243–258. *In* J.M. Marzluff et al. (ed.) Avian ecology and conservation in an urbanizing world. Kluwer Academic Publ., Boston, MA.

Jerzak, L. 2001. Synurbanization of the magpie in the Palearctic. p. 404–425. *In* J.M. Marzluff et al. (ed.) Avian ecology and conservation in an urbanizing world. Kluwer Academic Publ., Boston, MA.

Kenney, S.P., and R.L. Knight. 1992. Flight distances of black-billed magpies in different regimes of human density and persecution. Condor 94:545–547.

Marzluff, J.M. 2001. Worldwide urbanization and its effects on birds. p. 19–38. *In* J.M. Marzluff et al. (ed.) Avian ecology and conservation in an urbanizing world. Kluwer Academic Publ., Boston, MA.

Marzluff, J.M. 2005. Island biogeography for an urbanizing world: How extinction and colonization may determine biological diversity in human-dominated landscapes. Urban Ecosyst. 8:157–177.

McDonnell, M.J., and S.T.A. Pickett. 1990. The study of ecosystem structure and function along urban–rural gradients: An unexploited opportunity for ecology. Ecology 71:1231–1237.

McKinney, M.L. 2002. Urbanization, biodiversity, and conservation. Bioscience 52:883–890.

McKinney, M.L. 2006. Urbanization as a major cause of biotic homogenization. Biol. Conserv. 127:247–260.

Mennechez, G., and P. Clergeau. 2001. Settlement of breeding European starlings in urban areas: Importance of lawns vs. anthropogenic wastes. p. 275–288. *In* J.M. Marzluff et al. (ed.) Avian ecology and conservation in an urbanizing world. Kluwer Academic Publ., Boston, MA.

Miller, J.R., and R.J. Hobbs. 2002. Conservation where people live and work. Conserv. Biol. 16:330–337.

Mills, G.S., J.B. Dunning, Jr., and J.M. Bates. 1989. Effects of urbanization on breeding bird community structure in southwestern desert habitats. Condor 91:416–428.

Nolan, P.M., and G.E. Hill. 2004. Female choice for song characteristics in the house finch. Anim. Behav. 67:403–410.

Partecke, J., I. Schwabl, and E. Gwinner. 2006a. Stress and the city: Urbanization and its effects on the stress physiology in European blackbirds. Ecology 87:1945–1952.

Partecke, J., T.J. Van't Hof, and E. Gwinner. 2006b. Underlying physiological control of reproduction in urban and forest-dwelling European blackbirds *Turdus merula*. J. Avian Biol. 36:295–305.

Patricelli, G.L., and J.L. Blickley. 2006. Avian communication in urban noise: Causes and consequences of vocal adjustment. Auk 123:639–649.

Pierotti, R., and C. Annett. 2001. The ecology of Western Gulls in habitats varying in degree of urban influence. p. 307–329. *In* J.M. Marzluff et al. (ed.) Avian ecology and conservation in an urbanizing world. Kluwer Academic Publ., Boston, MA.

Rebele, F. 1994. Urban ecology and special features of urban ecosystems. Global Ecol. Biogeogr. Lett. 4:173–187.

Rheindt, F.E. 2003. The impact of roads on birds: Does song frequency play a role in determining susceptibility to noise pollution? J. Fur Ornithol. 144:295–306.

Rodewald, A.D., and D.P. Shustack. 2008. Consumer resource matching in urbanizing landscapes: Are synanthropic species over-matching? Ecology 89:515–521.

Rosenzweig, M.L. 2003. Win-win ecology. Oxford Univ. Press, Oxford, UK.

Sandstrom, U.G., P. Angelstam, and G. Mikusinski. 2005. Ecological diversity of birds in relation to the structure of urban green space. Landscape Urban Plan. 77:39–53.

Sauter, A., R. Bowman, S.J. Schoech, and G. Pasinelli. 2006. Does optimal foraging theory explain why suburban Florida scrub-jays (*Aphelocoma coerulescens*) feed their young human-provided food? Behav. Ecol. Sociobiol. 60:465–474.

Schoech, S.J., and R. Bowman. 2001. Variation in the timing of breeding between suburban and wildland Florida scrub-jays: Do physiologic measures reflect different environments? p. 289–306. *In* J.M. Marzluff et al. (ed.) Avian ecology and conservation in an urbanizing world. Kluwer Academic Publ., Boston, MA.

Sewell, S.R., and C.P. Catterall. 1998. Bushland modification and styles of urban development: Their effects on birds in south-east Queensland. Wildl. Res. 25:41–63.

Shochat, E. 2004. Credit or debit? Resource input changes population dynamics of city slicker birds. Oikos 106:622–626.

Shochat, E., S.B. Lerman, J.M. Anderies, P.S. Warren, S.H. Faeth, and C.H. Nilon. 2010. Invasion, competition, and biodiversity loss in urban ecosystems. Bioscience 60(3):199-208.

Shochat, E., S.B. Lerman, M. Katti, and D.B. Lewis. 2004. Linking optimal foraging behavior to bird community structure in an urban-desert landscape: Field experiments with artificial food patches. Am. Nat. 164:232–243.

Shochat, E., P.S. Warren, S.H. Faeth, N.E. McIntyre, and D. Hope. 2006. From patterns to emerging processes in urban evolutionary ecology. Trends Ecol. Evol. 21:186–191.

Sims, V.S., K.L. Evans, S.E. Newson, J.A. Tratalos, and K.J. Gaston. 2008. Avian assemblage structure and domestic cat densities in urban environments. Divers. Distrib. 14:387–399.

Slabbekoorn, H., and M. Peet. 2003. Birds sing at a higher pitch in urban noise—Great tits hit the high notes to ensure that their mating calls are heard above the city's din. Nature 424:267.

Sodhi, N.S. 1992. Comparison between urban and rural bird communities in prairie Saskatchewan—Urbanization and short-term population trends. Can. Field Nat. 106:210–215.

Sodhi, N.S., C. Briffett, L. Kong, and B. Yuen. 1999. Bird use of linear areas of a tropical city: Implications for park connector design and management. Landscape Urban Plan. 45:123–130.

Sorace, A. 2002. High density of bird and pest species in urban habitats and the role of predator abundance. Ornis Fenn. 79:60–71.

Tigas, L.A., D.H. Van Vuren, and R.M. Sauvajot. 2002. Behavioral responses of bobcats and coyotes to habitat fragmentation and corridors in an urban environment. Biol. Conserv. 108:299–306.

Valcarcel, A., and E. Fernández-Juricic. 2009. Anti-predator strategies of house finches: Are urban habitats safe spots from predators even when humans are around? Behav. Ecol. Sociobiol. 63:673–685

Wang, W.P., S.H. Chen, P.P. Jiang, and P. Ding. 2008. Black-billed Magpies (*Pica pica*) adjust nest characteristics to adapt to urbanization in Hangzhou, China. Can. J. Zool. 86:676–684.

Warren, P.S., M. Katti, M. Ermann, and A. Brazel. 2006. Urban bioacoustics: It's not just noise. Anim. Behav. 71:491–502.

Woods, M., R.A. McDonald, and S. Harris. 2003. Predation of wildlife by domestic cats *Felis catus* in Great Britain. Mammal Rev. 33:174–188.

Ziv, Y., Z. Abramsky, B.P. Kotler, and A. Subach. 1993. Interference competition and temporal and habitat partitioning in two gerbil species. Oikos 72:237–246.

5

Urban Mammals

Robert McCleery

Abstract

There is a clear relationship between mammalian diversity and the degree of urbanization of the landscape. As urbanization increases, mammalian diversity is lost. The loss of species in urban areas is generally attributed to habitat degradation and fragmentation, the loss of vegetation to impervious surface, and the simplification of vegetation. Mammals with large body sizes and corresponding large movement and range patterns, predators, interior species, and habitat specialists all appear to be sensitive to urbanization. On the other hand, a number of omnivores, generalists, and medium-size carnivores have been able to utilize the ample resources of the urban environment. Mammalian populations that have adapted to the urban environment commonly have higher densities, higher rates of survival and reproduction, and lower rates of dispersal than their rural counterparts. Additionally, urban populations also frequently show different behaviors from rural populations, including reduced responses to humans, altered activity periods, smaller ranges of movement, and different territorial behaviors. Urban mammals also have been shown to alter their diets to consume more anthropogenic foods and to use buildings, culverts, and bridges for shelter and the rearing of young. However, the concentrated nature of urban resources and the high population densities of urban adapted mammal population make them more susceptible to disease outbreaks and parasites. The moderately developed areas with anthropogenic food, high net primary production, and structures for shelter probably hold the most potential for urban conservation efforts of native mammals.

As the dominant ecological force on the planet, humans have altered more than 75% of the ice-free surface of the globe (Ellis and Ramankutty, 2008). The most intensive alterations to Earth's surface have come from the creation of cities. Cities play a vital role in human culture as centers for commerce, residence, entertainment, and social interactions. Currently, more than 50% of the planet's human population lives in cities and more than 65% of the population is projected to live in urban areas by 2045 (United Nations, 2008). Cities account for only 2.4% of the Earth's surface (Millennium Ecosystem Assessment, 2005), but because of the intensity and magnitude of this land use, it has a disproportionate influence on regional and global ecological systems and processes (Collins and Kinzig, 2000).

R. McCleery, Western Illinois University, Dep. of Biological Sciences. 1 University Circle, Macomb, IL 61455 (ra-mccleery@wiu.edu).

doi:10.2134/agronmonogr55.c5

The landscape, along with the biotic and abiotic features of urban environments, can vary immensely between cities, and even among neighborhoods in the same city. Nonetheless, we can characterize urban environments as areas of dense human populations, buildings, impermeable surfaces, and introduced vegetation with high concentrations of food, water, energy, materials, sewage, and pollution (McDonnell and Pickett, 1990; Pickett et al., 2001; Adams et al., 2006). From a landscape perspective urban systems are mosaics of residential, industrial, commercial, and infrastructure interspersed with green areas (Breuste et al., 2008). For mammals these important green areas often come in the form of a patchy network of lawn, parks, trails, golf courses, cemeteries, and a few areas of native vegetation.

How mammalian populations and communities respond to the urban environment, among other factors, is a function of the specific features of a particular urban environment and of the scale of the surrounding urbanization. Although oversimplified, a useful way to conceptualize a city is as a gradient of development intensity (McDonnell and Pickett, 1990). At the core of a city most vegetation has been replaced by impervious surfaces, there are high densities of inhabitants and buildings, and there is negligible net primary production. Moving away from the city core, impervious surfaces and the numbers of buildings and residences slowly decrease, and the landscape eventual changes to rural or wildlands. This simplified pattern of urbanization helps us to generalize how mammals respond to changing ecological processes and features along a changing gradient of land uses from the city center to rural areas. For example, in the city center mammals' food resources come from garbage, and they utilize buildings and culverts for den sites. Alternatively, further along the gradient in suburban areas, watered and fertilized laws provide mammals with ample food from vegetation and shelter in the form of trees and shrubs. Some of the other important features that change along the urban–rural gradient that are believed to influence mammal populations are roads, climate, and the density of predators.

Regardless of where along the urban–rural gradient we look, mammals can be found, occurring in every part of the urban environment. With one-half of the world's population living in urban areas, this is where most humans will observe and interact with other mammalian species. The interaction of humans with mammals has implications for health, quality of life, education, aesthetics, and conservation. In this chapter I will begin by exploring how mammalian communities change with urbanization. Then I will investigate how urbanization has changed the demographics, movements, behaviors, diet, physiology, and health of mammalian populations. I will conclude the chapter by showing how mammalian populations utilize the anthropogenic features of the urbanized environment and examining the potential for the conservation of mammals in urban environments.

Mammal Communities in Urban Areas

Urbanization has been detrimental to the mammalian diversity on local and globe scales. Urban development is potentially the leading cause of endangerment and expatriation of native mammal species (Czech et al., 2000; McKinney, 2002). In areas such as Australia less than one-half of the native mammal species still occur in and around urban areas (Ree and McCarthy, 2005; Tait et al., 2005).

Most researchers have shown a clear relationship between mammalian diversity or richness and the amount of urbanization—as urbanization increases mammalian diversity is lost (e.g., McKinney, 2008). The loss of species in urban areas is generally attributed to habitat degradation and fragmentation, the loss of vegetation to impervious surface, and the simplification of vegetation (McKinney, 2008). Moreover, the introduction of species and persecution of mammals has also played a role in the loss of mammalian diversity in urban areas (Matthiae and Sterns, 1981; Tait et al., 2005). Nonetheless, not all research has shown a pattern of declining diversity with increased urbanization. Racey and Euler (1982) found that small mammal diversity increased at moderate levels of development. They attributed this finding to the habitat heterogeneity and edge effects created by clearing forest lots for cottages.

The mammalian species that have been the most sensitive to urbanization generally have similar morphologies or life history strategies. These sensitive mammals often have large body sizes and corresponding large movement and range patterns, such as bison (*Bison bison*) or elk (*Cervus canadensis*). Similarly, large predators and those mammals that have been persecuted are rarely found in urban areas, including wolf (*Canis lupus*), bear (*Urus* spp.), and mountain lion (*Felis concolor*) (Matthiae and Sterns, 1981; Dickman, 1987; Vandruff et al., 1996). Additionally, interior species and habitat specialists, like fisher (*Martes pennant*) or lynx (*Lynx canadensis*), appear to be vulnerable to urbanization.

Although diversity often decreases with development, some mammals take advantage of favorable conditions within the matrix of lawns, houses, parks, gardens, and natural areas found in urban areas outside of the city core. Refuse, supplemental food, ornamental plants, and highly productive lawns often provide plentiful food resources for the mammals able to exploit them (McKinney, 2002). In fact, parks and lawns can have more net primary production than the neighboring wildlands of the region (Imhoff and Tucker, 2000).

The mammalian species found utilizing the lawns and other features of moderately developed areas are termed *urban adapters* (McKinney, 2002). Urban adapters are often generalist, and in nonurban areas they commonly thrive on edges or in savanna habitats that are structurally similar to many urban backyard, forest, and park matrices (McKinney, 2002). Urban adapters utilize many of the food resource available around human development, including ornamental plants, gardens, garbage, and supplemental food. Common urban adapted mammals include burrowing species—moles (*Talpidae*), groundhogs (*Marmota monax*), armadillos (*Dasypus novemcinctus*)—which are capable of finding refuge, avoiding human in their burrows, and finding shelter under porches and houses. They can also take advantage of rapidly growing grasses, ornamentals, and ample invertebrate populations found on urban lawns (Falk, 1976; McKinney, 2002). Medium-size omnivorous and carnivores, such as opossums (*Didelphis virginiana*), raccoons (*Procyon lotor*), foxes (*Vulpes* spp.), and coyotes (*Canis laterns*), have also been successful urban adapters, in part because of the elimination of large predators (Crooks and Soule, 1999). Omnivorous predators can also become highly adept at utilizing human food resources, such as trash, gardens, and bird seed (McKinney, 2002).

The mammals that are capable of living in even the most developed portion of the urban environment are termed *urban exploiters* or *synanthropes*. They are a homogenous group of early successional species that are rarely native to a region and highly adapted to urban environments. They are usually omnivorous, and their populations are dependent on human resources, with little reliance on local vegetative communities (Nilon and Vandruff, 1987; McKinney, 2002). Urban exploiters commonly found in the city core include house mice (*Mus mus*), black rats (*Rattus rattus*), feral cats (*Felis catus*), and Norway rats (*Rattus norvitecus*).

While the creation and expansion of cities has removed many native mammal species, invasive species have been filling the ecological vacuum to replace them. It has been shown that as urbanization increases, the portion and density of nonnative mammals also increases (Mackin-Rogalska et al., 1988; McKinney, 2002; Tait et al., 2005). This is helping to create a situation where the same few species can be found in the cores of cities worldwide (McKinney 2006).

Demographics

Populations of mammalian species that have been able to utilize the urban environment often show considerably different demographics than their rural counterparts. Most notably, many urban adapted species, including white-tailed deer (*Odocoileus virginianus*) (Coté et al., 2004), raccoons (Prange et al., 2003), gray squirrels (*Sciurus carolinensis*) (Hadidian et al., 1987), striped field mice (*Apodemus agrarius*) (Gliwicz et al., 1994), coyotes (Fedriani et al., 2001), and foxes (Harris and Smith, 1987), have all shown increased densities in urban areas. In some cases the densities of urban adapters have been extraordinary. For example, raccoons in nonurban areas rarely exceed 20 km^{-2}, but in urban environments they have recorded densities up to 333 km^{-2} (Lotze and Anderson, 1979; Riley et al., 1998). Similarly, gray squirrels usually do not exceed 3 ha^{-1} in nonurban areas, but densities greater than 50 ha^{-1} have been recorded in an urban park (Koprowski, 1994; Hadidian et al., 1987). Population densities of urban exploiters can also be astronomically high. The core areas of numerous cities worldwide have rat densities of more than 3000 km^{-2} (Keeling and Gilligan, 2000).

The specific demographic reasons for high densities of urban mammals vary by population and species but appear to be from a combination of increased survival and fecundity rates, and decreased rates of dispersal. What has been less clear, are the specific mechanisms causing the population demographics of urban populations to differ from their rural counterparts.

The survival rates of many mammalian populations have been shown to be higher in urban environments. One of the primary reasons given by researchers for this phenomenon is the decrease in potential sources of mortality and specifically a reduction in the risks of predation and hunting (Gliwicz et al., 1994; Etter et al., 2002; Prange et al., 2003; McCleery et al., 2008). For game mammals such as white-tailed deer, it is also logical to believe reductions in and restrictions to hunting in urban areas have helped to increase survival rates (Etter et al., 2002). Crooks and Soule (1999) showed that habitat patches without large predators (i.e., coyotes) had twice the number of mammalian mesopredators common in urban environments, presumably because coyotes weren't eating them or restricting their movements. Similarly, research has shown that urban fox squirrel (*Sciurus niger*) were less likely to be predated in urban environments (McCleery et al., 2008), and high density populations of urban striped field mice were less likely to be present in raptor and owl pellets in the inner city (Gliwicz et al., 1994). Nonetheless, it is also possible that some smaller rodents might see a reduction in survival in areas of high cat activity. Cat abundance has been shown to be negatively correlated with the densities of some small mammals (Baker et al., 2003), and in one study, 69% of the prey items brought home by house cats were mammals (Woods et al., 2003).

The sources of mortality for mammals are often considerably different for urban populations. The greatest source of mortality for many urban adapted mammals is road kill. Collisions with vehicles has been shown to be a major source of mortalities for urban white-tailed deer (Etter et al., 2002; Lopez et al., 2003), raccoon (Prange et al., 2003), fox squirrel (McCleery et al., 2008), coyote (Gehrt, 2007), and fox populations (Gosselink et al., 2007). Disease outbreaks are also a common cause of mortality in urban areas, likely due to the high densities found for many urban mammal populations (Gosselink et al., 2007; Prange et al., 2003).

There are at least two other interesting patterns that researchers examining urban mammals survival rates have found. In several cases the survival rates of rodents have been higher during the winter months, a time when harsh condition should reduce survival (Gliwicz et al., 1994; McCleery et al., 2008). One explanation is that the warmer microclimate and a plentiful supply of food found in urban areas helps to mitigate the effect of winter (Andrzejewski et al., 1978). Additionally, researchers have found that urban male white-tail deer (Lopez et al., 2003) and fox populations (Gosselink et al., 2007) had lower rates of survival than urban females. It is possible that the more extensive movement patterns of males makes them more susceptible to road kill, a major cause of mortally for both populations.

In addition to higher rates of survival, many urban adapted mammal populations have shown increased reproductive success. Urban black bears (*Ursus americanus*) and raccoons appear to have larger litter sizes than rural populations, with urban bears giving birth to up to three times more cubs than bears in wildlands (Beckmann and Berger, 2003; Prange et al., 2003). Both of these species feed on highly caloric garbage in urban environments, which might provide a mechanism for increased litter sizes. Urban fox squirrel populations appear to have litter sizes comparable to rural populations, but unlike rural populations, most females in an urban population have been shown to have more than one litter annually (McCleery, 2009a). Similarly, urban striped field mice extend their breeding season later into autumn in the urban environment and appear to reach sexual maturity quicker in urban areas (Andrzejewski et al., 1978; Gliwicz et al., 1994).

Many urban populations have even sex ratios or ratios that match rural populations. Nonetheless, some urban mammalian populations, such as striped field mice (Andrzejewski et al., 1978), foxes (Harris and Smith, 1987), and black bears (Beckmann and Berger, 2003), have shown a bias toward males. On the other hand, a population of urban white-tailed deer has been dominated by females; however, this is likely due to differential survival rates of the sexes because fetal sex ratios of the population have been skewed toward males (Lopez et al., 2003).

Increased rates of reproduction in urban mammal populations can lead to a surplus of juveniles if they survive. In general, juvenile mammals have an increased risk of mortality from weaning to adulthood. However, as with adults, survival of young mammals has increased with urbanization for fox squirrel (McCleery, 2009a), white-tailed deer (Peterson et al., 2004), and raccoon (Rosatte et al., 1991). Increased survival of juveniles leaves many urban mammal populations with a surplus of juveniles that can either be recruited into the population or can disperse to other populations.

Movements

Decreased rates of dispersal for surplus juveniles have commonly been cited as a reason for increased densities of mammal populations in urban areas (Etter et al., 2002; Prange et al., 2003; Gliwicz et al., 1994). Not only do there appear to be less-frequent dispersals from urban habitat patches for some mammals, but the distances dispersed might also be shorter in the urban environment (Etter et al., 2002; Harris and Trewhella, 1988). Limited dispersal or reciprocal site fidelity in urban mammals has been hypothesized to be a function of the harsh environment surrounding urban habitat patches (Etter et al., 2002) and the high quality food and resources consistently available in the urban environment (Prange et al., 2003). Undoubtedly, the dominant features of urban landscapes, roads and development, pose a barrier to the movements of many urban mammals. Roads have been shown to restrict the movements of small mammals, such as hedgehogs (*Erinaceinae*) (Rondinini and Doncaster, 2002), woodrats (*Neotoma* spp.) (McCleery et al., 2006a), and white-footed mice (*Peromyscus* spp.) (Merriam et al., 1989). Roads might present less of a barrier to medium-size mammals, but their movements might still be restricted by larger highways (Prange et al., 2004; Gehrt, 2005; Riley, 2006). Even large mammals such as white-tailed deer appear to avoid highways with more than eight lanes and dense development (Etter et al., 2002). Regardless of these barriers, urban mammals still disperse across the urban matrix and in at least two studies, researchers have found that dispersing juveniles in urban areas ran a higher risk of mortality due to roadkill than juveniles dispersing from rural areas (Beckmann and Berger, 2003; Etter et al., 2002).

In this manner the urban environment may pose an ecological trap for some urban mammal populations. Urban areas can have a consistent source of high quality food and reduced risks of predation, allowing mammals to have relatively high rates of reproduction. However, when this surplus of juveniles disperses out of these high quality patches, a disproportionate number of young can be lost to collisions with automobiles or to other hazards of the urban environment. Similarly, a number of authors have suggested that urban mammal populations may have source–sink population dynamics (Dickman and Doncaster, 1987; Harveson et al., 2004; Gosselink et al., 2007; McCleery, 2009a). The sources come from high quality urban patches, such as parks, remnant habitat patches, and restorations, and sinks can come in the form of lower quality urban areas, such as those lacking vegetation and having frequent disturbances and high levels of cat activity, or adjacent rural land with numerous predators and hunters.

Many researchers examining the ranges of mammals in urban environments have found that their ranges have been smaller and often more stable than nonurban populations (Etter et al., 2002; Beckmann and Berger, 2003; Prange et al., 2003; Harveson et al., 2007). Additionally, some urban mammals have been shown to have more complex and less uniform ranges, likely due to the heterogeneous nature of the urban landscape (Harrison, 1997). A possible explanation for the smaller ranges seen in urban mammals is that the patchy nature of the urban environment restricts the movements of urban mammals to areas with strict boundaries (Etter et al., 2002; Prange et al., 2004). Additionally, the clumped, abundant, high quality, and stable nature of resources in the urban environment may help to concentrate the animal into smaller areas (Beckmann and Berger, 2003; Prange et al., 2004). A common finding from research of urban mammals is that they congregate in areas of their preferred resource, such as areas of shrubs, woodlots, picnic sites, and where garbage can be found, and they avoid areas with limited resources, such as ball fields and paths (Gliwicz et al., 1994; Beckmann and Berger, 2003; Prange et al., 2004). The dichotomous nature of habitat selection in urban mammals can lead to the clumped and uneven distributions found in some populations of urban mammals (Dickman, 1987; Gliwicz et al., 1994; Prange et al., 2004).

Territoriality

Examining social behaviors and the partitioning of resources, researchers of urban mammals have noted that urban population appears to differ from non-urban populations (Berger and Beckmann, 2003; Prange et al., 2004; Riley, 2006). Food (quantity, quality, and distribution), density and distribution of refugia, density of conspecifics, habitat features, mates, and predation pressure all influence to what extent animals become territorial or use social hierarchies to partition resources (Maher and Lott, 2000). Only when resources are at intermediate levels do the benefits of procuring them via territoriality outweigh the costs of defending them (Krebs and Davies, 1993). Likewise, individuals respond to population densities by showing territoriality at lower and intermediate levels, while abandoning areas at high levels, possibly due to intruder pressure and the effort needed maintain the territory (Maher and Lott, 2000). Thus, it should not be surprising that high densities of mammals in urban areas have led to increased range overlap and decreased territoriality, especially when there are large quantities of high quality anthropogenic food available (Harris, 1980; Beckmann and Berger, 2003; Prange et al., 2004). However, there are also features of the urban environment that might lead to increased territoriality. Not all resources in the urban environment are abundant. For example a shortage in large dead trees in the urban environment may lead to increased territoriality (especially for females) in a population of urban fox squirrels (McCleery, unpublished data, 2007). Additionally, there is a positive relationship between habitat complexity and territoriality, possibly because the cost of territoriality has been reduced by clear boundaries of demarcation (Maher and Lott, 2000). For example, urban fox squirrels use building to delineate ranges (McCleery, unpublished data, 2007) and bobcats (*Lynx rufus*) use roads to define their territories (Riley, 2006), so the heterogeneous nature of the urban environments may lead to more territorial behavior in certain cases.

Behaviors

There are at least two characteristics of the urban environment that should influence the behavioral choices made by mammals when balancing predation risks with foraging. First, as human activity increases toward the city center, it creates an almost constant predator stimulus for urban mammals. Even without the risk of predation, these disturbances may negatively alter an animal's behavior by increasing vigilance behaviors (Berger et al., 1983; Frid and Dill, 2002). The second factor of urbanized environments that should affect mammals' behavioral choices is a reduction in the risk of predation along the urban–rural gradient due to the elimination of or avoidance of predators in most urbanized areas (Blumstein,

2002; Adams et al., 2005), which hypothetically allows mammals to use more of the urban environment without need for vigilance behaviors. Some studies have shown reductions in mammals' responses to humans corresponding to the increased exposure of mammals to humans in developed areas (Reimers and Sigurd, 2001; Magle et al., 2005; Harveson et al., 2007; McCleery, 2009b). One of the most common and highly plausible explanations given for reduced responses to humans is habituation (McCleery, 2009b). *Habituation,* a decreased responsiveness to repeated stimuli, seems to provide a mechanism for urban adapted mammals to cope with the constant human induced predator stimuli and take advantage of the resources in an environment with a reduced risk of predation. It has also been shown that once animals habituated to humans they may also transfer this reduced response to other predator stimuli. McCleery (2009b) reported that urban fox squirrels that showed a reduced response to humans also showed a reduced response to hawk and coyote calls when compared with rural and suburban populations.

Human activity in urban environments has altered mammal's behaviors in other ways. A number of populations in urban environments have been shown to be active at different times than rural populations (Tigas et al., 2002; Riley et al., 2003; Gehrt, 2007). Coyotes, bobcats, and javelina (*Pecari tajacu*) showed reduced activity during daylight hours and increased nocturnal activity in areas of human activity (Riley et al., 2003; Gehrt, 2007; Ticer et al., 1998), likely to avoid contacts with humans. On the other hand, the striped field mouse, which has been strictly nocturnal in rural environments, has become more active during daylight hours in urban areas (Gliwicz et al., 1994).

Diet

A number of species of urban adapted mammals also have been shown to alter their diets in urban environments to consume anthropogenic foods (Ditchkoff et al., 2006). Midsize omnivores appear to be especially adept at utilizing human foods and trash (Fedriani et al., 2001; Prange et al., 2004; Contesse et al., 2004). For example, 14 to 25% of the diet of urbanized coyotes and more than 50% of the stomach contents of urban foxes have been comprised of anthropogenic foods (Fedriani et al., 2001; Contesse et al., 2004). Coyotes have even been shown to feed on house cats. In one study cats appeared in more than 13.6% of coyote's scat (Shargo, 1988). Other shifts in mammals' diets have been observed in urbanized areas. Gray foxes in urban areas consumed more birds and mammals and less plant matter in urban areas, and squirrels have been shown to alter their diet to exploit the exotic plants common in urban environments (Harrison, 1997; Jodice and Humphrey, 1992).

Physiological Differences

Differences in the diets of urban mammals are often cited to explain physiological differences between urban and nonurban mammal populations. Individuals in urban populations have commonly been larger or had a greater body mass than individuals from less developed landscapes (Andrzejewski et al., 1978; Harrison, 1997; Beckmann and Berger, 2003; Harveson et al., 2007). Similarly, body conditions for most urban mammals have been better or equal to nonurban populations (Cypher and Frost, 1999; Grinder and Krausman, 2001), although at exceedingly high densities, body conditions in some urban populations have shown deterioration (Etter et al., 2002; Hadidian et al., 1987). Urban mammals consuming anthropogenic food also can show differences in weight fluctuations. Urban black bears feeding on garbage gained weight over most the winter, while most bears in nonurban areas hibernate for the season and lose a considerable amount of body weight (Beckmann and Berger, 2003). On the other hand, raccoons that foraged around picnic areas lost more weight than rural raccoons over the winter when the garbage and table scraps were unavailable during the season (Prange et al., 2003).

Some of the most interesting comparisons between urban and rural population of mammals were conducted on striped field mice in Poland. Not only did urban field mice have larger body masses, but they also had different ratios of internal organs to body mass (Gliwicz et al., 1994). Research also found lower levels of white blood cells and a reduced oxygen-carrying capacity in urban striped field, possibly due to the pollutant levels found in an urban setting (Gliwicz et al., 1994). Furthermore, examining morphometrics, researchers found significant differences in skull and bone measurements between urban and rural populations of striped field mice (Gliwicz et al., 1994), possibly indicating the divergent evolution of a population of urban mammal.

Parasites and Disease

The population densities achieved and the clumped nature of urban resources make many urban mammal populations susceptible to disease outbreaks and parasites (Bradley and Altizer, 2007). Disease outbreaks have been shown to be a major cause of mortality in urban adapted mammals such as raccoon (Prange et al., 2003) and red fox (Gosselink et al., 2007). Additionally, the prevalence of disease, such as chronic wasting disease, has appeared to increase with urban development (Farnsworth et al., 2005). Parasites may also be a problem in urban mammal populations. Not only have urban populations of striped field mice shown higher concentrations of ectoparasites, but they have also been found with ectoparasites specific to cats and dogs that have never been found in rural individuals (Gliwicz et al., 1994).

Larger populations of urban adapters are not the only mammals at risk to disease. Rarer species that persist in urban environment can also be susceptible to pathogens from abundant urban adapted species, introduced species and domestic animals. Threatened Allegany woodrats (*Neotoma magister*) are highly susceptible to the common raccoon parasite *Baylisascris procyonis*, especially in areas of high raccoon density that can be found around development (Logiudice, 2003). In the United Kingdom, the transfer of paramyxovirus from introduced North American gray squirrels to native red squirrels (*Sciurus vulgaris*) has played a role in the species' decline (Bradley and Altizer, 2007). Moreover, contact with dogs and cats appear to have increased the exposure of urban gray foxes and bobcats to pathogens common in domestic pets (Riley et al., 2004).

The increase of pathogens, metals, nutrients, and novel carbons (Chapter 15, Steele et al., 2010, this volume) along the urban portion of the urban–rural gradient clearly poses a threat to the health of urban mammals. The continued exposure to these pathogens, elements, and compounds is likely to make urban mammals more susceptible to disease, reduce survival, and impair reproduction (Ditchkoff et al., 2006; Bradley and Altizer, 2007). An excellent example of this was found in sea otters (*Enhydra lutris nereis*) off the coast of California. Otters in areas of high runoff from urban development had infection rates three times higher than otters off the coast of less developed areas (Miller et al., 2002). Urban mammals have also been shown to have increased levels of heavy metals, such as lead (Raymond and Forbes, 1975) and polychlorinated biphenyls (PCBs) (Dip et al., 2003), both believed to have adverse health consequences for individuals. Additionally, exposure of urban bobcats to anticoagulant rodenticides has been strongly linked to mange-associated mortality (Riley et al., 2007).

Use of Anthropogenic Structures in the Environment

The urban environment is dominated by different types of structures, a surprising number of which provide shelter and dens for urban mammals. Small- and medium-sized mammals, such as squirrels, brushtailed possums (*Trichosurus vulpecula*), stone martins (*Martes fonica*), and raccoons, commonly use attics and buildings for shelter and raising young (Adams et al., 2005; Hill et al., 2007). Moreover, urban red foxes, skunks, virginal opossums, Eurasian badgers (*Meles meles*), woodchucks (*Marmota momax*), and armadillos (*Dasypodidae*) make extensive use of the areas under houses for denning (Adams et al., 2005; Harris, 1981; Lariviere et al., 1999; Davison et al., 2008). In some cases urban adapted wildlife almost exclusively use anthropogenic structures for denning. A study of urban stone martins found that 97% of their dens were found in buildings, and inhabited buildings were selected for during winter months, presumable for their warmth (Herr et al., 2009)

Culverts and refuse are two other features of the urban environment commonly used by urban mammals for shelter. Coyotes, foxes, raccoons, skunks (*Mephitis* spp.), black bears, and even spotted hyenas (*Crocuta crocuta*) are just some of the mammals known to use culverts for dens (Barnes and Bray, 1966; Weller and Pelton, 1987; Reese et al., 1992; Adams, 1994; Gosselink et al., 2007; Pokines and Kerbis Peterhans, 2007; Grubbs and Krausman, 2009). Garbage dumps can provide food and shelter for nonnative rats (*Ratus* spp.) and house mice, as well as indigenous chipmunks (*Tamias striatus*) and mice (*Peromyscus* spp.) (Courtney and Fenton, 1976). In Oxford, England, native small mammals used building refuse and garbage for nesting (Dickman, 1987) and North American woodrats (*Neotoma* spp.) have commonly been found nesting in trash and building materials (McCleery et al., 2006b).

Bats as a group are very adept at using anthropogenic structures for roosting and hibernacula. Bats commonly enter houses, roosting in attics and basements (Adams, 1994), and numerous bat species have used bridges and culverts associated with highways for roosting. In fact, 24 of the United States' 45 bat species have been recorded using bridges or culverts as roosts (Keeley and Tuttle, 1999). In Texas, large colonies of Mexican free-tailed bats (*Tadarida braziliensis*) can be found in the center of several cities roosting on building, bridges, and stadiums (Adams et al., 2006; Scales and Wilkins, 2007).

Conservation

Urbanization is possibly the greatest threat to the survival of sensitive and imperiled mammal species (Czech et al., 2000). Most of the mammalian species that occur in or around urban areas are widely distributed, if not ubiquitous species. Nonetheless, there are a few notable exceptions where portions of the developed landscape appear to have benefited mammalian species at risk of extinction. The

Big Cypress fox squirrel (*Sciurus niger avicennia*), listed as a threatened species by the state of Florida, appeared to reach higher densities around golf courses than in natural and protected areas (Jodice and Humphrey, 1992). Golf courses provided the squirrels with a diverse and stable food source (Jodice and Humphrey, 1992). Similarly, conservation efforts in urban areas have become critical to the recovery of the San Joaquin kit fox (*Vulpes macrotis mutica*). Urban kit foxes use golf courses, city parks, and school grounds, and they den in culverts and drainage pipes (Adams et al., 2006). Urban San Joaquin kit foxes likely benefit from their utilization of anthropogenic food, which may explain the increased weight of urban juveniles (Cypher and Frost, 1999). Finally, the federally endangered Florida key deer (*Odocolius virginiana clavium*), like other white-tailed deer, appear to have benefited from moderate levels of urbanization. Urbanization has resulted in more habitat and food resource available for the deer, in turn leading to increases in the size of the population (Adams et al., 2006). Moreover, urbanization also has increased the weights and survival rates of key deer (Harveson et al., 2007). All three of these species appear to have benefited from the increased resources in areas of moderate development commonly found surrounding the core urban areas. These moderately developed areas with anthropogenic food, high net primary production, and structures for shelter probably hold the most potential for urban conservation efforts directed at populations of native mammals.

Summary

Urbanization continues to threaten mammalian species and alter the distribution of mammalian communities across the globe. Nonetheless, some mammals have been able to cope with the anthropogenic structures, altered vegetative communities, and human activity of urban environments. These urban adapted populations of mammals have been able to utilize the food resource and structures of urban environment and for the most part face a minimal risk of predation. This has helped lead to dense and clumped population of urban mammals, often with increased rates of survival and reproduction. Urban populations commonly differ from nonurban populations in their diet, behavior, physiology, and exposure to disease. It is these urbanized populations of mammals that most of the world's human population will encounter and interact with on a daily basis. The proximity of humans to mammals provides an excellent opportunity for ecological education and research, which are critical for the management and conservation of our planet's mammalian species. It is imperative that we seize the opportunity to further our understanding of how urbanization will continue to alter the composition of mammalian communities, as well the demographics, behaviors, and physiology of mammalian populations.

References

Adams, C.E., K.J. Lindsey, and S.J. Ash. 2006. Urban wildlife management. Taylor and Francis, Boca Raton.

Adams, L.W. 1994. Urban wildlife habitats; a landscape perspective. Univ. of Minnesota Press, Minneapolis.

Adams, L.W., L.W. Vandruff, and M. Luniak. 2005. Managing urban habitats and wildlife. p. 714–739. *In* C.E. Braun (ed.) Techniques for wildlife investigations and management. Allen Press, Lawrence, MA.

Andrzejewski, R., J. Babinska-Werka, J. Gliwicz, and J. Goszczynski. 1978. Synurbanization process in a population of *Apodemus agrarius*. Acta Theriol. (Warsz.) 23:341–358.

Baker, P.J., R.J. Ansell, P.A.A. Dodds, C.E. Webber, and S. Harris. 2003. Factors affecting the distribution of small mammals in an urban area. Mammal Rev. 33:95–100.

Barnes, V.G., Jr., and O.E. Bray. 1966. Black bears use drainage culverts for winter dens. J. Mammal. 47:712–713.

Beckmann, J.P., and J. Berger. 2003. Using black bears to test ideal-free distribution models experimentally. J. Mammal. 84:594–606.

Berger, J., D. Daneke, J. Johnson, and S.H. Berwick. 1983. Pronghorn foraging economy and predator avoidance in a desert ecosystem: Implications for the conservation of large mammalian herbivores. Biol. Conserv. 25:193–208.

Blumstein, D.T. 2002. Moving to suburbia: Ontogenetic and evolutionary consequences of life on predator-free islands. J. Biogeogr. 29:685–692.

Bradley, C.A., and S. Altizer. 2007. Urbanization and the ecology of wildlife diseases. Trends Ecol. Evol. 22:95–102.

Breuste, J., J. Niemelä, and R. Snep. 2008. Applying landscape ecological principles in urban environments. Landscape Ecol. 23:1139–1142.

Collins, J.P., and A. Kinzig. 2000. A new urban ecology. Am. Sci. 88:416–425.

Contesse, P., D. Hegglin, S. Gloor, F. Bontadina, and P. Deplazes. 2004. The diet of urban foxes (*Vulpes vulpes*) and the availability of anthropogenic food in the city of Zurich, Switzerland. Mamm. Biol. 69:81–95.

Coté, S.D., T.P. Rooney, J.-P. Tremblay, C. Dussault, and D.M. Waller. 2004. Ecological impacts of deer overabundance. Annu. Rev. Ecol. Evol. Syst. 35:113.

Courtney, P.A., and M.B. Fenton. 1976. The effects of a small rural garbage dump on populations of *peromyscus leucopus rafinesque* and other small mammals. J. Appl. Ecol. 13:413–422.

Crooks, K.R., and M.E. Soule. 1999. Mesopredator release and avifaunal extinctions in a fragmented system. Nature 400:563.

Cypher, B.L., and N. Frost. 1999. Condition of San Joaquin kit foxes in urban and exurban habitats. J. Wildl. Manage. 63:930–938.

Czech, B., P.R. Krausman, and P.K. Devers. 2000. Economic associations among causes of species endangerment in the United States. Bioscience 50:593.

Davison, J., M. Huck, R.J. Delahay, and T.J. Roper. 2008. Urban badger setts: Characteristics, patterns of use and management implications. J. Zool. (Lond.) 275:190–200.

Dickman, C.R. 1987. Habitat fragmentation and vertebrate species richness in an urban environment. J. Appl. Ecol. 24:337–351.

Dickman, C.R., and C.P. Doncaster. 1987. The ecology of small mammals in urban habitats. I. populations in a patchy environment. J. Anim. Ecol. 56:629–640.

Dip, R., D. Hegglin, P. Deplazes, O. Dafflon, H. Koch, and H. Naegeli. 2003. Age- and sex-dependent distribution of persistent organochlorine pollutants in urban foxes. Environ. Health Perspect. 111:1608–1612.

Ditchkoff, S., S. Saalfeld, and C. Gibson. 2006. Animal behavior in urban ecosystems: Modifications due to human-induced stress. Urban Ecosyst. 9:5–12.

Ellis, E.C., and N. Ramankutty. 2008. Putting people in the map: Anthropogenic biomes of the world. Front. Ecol. Environ. 6:439–447.

Etter, D.R., K.M. Hollis, T.R.V. Deelen, D.R. Ludwig, J.E. Chelsvig, C.L. Anchor, and R.E. Warner. 2002. Survival and movements of white-tailed deer in suburban Chicago, Illinois. J. Wildl. Manage. 66:500–510.

Falk, J.H. 1976. Energetics of a suburban lawn ecosystem. Ecology 57:141–150.

Farnsworth, M.L., L.L. Wolfe, N.T. Hobbs, K.P. Burnham, E.S. Williams, D.M. Theobald, M.M. Conner, and M.W. Miller. 2005. Human land use influences chronic wasting disease prevalence in mule deer. Ecol. Appl. 15:119–126.

Fedriani, J.M., T.K. Fuller, and R.M. Sauvajot. 2001. Does availability of anthropogenic food enhance densities of omnivorous mammals? An example with coyotes in southern California. Ecography 24:325–331.

Frid, A., and L.M. Dill. 2002. Human-caused disturbance stimuli as a form of predation risk. Conserv. Ecol. 6:11.

Gehrt, S.D. 2005. Seasonal survival and cause-specific mortality of urban and rural striped skunks in the absence of rabies. J. Mammal. 86:1164–1170.

Gehrt, S.D. 2007. Ecology of coyotes in urban landscapes. p. 303–311. In Proc. 12th Wildl. Damage Conf. Univ. of Nebraska, Lincoln.

Gliwicz, J., J. Goszcynski, and M. Luniak. 1994. Characteristic features of animal populations under synurbanization-the case of the blackbird and of the stripped field mouse. Mem. Zool. 49:237–244.

Gosselink, T.E., T.R. Van Deelen, R.E. Warner, and P.C. Mankin. 2007. Survival and cause-specific mortality of red foxes in agricultural and urban areas of Illinois. J. Wildl. Manage. 71:1862–1873.

Grinder, M., and P. Krausman. 2001. Morbidity–mortality factors and survival of an urban coyote population in Arizona. J. Wildl. Dis. 37:312–317.

Grubbs, S.E., and P.R. Krausman. 2009. Use of urban landscape by coyotes. Southwest. Nat. 54:1–12.

Hadidian, J., D. Manski, V. Flyger, C. Cox, and G. Hodge. 1987. Urban gray squirrel damage and population management: A case history. p. 219–227. In Proc. 3rd Eastern Wildlife Damage Control Conf. Univ. of Nebraska, Lincoln.

Harris, S. 1980. Home range and patterns of distribution of foxes (Vulpes vulpes) in an urban area as revealed by radio tracking. p. 685–690. In C.J. Amlaner and D.W. Macdonald (ed.) A handbook on biotlemetry and radio tracking. Pergamon Press, Oxford, UK.

Harris, S. 1981. An estimation of the number of foxes (Vulpes vulpes) in the city of Bristol, and some possible factors affecting their distribution. J. Appl. Ecol. 18:455–465.

Harris, S., and G.C. Smith. 1987. Demography of two urban fox (Vulpes vulpes) populations. J. Appl. Ecol. 24:75–86.

Harris, S., and W.J. Trewhella. 1988. An analysis of some of the factors affecting dispersal in an urban fox (Vulpes vulpes) population. J. Appl. Ecol. 25:409–422.

Harrison, R.L. 1997. A comparison of gray fox ecology between residential and undeveloped rural landscapes. J. Wildl. Manage. 61:112–122.

Harveson, P.M., R.R. Lopez, B.A. Collier, and N.J. Silvy. 2007. Impacts of urbanization on Florida Key deer behavior and population dynamics. Biol. Conserv. 134:321–331.

Harveson, P.M., R.R. Lopez, N.J. Silvy, and P.A. Frank. 2004. Source–sink dynamics of Florida key deer on big pine key, Florida. J. Wildl. Manage. 68:909–915.

Herr, J., L. Schley, E. Engel, and T.J. Roper. 2009. Den preferences and denning behaviour in urban stone martens (Martes foina). Mammal. Biol. 75:138–145.

Hill, N.J., K.A. Carbery, and E.M. Deane. 2007. Human–possum conflict in urban Sydney, Australia: Public perceptions and implications for species management. p. 101–113. In Human dimensions of wildlife. Taylor and Francis, Oxford.

Imhoff, M.L., and C.J. Tucker. 2000. The use of multisource satellite and geospatial data to study the effect of urbanization on primary productivity in the United States. IEEE Trans. Geosci. Rem. Sens. 38:2549.

Jodice, P.G.R., and S.R. Humphrey. 1992. Activity and diet of an urban population of Big Cypress fox squirrels. J. Wildl. Manage. 56:685–692.

Keeling, M.J., and C.A. Gilligan. 2000. Metapopulation dynamics of bubonic plague. Nature 407:903–906.

Keeley, B.W., and M.D. Tuttle. 1999. Bats in American bridges. Resource Publ. 4. Bat Conservation Int., Austin, TX.

Krebs, J.R., and N.B. Davies. 1993. An introduction to behavioural ecology. 3rd ed. Blackwell Scientific Publ., Oxford.

Koprowski, J.L. 1994. Sciurus carolinensis. Mamm. Species 480:1–9.

Lariviere, S., L.R. Walton, and F.O. Messier. 1999. Selection by striped skunks (Mephitis mephitis) of farmsteads and buildings as denning sites. Am. Midl. Nat. 142:96–101.

Logiudice, K. 2003. Trophically transmitted parasites and the conservation of small populations: Raccoon roundworm and the imperiled allegheny Woodrat. Conserv. Biol. 17:258–266.

Lopez, R.R., M.E.P. Vieira, N.J. Silvy, P.A. Frank, S.W. Whisenant, and D.A. Jones. 2003. Survival, mortality, and life expectancy of Florida key deer. J. Wildl. Manage. 67:34.

Lotze, J., and S. Anderson. 1979. *Procyon lotor.* Mamm. Species 119:1–8.

Mackin-Rogalska, R., J. Pinowski, J. Solon, and Z. Wojcik. 1988. Changes in vegetation, avifauna, and small mammals in a suburban habitat. Pol. Ecol. Stud. 14:293–330.

Magle, S., J. Zhu, and K.R. Crooks. 2005. Behavioral responses to repeated human intrusion by black-tailed prairie dogs (*Cynomys ludovicianus*). J. Mammal. 86:524–530.

Maher, C.R., and D.F. Lott. 2000. A review of ecological determinants of territoriality within vertebrate species. Am. Midl. Nat. 143:1–29.

Matthiae, P.E., and F. Sterns. 1981. Mammals in forest islands of southeastern Wisconsin. p. 55–66. *In* R.L. Burgess and D.M. Sharpe (ed.) Forest islands dynamics in man-dominated landscapes. Springer-Verlag, New York.

McCleery, R.A. 2009a. Reproduction, juvenile survival and retention in an urban fox squirrel population. Urban Ecosyst. 12:177–184.

McCleery, R.A. 2009b. Changes in fox squirrel anti-predator behaviors across the urban–rural gradient. Landscape Ecol. 24:483–493.

McCleery, R.A., R.R. Lopez, and N.J. Silvy. 2006a. Movements and habitat use of the Key Largo woodrat. Southeast. Nat. 5:725–736.

McCleery, R.A., R.R. Lopez, N.J. Silvy, P.A. Frank, and S.B. Klett. 2006b. Population status and habitat selection of the endangered Key Largo woodrat. Am. Midl. Nat. 155:197–209.

McCleery, R.A., R.R. Lopez, N.J. Silvy, and D.L. Gallant. 2008. Fox squirrel survival in urban and rural environments. J. Wildl. Manage. 72:133–137.

McDonnell, M.J., and S.T.A. Pickett. 1990. Ecosystem structure and function along urban–rural gradients: An unexploited opportunity for ecology. Ecology 71:1232–1237.

McKinney, M.L. 2008. Effects of urbanization on species richness: A review of plants and animals. Urban Ecosyst. 11:161–176.

McKinney, M.L. 2006. Urbanization as a major cause of biotic homogenization. Biol. Conserv. 127:247–260.

McKinney, M.L. 2002. Urbanization, biodiversity, and conservation. Bioscience 52:883.

Merriam, G., M. Kozakiewicz, E. Tsuchiya, and K. Hawley. 1989. Barriers as boundaries for metapopulations and demes of *Peromyscus leucopus* in farm landscapes. Landscape Ecol. 2:227–235.

Millennium Ecosystem Assessment. 2005. Ecosystems and human well-being: Current state and trends: Findings of the condition and Trends Working Group. Available at http://www.millenniumassessment.org (verified 12 Feb. 2010). Island Press, Washington, DC.

Miller, M.A., I.A. Gardner, C. Kreuder, D.M. Paradies, K.R. Worcester, D.A. Jessup, E. Dodd, M.D. Harris, J.A. Ames, A.E. Packham, and P.A. Conrad. 2002. Coastal freshwater runoff is a risk factor for *Toxoplasma gondii* infection of southern sea otters (*Enhydra lutris nereis*). Int. J. Parasitol. 32:997–1006.

Nilon, C.H., and L.W. Vandruff. 1987. Analysis of small mammal community data and applications to management of urban greenspaces. p. 53–59. *In* Proc. Natl. Symp. Urban Wildlife 2.

Peterson, M.N., R.R. Lopez, P.A. Frank, B.A. Porter, and N.J. Silvy. 2004. Key deer fawn response to urbanization: Is sustainable development possible? Wildl. Soc. Bull. 32:493–499.

Pickett, S.T.A., M.L. Cadenasso, J.M. Grove, C.H. Nilon, R.V. Pouyat, W.C. Zipperer, and R. Costanza. 2001. Urban ecological systems: Linking terrestrial ecological, physical, and socioeconomic components of metropolitan areas. Annu. Rev. Ecol. Syst. 32:127–157.

Pokines, J.T., and J.C. Kerbis Peterhans. 2007. Spotted hyena (*Crocuta crocuta*) den use and taphonomy in the Masai Mara National Reserve, Kenya. J. Arch. Sci. 34:1914–1931.

Prange, S., S.D. Gehrt, and E.P. Wiggers. 2003. Demographic factors contributing to high raccoon densities in urban landscapes. J. Wildl. Manage. 67:324–333.

Prange, S., S.D. Gehrt, E.P. Wiggers, and T.J. O'Shea. 2004. Influences of anthropogenic resources on raccoon (*procyon lotor*) movements and spatial distribution. J. Mammal. 85:483–490.

Racey, G.D., and D.L. Euler. 1982. Small mammal and habitat response to shoreline cottage development in central Ontario. Can. J. Zool. 60:865–880.

Raymond, R.B., and R.B. Forbes. 1975. Lead in hair of urban and rural small mammals. Bull. Environ. Contam. Toxicol. 13:551–553.

Ree, R., and M.A. McCarthy. 2005. Inferring persistence of indigenous mammals in response to urbanization. Anim. Conserv. 8:309–319.

Reese, E.A., W.G. Standley, and W.H. Berry. 1992. Habitat, soils, and den use of San Joaquin kit fox (*Vulpes velox macrotis*) at Camp Roberts Army National Guard Training Site, California. Rep. EGG-10617-2156.

Reimers, E., and S. Sigurd. 2001. Vigilance behavior in wild and semi-domestic reindeer in Norway. Alces 37:303–313.

Riley, S.P.D. 2006. Spatial ecology of bobcats and gray foxes in urban and rural zones of a national park. J. Wildl. Manage. 70:1425–1435.

Riley, S.P.D., C. Bromley, R.H. Poppenga, F.A. Uzal, L. Whited, and R.M. Sauvajot. 2007. Anticoagulant exposure and notoedric mange in bobcats and mountain lions in urban southern California. J. Wildl. Manage. 71:1874–1884.

Riley, S.P.D., J. Foley, and B. Chomel. 2004. Exposure to feline and canine pathogens in bobcats and gray foxes in urban and rural zones of a national park in California. J. Wildl. Dis. 40:11–22.

Riley, S.P.D., J. Hadidian, and D.A. Manski. 1998. Population density, survival, and rabies in raccoons in an urban national park. Can. J. Zool. 76:1153–1164.

Riley, S.P.D., R.M. Sauvajot, T.K. Fuller, E.C. York, D.A. Kamradt, C. Bromley, and R.K. Wayne. 2003. Effects of urbanization and habitat fragmentation on bobcats and coyotes in southern California. Conserv. Biol. 17:566–576.

Rondinini, C., and C.P. Doncaster. 2002. Roads as barriers to movement for hedgehogs. Funct. Ecol. 16:504–509.

Rosatte, R.C., M.J. Power, and C.D. MacInnes. 1991. Ecology of urban skunks, raccoons, and foxes in metropolitan environments. p. 31–38. *In* L.W. Adam and D.L. Leedy (ed.) Wildlife conservation in metropolitan environments. Nation Institute for Urban Wildlife, Columbia, MD.

Scales, J.A., and K.T. Wilkins. 2007. Seasonality and fidelity in roost use of the Mexican free-tailed bat, *Tadarida brasiliensis*, in an urban setting. West. N. Am. Nat. 67:402–408.

Shargo, E.S. 1988. Home range movements and activity patterns of coyotes in a Los Angeles suburb. Ph.D. diss. University of California, Los Angeles.

Steele, M.K., W.H. McDowell, and J.A. Aitkenhead-Peterson. 2010. Chemistry of urban, suburban, and rural surface waters. p. 297–340. *In* J. Aitkenhead-Peterson and A. Volder (ed.) Urban ecosystem ecology. Agron. Monogr. 55. ASA, CSSA, and SSSA, Madison, WI.

Tait, C.J., C.B. Daniels, and R.S. Hill. 2005. Changes in species assemblages within the Adelaide metropolitan area, Australia. Ecol. Appl. 15:346–359.

Ticer, C.L., R.A. Ockenfels, J.C. Devos, and T.E. Morrell. 1998. Habitat use and activity patterns of urban-dwelling javelina. Urban Ecosyst. 2:141–151.

Tigas, L.A., D.H. Van Vuren, and R.M. Sauvajot. 2002. Behavioral responses of bobcats and coyotes to habitat fragmentation and corridors in an urban environment. Biol. Conserv. 108:299–306.

United Nations. 2008. World urbanization prospects. The 2007 revision. Highlights. Available at http://www.un.org/esa/population/publications/wup2007/2007WUP_Highlights_web.pdf (verified 12 Feb. 2010). United Nations, New York.

Vandruff, L.W., E.G. Bolen, and G.J. San Julian. 1996. Management of urban wildlife. p. 507–553. *In* T.A. Bookhout (ed.) Research and management techniques for wildlife and habits. Allen Press, Lawrence, MA.

Woods, M., R.A. Mcdonald, and S. Harris. 2003. Predation of wildlife by domestic cats *Felius catus* in great Britain. Mammal Rev. 33:174–188.

Weller, D.M.G., and M.R. Pelton. 1987. Denning characteristics of striped skunks in Great Smoky Mountains National Park. J. Mammal. 68:177–179.

6

Habitat Function in Urban Riparian Zones

Joan G. Ehrenfeld
Emilie K. Stander

Abstract

Several of the features of riparian zones are affected by surrounding urban land use, which has important implications for the structure and function of biotic communities within these zones. Both native and invasive plant species are discussed. Urban riparian zone animals include mammals, birds, fish, invertebrates, and herpetofauna. The role of urban riparian zones for species movement is also examined.

Plant Communities

Plant communities in urban riparian zones reflect both the particular environmental conditions of riparian areas in general and the specific effect of surrounding urban land use on these conditions. Riparian zones are known for their high diversity of plants, both native and nonnative, and have received extensive attention (Naiman and Decamps 1997; Stohlgren et al., 1998; Brown and Peet, 2003; Naiman et al., 2005; Tabacchi and Planty-Tabacchi, 2005; Dwire and Lowrance, 2006; Richardson et al., 2007). Several features of riparian zones have important implications for the structure and function of biotic communities, and also are affected by surrounding urban land use. First, riparian zones tend to be linear, narrow landscape features, in accord with their geomorphic position along river corridors (Chapter 13, Stander and Ehrenfeld, 2010, this volume). Second, they are strongly influenced by water flow in many ways. Flood regimes determine the amount, frequency, depth, and duration of flooding. Flood waters deposit, as well as remove, sediments and their associated nutrients, propagules, and coarse and fine woody debris. Flood regimes also create topographic variation across the floodplain, in turn creating diverse microhabitats for both plants and animals. The riparian zone ranges from the channel shelf and banks of the river to the upper terraces that are rarely flooded. Third, their connection to rivers implies a high degree of linear connectivity within and between stands (Nilsson and Berggren 2000; Naiman et al., 2005; Gurnell et al., 2008). Finally, the association with flowing water also creates disturbance regimes with varied intensities and periodicities. These various factors result in a high degree of physical, chemical,

J.G. Ehrenfeld, Dep. of Ecology, Evolution, and Natural Resources, Rutgers University, 14 College Farm Rd., New Brunswick, NJ (ehrenfel@rci.rutgers.edu); E.K. Stander, USEPA, Urban Watershed Management Branch, 2890 Woodbridge Ave., MS-104, Edison, NJ (stander.emilie@epa.gov)

doi:10.2134/agronmonogr55.c6

and biotic heterogeneity, which in turn generates very high species densities and makes them among the most biodiverse of terrestrial ecosystems (Naiman et al., 1993). Because of the narrow, linear nature of the habitat and their dependence on riverine flow regimes, however, riparian areas are strongly influenced by urban development, and indeed, human land use in general (Richardson et al., 2007).

Urban riparian plant communities differ most markedly from nonurban communities in the high degree of invasion of exotic species (Table 6–1). It is perhaps not surprising that riparian zones are subject to high invasion by nonnative species, since these species are strongly associated with physical disturbance, both natural and anthropogenic (Planty-Tabacchi et al., 1996; Stohlgren et al., 1998; D'Antonio et al., 1999; Prieur-Richard and Lavorel, 2000; Alston and Richardson, 2006; Richardson et al., 2007; Schnitzler et al., 2007). This pattern is in accord with patterns widely documented for urban patches of natural vegetation in general (Kowarik, 1995, 2005; Pysek, 1998; Pickett et al., 2001; Alberti, 2005; McKinney, 2006). However, patterns of exotic invasion in urban riparian communities are more complex than a uniform increase in abundance of all exotic species in all urban riparian systems. For example, in a study of riparian forests along the Assiniboine River in southern Manitoba, Canada (including sites in downtown Winnipeg), Moffatt et al. (2004) found that although urban sites did indeed have a higher proportion of exotic species in the flora than suburban, rural, and reference sites, this resulted from the fact that there were considerably fewer native species in the urban sites compared with the others and therefore a higher proportion of exotics, although the absolute number of exotic species was similar across the rural–urban gradient. These researchers also showed that within the herbaceous plant community and also within seed banks, the most urban sites were characterized by a suite of opportunistic species that included both native and exotic species (Moffatt and McLachlan, 2004; Moffatt et al., 2004). Thus, the prominence of exotic species in urban riparian communities may be a side effect of native species loss as much as a response of exotic species to available habitat. Lavoie et al. (2003), in a longitudinal study of riparian areas along 560 km of the St. Lawrence River, found that the frequency of exotic species was more strongly related to salinity than urban development, with lower frequencies of exotics in the more estuarine areas. This study suggests that environmental factors other than the amount of urban land may override the effects of urban disturbance in determining the presence of exotic species. However, Aronson et al. (2004), in a series of riparian forested wetlands along a 60-km urbanized section of the Passaic River in New Jersey, found that exotic plants represented between 5 and 20% of the site flora, but there was no apparent pattern to explain which sites were minimally invaded and which were more heavily invaded. Thus, exotic invasion of urban riparian sites, while generally observed, is variable in pattern along rivers and among geographic regions.

While most studies of urban riparian communities have compared those within urban regions with those within nonurban land-use settings, a few studies have specifically examined the effects of differing types of urban land use on exotic species invasions in riparian communities. In a study of riparian areas in North Carolina, Vidra and Shear (2008) found that the structure of vegetation within the human-occupied landscape adjacent to riparian areas affected the extent of exotic species invasion. In residential areas, large trees were present above a sparse or open understory (i.e., lawns beneath tree canopies in backyards),

Table 6–1. Characteristics of plant communities in urban riparian systems.

General pattern	Associated factors	Locations†
Nonnative species more abundant than in nonurban riparian communities	• More physical disturbance, trash, trails • Higher percentage impervious in watershed • Curvilinear relationship of exotic richness to distance from urban edge • Vegetation structure—little understory, bare ground, sparse overstory; small patch size; sandy soils; lower flooding frequencies	Cleveland, OH (1); western USA (2); Spanish Mediterranean(3); Georgia Piedmont (4,5); Cape Town, South Africa (6); Winnipeg, Canada (9, 17); Cincinnati, OH (10); southern CA (11); Wisconsin River, WI (14); Georgia, USA (16)
No difference in exotic presence	• No overall patterns, although some particular species more common in urban sites	Winnipeg, Canada (17); Raleigh, NC (18)
Many large trees; few small trees	• Vegetation management	Georgia Piedmont (4,5), Cincinnati, OH (10)
No difference in species richness, urban and nonurban		Georgia Piedmont (5)
Sparse understory	• Flowing water disturbance • Steep bank limiting colonization	South Africa (6); Cincinnati, OH (10)
Extensive areas of bare ground	• Paths, evidence of human disturbance	South Africa (6)
Patch size decreases	• Land-use history • River fragmented by flow regulation • River geomorphic changes (natural)	Quebec, Canada (7)
Forest vegetation (vs. herbaceous, grassy)	• Wider, more shallow stream corridors, less eroded	Southeastern Pennsylvania (8)
Herbaceous layer species composition distinct	• Soil properties (i.e., clay and sand content) • "Urban avoiders" vs. "opportunists"	Winnipeg, Canada (9)
Increased number of dead trees (snags)	• Built area adjacent to riparian corridor	Cincinnati, OH (10)
Some species occurrences unrelated to urban land cover	• Generalists • Preexisting environmental conditions	Winnipeg, Canada (9); southern CA (11)
Dispersal ecology	• Berry-dispersed species predominate • Water-mediated dispersal important in dispersal from gardens	Winnipeg, Canada (9); central Europe (12)
Presence of anthropogenic vegetation (e.g., lawns) within natural vegetation matrix	• Associated with trails, sewer lines	Raleigh, NC (13)
Native diversity	• No patterns with urbanization • Variable effects of exotics	St. Lawrence, Canada (15)
Flood-tolerance of plants	• Exotics have lower tolerance	Wisconsin River, WI (14); Winnipeg, CA (17); New Jersey (22, 23)

Table continued.

Table 6–1. Continued.

General pattern	Associated factors	Locations†
Soil properties	• Drier, more alkaline in urban areas	Winnipeg, CA (17)
Physiographic province structures vegetation	• Proxy for differences in climate, topography, hydrology	Wisconsin River, WI (14)
Floodplain topography affects species composition	• Topographic complexity • Relationship to river determine species richness • Community structure	Potomac River, Washington, DC (19); St. Lawrence, CA (7)
Dominance of facultative or upland plant species (and lack of obligate wetland plants)	• Result of altered hydrology—lower water tables, less flooding	New Jersey (21, 22)

† References: (1)Wolin and MacKeigan, 2005; (2) Ringold et al., 2008; (3) Aguiar et al., 2007; (4) Burton and Samuelson, 2008; (5) Burton et al., 2005; (6) Alston and Richardson, 2006; (7) Charron et al., 2008; (8) Hession et al., 2000; (9) Moffat and McLachlan, 2004; (10) Pennington et al., 2008; (11) Oneal and Rotenberry, 2008; (12) Kowarik, 2005; (13) Mason et al., 2007; (14) Predick and Turner, 2008; (15) Lavoie et al., 2003; (16) Loewenstein and Loewenstein, 2005; (17) Moffat et al., 2004; (18) Vidra and Shear, 2008; (19) Pyle, 1995; (20) Everson and Boucher, 1998; (21) Ehrenfeld et al., 2003; (22) Ehrenfeld, 2005.

whereas in commercial areas, few large trees were found. These differences were correlated with higher levels of exotic plant invasions in riparian zones bordered by residential land use versus commercial land use, despite the larger width of riparian zones in residential neighborhoods. Vidra and Shear (2008) speculated that the presence of mature canopy trees and intact canopy coverage provides good habitat for berry-dispersing birds. Because there is a predominance of bird-dispersed species within the exotic flora, the adjacent habitat structure evidently promotes more extensive invasion within riparian areas with large canopy trees. Ehrenfeld (2008) and Cutway and Ehrenfeld (2009) also found that exotic species invasions were lower in wetland areas, both riparian and nonriparian, bordered by industrial land use than by residential land use. They associated the difference in invasion with the structure of the wetland edge: in residential areas, the boundaries of the riparian areas are open, and grade into backyards and lawns, whereas in industrial regions, there is a dense wall of shrubs and vines that separates the riparian zone from any adjacent upland habitat. This difference in edge structure may affect seed dispersal into the wetland, and indeed is correlated with differences in the presence of exotic seeds in the soil seed bank (Cutway, 2004). Because riparian zones have high perimeter/area ratios due to their occurrence as long, narrow bands, edge structure may be a component of vegetation structure that deserves further research (Devlaeminck et al., 2005; Hamberg et al., 2009).

Exotic invasion into urban riparian zones is associated with several aspects of the physical environment (Table 6–1). While geomorphic and geological setting structure riparian communities in general (Brinson, 1993; Brinson et al., 1995; Hauer and Smith, 1998), hydrogeomorphic conditions appear to be less well correlated with wetland communities in urban than nonurban environments (Ehrenfeld et al., 2003; Ehrenfeld, 2005). Within the constraints set by physiography and geomorphology, however, the main effects of urban land use are on the patterns of flooding and on soil characteristics. Urban streams have altered hydrologic regimes, with increased frequency of floods but shorter durations and the down-cutting or entrenchment of stream channels, resulting in lower frequencies of floodplain inundation, shorter hydroperiods for periods of flooding, and less interaction of surface waters with riparian soils. These drier conditions lead to greater degrees of invasion (Pyle, 1995; Ehrenfeld et al., 2003; Predick and Turner, 2008) because the large majority of exotic species are not adapted to wetland conditions. However, in some circumstances, stream entrenchment can lead to steep banks that restrict the opportunity for water-mediated dispersal of exotic plants and therefore results in less colonization of riparian areas (Alston and Richardson, 2006). Soils in urban riparian areas tend to have sandier textures than nonurban riparian areas (Moffatt et al., 2004; Predick and Turner, 2008). Ehrenfeld et al. (2003) also found that urban riparian sites tended to have sandier soils, with less organic matter, higher matrix chroma, and a lower frequency of redoximorphic features than other types of wetlands in urban areas, and these differences paralleled a lower percentage of obligate wetland plants than the other wetland types. Groffman et al. (2003) similarly reported that urban riparian areas have a lower frequency of wetland species, and conversely a higher frequency of upland species, than comparable riparian areas in forested landscapes. Another factor that has been associated with exotic invasion in urban riparian areas is nutrient enrichment (Wolin and MacKeigan, 2005; Aguiar et al., 2007). Both nitrogen (Stohlgren et al., 1998), in a study of nonurban riparian areas, and phosphorus (Vidra et al., 2006) have been implicated as causes of nutrient enrichment in riparian soils, and a possible explanation for the high degree of exotic species invasion in these communities.

Exotic invasion in urban riparian zones is also widely associated with large amounts of disturbance from both natural (flooding, natural tree mortality) and anthropogenic (trampling, foot and bicycle paths) sources (Moffatt and McLachlan, 2004; Moffatt et al., 2004; Burton et al., 2005; Aguiar et al., 2007; Burton and Samuelson, 2008; Oneal and Rotenberry, 2008). Indeed, Ehrenfeld et al. (2003) found that urban riparian wetlands were more heavily invaded than nonriparian urban wetlands, presumably reflecting both less continuous flooding and more physical disturbance from flowing waters during overbank flow events. Although most studies of urban riparian communities associated invasion with high levels of physical disturbance (Table 6–1), there have been few studies that separately examine the various types of disturbance as causes of exotic invasion. For example, the presence of walking trails, while a prominent feature of many urban riparian sites, is not always clearly associated with exotic invasion (Cutway, 2004; Ehrenfeld, 2008). There are also few data that clearly quantify the difference in flooding-generated disturbance between urban and nonurban riparian areas. Since flood-generated disturbance is a key feature of all riparian systems (Gregory et al., 1991; Naiman and Decamps, 1997; Naiman et al., 2005), some level of physical disturbance is to be expected. However, the extent to which disturbance is increased by direct human activities (e.g., trampling, trail usage, dumping of waste) vs. indirect human impacts (e.g., changing hydrology and changes to stream morphology and flooding regime) has not been clearly addressed. Moreover, the relationships between different types of disturbance and exotic invasion are also not clear. While disturbances that result in higher light availability through destruction of canopy or subcanopy trees and/or higher nutrient availability through the deposition of nutrient-rich sediments are likely to be strongly associated with exotic invasions, disturbances that do not directly affect resource availability (e.g., the presence of walking paths, the deposition of trash) may not necessarily enhance exotic invasion. Moreover, there are no data available to examine the association of disturbance regimes with differing types of surrounding urban land use (e.g., residential vs. industrial vs. commercial). Finally, it is likely that associations among modes of disturbance in riparian ecosystems, urban land use, and exotic invasion will vary greatly with stream order, geomorphic setting, and the socioeconomic descriptors of the urban region.

In summary, urban riparian communities are often, but not always, more extensively invaded by exotic plants than either nonurban riparian communities or nonriparian wetlands within urban landscapes. The relatively high susceptibility to invasion derives from both the factors intrinsic to riparian systems and the augmentation of these factors by the urban environment (Fig. 6–1). All riparian communities experience high rates of disturbance as a function of flowing water, soils often have high nutrient availability as a function of the fresh deposition of sediments (Chapter 13, Stander and Ehrenfeld, 2010, this volume), and disturbances open both understory and overstory canopies to allow high light availability in the understory. Urban land use within a watershed acts to exacerbate the intensity and frequency of flood-based disturbances, introduces novel disturbances from direct human activity within the community, and introduces a large novel source of exotic plant propagules from horticulture and from the presence of weedy upland areas. However, there remains a high degree of variability in the presence and species composition of exotic species within riparian sites. This variability may be explained, in part, by strongly individualistic responses of exotic species to environmental factors, so that different groups of species respond to environmental factors, and the effects of urban land use on these factors, in different ways (Ehrenfeld, 2008; Oneal and Rotenberry, 2008). Thus, exotic invasion does not increase with urban disturbance and human population size in a simple, linear way. Rather, as McDonnell and Hahs (2008) have demonstrated, urban land use creates a complex gradient that affects different

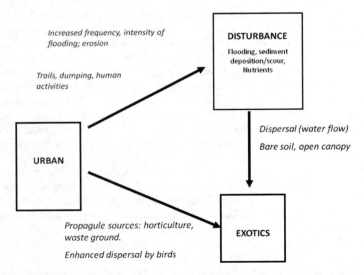

Fig. 6–1. Relationships between urbanization and natural riparian characteristics that explain the high degree of exotic species invasion in urban riparian communities.

species and different species functional groups in varying ways depending on the local context.

There have been fewer studies of vegetation structure within riparian communities, and, as with the data on exotic species, the available literature presents conflicting evidence. These studies often find that urban riparian areas tend to have more large trees, and fewer small-diameter trees or understory plants than nonurban areas. This may reflect human management to produce a park-like environment, with wide and open vistas (Burton et al., 2005; Alston and Richardson, 2006; Burton and Samuelson, 2008; Pennington et al., 2008). There may also be more dead trees in urban riparian zones than in nonurban sites (Pennington et al., 2008). However, other studies (e.g., Ehrenfeld, 2005) have found that metrics of vegetation structure, including mean tree diameter, basal area, tree and shrub densities, and native species diversity, are comparable to studies of nonurban wetlands of the same general types.

The size of patches of urban riparian vegetation is also frequently found to be smaller than nonurban patches, leading to a greater degree of isolation among the stands (e.g., Pyle, 1995; Ehrenfeld, 2000; Moffatt et al., 2004; Fullerton et al., 2006; Predick and Turner, 2008). Fragmentation is thus perhaps the most widely established characteristic of urban riparian systems.

Ecosystem Functions of Plant Communities

Although the vegetation of urban riparian wetlands has been examined from the point of view of its suitability as habitat for birds and other animals, and its effects on streams and stream biota, there has been virtually no examination of urban riparian vegetation with respect to biogeochemical functions (Chapter 13, Stander and Ehrenfeld, 2010, this volume) or other ecosystem functions. The role of vegetation in urban regions in mediating temperature, noise, air pollution, and as areas that promote infiltration of precipitation rather than direct runoff has been widely reviewed (e.g., Bolund and Hunhammar, 1999; Pickett et al., 2001; Kowarik, 2005). Similarly, wetlands have been discussed in general terms with respect to these functions, and also their ability to carry out the important function of maintaining water quality or removing pollutants, and their aesthetic valuation by urban residents (Doss and Taff, 1996; Brander et al., 2006). Although there is evidence that urban residents place value on riparian zones for their ability to improve water quality and provide wildlife habitat (Wagner, 2008), there have been no published studies that we are aware of that specifically evaluate urban riparian and floodplain ecosystems for their contributions to these functions, or for their contribution to functions of emerging importance, such as carbon storage, separate from the rest of the urban plant community, such as street trees and upland parks. Carbon storage and sequestration is currently a prominent ecosystem service that is becoming increasingly important in wetland management. While wetlands in general account for as much as 33% of the soil carbon pool (Bridgham et al., 2006), most of this carbon stock, and the associated sequestration of carbon, is associated with wetlands with organic soils, rather than the mineral soils typical of riparian systems. While riparian plant communities often support large trees, this carbon pool, however, has not been examined for its role in carbon storage or as a potential source of continuing carbon sequestration. This represents an important knowledge gap and a potential area for significant research advances.

Animal Communities
of Riparian Areas and Associated Streams
Invertebrates, Fish, and Herpetofauna

The primary functions for which urban riparian communities have been evaluated are as habitat for wildlife and as corridors for animal dispersal within highly developed areas, as well as modifiers of the stream environment (Chapter 13, Stander and Ehrenfeld, 2010, this volume). Several studies have compared amphibian and reptile communities in urban vs. nonurban riparian zones. Barrett and Guyer (2008) studied a series of riparian areas along second- and third-order streams in southwestern Georgia, and found that amphibians and reptiles responded quite differently to the effects of urban land use. Amphibian species richness declined; indeed only one species of salamander (*Eurycea cirrigera*) was found in the urban sites. Similar decreases in salamander diversity and abundance were documented for a set of sites in North Carolina (Price et al., 2006) and throughout the upper Mississippi basin (Knutson and Klaas, 1998). In contrast, in the Georgia study, reptile diversity increased in urban sites. Barrett and Guyer (2008) related these trends to changes in riparian plant community structure and changes in hydrology. Closed-forest communities, with ample wet areas and pools on the forest floor supported amphibian populations but provided little habitat for most snakes and turtles in nonurban sites. In contrast, in the urbanized sites, open canopies from sparse woody vegetation and less flooding fostered dry forest floors and warmer streams, which allowed several species of turtles and snakes associated with larger, open, warm streams to thrive, but eliminated the amphibia. Pillsbury and Miller (2008) identified the loss of connectivity among habitat patches, and the presence of adjacent suitable upland habitat, as primary factors affecting anurans in a study of urban nonriparian wetlands, and this factor was also identified by Hamer and McDonnell (2008) in their review of the responses of amphibians to urbanization in general. In contrast, both Willson and Dorcas (2003) and Miller et al. (2007) found that stream-breeding salamanders responded most strongly to watershed-scale land characteristics and were not strongly affected by the quality of the riparian communities. These few studies suggest that the structure of the plant community within riparian zones may affect some species of amphibian and reptiles, but the small size (i.e., fragmentation) of riparian communities, the lack of contiguous upland habitat, and other effects operating at the watershed scale may be more important to amphibian communities than the riparian vegetation itself.

Studies of stream organisms have similarly shown that while the structure of riparian vegetation affects the quantity and composition of stream communities, riparian buffers along streams do not completely mitigate the effects of urbanization within whole watersheds. For example, in a study of streams in Georgia (USA), Roy et al. (2005, 2007) found that although the communities of fish in streams with forested or grassy buffers were very different in composition, both types of communities could have high diversity and integrity, and the land use of the entire watershed had a larger effect on the community than did the vegetation composition of the riparian buffers. Fullerton et al. (2006) examined riparian area characteristics in the Pacific Northwest with respect to the suitability of streams as salmonid habitat. For these systems, riparian zones function to protect bank stability, prevent excess sedimentation and nutrient loading, moderate water temperatures, and supply coarse woody debris to the streams, an essential component of fish habitat (Broadmeadow and Nisbet, 2004). Within the Columbia River basin, Fullerton et al. (2006) found that urbanization (and agricultural use) removed about 50% of the riparian habitat on floodplain reaches. They concluded that riparian community patch size (width and length along the stream channel) was the critical parameter that determined whether the riparian community protected salmonid fish in a given reach. In a detailed comparison of the role of riparian vegetation with the effects of watershed urbanization in southeastern Pennsylvania, Horwitz et al. (2008) found that both the type of vegetation (forested vs. nonforested) and the nature of the watershed (highly urbanized vs. not urbanized) affected species composition and abundance of individual fish species, often in complex ways. While it was clear that in forested reaches, streams are narrower, more shaded, and have more riffle habitat, and that urban fish communities, regardless of riparian vegetation, are dominated by generalist species tolerant of a wide range of water conditions, there were variable vegetation × urbanization interactions for individual fish species. Horwitz et al. (2008) concluded that there was no simple relationship between urbanization and riparian cover—riparian vegetation does not provide much protection against the effects of watershed urbanization, but urbanization does not completely override the effects of differing riparian vegetation.

Studies of the effect of riparian plant communities on invertebrates in urban streams have similarly found that both the adjacent riparian community and the land use characteristics of the upgradient watershed affect these organisms. In a study of stream invertebrates in Australia, Walsh et al. (2005a) found that although urban reaches had a greater degree of canopy cover for the streams (because of the conversion of open agricultural land to forested or wooded stream corridors), the stream invertebrate community was not protected by these buffers, but rather reflected the overall percentage of impervious surface within the watershed. Carroll and Jackson (2008) similarly compared urbanization and riparian cover with respect to stream invertebrates and their leaf litter food source. They found that while riparian vegetation did control the input of litter to streams as expected, litter retention, and therefore the effects on the invertebrate fauna, appeared to be more strongly affected by watershed-level conditions, including litter inputs from stormwater inputs. While the richness of shredder invertebrate species was lower in urbanized reaches, their abundance was not affected.

Riparian Bird Communities

Birds have attracted the most attention in terms of the function of riparian communities as an animal habitat, in part because riparian zones support a high diversity of breeding species (Lussier et al., 2006). Their prominence has led to extensive studies of the effects of urbanization (Chace and Walsh, 2006). However, relatively few of these studies target urban riparian environments specifically. As in many other studies of bird communities, without respect to landscape location or watershed urbanization, the vertical structure of the plant community, and the connectivity to other habitat types, emerge as primary variables affecting community composition and bird abundance. Thus, the effects of urbanization on riparian vegetation structure, and the effects of fragmentation and loss of adjacent upland habitat are the primary factors affecting bird communities, and these effects vary regionally (Chapter 4, Shochat et al., 2010, this volume). For example, in California, and the arid American Southwest in general, riparian habitat is the most productive and diverse community type, at least for birds (Luther et al., 2008). In a study in southern California, Luther et al. (2008) found that both the structure of the riparian vegetation (the amount of tree cover and the richness of shrub species) and the amount of urbanization in the area around each observation point were needed to explain avian abundance, whereas the diversity of birds in general and the diversity of riparian-specific birds was best explained by the structure of the riparian vegetation (shrub species richness and percentage tree cover). They also found that some species, including Allen's hummingbird (*Selasphorus sasin*) and the cedar waxwing (*Bombycilla cedrorum*) were only found in urban sites and sites within vineyards. In a similar study in Cincinnati, OH within a forested biome, Pennington et al. (2008) also found that both the structure of riparian forests, measured as the abundance and height of native tree species and the amount of tree cover, the surrounding matrix (the number of buildings within 250 m of the stream), and the width of the riparian zone were significant factors explaining bird community structure. Pennington et al. (2008) noted, however, that different guilds of birds responded to different components of the riparian environment (width, species composition, surrounding matrix) and that different sets of factors could affect bird density, richness, and evenness within each guild of birds. Nevertheless, they concluded that landscape context, riparian area width, and vegetation structure within the riparian zone were all important variables determining bird communities. Similar patterns have been demonstrated in a variety of other studies (Germaine et al., 1998; Rodewald and Bakermans, 2006; Mason et al., 2007; Palmer et al., 2008).

Urbanization, of course, has direct effects on those birds that do inhabit riparian communities. Human structures, such as lawns, buildings, and paths, within riparian corridors result in loss of disturbance-intolerant groups of bird species and increases in nonnative birds and disturbance-tolerant species (Chapter 4, Shochat et al., 2010, this volume; Mason et al., 2007). Paradoxically, in some cases adjacent land use can augment population size. Warmer winter temperatures (i.e., heat island effect) and the presence of bird feeder–based food subsidies have been implicated as causative factors in this effect (Atchison and Rodewald, 2006; Crosbie et al., 2006; Leston and Rodewald, 2006). However, not surprisingly, this is not the case for all birds. Rodewald and Shustack (2008a) showed that Acadian flycatchers (*Empidonax virescens*) experienced lower reproductive productivity, reflecting a later breeding season, more nest turnover, and fewer breeding attempts in urban than in nonurban riparian forests. In a related study, Rodewald and Shustack (2008b) found that American cardinals (*Cardinalis cardinalis*) had equivalent survivorship and productivity in urban and nonurban riparian forests, suggesting that higher densities of birds in the urban sites simply reflected higher resource availability. Also, while urban riparian communities may provide poor habitat for nesting species, they are important as stop-over sites during migrations (Pennington et al., 2008). Thus, urban riparian corridors may have neutral, positive, or negative effects on the population biology of individual bird species, a result of differences in the specific life-history requirements of individual species.

As with fish, amphibians, and stream invertebrates, the surrounding urban matrix also affects bird communities, separately from the structure of the riparian vegetation. Sundell-Turner and Rodewald (2008), for example, found that an index of environmental quality based on the composition of the breeding bird community was most strongly related to the amount of forested land within a 1-km radius of a site, and secondarily to the area of roads within that radius, when examining a sample of urban wetlands in Ohio. Lussier et al. (2006) similarly found landscape structure more important than riparian vegetation characteristics for birds in a set of sites in Rhode Island. However, they observed that the dominance of bird species tolerant of urban disturbance in sites within predominantly residential neighborhoods was strongly correlated with a high percentage of exotic plant species. Lussier et al. (2006) further suggested that exotic species provide poor food resources and poor habitat quality for nesting for most intolerant birds; these factors, combined with the more structurally simple forests in the highly urban sites, explain the lack of most guilds of birds aside from the tolerant species. These patterns have been repeated elsewhere, in Portland, OR (Hennings and Edge, 2003) and for riparian sites in Colorado (Miller et al., 2003). While the specific factors identified in studies comparing local and landscape factors affecting riparian bird communities vary with the geographic region in which the studies are conducted, they uniformly find two patterns: first, the nature of adjacent urban land use is at least as important, if not more important, than the composition of the riparian plant community, and second, the increasing fraction of exotic species in the flora is strongly positively correlated with nonnative bird species and the guild of "tolerant" native species.

Urban Riparian Systems as Corridors for Species Movements

Finally, riparian systems function as corridors of natural habitat within urban areas, which may be important for both plant and animal dispersal and movement across an otherwise inhospitable landscape. Indeed, the longitudinal connectivity either along intact riverine systems, or between remnant patches of riparian vegetation along disturbed rivers is a widely recognized function of riparian systems (Nilsson and Berggren, 2000; Nilsson and Svedmark, 2002). While many studies have addressed the characteristics, such as width, vegetation structure, and adjacent land use, that affect site occupancy by birds, other fauna, and exotic species (e.g., Mason et al., 2007), there have been fewer that have directly measured the function of riparian corridors in urban areas for dispersal and migration of species. The value of riparian corridors for migrating birds was demonstrated by Pennington et al. (2008), who showed, for riparian corridors in Cincinnati, OH, that migrating birds are more tolerant of habitat disturbance (e.g., narrow width, adjacent development, presence of exotic plant species) than nesting birds, and that therefore riparian corridors within urban areas are important habitat for them. However, Rodewald and Matthews (2005) found that spring migrants moving through Columbus, OH preferentially used upland forests, in comparison with riparian areas, reflecting a more complex shrub and understory structure in the upland forests.

Nevertheless, in recognition of the importance of habitat corridors to maintain connectivity among patches of natural vegetation, riparian corridors have been targeted in urban planning efforts. For example, a study of Shellharbor Local Government Area of New South Wales, Australia used criteria of patch size, vegetative cover, structural diversity, and plant species composition, in addition to land use and land ownership criteria, to design a network of elements across the highly urbanized city center that would promote connectivity between two nature reserves on either side of the urban area (Parker et al., 2008). Networks that relied heavily on riparian habitat were found to be optimal in terms of estimated continuity of habitat and quality of habitat for a variety of organisms. In Vienna, a series of habitats were constructed along the Danube River in conjunction with a river restoration effort in an effort to improve connectivity along the river (Chovanec et al., 2002). Analysis of fish, amphibian, and dragonfly communities, in comparison with upstream reference areas, showed that these "stepping stone" habitats did allow a variety of species to move downstream across a highly disturbed section of the river and to establish a new set of populations along the restored river section. However, urban riparian systems also can be important as corridors permitting the dispersal of exotic species (Thébaud and Debussche, 1991).

Notice

The U.S. Environmental Protection Agency, through its Office of Research and Development, funded and managed, or partially funded and collaborated in, the research described herein. It has been subjected to the Agency's peer and administrative review and has been approved for external publication. Any opinions expressed in this paper are those of the author(s) and do not necessarily reflect the views of the Agency; therefore, no official endorsement should be inferred. Any mention of trade names or commercial products does not constitute endorsement or recommendation for use.

References

Aguiar, F.C., M.T. Ferreira, A. Albuquerque, and I. Moreira. 2007. Alien and endemic flora at reference and non-reference sites in Mediterranean-type streams in Portugal. Aquat. Conserv.: Mar. Freshwat. Ecosyst. 17:335–347.

Alberti, M. 2005. The effects of urban patterns on ecosystem function. Int. Reg. Sci. Rev. 28:168–192.

Alston, K.P., and D.M. Richardson. 2006. The roles of habitat features, disturbance, and distance from putative source populations in structuring alien plant invasions at the urban/wildland interface on the Cape Peninsula, South Africa. Biol. Conserv. 132:183–198.

Aronson, M.F.J., C.A. Hatfield, and J.M. Hartman. 2004. Plant community patterns of low-gradient forested floodplains in a New Jersey urban landscape. J. Torrey Bot. Soc. 131:232–242.

Atchison, K.A., and A.D. Rodewald. 2006. The value of urban forests to wintering birds. Nat. Areas J. 26:280–288.

Barrett, K., and C. Guyer. 2008. Differential responses of amphibians and reptiles in riparian and stream habitats to land use disturbances in western Georgia, USA. Biol. Conserv. 141:2290–2300.

Bolund, P., and S. Hunhammar. 1999. Ecosystem services in urban areas. Ecol. Econ. 29:293–301.

Brander, L.M., R.J.G.M. Florax, and J.E. Vermaat. 2006. The empirics of wetland valuation: A comprehensive summary and a meta-analysis of the literature. Environ. Resour. Econ. 33:223–250.

Bridgham, S.D., J.P. Megonigal, J.K. Keller, N.B. Bliss, and C. Trettin. 2006. The carbon balance of North American wetlands. Wetlands 26:889–916.

Brinson, M.M. 1993. A hydrogeomorphic classification for wetlands. Tech. Rep. WRP-DE-4. U.S. Army Engineer Waterways Exp. Stn., Vicksburg, MS.

Brinson, M.M., R.D. Rheinhardt, R.F. Hauer, L.C. Lee, W.L. Nutter, R.D. Smith, and D.F. Whigham. 1995. A guidebook for application of hydrogeomorphic assessments to riverine wetlands. Wetlands Research Program Tech. Rep. WRP-DE-11. U.S. Army Corps of Engineers, Washington, DC.

Broadmeadow, S., and T.R. Nisbet. 2004. The effects of riparian forest management on the freshwater environment: A literature review of best management practice. Hydrol. Earth Syst. Sci. 8:286–305.

Brown, R.L., and R.K. Peet. 2003. Diversity and invasibility of southern Appalachian plant communities. Ecology 84:32–39.

Burton, M.L., and L.J. Samuelson. 2008. Influence of urbanization on riparian forest diversity and structure in the Georgia Piedmont, US. Plant Ecol. 195:99–115.

Burton, M.L., L.J. Samuelson, and S. Pan. 2005. Riparian woody plant diversity and forest structure along an urban–rural gradient. Urban Ecosyst. 8:93–106.

Carroll, G.D., and C.R. Jackson. 2008. Observed relationships between urbanization and riparian cover, shredder abundance, and stream leaf litter standing crops. Fund. Appl. Limnol. 173:213–225.

Chace, J.F., and J.J. Walsh. 2006. Urban effects on native avifauna: A review. Landsc. Urban Plan. 74:46–69.

Charron, I., O. Lalonde, A.G. Roy, C. Boyer, and S. Turgeon. 2008. Changes in riparian habitats along five major tributaries of the Saint Lawrence River, Quebec, Canada: 1964–1997. River Res. Appl. 24:617–631.

Chovanec, A., F. Schiemer, H. Waidbacher, and R. Spolwind. 2002. Rehabilitation of a heavily modified river section of the Danube in Vienna (Austria): Biological assessment of landscape linkages on different scales. Int. Rev. Hydrobiol. 87:183–195.

Crosbie, S.P., D.A. Bell, and G.M. Bolen. 2006. Vegetative and thermal aspects of roost-site selection in urban Yellow-Billed Magpies. Wilson J. Ornithol. 118:532–536.

Cutway, H.B. 2004. The effects of urban land use and human disturbance on forested wetland invasibility. Ph.D. diss. Graduate Program in Ecology and Evolution, Rutgers Univ., New Brunswick, NJ.

Cutway, H.B., and J.G. Ehrenfeld. 2009. Exotic plant invasions in forested wetlands: Effects of adjacent urban land use type. Urban Ecosyst. doi:10.1007/s11252-009-0088-9.

D'Antonio, C.M., T.L. Dudley, and M.C. Mack. 1999. Disturbance and biological invasions: Direct effects and feedbacks. p. 413–452. In L.R. Walker (ed.) Ecosystems of disturbed ground. Elsevier, Amsterdam.

Devlaeminck, R., B. Bossuyt, and M. Hermy. 2005. Inflow of seeds through the forest edge: Evidence from seed bank and vegetation patterns. Plant Ecol. 176:1–17.

Doss, C.R., and S.J. Taff. 1996. The Influence of wetland type and wetland proximity on residential property values. J. Agric. Resour. Econ. 21:120–129.

Dwire, K.A., and R.R. Lowrance. 2006. Riparian ecosystems and buffers—Multiscale structure, function, and management: Introduction. J. Am. Water Resour. Assoc. 42:1–4.

Ehrenfeld, J.G. 2000. Evaluating wetlands within an urban context. Urban Ecosyst. 4:69–85.

Ehrenfeld, J.G. 2005. Vegetation of forested wetlands of urban and suburban landscapes in New Jersey. J. Torrey Bot. Soc. 132:262–279.

Ehrenfeld, J. 2008. Exotic invasive species in urban wetlands: Environmental correlates and implications for wetland management. J. Appl. Ecol. 45:1160–1169.

Ehrenfeld, J.G., H.B. Cutway, R. Hamilton IV, and E. Stander. 2003. Hydrologic description of forested wetlands in northeastern New Jersey, USA—An urban/suburban region. Wetlands 23:685–700.

Everson, D.A., and D.H. Boucher. 1998. Tree species-richness and topographic complexity along the riparian edge of the Potomac River. For. Ecol. Manage. 109:305–314.

Fullerton, A.H., T.J. Beechie, S.E. Baker, J.E. Hall, and K.A. Barnas. 2006. Regional patterns of riparian characteristics in the interior Columbia River basin, Northwestern USA: Applications for restoration planning. Landscape Ecol. 21:1347–1360.

Germaine, S.S., S.S. Rosenstock, R.E. Schweinsburg, and W.S. Richardson. 1998. Relationships among breeding birds, habitat, and residential development in Greater Tucson, Arizona. Ecol. Appl. 8:680–691.

Gregory, S.V., F.J. Swanson, W.A. McKee, and K.W. Cummins. 1991. An ecosystem perspective of riparian zones. Bioscience 41:540–551.

Groffman, P.M., D.J. Bain, L.E. Band, K.T. Belt, G.S. Brush, J.M. Grove, R.V. Pouyat, I.C. Yesilonis, and W.C. Zipperer. 2003. Down by the riverside: Urban riparian ecology. Front. Ecol. Environ. 1:315–321.

Gurnell, A., K. Thompson, J. Goodson, and H. Moggridge. 2008. Propagule deposition along river margins: Linking hydrology and ecology. J. Ecol. 96:553–565.

Hamberg, L., S. Lehvävirta, and D.J. Kotze. 2009. Forest edge structure as a shaping factor of understorey vegetation in urban forests in Finland. For. Ecol. Manage. 257:712–722.

Hamer, A.J., and M.J. McDonnell. 2008. Amphibian ecology and conservation in the urbanising world: A review. Biol. Conserv. 141:2432–2449.

Hauer, F.R., and R.D. Smith. 1998. The hydrogeomorphic approach to functional assessment of riparian wetlands: Evaluating impacts and mitigation on river floodplains in the U.S.A. Freshwater Biol. 40:517–530.

Hennings, L.A., and W.D. Edge. 2003. Riparian bird community structure in Portland, Oregon: Habitat, urbanization, and spatial scale patterns. Condor 105:288–302.

Hession, W.C., T.E. Johnson, D.F. Charles, D.D. Hart, R.J. Horwitz, D.A. Kreeger, J.E. Pizzuto, D.J. Velinsky, J.D. Newbold, C. Cianfrani, T. Clason, A.M. Compton, N. Coulter, L. Fuselier, B.D. Marshall, and J. Reed. 2000. Ecological benefits of riparian reforestation in urban watersheds: Study design and preliminary results. Environ. Monit. Assess. 63:211–222.

Horwitz, R.J., T.E. Johnson, P.F. Overbeck, T.K. O'Donnell, W.C. Hession, and B.W. Sweeney. 2008. Effects of riparian vegetation and watershed urbanization on fishes in streams of the mid-Atlantic piedmont (USA). J. Am. Water Resour. Assoc. 44:724–741.

Knutson, M.G., and E.E. Klaas. 1998. Floodplain forest loss and changes in forest community composition and structure in the Upper Mississippi river: A wildlife habitat at risk. Nat. Areas J. 18:138–150.

Kowarik, I. 1995. On the role of alien species in urban flora and vegetation. p. 85–103. In P. Pysek et al. (ed.) Plant invasions—General aspects and special problems. SPB Academic Publ., Amsterdam, The Netherlands.

Kowarik, I. 2005. Wild urban woodlands: Towards a conceptual framework. p. 1–32. In I. Kowarik and S. Körner (ed.) Wild urban woodlands. Springer-Verlag, Heidleburg, Germany.

Lavoie, C., M. Jean, F. Delisle, and G. Létourneau. 2003. Exotic plant species of the St. Lawrence River wetlands: A spatial and historical analysis. J. Biogeogr. 30:537–549.

Leston, L.F.V., and A.D. Rodewald. 2006. Are urban forests ecological traps for understory birds? An examination using Northern cardinals. Biol. Conserv. 131:566–574.

Loewenstein, N.J., and E.F. Loewenstein. 2005. Non-native plants in the understory of riparian forests across a land-use gradient in the Southeast. Urban Ecosyst. 8:79–91.

Lussier, S.M., R.W. Enser, S.N. Dasilva, and M. Charpentier. 2006. Effects of habitat disturbance from residential development on breeding bird communities in riparian corridors. Environ. Manage. 38:504–521.

Luther, D., J. Hilty, J. Weiss, C. Cornwall, M. Wipf, and G. Ballard. 2008. Assessing the impact of local habitat variables and landscape context on riparian birds in agricultural, urbanized, and native landscapes. Biodivers. Conserv. 17:1923–1935.

Mason, J., C. Moorman, G. Hess, and K. Sinclair. 2007. Designing suburban greenways to provide habitat for forest-breeding birds. Landsc. Urban Plan. 80:153–164.

McDonnell, M.J., and A.K. Hahs. 2008. The use of gradient analysis studies in advancing our understanding of the ecology of urbanizing landscapes: Current status and future directions. Landscape Ecol. 23:1143–1155.

McKinney, M.L. 2006. Urbanization as a major cause of biotic homogenization. Biol. Conserv. 127:247–260.

Miller, J.R., J.A. Wiens, N.T. Hobbs, and D.M. Theobald. 2003. Effects of human settlement on bird communities in lowland riparian areas of Colorado (USA). Ecol. Appl. 13:1041–1059.

Miller, J.E., G.R. Hess, and C.E. Moorman. 2007. Southern two-lined salamanders in urbanizing watersheds. Urban Ecosyst. 10:73–85.

Moffatt, S.F., and S.M. McLachlan. 2004. Understorey indicators of disturbance for riparian forests along an urban–rural gradient in Manitoba. Ecol. Indicators 4:1–16.

Moffatt, S.F., S.M. McLachlan, and N.C. Kenkel. 2004. Impacts of land use on riparian forest along an urban-rural gradient in southern Manitoba. Plant Ecol. 174:119–135.

Naiman, R.J., H. Decamps, and M. Pollock. 1993. The role of riparian corridors in maintaining regional biodiversity. Ecol. Appl. 3:209–212.

Naiman, R.J., and H. Decamps. 1997. The ecology of interfaces: Riparian zones. Annu. Rev. Ecol. Syst. 28:621–658.

Naiman, R., H. Décamps, and M.E. McClain. 2005. Riparia. Ecology, conservation and management of streamside communities. Elsevier Academic Press, San Diego, CA.

Nilsson, C., and K. Berggren. 2000. Alterations of riparian ecosystems caused by river regulation. Bioscience 50:783–792.

Nilsson, C., and M. Svedmark. 2002. Basic principles and ecological consequences of changing water regimes: Riparian plant communities. Environ. Manage. 30:468–480.

Oneal, A.S., and J.T. Rotenberry. 2008. Riparian plant composition in an urbanizing landscape in southern California, USA. Landscape Ecol. 23:553–567.

Palmer, G.C., J.A. Fitzsimons, M.J. Antos, and J.G. White. 2008. Determinants of native avian richness in suburban remnant vegetation: Implications for conservation planning. Biol. Conserv. 141:2329–2341.

Parker, K., L. Head, L.A. Chisholm, and N. Feneley. 2008. A conceptual model of ecological connectivity in the Shellharbour Local Government Area, New South Wales, Australia. Landscape Urban Plan. 86:47–59.

Pennington, D.N., J. Hansel, and R.B. Blair. 2008. The conservation value of urban riparian areas for landbirds during spring migration: Land cover, scale, and vegetation effects. Biol. Conserv. 141:1235–1248.

Pickett, S.T.A., M.L. Cadenasso, J.M. Grove, C.H. Nilon, R.V. Pouyat, W.C. Zipperer, and R. Costanza. 2001. Urban ecological systems: Linking terrestrial, ecological, physical and socioeconomic components of metropolitan areas. Annu. Rev. Ecol. Syst. 32:127–157.

Pillsbury, F.C., and J.R. Miller. 2008. Habitat and landscape characteristics underlying anuran community structure along an urban–rural gradient. Ecol. Appl. 18:1107–1118.

Planty-Tabacchi, A.-M., E. Tabacchi, R.J. Naiman, C. Deferrari, and H. Decamps. 1996. Invasibility of species-rich communities in riparian zones. Conserv. Biol. 10:598–607.

Predick, K.I., and M.G. Turner. 2008. Landscape configuration and flood frequency influence invasive shrubs in floodplain forests of the Wisconsin River (USA). J. Ecol. 96:91–102.

Price, S.J., M.E. Dorcas, A.L. Gallant, R.W. Klaver, and J.D. Willson. 2006. Three decades of urbanization: Estimating the impact of land-cover change on stream salamander populations. Biol. Conserv. 133:436–441.

Prieur-Richard, A.H., and S. Lavorel. 2000. Invasions: The perspective of diverse plant communities. Austral Ecol. 25:1–7.

Pyle, L.L. 1995. Effects of disturbance on herbaceous exotic plant-species on the floodplain of the Potomac River. Am. Midl. Nat. 134:244–253.

Pysek, P. 1998. Alien and native species in Central European urban floras: A quantitative comparison. J. Biogeogr. 25:155–163.

Richardson, D.M., P.M. Holmes, K.J. Esler, S.M. Galatowitsch, J.C. Stromberg, S.P. Kirkman, P. Pysek, and R.J. Hobbs. 2007. Riparian vegetation: Degradation, alien plant invasions, and restoration prospects. Divers. Distrib. 13:126–139.

Ringold, P.L., T.K. Magee, and D.V. Peck. 2008. Twelve invasive plant taxa in U.S. western riparian ecosystems. J. North Am. Benthol. Soc. 27:949–966.

Rodewald, A.D., and M.H. Bakermans. 2006. What is the appropriate paradigm for riparian forest conservation? Biol. Conserv. 128:193–200.

Rodewald, P.G., and S.N. Matthews. 2005. Landbird use of riparian and upland forest stopover habitats in an urban landscape. Condor 107:259–268.

Rodewald, A.D., and D.P. Shustack. 2008a. Consumer resource matching in urbanizing landscapes: Are synanthropic species over-matching? Ecology 89:515–521.

Rodewald, A.D., and D.P. Shustack. 2008b. Urban flight: Understanding individual and population-level responses of nearctic–neotropical migratory birds to urbanization. J. Anim. Ecol. 77:83–91.

Roy, A.H., C.L. Faust, M.C. Freeman, and J.L. Meyer. 2005. Reach-scale effects of riparian forest cover on urban stream ecosystems. Can. J. Fish. Aquat. Sci. 62:2312–2329.

Roy, A.H., B.J. Freeman, and M.C. Freeman. 2007. Riparian influences on stream fish assemblage structure in urbanizing streams. Landscape Ecol. 22:385–402.

Schnitzler, A., B.W. Hale, and E.M. Alsum. 2007. Examining native and exotic species diversity in European riparian forests. Biol. Conserv. 138:146–156.

Shochat, E., S. Lerman, and E. Fernández-Juricic. 2010. Birds in urban ecosystems: Population dynamics, community structure, biodiversity, and conservation. p. 75–86. In J. Aitkenhead-Peterson and A. Volder (ed.) Urban ecosystem ecology. Agron. Monogr. 55. ASA, CSSA, and SSSA, Madison, WI.

Stander, E.K., and J.G. Ehrenfeld. 2010. Urban riparian function. p. 253–276. In J. Aitkenhead-Peterson and A. Volder (ed.) Urban ecosystem ecology. Agron. Monogr. 55. ASA, CSSA, and SSSA, Madison, WI.

Stohlgren, T.J., K.A. Bull, Y. Otsuki, C.A. Villa, and M. Lee. 1998. Riparian zones as havens for exotic species in the central grasslands. Plant Ecol. 138:113–125.

Sundell-Turner, N.M., and A.D. Rodewald. 2008. A comparison of landscape metrics for conservation planning. Landscape Urban Plan. 86:219–225.

Tabacchi, E., and A.M. Planty-Tabacchi. 2005. Exotic and native plant community distributions within complex riparian landscapes: A positive correlation. Ecoscience 12:412–423.

Thébaud, C., and M. Debussche. 1991. Rapid invasion of Fraxinus ornus L. along the Herault River system in southern France: The importance of seed dispersal by water. J. Biogeogr. 18:7–12.

Vidra, R.L., and T.H. Shear. 2008. Thinking locally for urban forest restoration: A simple method links exotic species invasion to local landscape structure. Restor. Ecol. 16:217–220.

Vidra, R.L., T.H. Shear, and T.R. Wentworth. 2006. Testing the paradigms of exotic species invasion in urban riparian forests. Nat. Areas J. 26:339–350.

Wagner, M.M. 2008. Acceptance by knowing? The social context of urban riparian buffers as a stormwater best management practice. Soc. Nat. Resour. 21:908–920.

Walsh, C.J., T.D. Fletcher, and A.R. Ladson. 2005a. Stream restoration in urban catchments through redesigning stormwater systems: Looking to the catchment to save the stream. J. North Am. Benthol. Soc. 24:690–705.

Willson, J.D., and M.E. Dorcas. 2003. Effects of habitat disturbance on stream salamanders: Implications for buffer zones and watershed management. Conserv. Biol. 17:763–771.

Wolin, J.A., and P. MacKeigan. 2005. Human influence past and present—Relationship of nutrient and hydrologic conditions to urban wetland macrophyte distribution. Ohio J. Sci. 105:125–132.

7

Chemical, Physical, and Biological Characteristics of Urban Soils

Richard V. Pouyat
Katalin Szlavecz
Ian D. Yesilonis
Peter M. Groffman
Kirsten Schwarz

Abstract

Urban soils provide an array of ecosystem services to inhabitants of cities and towns. Urbanization affects soils and their capacity to provide ecosystem services directly through disturbance and management (e.g., irrigation) and indirectly through changes in the environment (e.g., heat island effect and pollution). Both direct and indirect effects contribute to form a mosaic of soil conditions. In the Baltimore Ecosystem Study (BES), we utilized the urban mosaic as a series of "natural experiments" to investigate and compare the direct and indirect effects of urbanization on soil chemical, physical, and biological properties at neighborhood, citywide, and metropolitan scales. In addition, we compared these results with those obtained from other metropolitan areas to assess the effects at regional and global scales and to assess the generality of these results. Our overall results suggest that surface soils of urban landscapes have properties that can vary widely, making it difficult to define or describe a typical "urban" soil or soil community. Specifically, we conclude that (i) urban effects on soils occur at multiple scales; (ii) management effects are greater than environment effects, although environmental effects are more widespread reaching beyond the boundary of most urban areas; (iii) urban landscapes are biologically active in pervious areas and have a high potential for carbon storage and nitrogen retention; (iv) the importance of urban and native factors depends on the property being measured; and (v) cross-city comparisons support in part the biotic homogenization and urban ecosystem convergence hypothesis.

Soils in urban landscapes are generally thought of as highly disturbed and heterogeneous, with little systematic pattern in their characteristics. As such, most studies have focused on human-constructed soils along streets and in

R.V. Pouyat (rpouyat@fs.fed.us) and I.D. Yesilonis (iyesilonis@fs.fed.us), U.S. Forest Service, North Research Stn., c/o Baltimore Ecosystem Study, 5200 Westland Blvd., Baltimore, MD 21227; K. Szlavecz, Dep. of Earth and Planetary Sciences, The Johns Hopkins Univ., 3400 N. Charles Street, Baltimore, MD 21218 (szlavecz@jhu.edu); P.M. Groffman, Cary Inst. of Ecosystem Studies, Millbrook, NY 12545 (groffmanp@caryinstitute.org); K. Schwarz, Dep. of Ecology, Evolution and Natural Resources, Rutgers Univ., New Brunswick, NJ 08901, currently Cary Inst. of Ecosystem Studies, Box AB, Millbrook, NY 12545-0129 (schwarzk@caryinstitute.org).

doi:10.2134/agronmonogr55.c7

highly impacted areas (e.g., Craul and Klein, 1980; Patterson et al., 1980; Short et al., 1986a; Jim, 1993, 1998). As a result, "urban soils" have been viewed as drastically disturbed and of low fertility (Craul, 1992). However, observations of entire landscapes have shown that the chemical, physical, and biological response of soils to urban land use is complex and variable, such that soils that are largely undisturbed or of high fertility also have been identified in urban areas (e.g., Schleuss et al., 1998; Hope et al., 2005; Pouyat et al., 2007a).

Unfortunately, the response of soil to urban land use is considered by many soil taxonomists to diverge from natural soil formation, and as a consequence, changes in soil characteristics resulting from urban development have received limited attention in the current U.S. soil taxonomy (Fanning and Fanning, 1989; Effland and Pouyat, 1997; Evans et al., 2000). Recent efforts, however, have made progress in developing a taxonomic system for highly disturbed soils (Lehmann and Stahr, 2007; Rossiter, 2007; International Committee on Anthropogenic Soils, 2007), yet the challenge remains to identify a systematic pattern of soil responses to various disturbances, management activities, and environmental changes that typically occur in urban landscapes.

Although the current U.S. soil taxonomy neglects soils altered by urban land use, by definition the taxonomy states that a soil is "... a collection of natural bodies on the earth's surface, *in places modified or even made by man of earthy materials,* containing living matter and supporting or capable of supporting plants out-of-doors" (Soil Survey Staff, 1975), which suggests that soils of urban landscapes should be considered taxonomically with nonurban soils (Effland and Pouyat, 1997; Pouyat and Effland, 1999). Even without a taxonomic designation, definitions of soils associated with urban and urbanizing landscapes have been proposed in the literature. For example, Craul (1992) modified the definition of Bockheim (1974) and defined urban soil as "a soil material having a non-agricultural, man-made surface layer more than 50 cm thick that has been produced by mixing, filling, or by contamination of land surface in urban and suburban areas." As another definition, Evans et al. (2000) suggested the term *anthropogenic soil,* which places urban soils in a broader context of humanly altered soils rather than limiting the definition to urban areas alone. Similarly, Pouyat and Effland (1999) and more recently Lehmann and Stahr (2007) more broadly defined urban soils to include not only those soils that are physically disturbed (e.g., old industrial sites and landfill) but also those that are undisturbed and altered by urban environmental change (e.g., temperature or moisture regimes).

In this chapter, we report on the broad array of effects of urban land use on the physical, chemical, and biological responses of soil, drawing from our research in the Baltimore Ecosystem Study (BES, http://beslter.org, verified 19 Feb. 2010), one of two urban Long Term Ecological Research (LTER) sites funded by the National Science Foundation, as well as research reported in the literature. We begin by introducing soil as the "brown infrastructure" of human settlements and discussing the ecosystem services provided by soils to the inhabitants of urban and exurban areas. We next provide a conceptual framework to incorporate the wide-ranging spatial and temporal effects of urban land uses on soil formation and describe the "urban soil mosaic" as a template to study urban soils. We conclude by presenting case studies of soil responses measured at various scales in landscapes altered by urban and exurban development. We include responses related to the physical, chemical, and biological (i.e., soil fauna and nutrient cycling) characteristics of soils with the ultimate goal of identifying a systematic pattern of these responses with respect to urban land use.

Importance in Urban Ecosystems

Soils form the foundation for many ecological processes and interactions, such as nutrient cycling, distribution of plants and animals, and ultimately location of human habitation (Brady and Weil, 1999). Although soils in urban and urbanizing landscapes are predominately altered by human activity, they provide many of the same ecosystem services as unaltered soils (Effland and Pouyat, 1997). As such, soils can function in urban landscapes by reducing the bioavailability of pollutants, storing carbon and mineral nutrients, serving as habitat for soil and plant biota, and moderating the hydrologic cycle through absorption, storage, and supply of water (Bullock and Gregory, 1991; De Kimpe and Morel, 2000; Lehmann and Stahr, 2007; Pouyat et al., 2007a,b).

In providing these services, soil plays a unique role as the brown infrastructure of urban ecological systems, much in the same way urban vegetation is thought of as green infrastructure (Pouyat et al., 2007a; Heidt and Neef, 2008). Whereas green infrastructure provides services attributed to vegetation, such as the moderation of energy fluxes by tree canopies (Akbari, 2002; Heidt and Neef, 2008), brown infrastructure provides ecosystem services attributed to soil, such as storm water infiltration and purification, and as a support medium for built structures (De Kimpe and Morel, 2000; Lehmann and Stahr, 2007; Pouyat et al., 2007b).

Habitat and Medium for Animals and Plants

On regional and global scales the conversion of native habitats to urban land uses greatly contributes to local extinction rates of plant and animal species (McKinney, 2002; Williams et al., 2009). Exasperating the effect of habitat loss, urban areas are epicenters of many introductions of aboveground and belowground nonnative species, some of which have become invasive or important pathogens or insect pests (Lilleskov et al., 2008; McKinney, 2008; Chapter 12, Reichard, 2010, this volume). The extinctions of native species and the naturalization of urban-adapted species have led to assemblages of species novel to urban areas (McIntyre et al., 2001; Korsós et al., 2002; Hornung and Szlavecz, 2003; Williams et al., 2009). The net result is a general pattern of nonnative species increasing and native species decreasing from outlying rural areas to urban centers.

Plants

Even with the depression of native species richness, the overall species richness of plants may be greater in urban than in rural habitats (McKinney, 2008). The higher species richness of plants in urban landscapes is due to the preferences of people and the naturalization of introduced species (Nowak, 2000; Williams et al., 2009). For instance, Nowak (2010) measured Shannon–Weiner Diversity Index values ranging from 3.0 to 3.8 across several cities in the eastern United States. These values are higher than the range of values found for eastern deciduous forests (1.9–3.1, Barbour et al., 1980). Likewise, Hope et al. (2003) found in the Phoenix metropolitan area that plant species richness was greater in developed areas than in the surrounding desert. Moreover, within the urban areas, species diversity was positively correlated to household income.

Therefore, it appears that soils of urban landscapes can support a greater number of plant species than the native soils they replaced, albeit with some species requiring supplements of water and nutrients. Even without supplements, urban soils appear to have sufficient resources to support plant growth. For example, observations of surface soils in Baltimore City have shown that chemical and physical characteristics fall within the range of requirements for most plants (Pouyat et al., 2007a). Specifically, only 10% of the sampled locations had bulk density measurements greater than 1.4 Mg m^{-3}, the level above which root growth is curtailed in silt loam soils. Moreover, concentrations of potassium, magnesium, and phosphorus were in most cases sufficient for plant growth, while concentrations of calcium exceeded the recommended ranges for horticultural plants in the region. The authors concluded that the apparent accumulation of Ca in these soils occurred due to the widespread use of concrete and gypsum as construction materials, which eventually degrade and get redistributed in the landscape (e.g., Lovett et al., 2000; Juknys et al., 2007).

Soil Fauna

Soil biota is an important component of the soil ecosystem, actively contributing to soil formation by altering its physicochemical properties. All major invertebrate taxa are represented in the soil, and in most terrestrial ecosystems the highest species diversity is found in the soil. Many soil taxa are poorly known, yet new species have actually been discovered and described in urban landscapes (Csuzdi and Szlavecz, 2002; Foddai et al., 2003; Kim and Byrne, 2006). Still less is known about the natural history and ecology of soil fauna than animals in the aboveground community, which is particularly true of urban ecosystems because these systems have been studied less than nonurban systems.

Most soil organisms are part of the decomposer food web, so their major ecosystem function is processing detritus and mobilizing nutrients (Chapter 18, Aitkenhead-Peterson et al., this volume). The soil food web is extremely complex and is currently an important focus of soil ecological research (Bardgett, 2005). Many soil ecologists consider the soil food web highly redundant (e.g., Andrèn et al., 1995; Laakso and Setälä, 1999) meaning that species can be replaced without major functional consequences. Whether functional redundancy exists or not in urban soil communities has important implications because of the presence in urban landscapes of nonnative species, which often occur more abundantly there than native species (Lilleskov et al., 2010).

The composition and abundance of urban soil fauna are determined by many interacting factors, both natural and anthropogenic, and will vary by taxon. Factors contributing to high species richness include the mosaic of land-use and cover types typically existing in urban landscapes and the likelihood of the introduction and establishment of nonnative species (McIntyre et al., 2001; Smith et al., 2006; Byrne et al., 2008; Lilleskov et al., 2008). Moreover, the occurrence of soil organisms in novel habitats, such as built structures, greenhouses, and green roofs, adds to the species richness of urban landscapes (Korsós et al., 2002; Schrader and Böning, 2006; Jordan and Jones, 2007; Csuzdi et al., 2008).

The proportion of nonnative species in an urban landscape is highly taxon dependent and varies with geographical region. Studies conducted in the

Baltimore metropolitan area showed that all carrion beetle species (*Silphidae*) are native (Wolf and Gibbs, 2004), while all terrestrial isopods (*Oniscidea*) were introduced (Hornung and Szlavecz, 2003). In addition, the proportion of nonnative earthworms in the Baltimore metropolitan area is roughly 50% (Szlavecz et al., 2006), and in the New York City metropolitan area no native earthworms occur (Steinberg et al., 1997). As a result, urban soils have fundamentally different soil faunal communities, with a higher proportion of introduced species when compared with their native soil counterparts (e.g., Spence and Spence, 1988; Pouyat et al., 1994; Bolger et al., 2000; McIntyre et al., 2001; Connor et al., 2002).

The successful adaptation of soil fauna to urban environmental conditions involves many physiological and behavioral traits. Individual organisms can cope with environmental stress through behavioral (e.g., migration, shift in food preferences, seeking more favorable microhabitats) or physiological (e.g., regulating absorption and storage of heavy metals) mechanisms (Ireland, 1976; Alikhan, 2003; Lev et al., 2008). Early studies of soil fauna inhabiting urban landscapes focused on pollution tolerance, particularly to contamination by heavy metals. A wealth of information is available on this topic, and results thus far have revealed a complex relationship between levels of contamination and physiology, behavior, and life history of these organisms (Gish and Christensen, 1973; Beeby, 1978; Ash and Lee, 1980; Pizl and Josens, 1995). For instance, responses by soil invertebrates to elevated levels of metals or other pollutants have varied by taxonomic group (Ireland, 1983; Lee, 1985; Beyer and Cromartie, 1987; Morgan and Morgan, 1993). Moreover, responses by individuals within the same population can vary by age, maturity, season, diet, and genetic differences (Ireland, 1983; Spurgeon and Hopkin, 2000).

In addition to the potential for pollution effects on individual soil organisms, the bioaccumulation of metals or other pollutants in urban soil fauna can subject higher trophic organisms, such as predators, to contaminants. Typically the accumulation of metals is magnified the higher the trophic level (Getz et al., 1977; Hopkin and Martin, 1985). In urban landscapes, the potential for the biomagnification of metals is especially true for predators of earthworms, such as birds, lizards, and mammals (Loumbourdis, 1997; Komarnicki, 2000).

Water Infiltration and Storage

Urban landscapes typically exhibit complex and variable soil drainage patterns and moisture regimes (Chapters 14 [Burian and Pomeroy, 2010] and 15 [Steele et al., 2010] this volume). This complexity is a result of a combination of urban factors that either increase or decrease the content of water in soils. For example, highly impacted urban soils often exhibit hydrophobic soil surfaces, surface crust formation, and high bulk densities that restrict infiltration rates (Craul, 1992). Moreover, under urban environment conditions such as heat stress, soil water is more likely to be depleted through higher rates of evapotranspiration. By contrast, soils in urban areas often are irrigated and have abrupt textural and structural interfaces that can restrict drainage resulting in higher soil water contents (Craul, 1992; Pouyat et al., 2007b). Additionally, urban landscapes often have surface drainage features that concentrate water flows (Tenenbaum et al., 2006; Pouyat et al., 2007b; Chapters 14 [Burian and Pomeroy, 2010] and 15 [Steele et al., 2010] this volume). Further complicating these effects are belowground infrastructures that can alter soil water through pressurized potable water distribution systems, which can leak water into adjacent soils, or through storm and sanitary water systems, which can drain soils of water through fractures in pipes (Band et al., 2004; Wolf et al., 2007).

The net result of these differing effects on soil water resources and ultimately the water cycle is unclear and represents an opportunity for future research (Mohrlok and Schiedek, 2007). Preliminary results in the Baltimore Ecosystem Study have revealed differences in soil moisture regimes between types of vegetation cover. In one study, a comparison between nonirrigated residential lawns and an adjacent remnant forest showed that the lawn soils had higher moisture levels at a 10-cm depth than the forest soil during the growing season. However, there were no differences between the two patch types after leaf drop. Moisture differences in the summer were apparently due to higher transpiration rates of the broad-leaved trees (Pouyat et al., 2007b). At a greater soil depth a more complex relationship was revealed in the long-term monitoring of forest and grass plots in the Baltimore metropolitan area. There were no consistent differences in soil moisture between grass and forest plots, although over the entire period from 2001 to 2005, moisture was significantly ($p < 0.05$) higher in grass than forest plots at a 50-cm depth (Groffman et al., 2009). There was an opposite pattern at a 30-cm depth, with significantly higher moisture in forest plots in 2001, 2002, and 2003.

In addition to the importance of cover, studies conducted at the scale of a watershed have shown impervious surfaces and soil disturbances can disrupt the relationship between topography and soil drainage that typically exist in unaltered landscapes. For example, in the Baltimore metropolitan area, Tenenbaum et al. (2006) compared a suburban watershed with a similar sized forested watershed and showed that the developed watershed lacked the typically strong relationship between topographic position and soil moisture. The authors concluded the poor relationship was primarily due to low infiltration and high runoff rates in the suburban watershed. Consequently, it may not be accurate to infer soil moisture or drainage when using topographic maps of urban landscapes.

In addition to a disconnect between soil moisture and topography, urban landscapes may have intact soils that are disturbed only at the surface and thus exhibit a restricted rate of infiltration. As a result, these soils do not hydrologically function as the same soil described in a nonurban context (Pitt and Lantrip, 2000). The potential of urban factors to restrict infiltration rates even in relatively intact soils is of particular importance because of the marked effect soil infiltration can have on stream flows during storm events (e.g., Holman-Dodds et al., 2003).

Sink for Trace Metals

For urban soils, elevated heavy metal concentrations are almost universally reported, although often with high variances (e.g., Wong et al., 2006: Table 7–1). Most of the heavy metal sources in urban landscapes have been associated with roadside environments (Van Bohemen and Janssen van de Laak, 2003; Zhang, 2006; Yesilonis et al., 2008), interior and exterior paint (Mielke, 1999), stack emissions (Govil et al., 2001; Walsh et al., 2001; Kaminiski and Landsberger, 2000), management inputs (Russell-Anelli et al., 1999), and industrial waste (Schuhmacher et al., 1997). Moreover, as heavy metals are emitted into urban environments, they may accumulate onto built surfaces and in the soil, which will vary depending on the source (Mielke, 1999). Cook and Ni (2007) found that relatively heavy

Table 7–1. Surface soil heavy metal elemental means of data reported in the literature.

Author	City	N	Extraction solution	Cd	Co	Cr	Cu	Mn	Ni	Pb	Zn
							mg kg^{-1}				
Aelion et al. (2008)[†]	South Carolina	60	HNO₃/HCl	–[‡]	–	7	3	86	2	12	–
Ikem et al. (2008)	New Madrid County, Missouri	62	HF, HClO₄, HNO₃	1.6 ± 1.38	10 ± 5.0	25 ± 10.6	18 ± 16.7	298 ± 172	16 ± 4.8	49 ± 39.8	96 ± 117
Kay et al. (2008)	Chicago, Illinois	57	USGS procedures	–	11 ± 3.7	71 ± 49.6	151 ± 373	584 ± 511	36 ± 23.5	395 ± 494	397 ± 411
Lee et al. (2006)	Hong Kong	236	strong acid digestion	0.36 ± 0.16	4 ± 1.6	18 ± 5.92	16 ± 22.6	–	4 ± 2.5	88 ± 62	103 ± 91.3
Odewande and Abimbola (2008)	Ibadan, Nigeria	106	HNO₃/HCl	8.4 ± 19.8	–	–	47 ± 44.1	1098 ± 522	20 ± 14.6	95 ± 127	228 ± 366
Ruiz-Cortes et al. (2005)[†]	Seville, Spain	52	three-step BCR, HNO₃/HCl	2.89	–	38	55	391	21	156	120
Tume et al. (2008)	Talcahuano, Chile	7	HNO₃/HCl	–	–	38 ± 16.7	–	–	23 ± 6.4	35 ± 43.3	333 ± 364
Yesilonis et al. (2008)[1]	Baltimore, Maryland	122	HNO₃/HCl	1.06	15	72	45	472	27	231	141
Zhang (2006)	Galway, Ireland	166	HF, HClO₄, HCl, HNO₃	–	6 ± 2.4	33 ± 16.3	33 ± 25.5	674 ± 780	21 ± 12.4	78 ± 72	99 ± 68.8
Zhao et al. (2007)	Wuxi, China	102	HNO₃/HCl, 3:l	0.14 ± 0.09	–	59 ± 16.2	40 ± 18.2	–	–	47 ± 25.8	113 ± 68.4
Zheng et al. (2008)	Beijing, China	773	HNO₃ and H₂O₂	0.15 ± 0.11	–	36 ± 13.9	24 ± 23.7	–	28 ± 8.7	29 ± 10.3	66 ± 29.8

† No measure of variation given (i.e., standard deviation).

‡ A dash (–) indicates no values were reported for the element.

particles in lead aerosols are deposited on or near roads, while relatively light particles are carried around structures by air currents. Particles of intermediate mass are carried by air currents and have an affinity for structural surfaces.

Soil acts as a sink for heavy metals through sorption, complexation, and precipitation reactions (Yong et al., 1992). These retention mechanisms are regulated by organic matter, pH, cation exchange capacity, and oxides of a soil. Therefore, the heavy metals that reach a soil will vary in their availability to plants, soil fauna, and humans on the basis of how these characteristics spatially vary in the urban landscape. Soils with relatively high amounts of organic matter and oxides and neutral to alkaline pH will generally have both a lowered availability to plants and animals and a lowered leaching rate to groundwater of heavy metals (Pizl and Josens, 1995; Brown et al., 2003). However, when soils are physically disturbed and lose organic matter and base elements, they lose their capacity to bind metals, resulting in an increase in the availability and mobility of metals (Farfel et al., 2005).

Retention and Storage of Carbon and Nitrogen

The most obvious impact on the retention and storage of carbon and nitrogen in urban landscapes is from physical disturbances of soil. Large volumes of soil are typically disturbed during construction, and disturbances continue to occur at finer scales once people inhabit the landscape. Using data collected at a commercial development site in the Baltimore metropolitan area (McGuire, 2004) Pouyat et al. (2007b) estimated that the amount of soil C that was disturbed during a development project of 2600 m^2 in area was roughly 2.7×10^4 kg. What happens to the pool of C disturbed during construction projects has not been reported in the literature to our knowledge. However, it is known that several factors are important to C retention, including the amount of C lost through erosion from the site, the amount of organic rich surface soil that is stockpiled or sold as topsoil, the proportion of the total C pool that is readily oxidized, and the amount of organic C buried during the grading process. These factors should also play a role in the loss of N as either nitrate or nitrous oxide.

The amount of C stored in soil over time, or C sequestered, is a balance between C input through net primary productivity (NPP) and loss through decay (soil heterotrophic respiration), both of which are controlled by environmental factors, including soil temperature and moisture and N availability. Carbon sequestration in urban soils is an important process that helps to mitigate the effects of increased emissions of greenhouse gases into the atmosphere. However, the gain or loss of C from soil can be greatly affected by urban land use and urban environmental change (Pouyat et al., 2002; Lorenz and Lal, 2009). For example, measurements taken in permanent forest and lawn plots of the Baltimore Ecosystem Study have shown that carbon dioxide fluxes from forest soils (a C loss) are increased under urban environmental conditions (Groffman et al., 2006), and the fluxes from managed lawns were as high or higher than the forested sites (Groffman et al., 2009).

Other soil–atmosphere exchanges of greenhouse gases, especially nitrous oxide and methane, are potentially altered by urban land use. Trace gas measurements taken in the Baltimore permanent plots indicate that urban forest and lawn soils have a reduced rate of methane uptake and increased nitrous oxide fluxes in comparison to rural forest soils (Groffman and Pouyat, 2009; Groffman et al., 2009). Likewise, Kaye et al. (2004) found that lawns in Colorado had reduced methane uptake and increased nitrous oxide fluxes relative to native shortgrass steppe. Goldman et al. (1995) found reduced methane uptake in urban versus rural forest patches in the New York City metropolitan area. In all these cases, changes in methane and nitrous oxide fluxes were associated with N inputs such as fertilization and N deposition that typically occur in urban landscapes (Groffman and Pouyat, 2009).

While the potential for losses of C and N in urban landscapes can be high, urban soils have the capacity to accumulate a surprising amount of C and N when compared with agricultural or native soils. Horticultural management efforts (e.g., fertilization and irrigation) tend to maximize plant productivity and soil organic matter accumulation for a given climate or soil type and thus increase the capacity of these soils to store C and N. This is particularly true of lawns, where soils are not regularly cultivated and turfgrass species typically grow through an extended growing season relative to most native grassland, forest, and crop ecosystems (Pouyat et al., 2002; Groffman et al., 2009). Indeed, lawn soils have shown a surprisingly high capacity to sequester C and cycle N (Pouyat et al., 2006; Raciti et al., 2008), although the net effect on C uptake may be somewhat lower if C emissions resulting from management activities are taken into account (Gordon et al., 1996; Pouyat et al., 2009b).

Due to the high amount of N inputs into urban landscapes, there is great interest in reducing exports of nitrate from urban soils to coastal receiving waters that are often N limited. There is particular interest in the ability of urban riparian soils to support denitrification, an anaerobic microbial process that converts nitrate into nitrous oxide and other N gases, thus serving as a sink for N leaching from upland soils. Urban riparian zones tend to be drier and more aerobic than riparian zones in agricultural or forested watersheds and therefore support less denitrification (Groffman et al., 2002, 2003; Chapter 13, Stander and Ehrenfeld, 2010, this volume). However, urban landscapes have areas of saturated soils in novel habitats such as stormwater detention basins and relict wetlands that have been shown to support high denitrification rates (Groffman and Crawford, 2003; Stander and Ehrenfeld, 2009).

Conceptual Framework for Urban Soil Formation

Urban Soil Formation

To increase our understanding of how urban land use alters soil formation and ultimately the characteristics of soils in urban landscapes, Pouyat (1991) suggested the use of the Factor Approach, a conceptual model first proposed by Jenny (1941) to describe the formation of soil at landscape scales. The Factor Approach posits that soil and ecosystem development is determined by a combination of state factors that include climate (cl), organisms (o), parent material (pm), relief (r), and time (t), where the characteristics of any given soil (or ecosystem), S, are the function:

$$S = f(cl, o, pm, r, t) \qquad [1]$$

Amundson and Jenny (1991) and Pouyat (1991) proposed that human effects can be incorporated into the factor approach by including a sixth or anthropogenic factor *a*, such that

$$S = f(a, cl, o, pm, r, t) \qquad [2]$$

To investigate the relative importance of individual factors, Jenny (1961), Vitousek et al. (1983), and Van Cleve et al. (1991) identified on landscapes "sequences" of soil bodies or ecosystems in which a single factor varies while the other factors are held constant, for example, a chronosequence where age *t* is the varying factor, or a toposequence where *r* is the varying factor. Likewise, Pouyat and Effland (1999) proposed that sequences can be used to investigate the effects of urban land use on soil characteristics and ecosystem processes. For example, when the anthropogenic factor *a* plays a role as in Eq. [2], situations where this factor varies over relatively short distances while the remaining factors are held constant would be considered an "anthroposequence" where:

$$S = f(a)_{cl,o,pm,r,t} \qquad [3]$$

Equation [3] represents a sequence, or study design, that can be used to investigate the effects of the *a* factor on a soil at landscape scales (i.e., tens to hundreds of kilometers). An anthroposequence also is suitable for comparing the effects of urban land use on soils at varying temporal scales (Pouyat and Effland, 1999). At one extreme, sequences can be identified where the *a* factor acts interdependently with the other factors for time scales in which pedogenic processes take place, and at the other extreme, sequences can be identified in which the *a* factor operates independently of the other factors (Fig. 7–1). For example, if the soil is disrupted, as in the case of a site that is graded to build a structure, the impact on that soil occurs independently of the other soil forming factors (Fig. 7–1a). In this case the temporal scale of the urban alteration is much shorter than the time frame in which most natural pedogenic processes operate. Here the material constituting the nonurban soil (*S* in Eq. [1]) predated the "new" modified soil (S2) and thus is considered the "new" parent material, so that

$$S2 = f(a, cl, o, S, r, t) \qquad [4]$$

Essentially, there is a new time zero from which pedogenesis takes place. Much in the same way that the till remaining from a retreating glacier is considered parent material, in an urban example transported material used to fill in a low-lying area

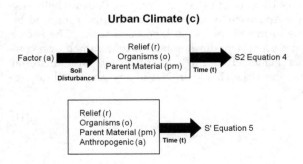

Fig. 7–1. Schematic representation of the temporal effects of the anthropogenic factor (*a*) on soil formation and the relationship of *a* with the natural soil forming factors; climate (*c*), relief (*r*), organisms (*o*), parent material (pm). (A) The anthropogenic factor works at a time frame independently of the other factors. (B) The anthropogenic factor works at a time frame that is interdependent of the other factors (modified from Pouyat and Effland, 1999).

is considered parent material (e.g., Short et al., 1986b). By contrast, factors of soil formation are interdependent, particularly in cases where urban effects occur on time scales similar to soil formation (usually tens to thousands of years, Fig. 7–1b). Under such conditions, interactions with other factors are more likely (Pouyat and Effland 1999; Pickett and Cadenasso, 2009; Pavao-Zuckerman, 2008). Here the profile of the soil remains largely intact, although various chemical, and to a lesser degree physical, properties may be altered through interactions between urban environmental factors and soil formation. Examples of when the a factor is interdependent with other soil forming factors include changes in water table depth (Groffman et al., 2003) and changes in soil temperature regimes (Savva et al., 2010) that typically occur in urban landscapes.

Even when a soil in an urban landscape is morphologically similar to a pre-existing condition, the soil may be functionally altered, such as when surface compaction inhibits the ability of an otherwise intact soil to infiltrate water (Pitt and Lantrip, 2000). In these circumstances, the soil parent material (pm) remains constant, but for a soil with an altered function (S'), such that

$$S' = f(a, cl, o, pm, r, t) \tag{5}$$

Therefore, there are two extreme circumstances, differentiated by temporal scales, in which anthroposequences can be defined and used for investigation of urban soils (Pouyat and Effland, 1999):

First, where

$$S2 = f(a_n)_{cl,o,S,r,t} \tag{6}$$

direct comparisons of different soil disturbances or management regimes associated with urban landscapes (a_1 vs. a_2 and so on) are possible between soils developing under similar environmental situations or parent materials of various origins. Moreover, soil formation can be compared among similar disturbance types along a chronosequence (e.g., Scharenbroch et al., 2005), where

$$S2 = f(t_n)_{a,cl,o,S,r} \tag{7}$$

In both cases, there is the opportunity to study and control factors (e.g., parent material or management regime) in the early stages of soil development that follows an urban disturbance (Evans et al., 2000).

Second, where

$$S' = f(a_n)_{cl,o,pm,r,t} \tag{8}$$

effects on soil formation of various urban environmental factors (a_1 vs. a_2 and so on) can be studied over similar soil types, e.g., forested soils along urban–rural environmental gradients (Pouyat et al., 1995; Carreiro et al., 2009). Such comparisons will be useful in delineating threshold responses of soil properties to deposition of atmospheric pollutants (e.g., Pouyat et al., 2008), or gradual changes in soil properties due to changes in temperature regimes (e.g., Savva et al., 2010).

In the first case (Eq. [6–7]), we consider the effects under these circumstances as direct; in the second case (Eq. [8]) we consider them as indirect. The following sections use the above conceptual framework to differentiate direct from indirect effects and the likely expression of these effects spatially and temporally in urban landscapes.

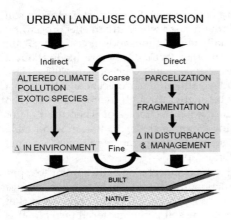

Fig.7–2. Conceptual diagram of the effect of urban land use conversion on native soils. As landscapes are urbanized, natural habitats are increasingly fragmented as parcels of land ownership become smaller. During this process humans introduce novel disturbance and management regimes that impact soil formation; these are known as *direct effects* (arrows on right). Concurrently there is a change in the environmental conditions in which soil formation takes place; these are *indirect effects* (arrows on left). The overlapping of the direct and indirect effects of anthropogenic factors on the native soil results in the "urban soil mosaic." (Modified from Pouyat et al., 2003 and Pouyat et al., 2007b.)

Urban Soil Mosaic

As land is converted to urban uses, direct and indirect factors affect soil chemical, physical, and biological characteristics (Fig. 7–2). Direct effects include those typically associated with urban soils, such as physical disturbances, incorporation of anthropic materials, and burial or coverage of soil by fill material and impervious surfaces (Craul, 1992; Jim, 1998; Schleuss et al., 1998; Galbraith et al., 1999). For most urban landscapes, these types of impacts are more pronounced during rather than after the land development process. Urban development of land typically includes clearing of existing vegetation, massive movements of soil, and building of structures. The extent and magnitude of these initial disturbances is dependent on topography (e.g., McGuire, 2004), infrastructure requirements, and other site limiting factors.

Soil management practices, such as fertilization and irrigation, that are introduced after the initial development disturbance are also considered direct effects. The spatial pattern of disturbance and management practices is largely the result of *parcelization*, or the subdivision of land by ownership, as landscapes are developed for human settlement. The parcelization of the landscape creates distinct parcels with characteristic disturbance and management regimes that will affect soils through time. The net result is a mosaic of soil patches, which will vary in size and configuration dependent on human population density, development patterns, and transportation networks, among other factors (Plate 7–1; see color images section). As mentioned in the previous section, these primarily physical disturbances often lead to "new" soil parent material from which soil develops (Eq. [4], Fig. 7–1a).

Indirect effects related to urban land use change involve changes in the abiotic and biotic environment, which can affect undisturbed soils. In our conceptual framework these effects work at temporal scales in which natural soil formation processes are at work (Eq. [8], Fig. 7–1b). Urban environmental factors include the urban heat island (Mount et al., 1999, Savva et al., 2010), soil hydrophobicity (White and McDonnell, 1988; Craul, 1992), introductions of exotic plant and animal species (Steinberg et al., 1997; Ehrenfeld et al., 2001), and atmospheric deposition of pollutants such as N and S (Lovett et al., 2000; Juknys et

Table 7–2. Fertilizer costs for different land uses in Maryland and turfgrass fertilization rates from the literature.

Land use	Nutrients and other application materials	Managed land	Costs per hectare	Literature values for application rates
		ha		kg N ha^{-1} yr^{-1}
Lawn care firms	$59,124,000	93,688	$631	217–289‡
Golf courses	$10,242,000	13,751	$745	168–239§
Detached single family homes	$248,872,000†	273,173	$911	65¶–120‡

† Costs include mowing equipment, turfgrass nutrients, and other related supplies.
‡ Morton et al. (1988).
§ Klein (1989).
¶ Yesilonis et al., unpublished data, 2004.

al., 2007), heavy metals (Orsini et al., 1986; Boni et al., 1988; Lee and Longhurst, 1992; DeMiguel et al., 1997; Mielke, 1999), and potentially toxic organic chemicals (Wong et al., 2004; Jensen et al., 2007; Zhang et al., 2006).

The net result is a diverse spatial mosaic that represents a variety of soil forming conditions that are useful in comparing the effects of urban land use change on soil characteristics and the distribution of soil fauna (Pouyat et al., 2009a). The spatial heterogeneity of natural soil forming factors may still underlie and constrain the effects of land-use and land-cover change. Additional heterogeneity is introduced from variations in human behavior and social structures that function at multiple scales (Grove et al., 2006; Pickett and Cadenasso, 2009). One example is variation related to fertilization practices among land owners (Table 7–2). The totality of this heterogeneity, both natural and human caused, represents a set of "natural experiments" to investigate the effects of the anthropogenic factor (a) on soil formation at different spatial and temporal scales (Eq. [6–8]).

Soil Chemical, Physical, and Biological Responses to Urban Land Use

The urban soil mosaic can be observed at several scales, with each scale of observation revealing a set of soil characteristics that can be related to various anthropogenic and nonanthropogenic soil forming factors (Pouyat et al., 2009a). For example, at the scale of a metropolitan area, the distribution of physically unaltered soil patches tends to increase in density, moving from the highly developed urban core to suburban and rural areas (Effland and Pouyat, 1997). However, these physically unaltered soils and their soil fauna can be affected by changes in environmental factors that are associated with urban land uses at a considerable distance from the urban core (Pouyat et al., 2008; Carreiro et al., 2009). At finer scales, such as the densely populated area of a city or town and the even finer scales of a neighborhood, subdivision, or residence, the characteristics of soils are expected to be more related to anthropogenic factors than to nonanthropogenic ones (Plate 7–1). Moreover, the finer the scale of observation, the more likely a particular human activity or alteration can be related to a specific soil response (e.g., Ellis et al., 2000).

In the next sections, we present several case studies that use the soil mosaic as a suite of "natural experiments" to investigate the response of soil characteristics to urban land use at multiple scales (Pouyat et al., 2009a). We begin with responses of undisturbed soils to environmental factors along urban–rural gradients at the scale of a metropolitan area (Eq. [8]). We then present studies that investigate soil responses at finer scales to different categories of land use and cover, each serving as an "experimental manipulation" of human management efforts, disturbances, and site histories (Eq. [6] and [7]). We finish with a comparison of selected soil characteristics at regional and global scales.

Urban Environmental Effects at the Scale of Metropolitan Areas

Studies using urban–rural gradients suggest that soils of remnant forests are altered by environmental changes occurring along the gradient. For instance, forest soils within or near urban areas often receive high amounts of heavy metals, organic compounds, and acidic compounds in atmospheric deposition. Lovett et al. (2000) quantified atmospheric N inputs over two growing seasons in oak forest stands along an urbanization gradient in the New York City metropolitan area. They found that urban remnant forests received up to a twofold greater N flux than rural forests. More interestingly, the input of N fell off in the stands 45 km from the urban core, which the authors suggest was due to the reaction of relatively small acidic anion aerosols with larger alkaline dust particles high in Ca^{2+} and Mg^{2+} that enabled the precipitation of N closer to the city. Similar results were found for the city of Louisville, KY; the San Bernardino Mountains in the Los Angeles metropolitan area, CA; the city of Oulu, Finland; and the city of Kaunas, Lithuania where N deposition rates, and in some cases S and base cation deposition rates, into urban and suburban forest patches were higher than in rural forest patches (Ohtonen and Markkola, 1991; Fenn and Bytnerowicz, 1993; Bytnerowicz et al., 1999; Juknys et al., 2007; Carreiro et al., 2009).

Evidence of a similar depositional pattern in the form of contents of metals in forest soils was found along urbanization gradients in the New York City, Baltimore, and Budapest, Hungary metropolitan areas. Pouyat et al. (2008) found up to a two- to threefold increase in contents of lead, copper, and nickel in urban forest remnants compared with suburban and rural counterparts. A similar pattern but with greater differences was found by Inman and Parker (1978) in the Chicago, IL metropolitan area, where levels of heavy metals were more than five times higher in urban than in rural forest patches. Other urbanization gradient studies have shown a similar pattern (Watmough et al., 1998; Sawicka-Kapusta et al., 2003), although smaller cities, or cities having more compact development patterns, exhibited less of a difference between urban and rural remnant forests (Pavao-Zuckerman, 2003; Pouyat et al., 2008; Carreiro et al., 2009). Besides accumulations of heavy metals, Wong et al. (2004) found a steep gradient of polycyclic aromatic hydrocarbons (PAH) concentrations in forest soils in the Toronto, Canada metropolitan area, with concentrations decreasing with distance from the urban center to the surrounding rural area. Similarly, Jensen et al. (2007) and Zhang et al. (2006) found significantly higher concentrations of PAHs in surface soils of Oslo, Norway and Hong Kong, China, respectively, than in surrounding rural areas. In all cases the total PAH concentrations were more than twofold higher in urban than in rural areas.

How these pollutants affect soil fauna or soil biological processes is uncertain, but results thus far suggest that the effects are variable and depend on the importance of various urban environmental factors (Pouyat et al., 2007b; Lorenz and Lal, 2009; Carreiro et al., 2009). For example, Inman and Parker (1978) found slower leaf litter decay rates in urban stands that were highly contaminated with Cu (76 mg kg^{-1}) and Pb (400 mg kg^{-1}) compared with unpolluted rural stands, suggesting a negative pollution effect in the Chicago metropolitan area. Similarly, Pouyat et al. (1994) found an inverse relationship between litter fungal biomass and fungivorous invertebrate abundances with heavy metal concentrations along an urbanization gradient in the New York City metropolitan area. By contrast, responses of soil invertebrates along urbanization gradients in Europe were more related to local factors, such as habitat connectivity or patch size, rather than environmental changes occurring along the gradient (Niemelä et al., 2002).

Where heavy metal contamination of soil is moderate to low relative to other atmospherically deposited elements such as N, biological activity may actually be stimulated. For example, decay rates, soil respiration, and soil N transformation increased in forest patches near or within major U.S. metropolitan areas in southern California (Fenn and Dunn, 1989; Fenn, 1991), Ohio (Kuperman, 1999), southeastern New York (McDonnell et al., 1997; Carreiro et al., 2009), and Maryland (Groffman et al., 2006; Szlavecz et al., 2006). However, where excessively high rates of S deposition also occur, several soil biological measurements were negatively affected in forests near an industrial city in northern Finland (Ohtonen, 1994).

In addition to environmental changes related to the inputs of chemicals into soils, urban areas are the foci for many introduced plant and animal species, some of which are invasive and can have large effects on soil processes (Lilleskov et al., 2010). In particular, invasive species can play a disproportionate role in controlling C and N cycles in terrestrial ecosystems (Ehrenfeld, 2003; Bohlen et al., 2004). Therefore, the relationship between invasive species abundances and urban land-use change has important implications for soil mediated ecosystem processes (Pouyat et al., 2007b). For example, in the northeastern and mid-Atlantic United States, where native earthworm species are rare or absent, urban areas are important foci of invasive earthworm introductions, especially Asian species from the genus *Amynthas*, which are expanding their range to outlying forested areas (Steinberg et al., 1997; Groffman and Bohlen, 1999; Szlavecz et al., 2006). Invasions by earthworms into forests have resulted in highly altered C and N cycling processes (Bohlen et al., 2004; Hale et al., 2005; Carreiro et al., 2009). Likewise, plant species invasions can impact C and N dynamics, which in some cases can facilitate the colonization of additional invasive species, such as earthworms, further exacerbating the turnover of N in the soil (Pavao-Zuckerman, 2008). Examples of plant invasions in urban metropolitan areas that have altered C and N cycles include species of the shrub *Berberis thunbergii* DC., the tree *Rhamnus cathartica* L., and the grass *Microstegium vimineum* (Trin.) A. Camus (Ehrenfeld et al., 2001; Heneghan et al., 2002; Kourtev et al., 2002).

Land Use and Cover at the City and Neighborhood Scale

Land use and cover can serve as an indicator of disturbance, site history, and management—factors that, as previously mentioned, have the potential to affect soil characteristics. As such, drawing a relationship between soil characteristics and land use and cover in an urban context should advance our understanding of anthropogenic effects of soils. Moreover, if these relationships are systematic, they will be useful in the development of mapping concepts for soils in urban landscapes. Therefore, an important task in urban soil research is to establish at multiple scales coherent relationships between soil characteristics (surface and subsurface) and land use and cover (Pouyat et al., 2007a, 2009a).

As population density increases, the parcels of landownership become smaller in area; thus, management and disturbance occur at finer scales. For this reason, to capture the heterogeneity inherent in the urban soil mosaic, the patches that make up the mosaic require delineation at relatively fine scales (Ellis et al., 2006; Pickett and Cadenasso, 2009). Management can vary significantly in urban mosaics (Table 7–2), which should result in a corresponding response by the affected soils, although the response will be tempered by other anthropo-genic factors and the characteristics of the native soil that remain. Therefore, an important question related to the formation of urban soils is the relative impor-tance of human alterations such as landscape management versus the importance of individual natural soil forming factors (i.e., factor a versus pm, r, o, cl, and t in Eq. [2]). We address this question and look for a systematic pattern in characteristics of surface soils in the following sections by comparing the spatial relationships of various soil variables with land use, cover, and the natural soil forming factor (pm) at the scale of a city, neighborhood, or subdivision (Plate 7–1).

City Scale

To investigate the relationship between urban land use and soil characteristics at a 0- to 10-cm depth, 130 plots of 0.04 ha were stratified by Anderson Level II land-use and cover classes as part of the Baltimore Ecosystem Study (Nowak et al., 2004; Pouyat et al., 2007a). Results showed a wide range of characteristics among all land-use and cover classes, although a subset of the soil variables measured (P, K, bulk density, and pH) were differentiated by the Anderson classes (Pouyat et al., 2007a). Differences were greatest between classes of land use and cover char-acterized by intensive land management (lawns) and the absence of management (forests). In particular, concentrations of P and K, both of which are components of most lawn fertilizers, and bulk density, an indirect measure of soil compaction, differentiated the forest from the grass cover plots.

Taking a subset of the forest and lawn cover-type plots and adding a set of agricultural plots, Groffman et al. (2009) measured potential N mineralization and nitrification rates among the different cover types. Relatively large differ-ences were found among the forested, lawn, and agricultural land-use types. Moreover, when these data were compared with data collected in forest soils along an urban–rural gradient in the Baltimore metropolitan area (Szlavecz et al., 2006), differences were much higher between land-use and cover (50-fold differ-ence) than between urban and rural forest remnants (10-fold difference) (Pouyat et al., 2009a). These results suggest that soil management associated with differ-ent land uses has a much greater effect on N cycling than environmental factors, such as N deposition and temperature changes, that occur along urban–rural gradients. In addition, Scharenbroch et al. (2005) showed that the differences in N cycling that may occur with respect to cover and management will increase as the soil ages after a site disturbance (i.e., a chronosequence). Research in the Phoenix metropolitan area also showed the importance of urban land use and management on soil N processes. Soils associated with urban land uses had sig-nificantly higher inorganic N pools than desert soils (Hope et al., 2005; Zhu et al., 2006), and soils associated with irrigated residential landscapes, or "mesic yards," had higher total N and available P than xeric landscaped yards and nonresidential spaces (Kaye et al., 2008).

Unlike N, P, or K, heavy metals such as Pb, Cu, and Zn did not differentiate the land-use and cover classes used in the Baltimore study, but rather were found to be related to major transportation corridors and the age of housing stock (Yesilonis et al., 2008). A similar relationship was found in Baltimore with an earlier investigation of garden soils (Mielke et al., 1983). Several investigations of other cities have found similar associations of these metals with transportation networks, suggesting the importance of vehicular traffic as a source of Pb, Cu, and Zn in urban environments (Bityukova et al., 2000; Facchinelli et al., 2001; Manta et al., 2002; Madrid et al., 2002; Li et al., 2004; Zhang, 2006). Indeed, as an additive to gasoline, Pb was an important component of automobile exhaust until 1986 in the United States (Mielke, 1999), while Cu and Zn contamination continues to occur from brake emissions and tire abrasion, respectively (Councell et al., 2004; Hjortenkrans et al., 2006).

An interesting and important result of the citywide analysis in Baltimore was the continued importance of parent material that differentiated plots by physiographic province (Atlantic Coastal Plain and Piedmont) for a subset of variables measured, primarily soil texture and the trace elements Al, Mg, V, Mn, Fe, and Ni, which are important constituents of the surface rock types found in the Piedmont province. These results show the continued importance of the effect of parent material on surface soils in urban landscapes (Pouyat et al., 2007a).

Neighborhood Scale

To relate soil characteristics at finer scales in the urbanized areas of the Baltimore metropolitan area, we delineated patches with higher categorical resolution at the scale of a neighborhood or subdivision and individual parcel. At this scale, soil responses can be related to individual patches with specific site histories and activities of individual land managers (Pouyat et al., 2009a). Two neighborhoods in the Baltimore metropolitan area, one medium-density (suburban) and the other high-density (urban) residential, were delineated using high resolution ecotope mapping (Ellis et al., 2006). The suburban neighborhood was situated just outside the city boundary and had households with a significantly higher mean income than households in the urban neighborhood. Similar to the citywide analysis described earlier, 80 plots of 0.04 ha were randomly stratified, but this time by ecotope classes of cover and use delineated in each neighborhood. Results showed that lower concentrations of Pb and Cu were found in the suburban neighborhood than in the urban one, and by contrast to the citywide analysis, these elements varied by use and cover classes delineated at a finer resolution, with the highest concentrations occurring in urban vacant and disturbed lots. Moreover, the proportion of plots with concentrations exceeding the USEPA's soil Pb screening level of 400 mg kg^{-1} was more than 16% in the urban neighborhood, while none of the plots in the suburban neighborhood exceeded the USEPA standard for Pb (I.D. Yesilonis and R.V. Pouyat, unpublished data, 2007).

Unlike the neighborhood-scale analysis, Pb concentrations were examined at the parcel level and were found to be consistent with the citywide results. A total of 60 residential properties in Baltimore City were intensively sampled for total Pb using field portable X-ray fluorescence. Thirty percent of sites had average Pb values that exceeded the USEPA screening level while 53% of sites exhibited lead

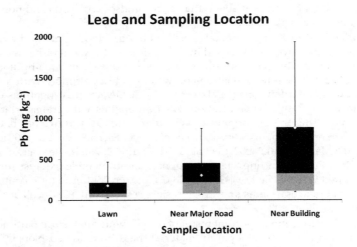

Fig. 7–3. Mean values (white diamonds) and error bars representing the lowest and highest values falling within 1.5 times the interquartile range. Observations outside the error bars are not shown; however, they were not removed from the dataset. Samples were collected from residential lots in Baltimore City. The "lawn" classification represents areas of the lawn not adjacent to a major roadway or building. The "near major road" classification represents samples closest to a major roadway (defined as the primary and secondary roads in the TIGER classification). The "near building" classification represents samples collected directly adjacent to a built structure (Schwarz, 2010).

levels >400 mg kg^{-1} in at least one area of the yard. The highest lead levels were found near buildings and major road networks (Fig. 7–3). Elevated lead levels were observed next to buildings regardless of building type, including brick and wood frame structures. Housing age was an important predictor of soil lead levels with none of the structures built after the ban on lead-based paint and leaded gasoline exhibiting soil lead levels above the EPA limit (Schwarz, 2010).

Using a subset of sample locations in the suburban neighborhood, soil concentrations of Ca, K, Mg, P, and organic matter were compared with N fertilizer rates applied by homeowners (Table 7–2). Both Ca and Mg concentrations showed a significant ($p < 0.01$) indirect response to fertilizer rates ($r^2 = 0.96$ and 0.65, respectively), while organic matter, K, and P did not show a trend. Similarly, springtail (Collembola) and mite (Acari) densities were indirectly related to fertilizer applications in the same neighborhood (Fig. 7–4). However, when we compared two subdivisions within this neighborhood that were built 10 yr apart, P and organic matter concentrations were significantly higher in the older subdivision (Fig. 7–5a). In addition, a comparison with an adjacent forest patch showed that both subdivisions had higher concentrations of P and K, which was consistent with the citywide results (Fig. 7–5b). Again, comparisons at a finer scale (neighborhood vs. citywide) enabled the association between an anthropogenic factor, in this case a management effect (N fertilization), on a soil response (Ca and Mg concentrations and mesofauna abundances), but also verified a relationship at a citywide scale between an intensively managed soil (lawn) versus an unmanaged forest soil.

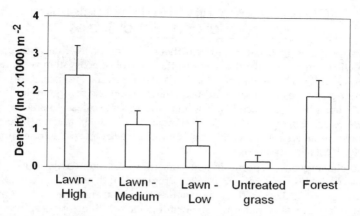

Fig. 7–4. Mean (SE) springtail (Collembola) densities by lawn maintenance level and for a nearby forest patch of a suburban neighborhood in Baltimore County, Maryland. Lawn maintenance level is based on number of times the lawn was fertilized during the growing season (K. Szlavecz, unpublished data, 2006).

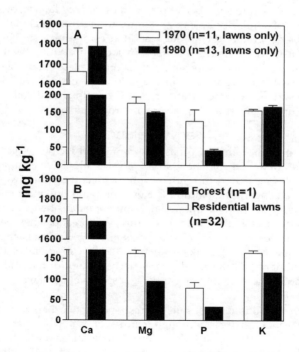

Fig. 7–5. Mean (SE) concentrations of Ca, Mg, P, and K for soils sampled to the 0- to 5-cm depth in a suburban neighborhood in Baltimore County, Maryland: (A) by year of development (1970 and 1980); (B) by land use and cover (Forest and Lawn) (I.D. Yesilonis, unpublished data, 2005).

Comparisons of Urban Soil Characteristics at Regional and Global Scales

A central principle in urban ecological theory presupposes that anthropogenic drivers will dominate natural drivers in the control of ecosystem response variables (Alberti, 1999; Kaye et al., 2006). If this assumption were true, it follows that ecosystem responses to urban land-use change should converge, relative to the native systems being replaced, at regional and global scales—this is termed the *urban ecosystem convergence hypothesis* (Pouyat et al., 2003), and with respect to plant and animal species assemblages, *biotic homogenization* (McKinney and Lockwood, 2001; McKinney, 2006).

The convergence hypothesis suggests that ecosystem responses, like soil response variables, will converge across regional and global scales as long as the anthropogenic drivers (i.e., management, disturbance, and environmental change) dominate over natural soil forming factors like topography and parent material (i.e., *r* and *pm* in Eq. [2]). The result is that soil characteristics will be more similar across urban landscapes at regional and global scales than the native soils they replaced (Pouyat et al., 2008). The biotic homogenization hypothesis suggests that urban land-use change results in local extinctions of regional soil fauna and that the movements of humans across geographical boundaries help facilitate the establishment and spread of urban adapted, or synanthropic, soil fauna. These synanthropic species thrive in urbanized landscapes, resulting in a high degree of similarity in species composition across regional and global scales. While biogeography theory suggests that community similarity should decrease with distance in "natural habitats" (Nekola and White, 1999), in human-dominated systems the decrease in similarity should be less or absent (McKinney, 2006). To address these hypotheses, in the following sections we compare soil responses of heavy metal concentrations, soil C densities, and soil invertebrate abundances to urban factors among several metropolitan areas.

Heavy Metals

To investigate the importance of urban environmental factors (*a*) versus parent material (*pm*) in Eq. [2] on soil chemistry, Pouyat et al. (2008) compared 15 soil response variables measured in remnant forests (deciduous hardwood) among urbanization gradients in the Baltimore, New York, and Budapest metropolitan areas. These metropolitan areas differed in population densities, size of area, and transportation systems. In the case of New York, the forest patches were situated on surface geology of the same type (Pouyat et al., 1995) and thus approximated an anthroposequence (Eq. [3]) along an urbanization gradient of 0 to125 km, whereas in Baltimore and Budapest, the surface geology differed along a 0- to 30-km gradient and a 0- to 20-km gradient, respectively (Szlavecz et al., 2006; Pouyat et al., 2008).

The authors found that forest soils are responding to urbanization gradients in all three metropolitan areas, although features of each city (spatial pattern of development, forms of transportation, parent material, and site history) influenced the soil chemical response. The authors suggested that the changes measured resulted from locally derived atmospheric pollution of Pb, Cu, and to a lesser extent Ca, which was more extreme at the urban end of the gradient, but extended beyond the political boundary of each city. In the case of Pb and

Cu, the soil concentrations were highly correlated to traffic volume and density of roads, confirming the importance of vehicle emissions with respect to heavy metal pollution. Moreover, in Baltimore and Budapest the soil chemical response was confounded by differences in parent material found along those urbanization gradients. A similar result showing the importance of parent material was found in regional analyses of pollution effects on soils in Tallinn, Estonia, and the Piedmonte Region of northwestern Italy (Bityukova et al., 2000; Facchinelli et al., 2001). Thus, the comparison of urban–rural gradients across three distinct metropolitan areas suggests that characteristics of the native parent material persist in the surface of urban soils. However, at the same time, the influence of urban environmental factors (deposition of Pb, Cu, and Ca) resulted in similar soil responses across the three metropolitan areas, which also occurred at distances beyond the boundaries of each city.

Soil Carbon

An important characteristic of urban land-use change with respect to the C cycle is the replacement of native cover types with lawn cover (Kaye et al., 2005; Milesi et al., 2005; Golubiewski 2006; Pouyat et al., 2009b). The estimated amount of lawn cover for the conterminous United States is 163,800 km^2 ± 35,850 km^2, which is approximately three times the area of other irrigated crops (Milesi et al., 2005). To manage this lawn cover, roughly half of residential and institutional managers appear to apply fertilizers (Law et al., 2004; Osmond and Hardy, 2004), with some applying at rates similar to or exceeding those of cropland systems, e.g., >200 kg ha^{-1} yr^{-1} (Morton et al., 1988; Table 7–2).

Due to the management efforts of landowners and the expansive use of lawn cover, there should be a significant accumulation of soil C in most urban landscapes (Pouyat et al., 2006; Huh et al., 2008). Qian et al. (2003) showed with simulations of the CENTURY model that N fertilization coupled with grass clipping replacement increased soil C accumulations by up to 59% in comparison with sites that were not fertilized and clippings were removed. Likewise, Golubiewski (2006) measured soil C stocks in 13 residential yards of different ages in the semiarid shortgrass prairie of Colorado and found that soil C recovered from the initial development disturbance after 20 yr and exceeded the semiarid prairie soil in 40 yr. Moreover, the effects of lawn care management on soil C accumulation exceeded the effects of other soil forming factors, such as elevation and soil texture.

The amount of effort to manage lawn or turfgrass systems, including the intensity of irrigation or nutrient applications, also should have an effect on soil C accumulation (Huh et al., 2008). Moreover, the effort should reflect the natural constraints on the turfgrass system (Pouyat et al., 2009b). For example, lawns growing in arid areas require more irrigation than lawns growing in temperate climates. Milesi et al. (2005) found that in the absence of irrigation and fertilization most species of turfgrass would not be able to grow and compete with native vegetation in most of the conterminous United States. Therefore, compensatory inputs of water and nutrients required to maintain turfgrass cover should result in a higher accumulation of soil C than in the previous dryland system (e.g., Golubiewski, 2006; Jenerette et al., 2006; Kaye et al., 2008). However, in more temperate regions of the United States, where turfgrasses may grow with minimal supplements, soil C in residential areas should be equivalent to or lower than the native soil, which will most likely be a forest soil (Pouyat et al., 2009b).

To address the potential for lawns and other urban land use and cover types to accumulate soil C, Pouyat et al. (2006) reviewed the literature and found that across different climate and soil types, older residential lawns had surprisingly similar C densities of 14.4 ± 1.2 kg m^{-2} m^{-1} depth. In addition, the authors conducted an analysis of available soil C data for urban and native soils, which showed remnant patches of native vegetation accounting for up to 34% of the stock of soil C of a city. However, when soils beneath impervious surfaces were excluded from the analysis, the estimated soil C densities rose substantially for the urban land-use and cover types, indicating the potential for soils in pervious areas of urban landscapes to sequester large amounts of C. The authors also compared the pre- and post-urban estimates of soil C stocks in six cities and found the potential for large decreases in soil C post-urban development for cities located in the northeastern United States, where native soils have inherently high soil C densities, but in drier climates with inherently low soil C densities, cities tended to have slightly higher soil C densities than the native soil types.

Soil Invertebrates

Biotic homogenization has been tested primarily for large organisms with relatively straightforward taxonomy, such as plants and birds (McKinney, 2006). However, large-scale zoogeographical analysis of invertebrate communities is challenging because of differences in collection techniques, difficulties identifying species, and confusion about nomenclature (Byrne et al., 2008). Investigating the spatial distribution of soil faunal communities along urban–rural gradients has been a focus of urban ecological research (e.g., Pouyat et al., 1994; Pizl and Josens, 1995; Steinberg et al., 1997; Szlavecz et al., 2006 Hornung et al., 2007). To make results from this approach comparable among many locations, a global network (GLOBENET) using standardized sampling methodology was initiated (Niemelä et al., 2000). Although GLOBENET initially focused on ground beetles (Carabidae) (Niemelä et al., 2002; Ishitani et al., 2003), other epigeic (surface-dwelling) arthropod taxa have also been analyzed (Hornung et al., 2007; Vilisics et al., 2007).

Preliminary results suggest that the degree of biotic homogeneity among soil fauna is not uniform and varies by taxa. For instance, international comparisons of urban carabid beetle assemblages at GLOBENET sites have shown a large degree of local differences (Niemelä et al., 2002). Other taxa, such as carrion beetles (Silphidae) have more specialized natural histories, and their species composition is largely determined by the regional species pool and the size and quality of urban forest habitats (Wolf and Gibbs, 2004). By contrast, earthworm species show a high degree of similarity among cities in the United States and Europe due to a high proportion of synanthropic, peregrine species making up earthworm assemblages (Table 7–3). These species have been carried accidentally or deliberately across continents and are now cosmopolitan in their distribution.

Acting against biotic homogenization is the survival of native species in remnant habitat patches or their adaptation to novel environments, as "urban adapters" (Johnston, 2001). The regional species pool varies geographically, and thus the subset adapting to urban environments will vary as well. Often these species are taxonomically similar and ecologically equivalent, resulting in urban vicariance. Two ecologically similar millipede species that have adapted to urban environments illustrate this phenomenon: *Brachyiulus bagnali* Roleman, a species

Table 7–3. The most commonly occurring earthworm species reported from cities surveyed in Europe and Baltimore, MD.†

Species	Baltimore	Budapest	Brussels (1)	Warsaw (2)	Greifswald (3)	Basel (4)	Bonn (5)	Dorsten (6)
Allolobophora chlorotica Savigny		x	x	x	x	x	x	x
Aporrectodea caliginosa Savigny	x	x	x	x	x	x	x	x
Aporrectodea longa Ude		x	x	x	x	x	x	x
Aporrectodea rosea Savigny	x	x	x	x	x	x	x	
Dendrobaena octaedra Savigny	x	x	x	x	x			x
Eiseniella tetraedra Savigny	x		x	x	x	x		
Lumbricus castaneus Savigny	x	x	x	x	x	x	x	x
Lumbricus rubellus Hoffmeister	x	x	x	x	x	x	x	x
Lumbricus terrestris L.	x	x	x	x	x	x	x	x
Octolasion cyaneum Savigny		x		x	x	x		x
Octolasion lacteum Örley	x	x	x	x	x	x		x
Total number of species	18	19	14	15	15	13	7	15

† References: (1) Pizl and Josens, 1995; (2) Pilipiuk, 1981; (3) Unger, 1999; (4) Glasstetter and Nagel, 2001; (5) Keplin and Broll, 1999; (6) Fründt and Ruszkowski, 1989.

of Continental Europe has a distribution limited to the eastern European city of Budapest, and *B. pusillus* Leach, an Atlantic European species, has been recorded more widely in cities located throughout Europe. *Brachyiulus bagnali*, while limited to Continental Europe is urban adapted and functionally similar to the *B. pusillus*.

Conclusions

Research on soils of urban landscapes has shown that properties of surface soils and their soil fauna can vary widely, making it difficult to define or describe a typical "urban" soil or soil community. Urban soils vary in characteristics depending on the nature and history of disturbance, management regime, and the effect of urban environmental changes. Moreover, the importance of natural soil forming factors (e.g., *pm* in Eq. [2]) continues to influence the spatial distribution of chemical and physical characteristics even when a significant proportion of the land area has been impacted by urban development.

Although urban landscapes are highly altered by development and human activities, urban soils have exhibited a surprising capacity to support plant growth and soil fauna; as a result they may have relatively high rates of biological activity and richness of species relative to the native systems they have replaced. With this capacity, urban soils have the potential to provide various ecosystem services to inhabitants of urban areas and settlements. However, frequent disturbance and many of the environmental changes occurring in urban landscapes can reduce the capacity of these soils to provide ecosystem services or support a diverse soil community, suggesting the importance of developing sustainable management practices that enhance ecosystem services of urban soils.

Some systematic patterns are emerging from our use of "natural experiments" (e.g., Eq. [5–8]) in the urban soil mosaic. Management, disturbance, site history, and environmental patterns all have been shown to have an impact on the spatial response of surface soils and soil fauna at multiple scales in the urban landscape. While physical and management impacts appear to be more pronounced, environmental factors have the potential to be more widespread, often having influence beyond the political boundary of most urban settlements.

The patterns discovered through our conceptual framework of urban soil formation should be useful in developing mapping concepts of soils in urban landscapes. For example, differences in surface soil properties among land-use and cover types could help determine mapping concepts in soil surveys of urban landscapes. Moreover, those soil properties associated with management, such as fertilizer applications and intensity of use, may be useful as surface diagnostic properties to differentiate human-altered soils associated with urban landscapes. Research on the patterning of the characteristics of soil at and substantially below the surface is needed in metropolitan areas that represent various-sized human settlements, cultural and economic factors, native soil types, and climates to generalize these patterns.

Acknowledgments

We thank R. Shaw for his comments on this manuscript. We thank Csaba Csuzdi, Liz Hornung, Zoltan Korsos, and Ferenc Vilisics for their expertise on soil fauna and many others for field and laboratory assistance. Funding support came from the U.S. Forest Service, Northern Global Change Program and Research Work Unit (NE-4952), Syracuse, NY; the Baltimore Ecosystem Study's Long Term Ecological Research grant from the National

Science Foundation (DEB 97-14835); NSF Undergraduate Mentoring in Environmental Biology Program (DEB99-75463); and the Center for Urban Environmental Research and Education, University of Maryland Baltimore County (NOAA Grants NA06OAR4310243 and NA07OAR4170518).

References

Aelion, C.M., H.T. Davis, S. McDermott, and A.B. Lawson. 2008. Metal concentrations in rural topsoil in South Carolina: Potential for human health impact. Sci. Total Environ. 402:149–156.

Aitkenhead-Peterson, J.A., M.K. Steele, and A. Volder. 2010. Services in natural and human dominated ecosystems. p. 373–390. In J. Aitkenhead-Peterson and A. Volder (ed.) Urban ecosystem ecology. Agron. Monogr. 55. ASA, CSSA, and SSSA, Madison, WI.

Akbari, H. 2002. Shade trees reduce building energy use and CO_2 emissions from power plants. Environ. Pollut. 116:S119–S126.

Alberti, M. 1999. Modeling the urban ecosystem: A conceptual framework. Environ. Plann. B 26:605–630.

Alikhan, M. 2003. The physiological consequences of metals and other environmental contaminants to terrestrial isopod species. p. 263–285. In A. Sfenthourakis et al. (ed.) The biology of terrestrial isopods. Crustaceana Monogr. 2. Brill, Leiden, the Netherlands.

Amundson, R., and H. Jenny. 1991. The place of humans in the state factor theory of ecosystems and their soils. Soil Sci. 151:99–109.

Andrèn, O., J. Bengtsson, and M. Clarholm. 1995. Biodiversity and species redundancy among litter decomposers. 141–152. In J.P. Collins (ed.) The significance and regulation of soil biodiversity. Kluwer, Dordrecht, the Netherlands

Ash, C., and D. Lee. 1980. Lead, cadmium, copper and iron from roadside sites. Environ. Pollut. 22:59–67.

Band, L., M. Cadenasso, S. Grimmond, J. Grove, and S. Pickett. 2004. Heterogeneity in urban ecosystems: Patterns and process. 257–258. In G. Lovett et al. (ed.) Ecosystem function in heterogeneous landscapes. Springer-Verlag, New York.

Barbour, M.G., J.H. Burk, and W.D. Pitts. 1980. Terrestrial plant ecology. Benjamin/Cummings, Menlo Park, CA.

Bardgett, R. 2005. The biology of soil: A community and ecosystem approach. Oxford Univ. Press, Oxford, UK.

Beeby, A. 1978. Interaction of lead and calcium uptake by the woodlouse Porcellio scaber (Isopoda, Porcellionidae). Oecologia 32:255–262.

Beyer, W., and E. Cromartie. 1987. A survey of lead, copper, zinc, cadmium, chromium, arsenic and selenium in earthworms and soil from diverse sites. Environ. Monit. Assess. 8:27–36.

Bityukova, L., A. Shogenova, and M. Birke. 2000. Urban geochemistry: A study of element distributions in the soils of Tallinn (Estonia). Environ. Geochem. Health 22:173–193.

Bockheim, J.G. 1974. Nature and properties of highly-disturbed urban soils. p. 161. In Agronomy Abstracts. ASA, CSSA, and SSSA, Madison, WI.

Bohlen, P.J., P.M. Groffman, T.J. Fahey, M.C. Fisk, E. Suarez, D.M. Pelletier, and R.T. Fahey. 2004. Ecosystem consequences of exotic earthworm invasion of north temperate forests. Ecosystems 7:1–12.

Bolger, D.T., A.V. Suarez, K.R. Crooks, S.A. Morrison, and T.J. Case. 2000. Arthropods in urban habitat fragments in southern California: Area, age, and edge effects. Ecol. Appl. 10:1230–1248.

Boni, C., E. Caruso, E. Cereda, G. Lombardo, G.M. Braga, and P. Redaelli. 1988. Particulate matter elemental characterization in urban areas: Pollution and source identification. J. Aerosol Sci. 19:1271–1274.

Brady, N.C., and R.R. Weil. 1999. The nature and properties of soils. Prentice Hall, Upper Saddle River, NJ.

Brown, S., R.L. Chaney, J.G. Hallfrisch, and Q. Xue. 2003. Effect of biosolids processing on lead bioavailability in an urban soil. J. Environ. Qual. 32:100–108.

Bullock, P., and P.J. Gregory. 1991. Soils: A neglected resource in urban areas. p. 1–4. *In* P. Bullock and P.J. Gregory (ed.) Soils in the urban environment. Blackwell Scientific Publ., Oxford, UK.

Burian, S.J., and C.A. Pomeroy. 2010. Urban impacts on the water cycle and potential green infrastructure implications. p. 277–296. *In* J. Aitkenhead-Peterson and A. Volder (ed.) Urban ecosystem ecology. Agron. Monogr. 55. ASA, CSSA, and SSSA, Madison, WI.

Byrne, L.B., M.A. Bruns, and K.C. Kim. 2008. Ecosystem properties of urban land covers at the aboveground–belowground interface. Ecosystems 11:1065–1077.

Bytnerowicz, A., M.E. Fenn, P.R. Miller, and M.J. Arbaugh. 1999. Wet and dry pollutant deposition to the mixed conifer forest. p. 235–269. *In* P.R. Miller and J.R. Mcbride (ed.) Oxidant air pollution impacts in the Montane Forests of southern California: A case study of the San Bernardino Mountains. Ecol. Stud. 134. Springer, New York.

Carreiro, M.M., R.V. Pouyat, C.E. Tripler, and W.X. Zhu. 2009. Carbon and nitrogen cycling in soils of remnant forests along urban–rural gradients: Case studies in the New York metropolitan area and Louisville, Kentucky. p. 308–328. *In* M.J. McDonnell et al. (ed.) Ecology of cities and towns: A comparative approach. Cambridge Univ. Press, Cambridge, UK.

Cook, R.D., and L.Q. Ni. 2007. Elevated soil lead: Statistical modeling and apportionment of contributions from lead-based paint and leaded gasoline. Ann. Appl. Stat. 1:130–151.

Connor, E., J. Hafernik, J. Levy, V. Moore, and J. Rickman. 2002. Insect conservation in an urban biodiversity hotspot: The San Francisco Bay Area. J. Insect Conserv. 6:247–259.

Councell, T.B., K.U. Duckenfield, E.R. Landa, and E. Callender. 2004. Tire-wear particles as a source of zinc to the environment. Environ. Sci. Technol. 38:4206–4214.

Craul, P.J. 1992. Urban soil in landscape design. John Wiley and Sons, New York.

Craul, P.J., and C.J. Klein. 1980. Characterization of streetside soils of Syracuse, New York. p. 88–101. *In* METRIA 3, Proc. of the Third Conference of the Metropolitan Tree Improvement Alliance. Rutgers, New Brunswick, NJ.

Csuzdi, C., T. Pavlicek, and E. Nevo. 2008. Is *Dichogaster bolaui* (Michaelsen, 1891) the first domicole earthworm species? Eur. J. Soil Biol. 44:198–201.

Csuzdi, C., and K. Szlavecz. 2002. *Diplocardia patuxentis*, a new earthworm species from Maryland, North America (Oligochaeta: Acanthodrilidae). Ann. Hist. Nat. Musei Natl. Hungarici 94:193–208.

De Kimpe, C.R., and J.L. Morel. 2000. Urban soil management: A growing concern. Soil Sci. 165:31–40.

De Miguel, E., J.F. Llamas, E. Chacon, T. Berg, S. Larssen, O. Royset, and M. Vadset. 1997. Origin and patterns of distribution of trace elements in street dust: Unleaded petrol and urban lead. Atmos. Environ. 31:2733–2740.

Effland, W.R., and R.V. Pouyat. 1997. The genesis, classification, and mapping of soils in urban areas. Urban Ecosyst. 1:217–228.

Ehrenfeld, J.G. 2003. Effects of exotic plant invasions on soil nutrient cycling processes. Ecosystems 6:503–523.

Ehrenfeld, J.G., P. Kourtev, and W. Huang. 2001. Changes in soil functions following invasions of exotic understory plants in deciduous forests. Ecol. Appl. 11:1278–1300.

Ellis, E.C., R.G. Li, L.Z. Yang, and X. Cheng. 2000. Changes in village-scale nitrogen storage in China's Tai Lake region. Ecol. Appl. 10:1074–1089.

Ellis, E.C., H.Q. Wang, H.S. Xiao, K. Peng, X.P. Liu, S.C. Li, H. Ouyang, X. Cheng, and L.Z. Yang. 2006. Measuring long-term ecological changes in densely populated landscapes using current and historical high resolution imagery. Remote Sens. Environ. 100:457–473.

Evans, C.V., D.S. Fanning, and J.R. Short. 2000. Human-influenced soils. p. 33–67. *In* R.J. Brown et al. (ed.) Managing soils in an urban environment. Agron. Monogr. 39.

Facchinelli, A., E. Sacchi, and L. Mallen. 2001. Multivariate statistical and GIS-based approach to identify heavy metal sources in soils. Environ. Pollut. 114:313–324.

Fanning, D.S., and M.C.B. Fanning. 1989. Soil morphology, genesis, and classification. John Wiley, New York.

Farfel, M.R., A.O. Orlova, R.L. Chaney, P.S.J. Lees, C. Rohde, and P.J. Ashley. 2005. Biosolids compost amendment for reducing soil lead hazards: A pilot study of Orgro® amendment and grass seeding in urban yards. Sci. Total Environ. 340:81–95.

Fenn, M.E. 1991. Increased site fertility and litter decomposition rate in high pollution sites in the San Bernardino Mountains. Forest Sci. 37:1163–1181.

Fenn, M.E., and A. Bytnerowicz. 1993. Dry deposition of nitrogen and sulfur to ponderosa and Jeffrey pine in the San Bernardino National Forest in southern California. Environ. Pollut. 81:277–285.

Fenn, M.E., and P.H. Dunn. 1989. Litter decomposition across an air-pollution gradient in the San Bernardino Mountains. Soil Sci. Soc. Am. J. 53:1560–1567.

Foddai, D., L. Bonato, L. Pereira, and A. Minelli. 2003. Phylogeny and systematics of the Arrupinae (Chilopoda: Geophilomorpha: Mecistocephalidae) with the description of a new dwarfed species. J. Nat. Hist. 37:1247–1267.

Fründt, H., and B. Ruszkowszki. 1989. Untersuchungen zur Biologie stadtischer Boden. 4. Regenwurmer, Asseln und Diplopoden. Verh. Ges. Okol. 18:193–200.

Galbraith, J.M., R.B. Bryant, and J. Russell-Anelli. 1999. Major kinds of humanly altered soils. p. 115–119. In J.M. Kimble et al. (ed.) Classification, correlation, and management of anthropogenic soils. Proc. Nevada and California Worksh. 21 Sept.–2 Oct. 1998. USDA-NRCS, National Soil Survey Center, Lincoln, NE.

Getz, L., L. Best, and M. Prather. 1977. Lead in urban and rural songbirds. Environ. Pollut. 12:335.

Gish, C., and R. Christensen. 1973. Cadmium, nickel, lead and zinc in earthworms from roadside soil. Environ. Sci. Technol. 7:1060–1062.

Glasstetter, M., and P. Nagel. 2001. Earthworm species as bioindicators in urban soils (City of Basel, Northwestern Switzerland). Verh. Ges. Okol. 31:190.

Goldman, M.B., P.M. Groffman, R.V. Pouyat, M.J. Mcdonnell, and S.T.A. Pickett. 1995. CH_4 uptake and N availability in forest soils along an urban to rural gradient. Soil Biol. Biochem. 27:281–286.

Golubiewski, N.E. 2006. Urbanization increases grassland carbon pools: Effects of landscaping in Colorado's front range. Ecol. Appl. 16:555–571.

Gordon, A.M., G.A. Surgeoner, J.C. Hall, J.B. Ford-Robertson, and T.J. Vyn. 1996. Comments on "The role of turfgrasses in environmental protection and their benefits to humans". J. Environ. Qual. 25:206–208.

Govil, P.K., G.L.N. Reddy, and A.K. Krishna. 2001. Contamination of soil due to heavy metals in the Patancheru industrial development area, Andhra Pradesh, India. Environ. Geol. 41:461–469.

Groffman, P.M., D.J. Bain, L.E. Band, K.T. Belt, G.S. Brush, J.M. Grove, R.V. Pouyat, I.C. Yesilonis, and W.C. Zipperer. 2003. Down by the riverside: Urban riparian ecology. Front. Ecol. Environ. 1:315–321.

Groffman, P.M., and P.J. Bohlen. 1999. Soil and sediment biodiversity cross-system comparisons and large-scale effects. Bioscience 49:139–148.

Groffman, P.M., N.J. Boulware, W.C. Zipperer, R.V. Pouyat, L.E. Band, and M.F. Colosimo. 2002. Soil nitrogen cycle processes in urban riparian zones. Environ. Sci. Technol. 36:4547–4552.

Groffman, P.M., and M.K. Crawford. 2003. Denitrification potential in urban riparian zones. J. Environ. Qual. 32:1144–1149.

Groffman, P.M., and R.V. Pouyat. 2009. Methane uptake in urban forests and grasslands. Environ. Sci. Technol. doi:10.1021/es803720h.

Groffman, P.M., R.V. Pouyat, M.L. Cadenasso, W.C. Zipperer, K. Szlavecz, I.D. Yesilonis, L.E. Band, and G.S. Brush. 2006. Land use context and natural soil controls on plant community composition and soil nitrogen and carbon dynamics in urban and rural forests. For. Ecol. Manage. 236:177–192.

Groffman, P.M., C.O. Williams, R.V. Pouyat, L.E. Band, and I.D. Yesilonis. 2009. Nitrate leaching and nitrous oxide flux in urban forests and grasslands. J. Environ. Qual. 38:1848–1860.

Grove, J.M., A.R. Troy, J.P.M. O'Neil-Dunne, W.R. Burch, M.L. Cadenasso, and S.T.A. Pickett. 2006. Characterization of households and its implications for the vegetation of urban ecosystems. Ecosystems 9:578–597.

Hale, C.M., L.E. Frelich, P.B. Reich, and J. Pastor. 2005. Effects of European earthworm invasion on soil characteristics in northern hardwood forests of Minnesota, USA. Ecosystems 8:911–927.

Heidt, V., and M. Neef. 2008. Benefits of urban green space for improving urban climate. p. 84–96. In M. Carreiro et al. (ed.) Ecology, planning, and management of urban forests. Springer-Verlag, New York.

Heneghan, L., C. Clay, and C. Brundage. 2002. Observations on the initial decomposition rates and faunal colonization of native and exotic plant species in an urban forest fragment. Ecol. Res. 20:108–111.

Hjortenkrans, D., B. Bergback, and A. Haggerud. 2006. New metal emission patterns in road traffic environments. Environ. Monit. Assess. 117:85–98.

Holman-Dodds, J.K., A.A. Bradley, and K.W. Potter. 2003. Evaluation of hydrologic benefits of infiltration based urban storm water management. J. Am. Water Resour. Assoc. 39:205–215.

Hope, D., W. Zhu, C. Gries, J. Oleson, J. Kaye, N. Grimm, and L. Baker. 2005. Spatial variation in soil inorganic nitrogen across an arid urban ecosystem. Urban Ecosyst. 8:251–273.

Hope, D., C. Gries, W. Zhu, W.F. Fagan, C.L. Redman, N.B. Grimm, A.L. Nelson, C. Martin, and A. Kinzig. 2003. Socioeconomics drive urban plant diversity. Proc. Natl. Acad. Sci. USA 100:8788–8792.

Hopkin, S., and M. Martin. 1985. Assimilation of zinc, cadmium, lead, copper and iron by the spider Dysdera crocata, a predator of woodlice. Bull. Environ. Contam. Toxicol. 34:183–187.

Hornung, E., and K. Szlavecz. 2003. Establishment of a Mediterranean isopod (Chaetophiloscia sicula Verhoeff, 1908) in a North American temperate forest. p. 181–189. In A. Sfenthourakis et al. (ed.) The biology of terrestrial isopods. Crustaceana Monogr. 2. Brill, Leiden, the Netherlands

Hornung, E., B. Tóthmérész, and T. Magura. 2007. Changes of isopod assemblages along an urban–rural gradient in Hungary. Eur. J. Soil Biol. 43:158–161.

Huh, K.Y., M. Deurer, S. Sivakumaran, K. Mcauliffe, and N.S. Bolan. 2008. Carbon sequestration in urban landscapes: The example of a turfgrass system in New Zealand. Aust. J. Soil Res. 46:610–616.

International Committee on Anthropogenic Soils. 2007. Urban soils. Available at http://clic.cses.vt.edu/icomanth/urban_soils.htm (verified 19 Feb. 2010).

Ikem, A., M. Campbell, I. Nyirakabibi, and J. Garth. 2008. Baseline concentrations of trace elements in residential soils from Southeastern Missouri. Environ. Monit. Assess. 140:69–81.

Inman, J.C., and G.R. Parker. 1978. Decomposition and heavy metal dynamics of forest litter in northwestern Indiana. Environ. Pollut. 17:34–51.

Ireland, M. 1976. Excretion of lead, zinc and cadmium in Dendrobaena rubida (Oligochaeta) living in heavy metal polluted sites. Soil Biol. Biochem. 8:347–350.

Ireland, M. 1983. Heavy metal uptake and tissue distribution in earthworms. p. 245–265. In J. Satchell (ed.) Earthworm ecology: From Darwin to vermiculture. Chapman and Hall, London.

Ishitani, M., D. Kotze, and J. Niemelä. 2003. Changes in carabid beetle assemblages across an urban–rural gradient in Japan. Ecography 26:481–489.

Jenerette, G.D., J. Wu, N.B. Grimm, and D. Hope. 2006. Points, patches, and regions: Scaling soil biogeochemical patterns in an urbanized arid ecosystem. Glob. Change Biol. 12:1532–1544.

Jenny, H. 1941. Factors of soil formation. McGraw-Hill, New York.

Jenny, H. 1961. Derivation of state factor equations of soils and ecosystems. Soil Sci. Soc. Am. Proc. 25:385–388.

Jensen, H., C. Reimann, T.E. Finne, R.T. Ottesen, and A. Arnoldussen. 2007. PAH-concentrations and compositions in the top 2 cm of forest soils along a 120 km long transect through agricultural areas, forests and the city of Oslo, Norway. Environ. Pollut. 145:829–838.

Jim, C.Y. 1993. Soil compaction as a constraint to tree growth in tropical and subtropical urban habitats. Environ. Conserv. 20:35–49.

Jim, C.Y. 1998. Soil characteristics and management in an urban park in Hong Kong. Environ. Manage. 22:683–695.

Johnston, R. 2001. Synanthropic birds in North America. p. 49–68. In J. Marzluff et al. (ed.) Avian ecology in an urbanizing world. Kluwer, Norwell, MA.

Jordan, K., and S. Jones. 2007. Invertebrate diversity in newly established mulch habitats in a Midwestern urban landscape. Urban Ecosyst. 10:87–95.

Juknys, R., J. Zaltauskaite, and V. Stakenas. 2007. Ion fluxes with bulk and throughfall deposition along an urban–suburban–rural gradient. Water Air Soil Pollut. 178:363–372.

Kaminski, M.D., and S. Landsberger. 2000. Heavy metals in urban soils of East St. Louis, IL. Part I: Total concentration of heavy metals in soils. J. Air Waste Manage. Assoc. 50:1667–1679.

Kay, R.T., T.L. Arnold, W.F. Cannon, and D. Graham. 2008. Concentrations of polycyclic aromatic hydrocarbons and inorganic constituents in ambient surface soils, Chicago, Illinois: 2001–2002. Soil Sediment Contam. 17:221–236.

Kaye, J.P., I.C. Burke, A.R. Mosier, and J.P. Guerschman. 2004. Methane and nitrous oxide fluxes from urban soils to the atmosphere. Ecol. Appl. 14:975–981.

Kaye, J.P., P.M. Groffman, N.B. Grimm, L.A. Baker, and R.V. Pouyat. 2006. A distinct urban biogeochemistry? Trends Ecol. Evol. 21:192–199.

Kaye, J.P., A. Majumdar, C. Gries, A. Buyantuyev, N.B. Grimm, D. Hope, G.D. Jenerette, W.X. Zhu, and L. Baker. 2008. Hierarchical Bayesian scaling of soil properties across urban, agricultural, and desert ecosystems. Ecol. Appl. 18:132–145.

Kaye, J.P., R.L. Mcculley, and I.C. Burke. 2005. Carbon fluxes, nitrogen cycling, and soil microbial communities in adjacent urban, native and agricultural ecosystems. Glob. Change Biol. 11:575–587.

Keplin, B., and G. Broll. 1999. Earthworms and dehydrogenase activity of urban biotopes. Soil Biol. Biochem. 29:533–536.

Kim, K., and L. Byrne. 2006. Biodiversity loss and the taxonomic bottleneck: Emerging biodiversity science. Ecol. Res. 21:794–810.

Klein, R.D. 1989. The relationship between stream quality and golf courses. Community and Environmental Defense Services, Maryland Line, MD.

Komarnicki, G. 2000. Tissue, sex and age specific accumulation of heavy metals (Zn, Cu, Pb, Cd) by populations of the mole (Talpa europaea L.) in a central urban area. Chemosphere 41:1593–1602.

Korsós, Z., E. Hornung, J. Kontschán, and K. Szlavecz. 2002. Isopoda and Diplopoda of urban habitats: New data to the fauna of Budapest. Ann. Hist. Nat. Musei Natl. Hungarici 94:193–208.

Kourtev, P.S., J.G. Ehrenfeld, and M. Haggblom. 2002. Exotic plant species alter the microbial community structure and function in the soil. Ecology 83:3152–3166.

Kuperman, R.G. 1999. Litter decomposition and nutrient dynamics in oak–hickory forests along a historic gradient of nitrogen and sulfur deposition. Soil Biol. Biochem. 31:237–244.

Laakso, J., and H. Setälä. 1999. Sensitivity of primary production to changes in the architecture of belowground food webs. Oikos 87:57–64.

Law, N.L., L.E. Band, and J.M. Grove. 2004. Nitrogen input from residential lawn care practices in suburban watersheds in Baltimore county, MD. J. Environ. Plann. Manage. 47:737–755.

Lee, C.S., X.D. Li, W.Z. Shi, S.C. Cheung, and I. Thornton. 2006. Metal contamination in urban, suburban, and country park soils of Hong Kong: A study based on GIS and multivariate statistics. Sci. Total Environ. 356:45–61.

Lee, D.S., and J.W.S. Longhurst. 1992. A comparison between wet and bulk deposition at an urban site in the U.K. Water Air Soil Pollut. 64:635–648.

Lee, K.E. 1985. Earthworms: Their ecology and relationships with soil and land use. Academic Press, Sydney, Australia.

Lehmann, A., and K. Stahr. 2007. Nature and significance of anthropogenic urban soils. J. Soils Sediments 7:247–260.

Lev, S., E. Landa, K. Szlavecz, R. Casey, and J. Snodgrass. 2008. Application of synchrotron methods to assess the uptake of roadway derived Zn by earthworms in an urban soil. Mineral. Mag. 72:33–37.

Li, X.D., S.L. Lee, S.C. Wong, W.Z. Shi, and L. Thornton. 2004. The study of metal contamination in urban soils of Hong Kong using a GIS-based approach. Environ. Pollut. 129:113–124.

Lilleskov, E., M. Callaham, R.V. Pouyat, J. Smith, M. Castellano, G. Gonzalez, D.J. Lodge, R. Arango, and F. Green. 2009. Invasive soil organisms and their impacts on below ground processes. In M. Dix and K. Britton (ed.) A dynamic invasive species research vission. General Tech. Rep. WO-83. USDA Forest Service, Washington, DC.

Lilleskov, E.A., W.J. Mattson, and A.J. Storer. 2008. Divergent biogeography of native and introduced soil macroinvertebrates in North America north of Mexico. Divers. Distrib. 14:893–904.

Lorenz, K., and R. Lal. 2009. Biogeochemical C and N cycles in urban soils. Environ. Int. 35:1–8.

Loumbourdis, N. 1997. Heavy metal contamination in a lizard, *Agama stellio*, compared in urban, high altitude and agricultural, low altitude areas of North Greece. Bull. Environ. Contam. Toxicol. 58:945–952.

Lovett, G.M., M.M. Traynor, R.V. Pouyat, M.M. Carreiro, W.X. Zhu, and J.W. Baxter. 2000. Atmospheric deposition to oak forests along an urban-rural gradient. Environ. Sci. Technol. 34:4294–4300.

Madrid, L., E. Diaz-Barrientos, and F. Madrid. 2002. Distribution of heavy metal contents of urban soils in parks of Seville. Chemosphere 49:1301–1308.

Manta, D.S., M. Angelone, A. Bellanca, R. Neri, and M. Sprovieri. 2002. Heavy metals in urban soils: A case study from the city of Palermo (Sicily), Italy. Sci. Total Environ. 300:229–243.

McDonnell, M.J., S.T.A. Pickett, P. Groffman, P. Bohlen, R.V. Pouyat, W.C. Zipperer, R.W. Parmelee, M.M. Carreiro, and K. Medley. 1997. Ecosystem processes along an urban-to-rural gradient. Urban Ecosyst. 1:21–36.

McGuire, M.P. 2004. Using DTM and LIDAR data to analyze human induced topographic change. ASPRS 2004 Fall Conf., Kansas City, MO. 12–16 Sept. 2004. Am. Soc. for Photogrammetry and Remote Sensing, Bethesda, MD.

McIntyre, N.E., J. Rango, W.F. Fagan, and S.H. Faeth. 2001. Ground arthropod community structure in a heterogeneous urban environment. Landsc. Urban Plan. 52:257–274.

McKinney, M.L. 2002. Urbanization, biodiversity, and conservation. Bioscience 52:883–890.

McKinney, M.L. 2006. Urbanization as a major cause of biotic homogenization. Biol. Conserv. 127:247–260.

McKinney, M.L. 2008. Effects of urbanization on species richness: A review of plants and animals. Urban Ecosyst. 11:161–176.

McKinney, M.L., and J. Lockwood. 2001. Biotic homogenization: A sequential and selective process. p. 1–18. In J. Lockwood and M. McKinney (ed.) Biotic homogenization. Springer Verlag, New York.

Mielke, H.W. 1999. Lead in the inner cities. Am. Sci. 87:62–73.

Mielke, H.W., J.C. Anderson, K.J. Berry, P.W. Mielke, R.L. Chaney, and M. Leech. 1983. Lead concentrations in inner-city soils as a factor in the child lead problem. Am. J. Public Health 73:1366–1369.

Milesi, C., S.W. Running, C.D. Elvidge, J.B. Dietz, B.T. Tuttle, and R.R. Nemani. 2005. Mapping and modeling the biogeochemical cycling of turfgrasses in the United States. Environ. Manage. 36:426–438.

Mohrlok, U., and T. Schiedek. 2007. Urban impact on soils and groundwater—From infiltration processes to integrated urban water management. J. Soils Sediments 7:68.

Morgan, J.E., and A.J. Morgan. 1993. Seasonal changes in the tissue metal (Cd, Zn and Pb) concentration in two ecophysiologically dissimilar earthworm species- pollution monitoring implications. Environ. Pollut. 82:1–7.

Morton, T.G., A.J. Gold, and W.M. Sullivan. 1988. Influence of overwatering and fertilization on nitrogen losses from home lawns. J. Environ. Qual. 17:124–130.

Mount, H., L. Hernandez, T. Goddard, and S. Indrick. 1999. Temperature signatures for anthropogenic soils in New York City. p. 137–140. In J.M. Kimble et al. (ed.) Classification, correlation, and management of anthropogenic soils. Proc. Nevada and California Worksh. 21 Sept.–2 Oct. 1998. USDA-NRCS, National Soil Survey Center, Lincoln, NE

Nekola, J., and P. White. 1999. The distance decay of similarity in biogeography and ecology. J. Biogeogr. 26:867–878.

Niemelä, J., D. Kotze, and S. Venn. 2002. Carabid beetle assemblages (Coleoptera, Carabidae) across urban–rural gradients: An international comparison. Landscape Ecol. 17:387–401.

Niemelä, J., J. Kotze, and A. Ashworth. 2000. The search for common anthropogenic impacts on biodiversity: A global network. J. Insect Conserv. 4:3–9.

Nowak, D. 2010. Urban biodiversity and climate change. p. 101–117. In N. Muller et al. (ed.) Urban biodiversity and design. Blackwell, Oxford, UK.

Nowak, D.J. 2000. The interactions between urban forests and global climate change. p. 31–44. In K.K. Abdollahi et al. (ed.) Global climate change and the urban forest. Franklin, Baton Rouge, LA.

Nowak, D.J., M. Kuroda, and D.E. Crane. 2004. Tree mortality rates and tree population projections in Baltimore, Maryland, USA. Urban For. Urban Green. 2:139–147.

Odewande, A.A., and A.F. Abimbola. 2008. Contamination indices and heavy metal concentrations in urban soil of Ibadan metropolis, southwestern Nigeria. Environ. Geochem. Health 30:243–254.

Ohtonen, R. 1994. Accumulation of organic matter along a pollution gradient: Application of Odum's theory of ecosystem energetics. Microb. Ecol. 27:43–55.

Ohtonen, R., and A.M. Markkola. 1991. Biological activity and amount of FDA mycelium in mor humus of Scots pine stands (Pinus sylvestris L.) in relation to soil properties and degree of pollution. Biogeochemistry 13:1–26.

Orsini, C.Q., M.H. Tabacniks, P. Artaxo, M.F. Andrade, and A.S. Kerr. 1986. Characteristics of fine and coarse particles of natural and urban aerosols of Brazil. Atmos. Environ. 20:2259–2269.

Osmond, D.L., and D.H. Hardy. 2004. Characterization of turf practices in five North Carolina communities. J. Environ. Qual. 33:565–575.

Patterson, J.C., J.J. Murray, and J.R. Short. 1980. The impact of urban soils on vegetation. METRIA: 3, Proc. of the Third Conference of the Metropolitan Tree Improvement Alliance. Rutgers, New Brunswick, NJ.

Pavao-Zuckerman, M.A. 2003. Soil ecology along an urban to rural gradient in the Southern Appalachians. Ph.D. diss. University of Georgia, Athens.

Pavao-Zuckerman, M.A. 2008. The nature of urban soils and their role in ecological restoration in cities. Restor. Ecol. 16:642–649.

Pickett, S.T.A., and M.L. Cadenasso. 2009. Altered resources, disturbance, and heterogeneity: A framework for comparing urban and non-urban soils. Urban Ecosyst. 12:23–44.

Pilipiuk, I. 1981. Earthworms (Oligichaeta, Lumbricidae) of Warsaw and Mazovia. Memorabilia Zoologia 34:69–78.

Pitt, R., and J. Lantrip. 2000. Infiltration through disturbed urban soil. p. 1–22. In W. James (ed.) Advances in modeling the management of stormwater impacts. Computational Hydraulics International, Guelph, Canada.

Pizl, V., and G. Josens. 1995. Earthworm communities along a gradient of urbanization. Environ. Pollut. 90:7–14.

Pouyat, R.V. 1991. The urban–rural gradient: An opportunity to better understand human impacts on forest soils. Proc. of the Society of American Foresters, 1990 Annual Conv., Washington, DC. Soc. of Am. Foresters, Washington, DC.

Pouyat, R.V., K. Belt, D. Pataki, P.M. Groffman, J. Hom, and L. Band. 2007b. Effects of urban land-use change on biogeochemical cycles. p. 45–58. In P. Canadell et al. (ed.) Terrestrial ecosystems in a changing world. Springer, New York.

Pouyat, R.V., M.M. Carreiro, P.M. Groffman, and M. Zuckerman. 2009a. Investigative approaches to urban biogeochemical cycles: New York metropolitan area and Baltimore as case studies. p. 329–352. In M. Mcdonnell et al. (ed.) Ecology of cities and towns: A comparative approach. Cambridge Univ. Press, Cambridge, UK.

Pouyat, R.V., and W.R. Effland. 1999. The investigation and classification of humanly modified soils in the Baltimore Ecosystem Study. p. 141–154. In J.M. Kimble et al. (ed.) Classification, correlation, and management of anthropogenic soils. Proc. Nevada and California Worksh. 21 Sept.–2 Oct. 1998. USDA-NRCS, National Soil Survey Center, Lincoln, NE.

Pouyat, R., P. Groffman, I. Yesilonis, and L. Hernandez. 2002. Soil carbon pools and fluxes in urban ecosystems. Environ. Pollut. 116:S107–S118.

Pouyat, R.V., M.J. Mcdonnell, and S.T.A. Pickett. 1995. Soil characteristics of oak stands along an urban–rural land use gradient. J. Environ. Qual. 24:516–526.

Pouyat, R.V., R.W. Parmelee, and M.M. Carreiro. 1994. Environmental effects of forest soil-invertebrate and fungal densities in oak stands along an urban–rural land use gradient. Pedobiologia 38:385–399.

Pouyat, R.V., J. Russell-Anelli, I.D. Yesilonis, and P.M. Groffman. 2003. Soil carbon in urban forest ecosystems. p. 347–362. In J.M. Kimble et al. (ed.) The potential of U.S. forest soils to sequester carbon and mitigate the greenhouse effect. CRC Press, Boca Raton, FL.

Pouyat, R.V., I.D. Yesilonis, and N.E. Golubiewski. 2009b. A comparison of soil organic carbon stocks between residential turfgrass and native soil. Urban Ecosyst. 12:45–62.

Pouyat, R.V., I.D. Yesilonis, and D.J. Nowak. 2006. Carbon storage by urban soils in the United States. J. Environ. Qual. 35:1566–1575.

Pouyat, R.V., I.D. Yesilonis, J. Russell-Anelli, and N.K. Neerchal. 2007a. Soil chemical and physical properties that differentiate urban land-use and cover types. Soil Sci. Soc. Am. J. 71:1010–1019.

Pouyat, R.V., I.D. Yesilonis, K. Szlavecz, C. Csuzdi, E. Hornung, Z. Korsos, J. Russell-Anelli, and V. Giorgio. 2008. Response of forest soil properties to urbanization gradients in three metropolitan areas. Landscape Ecol. 23:1187–1203.

Qian, Y.L., W. Bandaranayake, W.J. Parton, B. Mecham, M.A. Harivandi, and A.R. Mosier. 2003. Long-term effects of clipping and nitrogen management in turfgrass on soil organic carbon and nitrogen dynamics: The CENTURY model simulation. J. Environ. Qual. 32:1694–1700.

Raciti, S.M., P.M. Groffman, and T.J. Fahey. 2008. Nitrogen retention in urban lawns and forests. Ecol. Appl. 18:1615–1626.

Reichard, S.H. 2010. Inside out: Invasive plants and urban environments. p. 241–252. In J. Aitkenhead-Peterson and A. Volder (ed.) Urban ecosystem ecology. Agron. Monogr. 55. ASA, CSSA, and SSSA, Madison, WI.

Rossiter, D.G. 2007. Classification of urban and industrial soils in the world reference base for soil resources. J. Soils Sediments 7:96–100.

Ruiz-Cortes, E., R. Reinoso, E. Diaz-Barrientos, and L. Madrid. 2005. Concentrations of potentially toxic metals in urban soils of Seville: Relationship with different land uses. Environ. Geochem. Health 27:465–474.

Russell-Anelli, J.M., R. Bryant, and J. Galbraith. 1999. Evaluating the predictive properties of soil survey–soil characteristics, land practices and concentration of elements. p. 155–168. In J.M. Kimble et al. (ed.) Classification, correlation, and management of anthropogenic soils. Proc. Nevada and California Worksh. 21 Sept.–2 Oct. 1998. USDA-NRCS, National Soil Survey Center, Lincoln, NE.

Savva, Y., K. Szlavecz, R.V. Pouyat, P.M. Groffman, and G. Heisler. 2010. Land use and vegetation cover effects on soil temperature in an urban ecosystem. Soil Sci. Soc. Am. J. 74:469–480.

Sawicka-Kapusta, K., M. Zakrzewska, K. Bajorek, and J. Gdula-Argasinska. 2003. Input of heavy metals to the forest floor as a result of Cracow urban pollution. Environ. Int. 28:691–698.

Scharenbroch, B.C., J.E. Lloyd, and J.L. Johnson-Maynard. 2005. Distinguishing urban soils with physical, chemical, and biological properties. Pedobiologia 49:283–296.

Schleuss, U., Q.L. Wu, and H.P. Blume. 1998. Variability of soils in urban and periurban areas in Northern Germany. Catena 33:255–270.

Schrader, S., and M. Böning. 2006. Soil formation on green roofs and its contribution to urban biodiversity with emphasis on Collembolans. Pedobiologia 50:347–356.

Schwarz, K. 2010. The spatial distributin of lead in urban residential soil and correlations with urban land cover of Baltimore, Maryland. Ph.D. diss. Rutgers Univ., New Brunswick, NJ.

Schuhmacher, M., M. Meneses, S. Granero, J.M. Llobet, and J.L. Domingo. 1997. Trace element pollution of soils collected near a municipal solid waste incinerator: Human health risk. Bull. Environ. Contam. Toxicol. 59:861–867.

Short, J.R., D.S. Fanning, J.E. Foss, and J.C. Patterson. 1986a. Soils of the Mall in Washington, DC: I. Statistical summary of properties. Soil Sci. Soc. Am. J. 50:699–705.

Short, J.R., D.S. Fanning, J.E. Foss, and J.C. Patterson. 1986b. Soils of the Mall in Washington, DC: II. Genesis, classification and mapping. Soil Sci. Soc. Am. J. 50:705–710.

Smith, R., P. Warren, K. Thompson, and K. Gaston. 2006. Urban domestic gardens (VI): Environmental correlates of invertebrate species richness. Biodivers. Conserv. 15:2415–2438.

Soil Survey Staff. 1975. Soil taxonomy: A basic system of soil classification for making and interpreting soil surveys. USDA, Soil Conserv. Serv., Washington, DC.

Spence, J.R., and D.H. Spence. 1988. Of ground beetles and men: Introduced species and the synanthropic fauna of western Canada. Mem. Entomol. Soc. Can. 144:151–168.

Spurgeon, D., and S. Hopkin. 2000. The development of genetically inherited resistance to zinc in laboratory selected generations of the earthworm *Eisenia fetida*. Environ. Pollut. 109:193–201.

Stander, E.K., and J.G. Ehrenfeld. 2009. Rapid assessment of urban wetlands: Do hydrogeomorphic classification and reference criteria work? Environ. Manage. 43:725–742.

Stander, E.K., and J.G. Ehrenfeld. 2010. Urban riparian function. p. 253–276. In J. Aitkenhead-Peterson and A. Volder (ed.) Urban ecosystem ecology. Agron. Monogr. 55. ASA, CSSA, and SSSA, Madison, WI.

Steinberg, D.A., R.V. Pouyat, R.W. Parmelee, and P.M. Groffman. 1997. Earthworm abundance and nitrogen mineralization rates along an urban–rural land use gradient. Soil Biol. Biochem. 29:427–430.

Steele, M.K., W.H. McDowell, and J.A. Aitkenhead-Peterson. 2010. Chemistry of urban, suburban, and rural surface waters. p. 297–340. In J. Aitkenhead-Peterson and A. Volder (ed.) Urban ecosystem ecology. Agron. Monogr. 55. ASA, CSSA, and SSSA, Madison, WI.

Szlavecz, K., S.A. Placella, R.V. Pouyat, P.M. Groffman, C. Csuzdi, and I. Yesilonis. 2006. Invasive earthworm species and nitrogen cycling in remnant forest patches. Appl. Soil Ecol. 32:54–62.

Tenenbaum, D.E., L.E. Band, S.T. Kenworthy, and C.L. Tague. 2006. Analysis of soil moisture patterns in forested and suburban catchments in Baltimore, Maryland, using high-resolution photogrammetric and LIDAR digital elevation datasets. Hydrol. Processes 20:219–240.

Tume, P., J. Bech, B. Sepulveda, L. Tume, and J. Bech. 2008. Concentrations of heavy metals in urban soils of Talcahuano (Chile): A preliminary study. Environ. Monit. Assess. 140:91–98.

Unger, K. 1999. Untersuchungen zur Lumbricidenfauna der Stadt Greifswald. Z. Okologie Naturschutz 8:125–134.

USDA-NRCS. 1998. Soil survey of City of Baltimore, Maryland. Soil Surv. Rep. USDA-NRCS, Washington, DC.

Van Bohemen, H.D., and W.H. Janssen Van De Laak. 2003. The influence of road infrastructure and traffic on soil, water, and air quality. Environ. Manage. 31:50–68.

Van Cleve, K., F.S. Chapin III, C.T. Dyrness, and L.A. Viereck. 1991. Element cycling in taiga forests: State-factor control. Bioscience 41:78–88.

Vilisics, F., Z. Elek, G. Lovei, and E. Hornung. 2007. Composition of terrestrial isopod assemblages along an urbanisation gradient in Denmark. Pedobiologia 51:45–53.

Vitousek, P.M., K.V. Cleve, N. Balakrishnan, and D. Mueller-Dombois. 1983. Soil development and nitrogen turnover on recent volcanic substrates in Hawaii. Biotropica 15:268–274.

Walsh, D.C., S.N. Chillrud, H.J. Simpson, and R.F. Bopp. 2001. Refuse incinerator particulate emissions and combustion residues for New York City during the 20th century. Environ. Sci. Technol. 35:2441–2447.

Watmough, S.A., T.C. Hutchinson, and E.P.S. Sager. 1998. Changes in tree ring chemistry in sugar maple (*Acer saccharum*) along an urban-rural gradient in southern Ontario. Environ. Pollut. 101:381–390.

White, C.S., and M.J. McDonnell. 1988. Nitrogen cycling processes and soil characteristics in an urban versus rural forest. Biogeochemistry 5:243–262.

Williams, N.S.G., M.W. Schwartz, P.A. Vesk, M.A. McCarthy, A.K. Hahs, S.E. Clemants, R.T. Corlett, R.P. Duncan, B.A. Norton, K. Thompson, and M.J. McDonnell. 2009. A conceptual framework for predicting the effects of urban environments on floras. J. Ecol. 97:4–9.

Wolf, J., and J. Gibbs. 2004. Silphids in urban forests: Diversity and function. Urban Ecosyst. 7:371–384.

Wolf, L., J. Klinger, H. Hoetzl, and U. Mohrlok. 2007. Quantifying mass fluxes from urban drainage systems to the urban soil-aquifer system. J. Soils Sediments 7:85–95.

Wong, C.S.C., X.D. Li, and I. Thornton. 2006. Urban environmental geochemistry of trace metals. Environ. Pollut. 142:1–16.

Wong, F., T. Harner, Q. Liu, and M.L. Diamond. 2004. Using experimental and forest soils to investigate the uptake of polycyclic aromatic hydrocarbons (PAHs) along an urban–rural gradient. Environ. Pollut. 129:387–398.

Yesilonis, I.D., R.V. Pouyat, and N.K. Neerchal. 2008. Spatial distribution of metals in soils in Baltimore, Maryland: Role of native parent material, proximity to major roads, housing age and screening guidelines. Environ. Pollut. 156:723–731.

Yong, R.N., A.M.O. Mohamed, and B.P. Warkentin. 1992. Principles of contaminant transport in soils. Elsevier, New York.

Zhang, C.S. 2006. Using multivariate analyses and GIS to identify pollutants and their spatial patterns in urban soils in Galway, Ireland. Environ. Pollut. 142:501–511.

Zhang, H.B., Y.M. Luo, M.H. Wong, Q.C. Zhao, and G.L. Zhang. 2006. Distributions and concentrations of PAHs in Hong Kong soils. Environ. Pollut. 141:107–114.

Zhao, Y.F., X.Z. Shi, B. Huang, D.S. Yu, H.J. Wang, W.X. Sun, I. Oboern, and K. Blomback. 2007. Spatial distribution of heavy metals in agricultural soils of an industry-based peri-urban area in Wuxi, China. Pedosphere 17:44–51.

Zheng, Y.M., T.B. Chen, and J.Z. He. 2008. Multivariate geostatistical analysis of heavy metals in topsoils from Beijing, China. J. Soils Sediments 8:51–58.

Zhu, W.X., D. Hope, C. Gries, and N.B. Grimm. 2006. Soil characteristics and the accumulation of inorganic nitrogen in an arid urban ecosystem. Ecosystems 9:711–724.

8

Lawn Ecology

Thomas W. Cook
Erik H. Ervin

Abstract

Lawns make up a significant portion of the urban and suburban landscapes, with area estimates ranging from 58,000 to 163,800 km² nationwide, depending on assumptions (Vinlove and Torla, 1995; Milesi et al., 2005). Lawns are commonly viewed as a continuum of plant communities ranging from single species grass monocultures to more complicated polystand grass mixtures (Beard, 1973). Grass monocultures and polystands are commonly referred to as turf and are the most highly managed and manipulated lawn ecosystems. They depend largely on input from humans to maintain purity. Home lawns, parks, and other lower input general areas of the landscape are characterized by greater diversity of plant components and less control exercised by humans. As ecosystems these are surprisingly complex and may be quite stable through time. Our premise in this chapter is that the majority of lawns in the urban environment are the more complex and diverse ecosystems described above rather than pure monocultures or polystands of planted grasses found in intensively managed landscapes. Much of our discussion will focus on understanding the nature of lawn ecosystems in the 48 conterminous states in the USA, including regional differences in species, succession and development of diversity as it occurs through time, and the influence that cultural inputs have on species composition.

Technically, lawns are a subclimax of plants (Ashby, 1969) maintained as a result of regular mowing. This sub climax is further influenced by irrigation, fertility, shade, and pesticide use. Soil type, pH, and physical condition also influence species composition. Since lawns may persist for many decades, fluctuations in maintenance intensity due to changes in ownership, microenvironment, and available or allowable resources play an important role in lawn ecosystem development. Because these factors are not constant, lawn composition may fluctuate through time and may or may not develop a relatively stable subclimax of adapted species. In the post World War II era, lawns have been viewed almost exclusively as cool or warm season grasses from which all dicot species and unwanted warm season annual grasses are removed via herbicides or hand weeding. The available American literature discussing turfgrass ecology has emphasized short-term competition and succession among the planted grasses as influenced by such variables

T.W. Cook, Dep. of Horticulture, Oregon State University, 4017 Ag. and Life Sciences Bldg., Corvallis, OR 97331-7304 (cookt@hort.oregonstate.edu); E.H. Ervin, Dep. of Crop and Soil Environmental Sciences, Virginia Polytechnic Inst. & State University, 330 Smyth Hall, Blacksburg, VA 24061-0404 (eervin@vt.edu).

doi:10.2134/agronmonogr55.c8

that include seeding rates, mixture percentages, planting times, fertility management, and mowing heights (Schmidt and Blaser, 1969; Watschke and Schmidt, 1992; Danneberger, 1993). Since only a few dozen of the estimated 10,000 species of grass are adapted to regular mowing, most research has involved a very small number of species. Research from the United Kingdom has taken a broader view of ecology, emphasizing the interactions of a wider range of plant forms present in mature lawns (Gilbert, 1989; Thompson et al., 2004). An understanding of lawn ecology has to start with the planted grasses, the majority of which are introduced species to North America. Grasses used for lawns can be divided into groups based on regional climate zones in which species are listed according to adaptation in regions with similar climates (Ward, 1969; Madison, 1971). This approach creates a useful, though overly simplified, vision of grass adaptation.

Zones of Adaptation

Madison (1971) calculated effective day and night temperatures from standard weather data and developed a climatograph to demonstrate how summer temperature regimes influence competitive ability of grasses. Madison's climatograph identifies three basic turfgrass adaptation zones present in the United States. Warm season grasses (C4 metabolism), such as *Cynodon, Zoysia, Stenataphrum, Eremochloa,* and *Paspalum,* are most competitive in climates where summer nighttime effective temperatures are above 21°C and summer effective daytime temperatures are above 27°C. Cool season grasses (C3 metabolism), such as *Agrostis, Festuca, Lolium,* and *Poa,* are most competitive where effective summer nighttime temperatures are below 21°C and effective daytime temperatures are below 27°C. The gap area between the two distinct zones is termed a *transition zone,* where both cool and warm season grasses grow but warm season grasses struggle in winter and cool season grasses struggle in summer. Zones of adaptation can be further refined by natural precipitation and humidity patterns (Ward, 1969). Many different zone maps have been proposed by various authors (Dickinson, 1930; Wise, 1961; Musser, 1962; Ward, 1969; Madison, 1971). Turgeon (2008) adapted the work of Trewartha (1968) to develop global zone maps. Detailed zone maps have also been developed for individual states (Youngner et al., 1962). More recently, Beard and Beard (2005) proposed a scheme composed of 12 separate climatic zones for the continental United States. For this discussion of lawn ecology we have adapted the map of Beard and Beard (2005) to illustrate turf adaptation zones from our perspective (Plate 8–1, see color insert section).

Cool Humid Pacific Zone

This zone includes the coast, mountains, and valleys of Oregon, Washington, and British Columbia west of the Cascade Mountains. It moves south into California as a narrow corridor along the coast as far south as the west side of the San Francisco Bay and along the coast as a narrow strip to the Monterey area. Historically it has been grouped together with the Cool Humid Zone by other authors (Dickinson, 1930; Musser, 1962; Madison, 1971). Grasses in the Cool Humid Pacific Zone experience mild wet winters during which dormancy rarely occurs, prolonged spring-like weather, moderately warm dry summers, and mild falls leading up to the late fall rainy period. Midsummer relative humidity generally runs above 50%. The effective growing period for turf is typically more than 10 months per

year where irrigation is practiced. Without summer rainfall, all grasses eventually go dormant from mid July to the onset of fall rains, typically in October. Some
notable differences in climate patterns between the Cool Humid Pacific Zone and
Cool Humid Zone are illustrated in Fig. 8–1 and 8–2. The year-round mild climate
in the Cool Humid Pacific Zone fosters growth of a large palette of cool season
grasses, including bentgrasses (*Agrostis* spp.), bluegrasses (*Poa* spp.), fine fescues
(*Festuca* spp.), ryegrasses (*Lolium* spp.), tall fescue [*Lolium arundinaceum* (Schreb.)
S.J. Darbyshire], velvetgrasses (*Holcus* spp.), and rat-tail fescue [*Vulpia myuros* (L.)
C.C. Gmelin]. The actual proportions of grasses in mature lawns are most often
dependent on the intensity of irrigation and fertilization (Table 8–1).

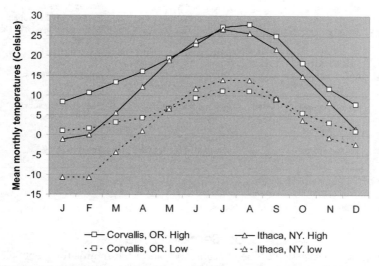

Fig. 8–1. Comparison of average high and low temperatures for representative sites in Cool Humid
and Cool Humid Pacific zones. Note the mild winter temperatures in the Cool Humid Pacific zone.

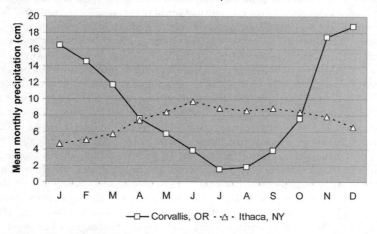

Fig 8–2. Comparison of precipitation patterns for representative sites in Cool Humid and Cool
Humid Pacific zones. Note the lack of summer rainfall in the Cool Humid Pacific zone.

Table 8–1. Long-term abundance of cool season grasses in lawns maintained under a variety of cultural intensities in the Cool Humid Pacific Zone.

	Irrigated, low fertility	Irrigated, high fertility	Nonirrigated
Abundant	Agrostis stolonifera L.	Poa annua L.†	Agrostis castellana Boiss. & Reut.
	Agrostis castellana Boiss. & Reut.	Agrostis stolonifera L.	Lolium arundinaceum Schreb. subsp. arundinacea
	Agrostis capillaris L.	Agrostis castellana Boiss. & Reut.	Vulpia myuros (L.) C.C. Gmel.‡
	Holcus lanatus L.	Agrostis capillaris L.	Poa trivialis L.
	Holcus mollis L.	Poa trivialis L.	Agrostis stolonifera L.
	Poa annua L.†	Holcus lanatus L.	Holcus lanatus L.
	Poa trivialis L.	Holcus mollis L.	Holcus mollis L.
	Festuca rubra L. subsp. rubra	Lolium perenne L.	Festuca rubra L. subsp. rubra
	Festuca rubra subsp. commutata Gaudin	Poa pratensis L.	Festuca trachyphylla (Hack.) Krajina
	Festuca trachyphylla (Hack.) Krajina	Festuca rubra L. subsp. rubra	Poa pratensis L.
	Lolium arundinaceum Schreb. subsp. arundinacea	Festuca rubra subsp. commutata Gaudin	Agrostis capillaris L.
	Lolium perenne L.	Festuca trachyphylla (Hack.) Krajina	Festuca rubra subsp. commutata Gaudin
	Poa pratensis L.	Lolium arundinaceum Schreb. subsp. arundinacea	Lolium perenne L.
Scarce	Vulpia myuros (L.) C.C. Gmel. ‡	Vulpia myuros (L.) C.C. Gmel. ‡	Poa annua L.§

† Perennial under regular irrigation.

‡ Winter annual grass.

§ Winter annual under nonirrigated conditions.

Cool Semiarid Zone

Starting on the east side of the Cascade and Sierra Mountains the Cool Semiarid Zone extends through Montana and Wyoming to the east. It reaches south into the central parts of Colorado, Utah, and the northern half of Nevada, as noted in Plate 8–1. Average summer relative humidity at midday is typically around 30%. Nonnative turfgrasses can only be grown with regular irrigation. The entire zone is cold in winter, and all grasses survive winters via dormancy. Growth begins in spring as temperatures rise above freezing, reaching a peak in June. Depending on location, midsummer may bring above-optimum temperatures and reduced growth. Fall provides a second optimum growth period before development of winter dormancy. The growing season is well suited to growth of cool season grasses, and adaptation is largely determined by the cold winter temperatures. Lack of a cool season grass soil seed bank means that generally only planted grasses occur in turf unless others are transported in via contaminated seed. With irrigation, this is excellent Kentucky bluegrass (*Poa pratensis* L.) country. Other cool season grasses common to the area include fine fescues, tall fescue, perennial ryegrass in milder areas, and annual bluegrass (*Poa annua* L.) in intensively managed turf. Native cool season grasses adapted to low-input turf in parts of the zone include western wheatgrass [*Pascopyrum smithii* (Rydb.) A. Löve], streambank wheatgrass (*Elymus lanceolatus* Scribn. and J.G. Sm. Gould), and crested wheatgrass [*Agropyrum cristatum* (L.) Gaertn.]. The abundance of common cool season grasses in the region under various cultural regimes is listed in Table 8–2.

Table 8–2. Long-term abundance of common grasses in lawns maintained under a variety of cultural intensities in the Cool Semiarid Zone.

	Irrigated, low fertility	Irrigated, high fertility	Minimal irrigation, low fertility
Abundant	Poa pratensis L.	Poa pratensis L.	Buchloe dactyloides (Nutt.) Engelm.†
	Festuca rubra L. subsp. rubra	Lolium perenne L.	Bouteloua gracilis (Kunth) Lag. ex Griffiths†
	Festuca rubra subsp. commutata Gaudin	Festuca rubra L. subsp. rubra	Agropyron cristatum (L.) Gaertn.
	Festuca trachyphylla (Hack.) Krajina	Festuca rubra subsp. commutata Gaudin	Pascopyrum smithii (Rydb.) Barkworth & D.R. Dewey†
	Lolium perenne L.	Festuca trachyphylla (Hack.) Krajina	Agropyron riparium Scribn. & J.G. Sm.†
	Lolium arundinaceum Schreb. subsp. arundinacea	Lolium arundinaceum Schreb. subsp. arundinacea	Elymus lanceolatus (Scribn. & J.G. Sm.) Gould
	Agrostis stolonifera L.	Poa annua L.‡	
	Poa annua L.‡	Agrostis stolonifera L.	
	Agrostis capillaris L.	Agrostis capillaris L.	
Scarce	Poa trivialis L.	Poa trivialis L.	

† Native species. Buchloe and Bouteloua are warm season grasses.

‡ Short to long lived perennial under irrigation.

Semi-cool Semiarid Zone

This zone runs from the lower third of North Dakota south through Nebraska and the western two-thirds of Kansas. It also extends into parts of Colorado, Utah, and Nevada as noted in Plate 8–1. Average summer relative humidity at midday runs approximately 45% in much of the zone. Grasses in the eastern part of the zone may survive without irrigation, but quality cool season turf requires regular irrigation. The entire zone is cold in winter, and all grasses survive winters via dormancy. Growth begins in spring as temperatures rise above freezing, reaching a peak in June. Depending on location, midsummer may bring above-optimum temperatures and reduced growth. Fall provides a second optimum growth period, followed by cold winter temperatures and dormancy. The growing season is well suited to growth of cool season grasses, and adaptation is largely determined by winter low temperatures. Lack of a cool season grass soil seed bank means that generally only planted grasses occur in turf unless others are transported in via contaminated seed. Commonly planted cool season grasses include Kentucky bluegrass, fine fescues, tall fescue, perennial ryegrass in milder areas, and annual bluegrass in intensively managed turf. Native cool season grasses adapted to low input turf in parts of the zone include western wheatgrass, streambank wheatgrass, and crested wheatgrass. The abundance of common cool season grasses in the region under various cultural regimes is listed in Table 8–3. Eastern parts of the Semi-cool Semiarid Zone also support the growth of American buffalograss [Buchloe dactyloides (Nutt.) Engelm.] and blue grama [Bouteloua gracillis (Willd. ex Kunth) Lag. ex Griffiths], which are cold-tolerant warm season natives adapted in the central plains. The southern part of the zone is suited for bermudagrass [Cynodon dactylon (L.) Pers.] and zoysiagrass (Zoysia japonica Steud.).

Table 8–3. Long-term abundance of cool season grasses in lawns maintained under a variety of cultural intensities in the Semi-cool Semiarid Zone.

	Irrigated, low fertility	Irrigated, high fertility	Nonirrigated
Abundant	*Poa pratensis* L.	*Poa pratensis* L.	*Buchloe dactyloides* (Nutt.) Engelm.†
	Festuca rubra L. subsp. *rubra*	*Lolium perenne* L.	*Bouteloua gracilis* (Kunth) Lag. ex Griffiths†
	Festuca rubra subsp. *commutata* Gaudin	*Festuca rubra* L. subsp. *rubra*	*Lolium arundinaceum* Schreb. subsp. *arundinacea*
	Festuca trachyphylla (Hack.) Krajina	*Festuca rubra* subsp. *commutata* Gaudin	*Agropyron cristatum* (L.) Gaertn.
	Lolium perenne L.	*Festuca trachyphylla* (Hack.) Krajina	*Pascopyrum smithii* (Rydb.) Barkworth & D.R. Dewey‡
	Lolium arundinaceum Schreb. subsp. *arundinacea*	*Lolium arundinaceum* Schreb. subsp. *arundinacea*	*Agropyron riparium*‡
	Agrostis stolonifera L.	*Poa annua* L.§	*Elymus lanceolatus* (Scribn. & J.G. Sm.) Gould‡
	Poa annua L.	*Agrostis stolonifera* L.	*Festuca trachyphylla* (Hack.) Krajina
	Agrostis capillaris L.	*Agrostis capillaris* L.	*Festuca rubra* L. subsp. *rubra*
Scarce	*Poa trivialis* L.	*Poa trivialis* L.	*Agrostis capillaris* L.

† Warm season perennial native grasses.

‡ Cool season perennial native grasses.

§ Behaves as a perennial under intensive maintenance.

Cool Humid Zone

Located along the Great Lakes, this zone encompasses the northern half of Minnesota, Wisconsin, Michigan, and New York, as well as all of Vermont, New Hampshire, and Maine (Plate 8–1). As the name implies, average midday relative humidity throughout the region is high, running typically between 50 and 75%. Grasses in the Cool Humid Zone have a burst of growth in spring followed by a summer stress period of above-optimum temperatures, high humidity, and abundant summer rainfall. As humidity and temperatures drop in the fall, a second flush of growth occurs before development of winter dormancy. This zone experiences higher night temperatures during summer and significantly higher summer rainfall than the Cool Semiarid Zone. The effective growing season is approximately 7 mo. Grasses can survive without irrigation but are subject to dormancy under prolonged summer drought. Cold tolerance is important for survival. Common cool season grasses adapted to this zone include bluegrasses, bentgrasses, fine fescues, and tall fescue. Perennial ryegrass is marginally hardy in the coldest parts of the zone. Grass long-term abundance under varying cultural regimes is noted in Table 8–4.

Semi-cool Humid Zone

Ranging from the lower half of Minnesota down to the eastern part of Kansas, this zone extends through the lower portion of Wisconsin, Michigan, and New York through Massachusetts to the coast. The southern boundary runs from the northern half of Missouri to Pennsylvania, with a peninsula extending down through mountainous West Virginia (Plate 8–1). Average midday relative humidity throughout the region runs as high as 75%. Grasses have a burst of growth in spring followed by a summer stress period of above-optimum temperatures, high humidity, and abundant summer rainfall. As humidity and temperatures

Table 8–4. Long-term abundance of cool season grasses in lawns maintained under a variety of cultural intensities in the Cool Humid Zone.

	Irrigated, low fertility	Irrigated, high fertility	Nonirrigated
Abundant	*Poa pratensis* L.	*Poa pratensis* L.	*Lolium arundinaceum* Schreb. subsp. *arundinacea*
	Festuca rubra L. subsp. *rubra*	*Poa annua* L.†	*Festuca trachyphylla* (Hack.) Krajina
	Festuca rubra subsp. *commutata* Gaudin	*Lolium perenne* L.	*Festuca rubra* L. subsp. *rubra*
	Festuca trachyphylla (Hack.) Krajina	*Festuca rubra* L. subsp. *rubra*	*Festuca rubra* subsp. *commutata* Gaudin
	Agrostis stolonifera L.	*Festuca rubra* subsp. *commutata* Gaudin	*Poa pratensis* L.
	Lolium arundinaceum Schreb. subsp. *arundinacea*	*Festuca trachyphylla* (Hack.) Krajina	*Lolium perenne* L.
	Lolium perenne L.	*Lolium arundinaceum* Schreb. subsp. *arundinacea*	*Agrostis capillaris* L.
	Poa trivialis L.	*Poa trivialis* L.	*Agrostis stolonifera* L.
	Poa annua L.	*Agrostis stolonifera* L.	*Poa trivialis* L.
Scarce	*Agrostis capillaris* L.	*Agrostis capillaris* L.	*Poa annua* L.‡

† Behaves as a perennial under intensive maintenance.

‡ Behaves as a recurring annual under nonirrigated conditions.

drop in the fall, a second flush of growth occurs before development of winter dormancy. The effective growing period for much of the zone runs from 7 to 9 mo. Throughout the region, grasses can survive without irrigation but are subject to dormancy under prolonged summer drought. Cold tolerance is important, along with heat tolerance, for cool season grasses. Common cool season grasses adapted to this zone include bluegrasses, bentgrasses, fine fescues, ryegrasses, and tall fescue. Grass long-term abundance under varying cultural regimes is noted in Table 8–5.

Table 8–5. Long-term abundance of grasses in lawns maintained under a variety of cultural intensities in the Semi-cool Humid Zone.

	Irrigated, low fertility	Irrigated, high fertility	Nonirrigated
Abundant	*Lolium arundinaceum* Schreb. subsp. *arundinacea*	*Poa pratensis* L.	*Lolium arundinaceum* Schreb. subsp. *arundinacea*
	Festuca trachyphylla (Hack.) Krajina	*Lolium arundinaceum* Schreb. subsp. *arundinacea*	*Poa pratensis* L.
	Festuca rubra L. subsp. *rubra*	*Lolium perenne* L.	*Lolium perenne* L.
	Digitaria spp.†	*Zoysia japonica* Steud.‡	*Zoysia japonica* Steud.‡
	Poa pratensis L.	*Digitaria* spp.†	*Muhlenbergia schreberi* J.F. Gmel.‡
	Zoysia japonica Steud.‡	*Festuca trachyphylla* (Hack.) Krajina	*Digitaria* spp.†
	Muhlenbergia schreberi J.F. Gmel.‡	*Festuca rubra* L. subsp. *rubra*	*Festuca trachyphylla* (Hack.) Krajina
	Lolium perenne L.	*Muhlenbergia schreberi* J.F. Gmel.‡	*Festuca rubra* L. subsp. *rubra*
Scarce	*Poa trivialis* L.	*Poa trivialis* L.	*Poa trivialis* L.

† Summer annual warm season grasses.

‡ Perennial warm season grasses.

Humid Transition Zone

The Humid Transition Zone denotes a variable width band of the USA north and south of the juncture of the cool season grass zones and the warm season grass zones. As noted in Plate 8–1, the zone ranges as far north as Long Island and New Jersey, because of the moderating influence of the Atlantic Ocean on winter temperatures, and then cuts southward through Washington DC, Virginia, Kentucky, Missouri, and Kansas before ending in the eastern third of Oklahoma. Summers in this zone are well suited to warm season grasses, but temperatures are well above optimum for cool season grasses, which often suffer from summer diseases and physiological decline. Winters are suited to cool season grass survival but may cause significant freezing injury to warm season grasses. Cool season grasses typically grow well in the spring and early summer, thin out or dieback in summer, recover in the fall, and go at least partially dormant in winter. The most widely planted cool season grasses in this zone include tall fescue, Kentucky bluegrass, and perennial ryegrass. Tall fescue has the best heat and drought resistance and is most often the dominant component of irrigated or nonirrigated cool season lawns. Summer annual grasses such as crabgrass (*Digitaria* spp.) are ubiquitous throughout the region and are often quite conspicuous during summer when cool season grasses weaken and begin to fail. Warm season grasses coexist with cool season grasses, becoming increasingly prevalent or dominant in the southern half of the zone. The primary warm season grasses in this zone include bermudagrass and zoysiagrass. Nimblewill (*Muhlenbergia schreberi* Gmel.) is a weedy warm-season perennial that often invades moderately shaded areas of low maintenance lawns. Zoysiagrass is generally more winter hardy than bermudagrass throughout the zone, so it more often occurs as a contaminant in cool-season lawns in northern areas of the zone. This zone is one of the most difficult areas in the United States to grow healthy lawns and typically has more weed, disease, and insect problems than other zones. Grass abundance in this zone under varying cultural regimes is listed in Table 8–6.

Table 8–6. Long-term abundance of grasses in lawns maintained under a variety of cultural intensities in the Humid Transitional Zone.

	Irrigated, low fertility	Irrigated, high fertility	Nonirrigated
Abundant	*Lolium arundinaceum* Schreb. subsp. *arundinacea*	*Lolium arundinaceum* Schreb. subsp. *arundinacea*	*Zoysia japonica* Steud.†
	Zoysia japonica Steud.†	*Zoysia japonica* Steud.†	*Cynodon dactylon* (L.) Pers.†
	Cynodon dactylon (L.) Pers.†	*Cynodon dactylon* (L.) Pers.†	*Digitaria* spp.‡
	Digitaria spp.‡	*Digitaria* spp.‡	*Lolium arundinaceum* Schreb. subsp. *arundinacea*
	Poa pratensis L.	*Lolium perenne* L.	*Poa pratensis* L.
	Festuca trachyphylla (Hack.) Krajina	*Poa pratensis* L.	*Lolium perenne* L.
	Festuca rubra L. subsp. *rubra*	*Festuca trachyphylla* (Hack.) Krajina	*Festuca trachyphylla* (Hack.) Krajina
	Lolium perenne L.	*Festuca rubra* L. subsp. *rubra*	*Festuca rubra* L. subsp. *rubra*
	Poa trivialis L.	*Poa trivialis* L.	*Poa trivialis* L.
Scarce	*Muhlenbergia schreberi* J.F. Gmel.†	*Muhlenbergia schreberi* J.F. Gmel.†	*Muhlenbergia schreberi* J.F. Gmel.†

† Perennial warm season grasses.

‡ Annual warm season grasses.

Warm and Tropical Humid Zones

The Warm Humid and Tropical Zones encompass all of the Deep South states from the coastal portions of the Carolinas through the eastern third of Texas (Plate 8–1). In general, these two zones are characterized by a long, hot, humid growing season, with summer rainfall and generally mild winter temperatures ranging from 4°C in January in midsouthern Arkansas to 19°C in mid January in south Florida (Holt, 1969). Rainfall throughout the region averages approximately 125 cm yr⁻¹, with highest levels along the Gulf Coast and lowest levels at the western border of the zone. Warm season grasses common throughout these zones include bermudagrass and zoysiagrass, with centipedegrass [*Eremochloa ophiuroides* (Munro) Hack.], St. Augustinegrass [*Stenotaphrum secundatum* (Walt.) Kuntze], carpetgrass [*Axonopus fissifolius* (Raddi) Kuhlm.], and bahiagrass (*Paspalum notatum*) performing best in the southern portions of the zone (Ward, 1969). Bermudagrass dominates in full sun under all levels of maintenance and readily invades drought-weakened stands of other species. Monostands of St. Augustine-grass or centipedegrass are favored by summer irrigation and moderate shade, with centipedegrass being more persistent under low fertility. Bahiagrass, a prolific seeder, will often be a large component of unirrigated areas in the Humid Tropical Zone, while carpetgrass, a nurse-grass in centipedegrass plantings, is not competitive under high fertility or without irrigation. A recently improved species, seashore paspalum (*Paspalum vaginatum* O. Swartz.), is the most salt-tolerant of the group while also possessing excellent drought tolerance and improved shade tolerance relative to bermudagrass. While seashore paspalum most likely has a bright future in southern climate zones, its use has been restricted to golf and sports turf surfaces and is not currently planted or found in southern lawns. Tall fescue is the primary cool season grass in the region, but is marginally adapted primarily in the northern part of the Warm Humid Zone and at higher elevations. See Tables 8–7 and 8–8 for abundance of grasses in these zones under variable maintenance programs.

Table 8–7. Long-term performance of warm and cool season grasses in lawns maintained under a variety of cultural intensities in the Warm Humid Zone.

	Irrigated, low fertility	Irrigated, high fertility	Nonirrigated
Abundant	*Eremochloa ophiuroides* (Munro) Hack.	*Stenotaphrum secundatum* (Walter) Kuntze	*Paspalum notatum* Flüggé
	Axonopus spp.	*Cynodon* spp.	*Cynodon* spp.
	Stenotaphrum secundatum (Walter) Kuntze	*Zoysia* spp.	*Zoysia* spp.
	Cynodon spp.	*Eremochloa ophiuroides* (Munro) Hack.	*Stenotaphrum secundatum* (Walter) Kuntze
	Zoysia spp.	*Lolium arundinaceum* Schreb. subsp. *arundinacea*	*Eremochloa ophiuroides* (Munro) Hack.
	Muhlenbergia schreberi J.F. Gmel.†	*Muhlenbergia schreberi* J.F. Gmel.†	*Axonopus* spp.
	Lolium arundinaceum Schreb. subsp. *arundinacea*‡	*Axonopus* spp.	*Lolium arundinaceum* Schreb. subsp. *arundinacea*†
Scarce			*Muhlenbergia schreberi* J.F. Gmel.‡

† Warm season weedy perennial grass that invades in shade-thinned warm and cool season grass lawns.

‡ Cool season grass for partial shade sites in cooler northern portion of the zone.

Table 8–8. Long-term performance of warm and cool season grasses in lawns maintained under a variety of cultural intensities in the Warm Tropical Zone.

	Irrigated, low fertility	Irrigated, high fertility	Nonirrigated
Abundant	*Eremochloa ophiuroides* (Munro) Hack.	*Stenotaphrum secundatum* (Walter) Kuntze	*Paspalum notatum* Flüggé
	Paspalum notatum Flüggé	*Cynodon* spp.	*Cynodon* spp.
	Axonopus spp.	*Zoysia* spp.	*Zoysia* spp.
	Stenotaphrum secundatum (Walter) Kuntze	*Eremochloa ophiuroides* (Munro) Hack.	*Stenotaphrum secundatum* (Walter) Kuntze
	Paspalum vaginatum Sw.	*Paspalum notatum* Flüggé	*Eremochloa ophiuroides* (Munro) Hack.
	Cynodon spp.	*Axonopus* spp.	
Scarce	*Zoysia* spp.		

Warm Semiarid, Semiarid Transitional, and Warm Arid Zones

These zones begin in middle Oklahoma and Texas and move west through New Mexico, Arizona, California, and up past the Las Vegas area to Death Valley and west to Palm Springs. Maximum high temperatures in July average 35°C or above, with average minimum January temperatures generally greater than 6°C and total precipitation averaging between 25 and 50 cm yr^{-1}. Warm season grasses predominate, and summer irrigation is required for the maintenance of a reasonably uniform and persistent lawn. In full sun bermudagrass and buffalograss are best adapted, with bermudagrass being the most competitive in trafficked situations under either low or high fertility. *Zoysia japonica* and *Z. matrella* are more wear tolerant and shade tolerant than buffalograss, while also being adapted to low fertility. St. Augustinegrass is the best choice for moderately shaded lawns, but it requires more frequent irrigation and fertilization than the others to remain competitive. Seashore paspalum and saltgrass [*Distichlis spicata* (L.) Greene] are well adapted to alkaline, salt-affected soils or saline irrigation water common to these zones, although neither species is widely planted at present. Future improvements in quality attributes (e.g., color, texture) for saltgrass and cost-competitive availability of seeded cultivars will result in wide plantings of both species in these two zones. In cooler, high elevation areas within this zone tall fescue and the fine fescues can be successfully managed under low fertility if irrigated regularly. Persistence of high quality Kentucky bluegrass and perennial ryegrass lawns at higher elevations is also common, but requires more frequent fertilizer inputs. Grass abundance in these zones under a variety of cultural regimes is shown in Tables 8–9 through 8–11. A detailed map showing adaptation of grasses in California was presented by Youngner et al. (1962).

Table 8–9. Long-term performance of warm and cool season grasses in lawns maintained under a variety of cultural intensities in the Warm Semiarid Zone.

	Irrigated, low fertility	Irrigated, high fertility	Nonirrigated
Abundant	*Cynodon* spp.	*Cynodon* spp.	none
	Zoysia spp.	*Stenotaphrum secundatum* (Walter) Kuntze	
	Buchloe dactyloides (Nutt.) Engelm.	*Zoysia* spp.	
	Stenotaphrum secundatum (Walter) Kuntze	*Buchloe dactyloides* (Nutt.) Engelm.	
	Lolium arundinaceum Schreb. subsp. *arundinacea*†	*Lolium arundinaceum* Schreb. subsp. *arundinacea*†	
	Festuca rubra L. subsp. *rubra*†	*Poa pratensis* L.†	
Scarce		*Lolium perenne* L.†	

† Cool season grasses will only persist at cooler, higher elevations locations within this zone.

Cool Semiarid Pacific Zone

This unusual zone occupies the south coast of California from San Diego north to approximately Carmel. It extends inland to the western slope of the coastal mountains. It is strongly influenced by marine off-shore flow. Temperatures are mild year round, humidity is high, and fog is common. The growing season is effectively 12 mo. Important grasses include common and hybrid bermudagrasses, kikuyugrass (*Pennisetum clandestinum* Hochst. ex Chiov.), seashore paspalum, tall fescue, perennial ryegrass, and annual bluegrass (Table 8–12).

Table 8–10. Long-term performance of warm and cool season grasses in lawns maintained under a variety of cultural intensities in the Semiarid Transitional Zone.

	Irrigated, low fertility	Irrigated, high fertility	Nonirrigated
Abundant	*Buchloe dactyloides* (Nutt.) Engelm.	*Cynodon* spp.	none
	Zoysia spp.	*Lolium arundinaceum* Schreb. subsp. arundinacea†	
	Cynodon spp.	*Poa arachnifera*†	
	Lolium arundinaceum Schreb. subsp. arundinacea†	*Buchloe dactyloides* (Nutt.) Engelm.	
	Poa arachnifera†	*Zoysia* spp.	
	Paspalum notatum Flüggé	*Paspalum notatum* Flüggé	
Scarce	*Paspalum vaginatum* Sw.	*Paspalum vaginatum* Sw.	

† Perennial cool season grasses.

Table 8–11. Long-term performance of warm and cool season grasses in lawns maintained under a variety of cultural intensities in the Warm Arid Zone.

	Irrigated, low fertility	Irrigated, high fertility	Nonirrigated
Abundant	*Cynodon* spp.	*Lolium arundinaceum* Schreb. subsp. arundinacea	none
	Lolium arundinaceum Schreb. subsp. arundinacea	*Cynodon* spp.	
	Poa pratensis L.	*Poa pratensis* L.	
	Stenotaphrum secundatum (Walter) Kuntze	*Stenotaphrum secundatum* (Walter) Kuntze	
	Poa annua L.†	*Poa annua* L.†	
	Zoysia spp.	*Zoysia* spp.	
	Paspalum vaginatum Sw.	*Paspalum vaginatum* Sw.	
Scarce	*Buchloe dactyloides* (Nutt.) Engelm.	*Buchloe dactyloides* (Nutt.) Engelm.	

† Behaves as a winter annual in this zone.

Table 8–12. Long-term performance of warm and cool season grasses in lawns maintained under a variety of cultural intensities in the Cool Semiarid Pacific Zone.

	Irrigated, low fertility	Irrigated, high fertility	Nonirrigated
Abundant	*Lolium arundinaceum* Schreb. subsp. *arundinacea*	*Lolium arundinaceum* Schreb. subsp. *arundinacea*	*Poa annua* L.†
	Poa pratensis L.	*Poa pratensis* L.	*Pennisetum clandestinum* Hochst. ex Chiov.
	Pennisetum clandestinum Hochst. ex Chiov.	*Pennisetum clandestinum* Hochst. ex Chiov.	
	Cynodon spp.	*Cynodon* spp.	
	Zoysia spp.	*Zoysia* spp.	
	Buchloe dactyloides (Nutt.) Engelm.	*Buchloe dactyloides* (Nutt.) Engelm.	
	Paspalum vaginatum Sw.	*Paspalum vaginatum* Sw.	
Scarce	*Stenotaphrum secundatum* (Walter) Kuntze	*Stenotaphrum secundatum* (Walter) Kuntze	

† Behaves as a winter annual in this zone.

Ecosystem Development

Since humans play an integral role in lawn ecosystems, it is worth considering how key maintenance practices shape the development of lawns through time. Mowing, fertilization, irrigation, and weed control profoundly influence lawn ecosystem components. Variability in cultural practices can result in plant inventories ranging from single species monocultures to broad-based plant communities supporting a myriad of plant and animal life.

Mowing

Lawn ecosystems are defined by mowing. While mowing intensity ranges from daily to as little as four times per year, the most common frequency for amenity lawns is once per week during the growing season. Likewise, although mowing heights range from 2.5 to 10 cm, typical mowing heights fall in the range of 5 to 7.5 cm for most lawns. Regular mowing is a primary selection tool that eliminates all plants not suited to regular clipping. Specifically, grasses that normally grow with elevated apical meristems and/or with limited ability to produce lateral shoots fail to persist. Dicots that are erect growing with exposed apical meristems are eliminated, along with other annual plants unable to reproduce under regular mowing. In general terms, mowing favors low growing plants that can reproduce or spread vegetatively when frequently clipped.

Within the realm of plants tolerant of mowing, competition between plants is affected by mowing practices. Cutting erect growing grasses higher tends to reduce encroachment from summer annuals such as crabgrass (Dernoeden et al., 1993, 1998; Jagschitz and Ebdon, 1985) and annual bluegrass (Adams, 1980; Bush et al., 2000). Voigt et al. (2001) saw a dramatic increase in the percentage of crabgrass cover in tall fescue as the mowing height was lowered from 7.6 to 2.5 cm. Adams (1980) found dramatic increases in the annual bluegrass percentage of cover in a mixed stand of perennial ryegrass and colonial bentgrass as mowing height was lowered from 7.5 to 1.25 cm in the mild climate of the UK. A similar but less dramatic increase in white clover (*Trifolium repens* L.) was observed when mowing height was lowered from 7.6 to 2.5 cm in tall fescue (Voigt et al., 2001). It appears that for any given grass in any given climate there is a minimum mowing height at which it is most competitive. As the mowing height goes below that height, encroachment by other better adapted grasses and dicots will increase.

Fertility Levels

Many turfgrasses are highly responsive to N fertilizer, including bermudagrass, perennial ryegrass, Kentucky bluegrass, and annual bluegrass. Even grasses that perform well under low N fertility, such as bentgrass and fine fescues, often respond vigorously to N fertilizer applications. Nitrogen can be used as a tool to enhance turf density and increase competitive ability of planted grasses. In general, regular N applications favor grasses over dicots in mixed stands (Busey, 2003). Increasing rates of N fertilizer reduced dandelion (*Taraxacum officinale* Weber) in Kentucky bluegrass (Johnson and Bowyer, 1982) and in bermudagrass (Callahan and Overton, 1978). Voigt et al. (2001) found that tall fescue plots receiving N fertilizer had significantly less broadleaf cover than plots that were not fertilized.

Nitrogen fertility may also affect competition between grasses. In cool climates with minimal drought or temperature stress, increasing N rates increased percentage cover of annual bluegrass in mixed stands of cool season grasses (Adams, 1980; Lodge and Lawson, 1993). Under more severe climatic conditions, Gaussoin and Branham (1989) found no consistent effect of higher N levels on annual bluegrass encroachment in creeping bentgrass turf.

In mixed stands of cool and warm season grasses, N application timing has a strong impact on competition between species. In a transition zone climate, fertilizing Kentucky bluegrass in fall through spring, when it is most responsive to N, reduced encroachment from crabgrass, while fertilizing in summer increased crabgrass (Dunn et al., 1981). Zoysiagrass, a slow growing warm season grass, was more competitive with Kentucky bluegrass when fertilizer N was applied during its peak summer growth period as opposed to fall, which favored Kentucky bluegrass (Engel, 1974).

Under low N fertility, mixtures of cool season grasses with different N needs will generally shift toward the grasses tolerant of low N levels (Bradshaw, 1962). In the cool climate of Bradshaw's experiments, low fertility turf tended to have high populations of bentgrasses and fine fescues along with white clover. Regular clipping removal harvests N, P, and K contained in leaves. The net loss of N from the system reduces fertility, resulting in a need for supplemental N to maintain turf density and vigor. If adequate N is not put back into the system, grasses with high N requirements such as Kentucky bluegrass and perennial ryegrass lose vigor and thin. This facilitates encroachment by perennial weedy grasses, annual grasses and seed-borne dicots tolerant of low fertility conditions (Haley et al., 1985; Heckman et al., 2000; Harivandi et al., 2001). In addition to fertility effects associated with clippings, Gaussoin and Branham (1989) found that removing clippings reduced the encroachment rate of annual bluegrass in creeping bentgrass turf due to seed removal. For a detailed discussion of the effects of clipping management on grass and dicot encroachment into lawns see Busey (2003).

Irrigation and Drought

Irrigation is utilized to some degree in most areas where lawns are grown. In many parts of the United States irrigation intensity ranges from none to enough water to maintain stand density to excessive irrigation. This continuum influences species composition dramatically in climates where lawns can survive without supplemental irrigation. Unfortunately, there is little published research that addresses this issue. Long-term observations in the Pacific Coastal Zone indicate that lawns can exist without irrigation, but lawns that are not irrigated are characterized by distinct species shifts away from planted grasses toward a much more diverse collection of naturalized grasses and dicots.

Prolonged drought stress causes perennial grasses to go dormant, reduces stand density, and appears to favor drought-resistant dicots. With the onset of fall rains, dormant plants slowly resume growth, exposed seeds germinate, and open niches often become colonized by moss. With time, grass populations decline as dicot populations increase. This was demonstrated by Dernoeden et al. (1994), who found that drought-induced loss of turf was followed by an increase in weed colonization.

Regular but not excessive irrigation tends to maintain dense turf but does not stimulate excessive shoot growth. Excessive irrigation is often associated

with intensively managed landscapes and is common where automated irrigation systems are operated via timers. The tendency is to start irrigation earlier
than needed in spring, irrigate constantly through summer without periodically
adjusting runtimes, and then to leave systems running late into fall even though
turf water needs have decreased. This intensity of irrigation maximizes shoot
growth, is detrimental to root growth, and may have long-term negative impacts
on soil compaction and nutrient leaching. Little research has addressed issues
associated with excessive irrigation. Jiang et al. (1998) determined that neither
adequate nor excessive irrigation had any impact on crabgrass or common dandelion populations in perennial ryegrass in Kansas. Gaussoin and Branham (1989)
concluded that frequency of irrigation was more important than the amount of
water applied. In their trials, daily irrigation increased annual bluegrass populations more than irrigating three times per week, regardless of how much water
was applied.

Weed Control

Dicots such as common dandelion, broadleaf plantain (*Plantago major* L.), narrowleaf plantain (*Plantago lanceolata* L.), clover (*Trifolium* spp.), English daisy (*Bellis
perennis* L.), and dozens of others have long been components of lawns. Before the
introduction of selective herbicides, control of undesirable species in lawns was
a difficult task. In much of the United States crabgrass was a major component
of lawns, and clover was ubiquitous. Before development of effective selective
broadleaf herbicides, it was common to include clover in seeding mixes (Downing, 1859; Scott, 1870; Hunn and Bailey, 1906; Barron, 1906; Dickinson, 1930). More
recently grass–dicot mixtures have been developed to enhance lawn performance under minimal input conditions (Gibeault and Youngner, 1986; Rumball,
1989; Cook, 1993).

Emphasis on producing pure stands of grass for lawns is an outgrowth of the
development of selective herbicides, improved turfgrass cultivars, increased use
of fertilizers, and availability of irrigation water. Mainstream turfgrass mixtures
used for seed or sod today use improved cultivars, often assembled as blends.
Interspecies mixtures generally include species and cultivars adapted to both sun
and shade and/or low to high fertility. Herbicide use on lawns is now widespread
and focused on controlling both dicots and undesirable annual and perennial
grasses. The primary impact of herbicide use on lawns is reduced plant diversity,
giving rise to the notion that all lawns are grass monocultures.

Societal pressure to reduce use of pesticides in general is changing our
approach to lawn culture and may change our current vision of lawns. Many
municipalities have already quit using broadleaf herbicides on lawns. This will
ultimately result in an increase in dicot species and greater species diversity.

Regional Lawn Plant Communities

At first glance lawns appear to be simple monocultures or polystands of grasses.
When turf managers are asked what they are growing, the answer is almost
always a description of what they planted. The assumption is that whatever
was planted is what is actually there. While there may be cases where this is
true, often a dynamic transition occurs through time, producing regionally distinct plant communities. Lawn succession patterns for four sites throughout the

United States are detailed below, along with a discussion of the forces that shape the final mixture of plants in these communities.

Cool Humid Pacific Zone Example: Corvallis, Oregon

The most common lawn seed mixtures in this zone contain about 70% perennial ryegrass and 30% fine fescue by weight or, more recently, 100% tall fescue. Lawns can be planted in spring through fall, but the most common planting time is generally fall. If sod is used, it is likely to be 100% perennial ryegrass or, more recently, 100% tall fescue. Lawns planted before 1980 were likely to have been planted with mixtures of perennial ryegrass, fine fescue, Kentucky bluegrass, and colonial or dryland bentgrass. Lawns planted before 1960 were invariably mixtures of colonial (*Agrostis capillaris* L.) or dryland bentgrass (*A. castellana* Boiss. and Reut.) plus fine fescue (Schudel and Rampton, 1954).

Native soils in this area are typically clay loams of around pH 6.0. Imported soils are commonly loamy sands or sandy loams extracted from gravel mining sites throughout the area. Native soils maintain soil seed banks composed of creeping bentgrass (*Agrostis stolonifera* L.), colonial bentgrass, dryland bentgrass, velvetgrasses (*Holcus lanatus* L. and *H. mollis* L.), annual bluegrass, rough bluegrass (*Poa trivialis* L.), and rat tail fescue. Most of these are documented as producing persistent seed banks (Grime, 1980). Roxburgh and Wilson (2000) found a wide range of species, including many listed above, in a soil seed bank associated with a 30-yr-old lawn in New Zealand.

During the establishment year, lawns planted with perennial ryegrass–fine fescue mixtures that are regularly irrigated will generally appear to be nearly 100% ryegrass due to the inherently fast establishment characteristics of that grass. Off-type unplanted grasses are most likely to be annual bluegrass. During the next several years, depending on maintenance intensity, the lawn will increasingly be dominated by the fine fescue component. Rose-Fricker et al. (1997) found that under low maintenance conditions Chewings fescue (*Festuca rubra* L. subsp. *commutata* Gaudin) was the most aggressive of the fine fescues after 4 yr when mixed with Kentucky bluegrass, followed by creeping red fescue [*Festuca rubra* L. subsp. *arenaria* (Osbeck) F. Aresch.] and hard fescue [*Festuca trachyphylla* (Hack.) Krajina]. Any bentgrasses, rough bluegrass, or velvetgrass present will also become visible. By year 10 the lawn will often be dominated by these species that were not planted.

Irrigated, Low Fertility Sites

In general, under low fertility conditions with adequate but not excessive irrigation the dominant grass will be one or more bentgrasses (Taylor, 1974; Davis, 1958). Dryland bentgrass via strong rhizome growth and creeping bentgrass via stoloniferous growth are the most dominant and often grow intermixed throughout lawns. Rough bluegrass develops a distinct winter growth cycle. Showing up in fall with the arrival of persistent rains and cool temperatures, it then competes vigorously through winter and loses ground in summer when high temperatures and drought force dormancy. Common velvetgrass is often present as distinct patches but rarely as a dominant component (Bradshaw, 1962; Thorhallsdottir, 1990). German velvetgrass is less common, but where present it may become a dominant component due to extensive rhizome growth. Annual bluegrass will be present, but it is not as vigorous as bentgrass under low fertility conditions. All of these invasive grasses most likely come as contaminants in uncertified seed mixtures, from the local soil seed bank, or as transients on animals, equipment, or shoes. Remnants of the planted species will also be present but rarely as dominant components (Davis, 1958; Taylor, 1974). Ryegrass generally forms isolated clumps or occasionally dominant patches in wear areas. Chewings fescue will form isolated patches. Creeping red fescue may be diffusely integrated throughout the lawn area. Fine fescues are most likely to be found near tree bases or other drought stress areas.

Dicot components include common dandelion, false dandelion (*Hypochaeris radicata* L.), and white clover. There may also be subterranean clover (*Trifolium subterraneum* L.), English daisy, mousear chickweed [*Cerastium fontanum* subsp. *vulgare* (Hartman) Greuter and Burdet], heal all (*Prunella vulgaris* L.), common yarrow (*Achillea millefolium* L.), or Veronicas (*Veronica* spp.). In practice each site will have its own mix of dicots specific to local conditions. Invariably several bryophytes will be present, including *Brachythecium* spp., *Calliergonella cuspidata* (Hedw.) Loeske, *Eurhynchium praelongum* (Hedw.) Schimp. in B.S.G., *Rhytidiadelphus squarrosus* (Hedw.) Warnst., and *Scleropodium touretii* var. *colpophyllu* (Sull.) Lawt. ex Crum. Moss is a strong winter grower, along with rough bluegrass. Its growth cycle starts with the onset of fall rains followed by vigorous growth until drought and heat cause dormancy, generally by early May. These lawns may be 80 to 90% bentgrass during summer, with the remainder made up of velvetgrass, annual bluegrass, and dicots, primarily clover. Winters will see an increase in the proportion of rough bluegrass and various mosses. As long as N fertility stays low the resident population of annual bluegrass will remain small. The preceding observations are similar to those reported by Thompson et al. (2004) in their study of lawns in Sheffield, UK.

Irrigated, High Fertility Sites

Starting with a standard perennial ryegrass–fine fescue mix, the lawn will initially be dominated by perennial ryegrass. If the lawn is planted in late fall, as is common, annual bluegrass shows up immediately and by the following spring may constitute 10 to 50% of the stand, depending on the individual site seed bank. Over a period of several years the fine fescue component may increase and eventually dominate the perennial ryegrass, but will generally be out-competed by annual bluegrass (Adams, 1980) and bentgrass if they are present in the soil seed bank. Rough bluegrass grows primarily in winter and becomes part of the final subclimax lawn. Ultimately, the stand will be dominated by bentgrass, rough bluegrass, and annual bluegrass on most native soil sites. On imported soils free of the normal soil seed bank, the primary invasive species will be annual bluegrass and possibly rough bluegrass via contaminated seed.

Dicots will be less common due to competition from grasses responding to higher fertility. Common components may include mousear chickweed, common dandelion, false dandelion, and limited amounts of clover. Bryophytes will be much less common and will only develop in weak areas in the lawn, most commonly in thin turf in shade.

All things considered, these lawns tend to have fewer dicots and mosses than other lawns. They typically have more annual bluegrass and higher levels of the planted grasses. Ultimately, they are most likely to develop into a bentgrass–annual bluegrass subclimax.

Nonirrigated Sites

Nonirrigated lawns are common in this zone. They basically are dependent entirely on fall through spring rains. Summer is most often a severe drought stress period, during which perennial grasses survive via dormancy and annual grasses survive as seed. Most nonirrigated lawns are older lawns that have already developed bentgrass–rough bluegrass–velvetgrass–tall fescue subclimax status. If the lawns are recently planted to perennial ryegrass–fine fescue, the pattern will be similar to irrigated sites during establishment. Once drought develops in the first summer, both fine fescue and perennial ryegrass typically stay green through midsummer until drought dormancy sets in. If the summer is very dry, ryegrass will suffer partial die-out and thin to individual clumps. Creeping red fescue will eventually fill in and dominate after the first year. Any bentgrasses on site will slowly expand and fill gaps. Dryland bentgrass will normally become the dominant bentgrass in this scenario. After several years, ryegrass, hard fescue, Chewings fescue, velvetgrass, and tall fescue will form isolated clumps, with dryland or colonial bentgrass filling in the open areas. If Kentucky bluegrass is present, it will show up as emerging rhizomes once fall rains begin, but it rarely develops beyond a few scattered plants. Eventually, most sites will be invaded by rattail fescue or silvery hairgrass (*Aira caryophyllea* L.), which are winter annuals. These grasses germinate in fall and often develop very dense ground cover before flowering in mid spring. By late June both grasses will set seed and die. Annual bluegrass is rarely found in these lawns, but when it is present, it behaves distinctly as a winter annual.

The dominant dicots include false dandelion, common dandelion, white clover, subterranean clover, common yarrow, wild carrot (*Daucus carota* L.), English daisy, heal-all, mousear chickweed, and yellow spring bedstraw (*Galium verum* L.). Moss species will include *Brachythecium* spp., *Calliergonella cuspidata* (Hedw.) Loeske, *Eurhynchium praelongum* (Hedw.) Schimp. in B.S.G., *Rhytidiadelphus squarrosus* (Hedw.) Warnst., and *Scleropodium touretii* var. *colpophyllum* (Sull.) Lawt. ex Crum.

In most cases the dominant perennial grasses include dryland bentgrass, tall fescue (if it is present initially), and fine fescues. The dominant dicot will nearly always be false dandelion, which may achieve 50 to 80% ground cover year round. When present, rattail fescue may constitute more than 50% of the ground cover by mid winter. Moss will become a strong winter component, often covering up to 50% of the area by winter's end. These lawns have a strong seasonal character, with primary growth occurring from late fall through mid spring. Grass dormancy or die-out in summer leaves only drought-resistant dicots. Yellow spring bedstraw generally stays green and dense through the entire summer while false dandelion will eventually go dormant in most years, along with common yarrow and white clover. Many of the most persistent dicots are fall germinators, as typified by common and false dandelion, mousear chickweed, and various clovers. Summer annual grasses such as crabgrass (*Digitaria* spp.) and foxtails (*Setaria* spp.) are scarce in this climatic zone and are rarely observed in lawns under any level of maintenance. This may be a reflection of the cool spring weather, which is ideal for cool season grasses, and the extreme drought in summer, which does not facilitate germination and development of warm season annual grasses.

Cool Semiarid Zone Example: Boise, Idaho

Throughout this zone lawns cannot exist without regular irrigation. The most widely planted grass is Kentucky bluegrass, often sold alone or in mixtures with perennial ryegrass or one or more fine fescues. In recent years 100% tall fescue lawns have become more popular, but they still make up a relatively small portion of lawns in this zone. Most commercial sod is either straight Kentucky bluegrass or straight tall fescue. A few farms grow Kentucky bluegrass–perennial ryegrass mixtures. Old lawns planted from seed before about 1960 often contained bentgrasses mixed with Kentucky bluegrass and fine fescue. Common unplanted grasses include quackgrass [*Elymus repens* (L.) Gould] and annual bluegrass in localized areas. Common warm season annual grasses include crabgrass, foxtail, and barnyardgrass [*Echinochloa crusgalli* (L). Beauv.]. Bermudagrass is found as a contaminant in warmer microclimates and regions of this zone.

Soils throughout the region are variable, but many are silt loams. Imported soils are often sandy loams. Soil pH ranges from 6 to >8, depending on local conditions. Unlike the mild climate of the Pacific Coastal Zone, this area does not support persistent grass soil seed banks, so few cool season grass contaminants are present unless introduced via seed or sod.

Irrigated, Low or High Fertility Sites

Kentucky bluegrass lawns in this zone are likely to stay mostly pure for decades or longer as long as irrigation is adequate to sustain the stand. To the extent that contaminants are present, quackgrass is competitive but rarely dominates the stand. Bentgrass, when present, will dominate all other grasses. Old lawns that were likely to have been planted with bentgrass in mixture with other grasses often end up 90 to 95% bentgrass. An example of this can be found in Til-Taylor Park in Pendleton, OR and in older sections of the landscape around the state capitol in Boise, ID. Long-term persistence of tall fescue has not been determined, but in pure plantings it appears to be persistent. Rough bluegrass may occur as a contaminant in both seed and sod.

Annual bluegrass is rarely a significant component of low fertility lawns except in heavy wear areas or areas where turf has died out due to drought or other stresses. Under high fertility, lawn composition will be similar to low fertility but may contain more annual bluegrass and fewer dicots. Crabgrass is found throughout most of the zone, but may be absent from some locations, such as the high desert area in Central Oregon.

Common dicots in lawns throughout the Cool Semiarid Zone include common dandelion, narrowleaf plantain, broadleaf plantain, white clover, black medic (*Medicago lupulina* L.), common mallow (*Malva neglecta* Wallr.), field bindweed (*Convolvulus arvensis* L.), prostrate knotweed (*Polygonum aviculare* L.), lawn violet (*Viola* spp.), and ground ivy (*Glechoma hederacea* L.). Mosses are rare in this climate, generally occurring only in shady or wet sites.

Semi-cool Humid Zone Example: Wooster, Ohio

Dominant planted grasses in this zone include Kentucky bluegrass, fine fescues, perennial ryegrass, and tall fescue. Unplanted invasive cool season grasses include annual bluegrass, rough bluegrass, and, to a lesser degree, creeping bentgrass, quackgrass, and orchardgrass (*Dactylis glomerata* L.). Warm season perennial volunteer grasses include zoysiagrass in full sun and nimblewill in partial shade. Warm season summer annual grasses are represented by crabgrass, foxtails, and goosegrass [*Eleusine indica* (L.) Gaertn.].

Cheng et al. (2008) surveyed for 40 dicot species and found that dominant species in the Wooster, OH area include common dandelion, white clover, ground ivy, and broadleaf and narrow leaf plantains. Prostrate knotweed, heal all, woodsorrel (*Oxalis* spp.), and speedwell (*Veronica* spp.) are also common in Wooster lawns (Whitney, 1985).

Humid Transition Zone Example: Washington, DC

The dominant cool season grass planted in this zone is tall fescue, while from the middle portions of the zone and to the south zoysiagrass, then bermudagrass, can be planted and maintained in full sun. Soil pH generally ranges from 5 to 7, with sandy or silty loams overlaying clay. Commonly planted sun and shade mixtures contain 50 to 80% tall fescue, 10 to 20% perennial ryegrass, 10 to 20% fine fescues, and 10 to 20% Kentucky bluegrass. When the plant components of an unfertilized and unirrigated 10-yr-old Washington, DC lawn were surveyed, tall fescue was the majority species at 37%, with Kentucky bluegrass at 11%, and perennial ryegrass at less than 2% (Falk, 1980). Along with 10% bare ground, grass invaders consisted of crabgrass (14%), orchardgrass (8%), and bermudagrass (5%). In comparison, tall fescue made up almost 70% of a nearby lawn that had received merely one spring fertilizer application per year (~25 kg N ha^{-1}) and regular summer irrigation. On this lawn a denser stand of tall fescue had decreased summer crabgrass cover to 5% and completely eliminated bermudagrass invasion (Falk, 1980). Mixtures of cool and warm season grasses in low maintenance transition zone lawns are quite common, often resulting in a splotchy cold weather appearance due to the winter dormancy of the warm season components.

We present a description of the components of one author's yard (Ervin, personal observations, 2009) as an example of the diversity that can be present in an unirrigated northern transition zone location, Blacksburg, VA. On the north-facing, front lawn, which is not tree shaded, tall fescue and hard fescue have periodically been fall-seeded into drought-thinned Kentucky bluegrass and perennial ryegrass during the last 8 yr. The result is a lawn that is a fairly even mix of all four species, along with less than 5% each of common bermudagrass and nimblewill. In the lawn area immediately behind the house, a large block of *Zoysia japonica* was sodded and is persisting relatively weed free in the full sun. The far back portion of this landscape consists of an ordered stand of mature white pines (*Pinus strobus* L.), presenting a mix of dense to moderate to light shade. English ivy (*Herdera helix* L.) and periwinckle (*Vinca minor* L.) dominate the unmowed shaded understory, while moderately shaded areas consist of thin to thick stands of primarily *Poa trivialis* and secondarily fine fescues. The final component of this yard is a full sun to light shade area that was sodded 8 yr ago to tall fescue, mowed regulary at 7.5 cm, and left unfertilized. This block is now a blotchy mix of tall fescue, bermudagrass, and nimblewill. Similar species diversity is commonly observed across low maintenance lawns in the transition zone.

Looking Closer at Lawn Ecosystems

Lawns are unusual in that the crop is what is left after clipping harvest. While it would seem that net primary productivity would vary widely among different types of lawns, Falk (1980) found little difference between productivity of two temperate zone lawns maintained under dramatically different regimes and varying in species richness. The productivity of both lawns was in the range of 28 MJ m², which was equal to crops such as maize (*Zea mays* L.) and wheat (*Triticum aestivum* L.).

Lawns also produce a relatively persistent layer of debris (thatch) that accumulates beneath the green vegetation and above the soil level (Fig. 8–3). The thatch layer has a significant impact on performance of the turf and may influence the composition of the soil macro- and microorganisms in the ecosystem. Thatch also increases carbon and nutrient storage in the lawn system (Kaye et al., 2008). Recent studies have demonstrated that lawn ecosystems function as efficient carbon sinks, with mature lawns (>30 yr old) accumulating soil organic carbon at levels equivalent to urban forest remnants and at levels 40% higher than shortgrass prairie and rural forests (Pouyat et al., 2009).

Lawns are also efficient N sinks. Raciti et al. (2008) reported that N appears to be tightly cycled in lawns and forests, with small pools of available NO_3 and NH_4 and rapid turnover times, suggesting low leaching potential in both systems. Applications of water and nutrients beyond those required for a sustainable lawn may, at times, negate these benefits, but numerous references suggest that adherence to lawn care based on best management practices and integrated pest management functions to preserve these ecosystem benefits (Barton and Colmer, 2006; Alumai et al., 2009).

As noted by Rebele (1994), the conventional approach to ecology focuses on natural systems and is somewhat inadequate for describing ecological systems in the human-constructed urban environment. Brady et al. (1979) tried to

Foliage: Photosynthesizing young tillers, stolons, and emerging rhizomes of grasses, dicots, and bryophytes. Individual grass tillers rarely have more than three functional leaf blades. Older lower leaf blades and sheaths slough off and accumulate at the base of individual shoots forming a pseudo thatch layer.

Thatch: Organic debris composed of living and dead stems and roots. Live roots grow through thatch to reach soil and shoots develop at the top of the thatch zone. Thatch composition varies from recognizable leaf blade and sheath tissue at top to unrecognizable remnants of internodes, nodes and roots high in lignin.

Root zone: Growing media that anchors the plants and provides nutrients and water for growth. Lawn soils are variable ranging from sand to clay and are often composed of construction sub soils or artificial mixtures layered over existing soils. Lawn soil pH ranges from 4-8 throughout much of the United States.

Fig. 8–3. Lawns are a constructed ecosystem composed of a foliage, thatch, and root zone. Together they support a distinctive community of macro- and microflora and fauna that provides a wide array of ecosystem services in the urban environment.

relate urban ecosystems to larger biogeographical landscape units by proposing a schema to assign names to urban ecosystem components based on natural equivalents. Lawns were termed *Urban/Savannah*, which seems to equate this constructed ecosystem to an extension of a natural system. This leads to comparisons between lawns and natural grasslands that may not be warranted (Whitney, 1985; Cheng et al., 2008). In another example, Roxburgh and Wilson (2000) used a mature lawn to test the validity of community matrix theory in determining stability of natural ecosystems. They concluded that the theory developed for natural systems actually has a better fit for simple systems such as lawns than for more complex natural systems—in other words, lawns are different from natural systems. At some point a different approach to interpreting lawn ecosystems is needed.

The earliest conceptual model of a temperate zone turfgrass community ecosystem invertebrate food web was developed by Streu (1973). His model was created to show the interconnected nature of the macro- and microfauna in the system and the potential problems that might occur if the food web is disrupted by pesticides. It included numerous arthropods active in all three turf profile zones (Fig. 8–3) and other invertebrates that are part of the grazing food web as well as macro-, meso-, and microfauna associated with the detritus food web. His model is similar to that outlined by Petersen and Luxton (1982).

Earthworms are a key macrofauna component of the invertebrate detritus food web because they facilitate organic matter decomposition by comminuting organic debris and mixing microbes with organic matter via casts (Chapter 7, Pouyat et al., 2010, this volume). Lawns with large earthworm populations generally have less accumulated thatch than lawns with weak populations. Much of what we know about earthworm activity in lawns comes from research conducted in the United Kingdom (Ferro, 1937; Evans, 1947; Jefferson, 1955, 1956, 1958, 1961; Binns et al., 1999). More recently, Smetak et al. (2007) studied earthworms in Moscow, ID on young and old lawns and found the four most common species included *Lumbricus terrestris* L., *L. rubellus* Hoff., *Apporectodea longa* Ude., and *A. trapezoides* Duges. The dominant species across all sites were *L. terrestris* and *A. longa*. These findings are similar to surveys in England (Binns et al., 1999; Smith et al., 2006; Bartlett et al., 2008a). Smetak et al. (2007) noted that all observed earthworms were exotic species and that this was consistent throughout the region in both agricultural and grassland settings and not specific to lawns. High soil organic matter, periodic fertilizer applications, and soils kept moist via regular irrigation fostered higher species diversity and density. Smetak et al. (2007) also noted lower populations of worms in young versus older lawns. Due to the nature of lawn construction, which often includes utilization of on-site subsoils low in organic matter, importation of fill soils, and thorough tillage, earthworm populations may take time to build up as they colonize from surrounding areas. In pasture studies where earthworms were introduced at 10-m intervals to sites lacking worms, it took 7 yr for worms to completely colonize the trial areas (Stockdill, 1966, 1982).

The critical role earthworms play in the thatch decomposition process and the impacts of insecticides on both earthworms and decomposition in the United States have been discussed by Randell et al. (1972) and Potter et al. (1990). These studies found that any treatment that killed or excluded earthworms led to increased thatch accumulation or decreased thatch decomposition.

In recent years ecologists have used nematode population and community structure to determine the characteristics of the detritus food web. The idea is that nematodes feed on bacteria and fungi so the assemblage of microbial populations will be reflected in the makeup of the nematode population (Ferris and Matute, 2003). The impact of long-term turfgrass management practices on soil nematode and nutrients pools was reported by Cheng et al. (2008). They considered turfgrass soil food webs to be highly enriched but poorly to moderately structured compared with undisturbed natural grasslands that are usually highly structured but poorly enriched. On the basis of the turfgrass soil nematode community in this study they concluded that turfgrass systems are highly resistant to potential negative effects of a range of turfgrass pesticides due to generally high soil organic matter levels and high microbial activity. Fertilizer inputs increased the enrichment index in this study. Fertilizer source, whether inorganic or organic, had no impact on the nematode community. Ferris and Matute (2003) noted that any disturbance such as tillage or application of organic materials with high N content (low C/N ratio) increases the enrichment opportunist nematodes because of stimulation of bacterial growth by the organic source material.

Bacterial communities associated with turf vary significantly with the root zone media (Karp and Nelson, 2004; Bartlett et al., 2007, 2008b). In these studies, soil-based root zones were dominated by gram positive bacteria while sand-based root zones were dominated by gram negative bacteria. Zuberer (2005) reported high populations of fungi and bacteria in soil- and sand-based sports fields in Texas. Bartlett et al. (2008b) found that microbial biomass was concentrated in the top 0- to 7.5-cm zone in golf course turf. Fairway and rough areas, which are generally analogous to lawn turf, had greater microbial biomass than more intensively maintained greens and tees. Regardless of root zone type, microbial activity was higher near the surface where organic matter levels were enriched.

The picture emerging from the limited research on lawn community ecosystems indicates a system with a range of flora and fauna, grazing and detritus food webs that are well developed but skewed toward high enrichment and low structure, and a key role for earthworms in system development. For a system that is unstable according to classical ecological definitions, a lawn ecosystem appears to be remarkably stable, perhaps due to ongoing human inputs. It is clear that a great deal more research is needed to adequately explore all of the variability in lawn ecosystems. Most reported lawn ecology work has focused on cool season grasses in temperate climates, so more research is needed in warm season climates. Finally, turfgrass researchers need to become more involved in ecological research.

Acknowledgments

We thank Pat Donovan for preparing the Turfgrass Climate Zones figure and Mike Goatley, Jr., Paul Johnson, Bernd Leinauer, and Mark Mahady for their insights on grasses present in various zones.

References

Adams, W.A. 1980. Effects of nitrogen fertilization and cutting height on the shoot growth, nutrient removal, and turfgrass composition of an initially perennial ryegrass dominant sports turf. p. 343–350. In J.B. Beard (ed.) Proc. 3rd Int. Turfgrass Res. Conf., Munich, West Germany. 11–13 July 1977. Int. Turfgrass Soc. and ASA, CSSA, SSSA, Madison, WI.

Alumai, A., S.O. Salimen, D.S. Richmond, J. Cardina, and P.S. Grewal. 2009. Comparative evaluation of aesthetic, biological, and economic effectiveness of different lawn management programs. Urban Ecosyst. 12:127–144.

Ashby, M. 1969. Introduction to ecology. 2nd ed. Macmillan and Co. Ltd.

Barron, L. 1906. Lawns and how to make them. Doubleday, Page and Co. New York.

Bartlett, M., I. James, J. Harris, and K. Ritz. 2007. Interactions between microbial community structure and soil environment found on golf courses. Soil Biol. Biochem. 39:1533–1544.

Bartlett, M., I. James, J. Harris, and K. Ritz. 2008a. Earthworm community structure on five English golf courses. Appl. Soil Ecol. 39:336–341.

Bartlett, M., I. James, J. Harris, and K. Ritz. 2008b. Size and phenotypic structure of microbial communities within soil profiles in relation to play surfaces on a UK golf course. Eur. J. Soil Sci. 59:1013–1019.

Barton, L., and T.D. Colmer. 2006. Irrigation and fertilizer strategies for minimizing nitrogen leaching from turfgrass. Agric. Water Manage. 80:160–175.

Beard, J.B. 1973. Turfgrass science and culture. Prentice Hall, Englewood Cliffs, NJ.

Beard, J.B., and H.T. Beard. 2005. Beard's turfgrass encyclopedia. Michigan State Univ. Press, East Lansing.

Binns, D.J., S.W. Baker, and T.G. Piearce. 1999. A survey of earthworm populations on golf course fairways in Great Britain. J. Turf. Sci. 75:36–44.

Bradshaw, A.D. 1962. Turfgrass species and soil fertility. J. Sports Turf Res. Inst. 38:372–384.

Brady, R.F., T. Tobias, P.F.J. Eagles, R. Ohrner, J. Micak, B. Veale, and R.S. Dorney. 1979. A typology for the urban ecosystem and its relationship to larger biogeographical landscape units. Urban Ecol. 4:11–28.

Busey, P. 2003. Cultural management of weeds in turfgrass: A review. Crop Sci. 43:1899–1911.

Bush, E.W., A.D. Owings, D.P. Shepard, and J.N. McCrimmon. 2000. Mowing height and nitrogen rate affect turf quality and vegetative growth of common carpetgrass. HortScience 35:760–762.

Callahan, L.M., and J.R. Overton. 1978. Effects of lawn management practices on bermudagrass turf. p. 37–40. In Tennessee Farm and Home Science 108. Tennessee Agric. Exp. Stn., Knoxville.

Cheng, Z., D.S. Richmond, S.O. Salminen, and P.S. Grewal. 2008. Ecology of urban lawns under three common management programs. Urban Ecosyst. 11:177–195.

Cook, T. 1993. Low maintenance turf. The Hardy Plant Society of Oregon. 9(1):9–15.

Danneberger, T.K. 1993. Turfgrass ecology and management. G.I.E. Inc., Cleveland, OH.

Davis, R.R. 1958. The effect of other species and mowing height on persistence of lawn grasses. Agron. J. 50:671–673.

Dernoeden, P.H., M.J. Carroll, and J.M. Krouse. 1993. Weed management and tall fescue quality as influenced by mowing, nitrogen, and herbicides. Crop Sci. 33:1055–1061.

Dernoeden, P.H., M.J. Carroll, and J.M. Krouse. 1994. Mowing of three fescue species for low-maintenance turf sites. Crop Sci. 34:1645–1649.

Dernoeden, P.H., M.A. Fidanza, and J.M. Krouse. 1998. Low maintenance performance of five Festuca species in monostands and mixtures. Crop Sci. 38:434–439.

Dickinson, L.S. 1930. The lawn. Orange Judd Publ., New York.

Downing, A.J. 1859. A treatise on the theory and practice of landscape gardening. Funk and Wagnalls, New York.

Dunn, J.H., C.J. Nelson, and R.D. Winfrey. 1981. Effects of mowing and fertilization on quality of ten Kentucky bluegrass cultivars. p. 293–301. In R.W. Sheard (ed.) Proc. 4th Int. Turfgrass Res. Conf., Guelph, ON, Canada. 19–23 July 1981. Ontario Agric. College, Univ. Guelph, and Int. Turfgrass Soc., Guelph, ON, Canada.

Engel, R.E. 1974. Influence of nitrogen fertilization on species dominance in turfgrass mixtures. p. 104–111. In E.C. Roberts (ed.) Proc. 2nd Int. Turfgrass Res. Conf., Blacksburg, VA. 19–21 June 1973. ASA and CSSA, Madison, WI.

Evans, A.C. 1947. Earthworms. J. Sports Turf Res. Inst. 23:49–54.

Falk, J. 1980. The primary productivity of lawns in a temperate environment. J. Appl. Ecol. 17(3):689–695.

Ferris, H., and M.M. Matute. 2003. Structural and functional succession in the nematode fauna of a soil food web. Appl. Soil Ecol. 23:93–110.

Ferro, R.B. 1937. Some research factors affecting earthworm activity in turf. J. Sports Turf Res. Inst. 17:86–98.

Gaussoin, R.E., and B.E. Branham. 1989. Influence of cultural factors on species dominance in a mixed stand of annual bluegrass/creeping bentgrass. Crop Sci. 29:480–484.

Gibeault, V.A and V.B. Youngner. 1986. Strawberry clover with bermudagrass for low maintenance turf. A report of the Elvenia J. Slosson fund for ornamental horticulture 1983–1986. Univ. of California Div. of Agric. and Nat. Res., Sacramento.

Gilbert, O.L. 1989. The ecology of urban habitats. Chapman and Hall, London.

Grime, J.P. 1980. An ecological approach to management. In I.H. Rorison and R. Hunt (ed.) Amenity grassland: An ecological perspective. John Wiley and Sons Ltd., Chichester, UK.

Haley, J.E., D.J. Wehner, T.W. Fermanian, and A.J. Turgeon. 1985. Comparison of conventional and mulching mowers for Kentucky bluegrass maintenance. HortScience 20:105–107.

Harivandi, M.A., W.L. Hagan, and C.L. Elmore. 2001. Recycling mower effects on biomass, nitrogen recycling, weed invasion, turf quality, and thatch. Int. Turfgrass Soc. Res. J. 9:882–885.

Heckman, J.R., H. Liu, W. Hill, M. DeMilia, and W.L. Anastasia. 2000. Kentucky bluegrass response to mowing practice and nitrogen fertility management. J. Sustain. Agric. 15(4):25–33.

Holt, E.C. 1969. Turfgrasses under warm, humid conditions. p. 513–527. In A.A. Hanson and F.V. Juska (ed.) Turfgrass science. Agron. Monogr. 14. ASA, Madison, WI.

Hunn, C.E., and L.H. Bailey. 1906. The practical garden book. 5th ed. Grosset and Dunlap, New York.

Jiang, H., J. Fry, and N. Tisserat. 1998. Assessing irrigation management for its effects on disease and weed levels in perennial ryegrass. Crop Sci. 38:440–445.

Jagschitz, J.A., and J.S. Ebdon. 1985. Influence of mowing, fertilizer and herbicide on crabgrass infestation in red fescue turf. p. 699–704. In F. Lemaire (ed.) Proc. 5th Int. Turfgrass Res. Conf. Avignon, France. 1–5 July 1985. INRA, Paris.

Jefferson, P. 1955. Studies on the earthworms of turf a) the earthworms of experimental turf plots. J. Sports Turf Res. Inst. 31:6–27.

Jefferson, P. 1956. Studies on the earthworms of turf b) earthworms and soil. J. Sports Turf Res. Inst. 32:166–179.

Jefferson, P. 1958. Studies on the earthworms of turf c) earthworms and casting. J. Sports Turf Res. Inst. 32:166–179.

Jefferson, P. 1961. Earthworms and turf culture. J. Sports Turf Res. Inst. 37:276–289.

Johnson, B.J., and T.H. Bowyer. 1982. Management of herbicide and fertility levels on weeds and Kentucky bluegrass turf. Agron. J. 74:845–850.

Karp, M.A., and E.B. Nelson. 2004. Bacterial communities associated with creeping bentgrass in soil and sand rootzones. USGA Turfgrass Environ. Res. Online. 3(24):1–19.

Kaye, J.P., A. Majumdar, C. Gries, A. Buyantuyev, N.B. Grimm, D. Hope, G.D. Jenerette, W.X. Zhu, and L. Baker. 2008. Hierarchical Bayesian scaling of soil properties across urban. agricultural, and desert ecosystems. Ecol. Appl. 18(1):132–145.

Lodge, T.A., and D.M. Lawson. 1993. The construction, irrigation and fertiliser nutrition of golf greens: Botanical and soil chemical measurements over 3 years of differential treatment. J. Sports Turf Res. Inst. 69:59–73.

Madison, J.H. 1971. Principles of turfgrass culture. Van Nostrand Reinhold Co., New York.

Milesi, C., S.W. Running, C.D. Elvidge, J.B. Dietz, B.T. Tuttle, and R.R. Nemani. 2005. Mapping and modeling the biogeochemical cycling of turfgrasses in the United States. Environ. Manage. 36(3):426–438.

Musser, H.B. 1962. Turfgrass management. McGraw-Hill Book Co., New York.

Petersen, H., and M. Luxton. 1982. A comparative analysis of soil fauna populations and their role in decomposition processes. Oikos 39(3):291–357.

Potter, D.A., A.J. Powell, and M.S. Smith. 1990. Degradation of turfgrass thatch by earthworms (Oligochaeta: Lumbricidae) and other soil invertebrates. J. Econ. Entomol. 83:205–211.

Pouyat, R.V., K. Szlavecz, I.D. Yesilonis, P.M. Groffman, and K. Schwarz. 2010. Chemical, physical, and biological characteristics of urban soils. p. 119–152. In J. Aitkenhead-Peterson and A. Volder (ed.) Urban ecosystem ecology. Agron. Monogr. 55. ASA, CSSA, and SSSA, Madison, WI.

Pouyat, R.V., I.D. Yesilonis, and N.E. Golubiewski. 2009. A comparison of soil organic carbon stocks between residential turfgrass and native soil. Urban Ecosyst. 12:45–62.

Raciti, S.M., P.M. Groffman, and T.J. Fahey. 2008. Nitrogen retention in urban lawns and forests. Ecol. Appl. 18(7):1615–1626.

Randell, R., J.D. Butler, and T.D. Hughes. 1972. The effect on thatch accumulation and earthworm populations in Kentucky bluegrass turf. HortScience 7(1):64–65.

Rebele, F. 1994. Urban ecology and special features of urban ecosystems. Global Ecol. Biogeograph. Lett. 4:173–187.

Rose-Fricker, C., M. Fraser, and W.A. Meyer. 1997. Competitive abilities and performance of cool season turfgrass species in mixtures in the Willamette Valley of Oregon, USA, under high and low maintenance turf conditions. Int. Turf. Soc. Res. J. 8:1330–1335.

Roxburgh, S.H., and J.B. Wilson. 2000. Stability and coexistence in a lawn community: Experimental assessment of the stability of the actual community. Oikos 88:409–423.

Rumball, W. 1989. Breeding progress towards wildflower lawns. p. 91–93. In H.Takatoh (ed.) Proc. 6th Int. Turfgrass Res. Conf. 31 July–5 Aug., 1989. Int. Turfgrass Soc. and Japanese Society of Turfgrass Sci., Tokyo, Japan.

Schmidt, R.E., and R.E. Blaser. 1969. Ecology and turf management. p. 217–233. In A.A. Hanson and F.V. Juska (ed.) Turfgrass science. Agron. Monogr. 14. ASA, Madison, WI.

Scott, F.J. 1870. The art of beautifying home grounds. D. Appleton and Co., New York.

Schudel, H.L., and H.H. Rampton. 1954. Home lawns for Oregon. Bull. 516. Agric. Exp. Stn., Oregon State College, Corvallis.

Smetak, K.M., J.L. Johnson-Maynard, and J.E. Lloyd. 2007. Earthworm population density and diversity in different aged urban systems. Appl. Soil Ecol. 37:161–168.

Smith, J., A. Chapman, and P. Eggleton. 2006. Baseline biodiversity surveys of the soil macrofauna of London's green spaces. Urban Ecosyst. 9:337–349.

Stockdill, S.M.J. 1966. The effect of earthworms on pastures. Proc. N.Z. Ecol. Soc. 13:68–75.

Stockdill, S.M.J. 1982. Effects of introduced earthworms on the productivity of New Zealand pastures. Pedobiologia 24:29–35.

Streu, H.T. 1973. The turfgrass ecosystem: Impact of pesticides. Bull. Entomol. Soc. Am. 19(2):89–90.

Taylor, D.K. 1974. Cultivar response in turfgrass species mixture trials with mowing at two heights. p. 48–54. In E.C. Roberts (ed.) Proc. 2nd Int. Turfgrass Res. Conf., Blacksburg, VA. 19–21 June 1973. ASA and CSSA, Madison, WI.

Thompson, K., J.G. Hodgson, R.M. Smith, P.H. Warren, and K.J. Gaston. 2004. Urban domestic gardens (III): Composition and diversity of lawn floras. J. Veg. Sci. 15:373–378.

Thorhallsdottir, T.H. 1990. The dynamics of five grasses and white clover in a simulated mosaic sward. J. Ecol. 78:909–923.

Trewartha, G.T. 1968. An introduction to climate. McGraw-Hill Book Co.

Turgeon, A.J. 2008. Turfgrass management. 8th ed. Pearson/Prentice Hall, Upper Saddle River, NJ.

Vinlove, F.K., and R.F. Torla. 1995. Comparative estimations of U.S. home lawn area. J. Turf Manage. 1(1):83–97.

Voigt, T.B., T.W. Fermanian, and J.E. Haley. 2001. Influence of mowing and nitrogen fertility on tall fescue turf. Int. Turfgrass Soc. Res. J. 9:953–956.

Ward, C.Y. 1969. Climate and adaptation. p. 27–79. In A.A. Hanson and F.V. Juska (ed.) Turfgrass science. Agron. Monogr. 14. ASA, Madison, WI.

Watschke, T.L., and R.E. Schmidt. 1992. Ecological aspects of turf communities p. 129–162. *In* D.V. Waddington et al. (ed.) Turfgrass. Agron. Monogr. 32. ASA, CSSA, and SSSA, Madison, WI.

Whitney, G.G. 1985. A quantitative analysis of the flora and plant communities of representative Midwestern U.S. town. Urban Ecol. 9:143–160.

Wise, L.N. 1961. The lawn book. W.R. Thompson, State College, Miss.

Youngner, V.B., J.H. Madison, M.H. Kimball, and W.B. Davis. 1962. Climatic zones for turfgrass in California. Calif. Turfgr. Cult 12(4):2–4.

Zuberer, D. 2005. Microbes in soil and sand-based root zones: A few of the basics. Sportsturf. 5 July, p. 8–12.

9

Urban Plant Ecology

Astrid Volder

Abstract

Vegetation plays an essential role in the functioning of urban ecosystems. To achieve maximum, continued benefits from urban ecosystems we need to improve our understanding of how vegetation and the urban physical environment interact. This chapter describes the various types of plant communities that are common to most urban ecosystems and assesses how urban environmental conditions may affect plants growing in the urban environment.

Ecologists have traditionally focused on the dynamics of undisturbed natural ecosystems, while largely ignoring systems that are modified by human influences (McDonnell, 1997). Starting approximately two decades ago, ecologists realized the importance of researching ecological changes along rural–urban gradients and started developing models of urban ecosystem functioning (Clergeau et al., 1998; McDonnell and Pickett, 1990; McDonnell et al., 1997; Pouyat et al., 1995; van Rensburg et al., 1997a,b). Urban environments represent a unique habitat that, if properly managed, could benefit human society for centuries to come. To achieve sustainable urban ecosystem management, a thorough understanding of the structure and function of these ecosystems is needed (Chapter 18, Aitkenhead-Peterson et al., 2010, this volume).

Vegetation in urban environments, for the most part, can be divided into three types; vegetation that is planted with a purpose, vegetation that has emerged spontaneously, and remnants of predevelopment vegetation. Vegetation that is planted, for example, in gardens, parks, around buildings, and along streets and roadways, tends to have a high proportion of exotics (Pemberton and Liu, 2009), some of which can be invasive (Chapter 12, Reichard, 2010, this volume). Quite often, planted vegetation in urban areas suffers from less than optimal growth conditions, such as compacted soils (Chapter 7, Pouyat et al., 2010, this volume), high radiation and temperature stress (Chapter 2, Heisler and Brazel, 2010, this volume), poor soil hydrology and frequent physical disturbance, particularly when surrounded by hardscape infrastructure and large impervious surface areas (Chapter 14, Burian and Pomeroy, 2010, this volume). In contrast to planted vegetation, vegetation that has arrived spontaneously likely has a

A. Volder, Dep. of Horticultural Sciences, Texas A&M University, TAMU 2133, College Station, TX 77843-2133 (a-volder@tamu.edu).

doi:10.2134/agronmonogr55.c9

successful parent plant close by and may be well suited for the urban environment. These are generally species that thrive on disturbance such as ruderals (Grime, 1977) and may contain a high proportion of invasive species, either native or nonnative. The third category, remnant vegetation, is well suited for predevelopment conditions, but may experience considerable stress from the surrounding urban development and altered microclimate, such as higher air pollution, higher temperatures, altered hydrology, and frequent disturbance (Chapters 2 [Heisler and Brazel, 2010], 3 [Santosa, 2010], and 14 [Burian and Pomeroy, 2010], this volume). Increasing fragmentation in urbanizing areas makes these remnants more vulnerable to invasive species and species loss, as fragmentation increases the proportion of edge habitat that is exposed to disturbance (Christie and Hochuli, 2005; Williams et al., 2005).

Vegetated areas provide important services to society and are a crucial part of urban ecosystem functioning beyond the obvious aesthetic benefits and provision of recreational opportunities (Bolund and Hunhammar, 1999; Li et al., 2005; Andersson, 2006; Tratalos et al., 2007). These ecosystem services (Chapter 18, Aitkenhead-Peterson et al., 2010, this volume) provide significant economic, environmental, and social benefits to society. Urban vegetation helps mitigate the urban heat island through increased albedo, shading, and evaporative cooling (Grimmond et al., 1996; Taha, 1997; Akbari et al., 2001). Urban vegetation improves hydrological performance via rainfall interception (Xiao and McPherson, 2002; Herbst et al., 2006; Guevara-Escobar et al., 2007) and as a mechanism to remove excess water through plant transpiration (Dussaillant et al., 2004; Carter and Jackson, 2007; Lemonsu et al., 2007). Urban vegetation reduces energy usage via the shading of buildings (Simpson and McPherson, 1998; Shashua-Bar and Hoffman, 2000; Akbari et al., 2001; Carver et al., 2004; Yu and Hien, 2006; Donovan and Butry, 2009). It also provides wildlife habitat to urban mammals and birds (Pickett et al., 2001; Savard et al., 2000; Sorace, 2001; Wilby and Perry, 2006; Chapters 4 [Shochat et al., 2010], 5 [McCleery, 2010], and 6 [Ehrenfeld and Stander, 2010], this volume). Urban vegetation generally reduces air pollution (Beckett et al., 1998; Currie and Bass, 2008; Jim and Chen, 2008; Escobedo and Nowak, 2009; Chapter 3, Santosa, 2010, this volume). Urban vegetation plays a crucial role in the functioning of urban riparian zones and wetlands (Moffatt et al., 2004; Burton et al., 2005; Urban et al., 2006; Chapter 13, Stander and Ehrenfeld, 2010, this volume), and it provides an opportunity for carbon sequestration (Zipperer, 2002; Golubiewski, 2006; Gratani and Varone, 2006; Pouyat et al., 2009). Finally, urban vegetation promotes feelings of well-being in humans (Chiesura, 2004).

In addition to the benefits that urban plants provide in terms of ecosystem services, we also need to consider how the urban environment, with its higher air temperatures, altered air chemistry, reduced open space, altered light environment, compacted and impoverished soils, altered precipitation patterns and modified hydrology, affects the performance of urban vegetation (Fig. 9–1). A variety of studies have measured plant responses along a rural–urban gradient (e.g., Guntenspergen and Levenson, 1997; van Rensburg et al., 1997b; Zipperer et al., 1997; Zedler and Leach, 1998; Barradas, 2000; Gregg et al., 2003, 2006; Ziska et al., 2003), but when compared with our knowledge of plant responses to their environment in natural ecosystems, relatively little is known about the effect of the urban environment on plants. Responses of plants to urban conditions can be measured at the community level via measures of species composition and

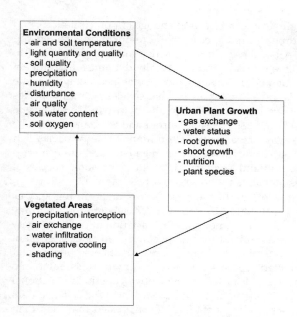

Environmental Conditions
- air and soil temperature
- light quantity and quality
- soil quality
- precipitation
- humidity
- disturbance
- air quality
- soil water content
- soil oxygen

Urban Plant Growth
- gas exchange
- water status
- root growth
- shoot growth
- nutrition
- plant species

Vegetated Areas
- precipitation interception
- air exchange
- water infiltration
- evaporative cooling
- shading

Fig. 9–1. Conceptual diagram linking environmental conditions, plant growth, and vegetation effects on environmental conditions in urban areas.

diversity or at the individual plant level by measuring growth and reproduction or physiological parameters, such as photosynthesis, respiration, transpiration, and water status.

There are many questions about vegetation that could be addressed in urban and suburban landscapes, including questions related to (i) physiological and growth responses of plants to air, water, and soil pollution; (ii) physiological and growth responses of plants to altered urban environments (e.g., higher temperatures, reduced radiation, increased vapor pressure deficit, soil compaction, drought or flooding); (iii) the structure and function of small (<1 ha) ecosystems and associated edge effects; (iv) potential facilitation and impact of invasive plants; and (v) the role of frequent anthropogenic disturbance (modified from McDonnell, 1997).

This chapter will focus on the role of urban vegetation in urban ecosystem functioning, as well as the effects that urban areas have on the performance of individual plants.

Urban Plant Communities

Trends along Urbanization Gradients and through Time

Vegetation in urban areas generally has high species diversity, mainly because urban areas have a wide variety of (artificially created) habitats on a relatively small surface area and include a wide range of planted ornamental species that are often nonnative. In a conceptual paper, Williams et al. (2009) argued that four major urban diversity filters add and remove species from the urban species pool, drastically influencing the resulting species composition. These filters are related to habitat transformation (i.e., the degree of urbanization), habitat fragmentation (i.e., size and shape of individual green areas and distance between green areas), environmental changes due to the urban environment (e.g., urban heat island, air pollution, light level changes, and changes in precipitation), and human preferences (i.e., adding desired species while removing unwanted species) (Williams et al., 2009). All of these filters are acting simultaneously and to differing degrees, resulting in a spatially and temporally dynamic environment where species composition may be in constant flux. A detailed analysis of the Brussels (Belgium) flora by land-use type showed that species richness, species rarity, and the number of exotic species varied greatly by urban land-use type as defined by building density and use type (Godefroid and Koedam, 2007). The size of woodlots in Brussels was also a major determinant of plant species richness, with larger woodlots containing more species in general and more species of conservation value (Godefroid and Koedam, 2003), although some species of conservation value were less prevalent in the largest forest fragment.

Land use and fragment size are not the only factors determining species composition, as Hope et al. (2003) found in Phoenix, AZ, where family income and housing age also explained a large amount of the variation in plant diversity found across the city. Greater household income generally led to greater plant diversity in the residential landscape, while newer residences also had greater species diversity. Increased plant species diversity in urban areas compared to rural areas does not necessarily lead to a greater phylogenetic or functional group diversity (Knapp et al., 2008). Based on a broad survey of the German flora, comparing rural and urban areas, Knapp et al. (2008) suggested that the urban environment tends to select for species that are closely related with similar functional characteristics that enable these plants to thrive under urban conditions. Thus, although species diversity is high in urban areas, the growth form and life history strategy of these species may all be very similar (e.g., mostly short-lived, ruderal species).

From city-wide surveys of vascular species growing along a rural–urban gradient near the British cities of Birmingham and Sheffield, Thompson and McCarthy (2008) found that successful urban plants generally were "robust plants of relatively fertile, dry, un-shaded, base rich habitats." Species richness in Brussels did not change when comparing a 1940 to 1971 survey with a 1990 to 1994 survey, but the turnover of species was high and 147 of the 731 original species were missing in the 1990 to 1994 survey, while 148 new species were recorded. The 1990 to 1994 survey included a greater proportion of alien species, more species that preferred habitats that are rich in nitrogen, and, contrary to the British study, more species that preferred shade (Godefroid, 2001). Interestingly, when existing native and alien species were compared, the latter were generally species that preferred high nitrogen, high light, drought, heat, and alkaline soils. Thus, native species newcomers had greater shade tolerance, while nonnative species newcomers preferred high light habitats. Contrary to the results in Brussels where species diversity did not decline with increasing urbanization, a survey comparing plant species diversity in the 1960s versus diversity in the 1990s in Plzen, Czech Republic showed a decline in species diversity with time, mainly due to a loss of archeophytes (alien species that immigrated before 1500 CE) from the urban area (Pysek et al., 2004).

Parks and Managed Urban Forests

Urban parks and urban forests serve an important function in the urban landscape. The vegetation in these parks provides green space for recreation, serves as wildlife habitat, retains stormwater runoff, cools the surrounding environment, and reduces air pollution. These parks and urban forests form the backbone of urban watersheds and are vital for maintaining high quality water sources (Long and Nair, 1999; Matteo et al., 2006). Vegetation in these areas generally is highly variable, ranging from fields planted with turfgrass for sports purposes to planted woody landscapes. Urban parks are a reservoir of biodiversity, although often containing a high proportion of exotic species (Koh and Sodhi, 2004; Li et al., 2006). Loeb (2006) found that a common urban park flora does not exist, even when the parks are located in a similar climate. This study of 10 urban parks located in Boston, New York, Baltimore, and Washington showed that less than 1% of the 1391 recorded species occurred in each park. Loeb (2006) also reported that there was a significant positive relationship between population density and nonnative species diversity, while native species diversity decreased as population density increased. Urban parks also provide microclimatic relief compared with the harsh conditions that planted street trees generally experience (Whitlow et al., 1992; Kjelgren and Clark, 1993; Georgi and Zafiriadis, 2006), thus allowing for a wider range of species. More information on urban forests and urban forestry can be found in Chapter 11 (Volder and Watson, 2010, this volume).

Gardens

Ornamental plants greatly contribute to an increase in plant species diversity in urban areas. They are used in parks, urban squares, green spaces around buildings, green strips, car parks, and private gardens. Residential gardens generally represent the largest planted area in cities (Gaston et al., 2005b; Loram et al., 2007) and can serve many functions, from providing home-grown food to serving as a private oasis for homeowners. Although individual gardens are small in size compared to the size of urban areas, the total combined area of residential gardens can be quite large. For example, in Sheffield, UK the total area of all garden areas combined was approximately 33 km^2, or 23% of the city area. These gardens housed 360,000 trees and a staggering estimated total of 52,000 domestic cats (Gaston et al., 2005b). The domestic gardens project in the UK has been a great source of information about domestic gardens (Thompson et al., 2003, 2005, 2004; Gaston et al., 2005a,b, 2007; Smith et al., 2005, 2006a,b,c; Loram et al., 2007, 2008a,b), revealing that they comprise 23 to 29% of undeveloped urban land. Outside of Europe, Mathieu et al. (2007) estimated that residential gardens comprised 36% of the Dunedin, New Zealand urban area. Garden areas generally contain a large proportion of exotic plants (McKinney, 2006, 2008; Loram et al., 2008b; Marco et al., 2008; Niinemets and Penuelas, 2008; Chapter 12, Reichard, 2010, this volume), and this may be a function of the degree of urbanization. Based on a survey in southern France, Marco et al. (2008) concluded that as rural gardens became more "urbanized," they contained a greater proportion of alien and invasive plant species. The generally smaller size of urban gardens may lead to more intensive garden practices per area of land, while rural gardens often have a large portion of unmanaged land. Homeowners are often encouraged to design residential gardens with the aim to improve overall biodiversity at a range of trophic scales. It is suggested that this can be done by planting a range of (native) plant species, creating different microclimates within a garden, creating a layered vegetation pattern from short stature plants to trees, and providing a wide variety of other landscape features.

By providing an assortment of microclimates, different textures (i.e., in soils and surface materials), and potential hiding places, a diverse group of wildlife, insects, plants, fungi, and microbes can potentially find a niche in which to thrive. Although the idea seems logical, using this technique did not work in Sheffield, where using artificial nests, ponds, dead wood, and patches of nettles to increase insect and fungal biodiversity did not lead to any noticeable increase in small-scale biodiversity (Gaston et al., 2005a). The relatively small size of the average garden may not provide enough habitat area to encourage and support biodiversity at all trophic levels. More studies are needed to study how garden design can be used to improve biodiversity at a range of spatial scales.

Remnant and Regenerated Vegetation

Remnant vegetation in urban areas is defined as patches of native vegetation that have never been cleared for urban use (Zipperer et al., 1997). Some of these sites may have been cleared in the past for agricultural purposes, but have never been built. Regenerated patches occur in sites where the land was initially cleared for urban development purposes, but the land was subsequently abandoned and recolonized by spontaneously growing vegetation. Both remnant and regenerated patches are ecologically and societally important and affect the urban environment strongly through the role these patches play in hydrologic processes, nutrient cycling, improvement of air and water quality, providing valuable wildlife habitat, and offering recreational opportunities. Regenerated patches may have lost a substantial amount of biodiversity compared with remnant vegetation because initially only the more opportunistic, "weedy," early successional species recolonize the site (Zipperer, 2002). Remnant patches are also unlikely to maintain presettlement species diversity because fragmentation often leads to losses of species from the system. A study in western Victoria, Australia clearly demonstrated that urbanization has a strong influence on species composition in remnant patches, disproportionally negatively affecting those species that are geophytes, hemicryptophytes with a flat rosette, and/or having wind- and ant-dispersed seeds (Williams et al., 2005). Due to the close proximity of nonnative species in urban areas, a large portion of the species that migrate into remnant and regenerated patches is often nonnative (Guntenspergen and Levenson, 1997; Zipperer, 2002), drastically altering species composition even in cases where species diversity is not much reduced.

Sites for spontaneous regeneration become available when existing vegetation is removed, dies, or when a disturbance occurs. The newly created open space can then be colonized by other individuals. Other sites where vegetation appears spontaneously are abandoned formerly inhabited sites, where building structures are torn down and not replaced, or just abandoned and left without maintenance. The size of these sites can vary from the site of a single dead tree to entire parks or abandoned building sites. The type of disturbance that created these new spaces determines the size and frequency with which these sites become available. Species available for colonization would be species growing nearby (within a mile) that are wind or bird dispersed, or species that grow from seed that was already present in the seed bank. A survey of seed banks in urban gardens in Sheffield, UK showed that the most abundant species in these seed banks were weedy plants and that most of the seeds present were of species that were known to have long-lived seeds (Thompson et al., 2005). Clonal species can move in if some individuals were already present before the disturbance and survived. New species can arrive on an abandoned site via humans, for example, through seeds for bird feeders or via transport on shoes, clothing, or car tires. The subsequent success of these species depends on how well these plants cope with the environmental conditions, such as light, water availability, nutrients, air pollution, and the degree and frequency of disturbance, how well they cope with the biotic environment (e.g., competition with other plants, herbivory, disease), and the life history of the species (e.g., lifespan and reproductive output).

Roadside Vegetation

Roadside plant communities are exposed to higher rates of air pollution, salt spray during the winter in northern climates, frequent mowing, wind from traffic, polluted runoff from the road during rainstorms, and altered hydrology. This type of habitat is ideal for plants using a ruderal strategy of high relative growth rates, large reproductive outputs, and short lifespans (Grime, 1977). For example, a survey of vegetation in Plzen demonstrated a strong increase in the proportion of ruderals as the area became increasingly urban between 1960 and 1990 (Pysek et al., 2004). In spite of the high disturbance rates along roadsides, these habitats can have significant conservation value (Ranta, 2008). The composition of roadside vegetation can be considerably affected by seed transport by vehicles, as suggested by von der Lippe and Kowarik (2007, 2008), who found that in Berlin, the seed rain contained many more seeds that were consistent with long-distance dispersal than with dispersal from nearby local vegetation. More than 50% of the seeds captured along the roadside were nonnative, and 19% of the seeds were from highly invasive species (Von der Lippe and Kowarik, 2007). In a followup study, Von der Lippe and Kowarik (2008) showed that the proportion of nonnative species was particularly high in the outbound lanes, suggesting that cities export nonnative, and sometimes invasive, species to more rural areas.

Restored Sites

Degraded urban sites have often lost or reduced some ecosystem functions, and restorative efforts are started to bring back key species and potentially restore preurban species diversity in the hopes of restoring the lost ecosystem functions. Urban wetlands and riparian areas are frequently a target for restoration efforts because they perform several crucial ecosystem services, such as water filtration, flood mitigation, provision of habitat for rare species (plants and animals), and recreational opportunities (Chapters 6 [Ehrenfeld and Stander, 2010] and 13 [Stander and Ehrenfeld, 2010], this volume). Altered hydrology in urbanizing areas make urban wetlands more vulnerable to invasion by nonnatives [e.g., by *Phragmites australis* (Cav.) Trin. ex Steud.] or expansion of other unwanted species such as the noxious weeds *Typha angustifolia* L. and *Phalaris arundinacea* L. (Choi and Bury, 2003). Reduced base flow through wetlands and riparian zones encourages encroachment and/or invasion by woody species such as *Salix* spp. and *Populus deltoides* W. Bartram ex Marshall (Choi and Bury 2003). Restoration of wetlands and riparian zones to their original state requires restoring the original hydrology, which is extremely difficult in urbanized areas where the hydrology is generally drastically altered through increased impervious surfaces, changes in vegetation, regrading, and other development.

Plant Responses to the Urban Environment

Effect of Soils and Belowground Conditions

Urban soils are often very different from rural soils (Craul, 1992; Effland and Pouyat, 1997; De Kimpe and Morel, 2000). Urban soils have been found to be more alkaline (Ware, 1990), warmer (Graves, 1994), higher in polycyclic aromatic carbons (PAH) and heavy metals (Miederer et al., 1995), more heterogeneous and more frequently disturbed (De Kimpe and Morel, 2000; Lorenz and Kandeler, 2005), and more compacted (Gregory et al., 2006) than rural soils. All these differences are heavily modified by site history, location, and soil age (Scharenbroch et al., 2005). Human activities such as sealing, infilling, removal of topsoil, compaction, and mixing strongly affect the soil structural properties, often reducing water and oxygen infiltration rates while increasing the resistance to root penetration. Reduced water infiltration rates due to compaction lead to an increase in the frequency of both drought and flooded soil conditions, while reduced soil oxygen availability reduces root growth and root soil exploration (Perata and Alpi, 1993; Bengough et al., 2006; Cook et al., 2007). These poor soil conditions induce great problems for the development of a healthy root system, often leading to stunted or abnormal tree development and an increased vulnerability to root diseases (Kozlowski, 1999). Generally, soils in highly urbanized areas are covered with an impervious surface and nutrient cycling from leaf litter is reduced. The impervious pavement reduces oxygen diffusion into the root zone, posing considerable problems for root growth (Beardsell, 1981; Balakina et al., 2005; Cook et al., 2007). For example, a combination of high vapor pressure deficits and reduced nutrient and water inputs likely reduced growth of *Liquidambar styraciflua* L. trees growing in an urban plaza compared with an urban park in Seattle (Kjelgren and Clark, 1993). These trees were experiencing a greater loss of water due to higher vapor pressure deficits while impervious pavement surrounding the trees was preventing water from infiltrating into the soil, thus creating exacerbated drought conditions. Environmental conditions such as found by Kjelgren and Clark in Seattle would be common in any urban area. The use of pervious pavement in pedestrian areas and other lightly used paved areas could help improve soil water infiltration (Volder et al., 2009).

In a study of urban soil temperatures, mean and maximum soil temperatures in urban paved areas were higher than in urban unpaved areas, particularly at depth (Graves and Dana, 1987). Higher soil temperatures can have strong detrimental effects on the growth of some species (Graves et al., 1991; Graves and Wilkins, 1991; Graves, 1994). Deposition of materials onto soils can also cause problems. Salts such as sodium can disrupt soil structure and change salt concentrations in the soil solution. Panno et al. (1999) found that deicing agents caused the originally diverse vegetation of fens in Illinois to be replaced by a more salt-tolerant species, *Typha angustifolia* L., thus resulting in a decrease in species richness in these fens. Salt can also affect streetscape vegetation. Trees planted in street planters in Geneva, IL were grown in low quality soils consisting of brick rubble, gravel, sand, and cinders with sodium levels high enough to classify the planter soils as sodic. Conditions such as these have strong adverse impacts on street tree survival, growth, and physiology (Kelsey and Hootman, 1990). Even when vegetation is not planted in planters, the presence of a complex network of drains, pipes, cables, and electrical conduits results in restricted root space and damage to root systems when regular maintenance of this belowground infrastructure is needed (Jim, 2001). On private land, poor gardening practices, such as excessive watering and fertilization, can cause anaerobic root stress and potentially toxic conditions as well (Huinink, 1998).

Effects of Air Quality, Light and Temperature

Plants in urban areas are exposed to a wide range of compounds such as CO_2, NO_x, ozone, dust, and PAH (Kuttler and Strassburger, 1999). These can have direct and indirect effects on plant growth, and the effects are variable. For example, elevated CO_2 levels generally benefit plant growth and plant water use efficiency, while ozone is phytotoxic to many plants, but some plants are more sensitive than others. Dust and smog (NO_x) can reduce light levels in urban areas, thus reducing rates of leaf photosynthesis. In addition, dust can cover the leaves and clog stomatal pores, which reduces rates of leaf photosynthesis as well. Besides dust, most of these compounds need to be absorbed by leaf stomates to cause leaf damage. Thus, the effect of air quality on urban vegetation is reduced in those species with lower rates of stomatal conductance. Similarly, effects of air pollution are reduced at night when stomates are closed, and during winter when the vegetation is leafless. Air quality can also affect the quality of food produced in urban areas. Carrots (*Daucus carota* L.) grown in an urban area in India showed reduced growth and reduced phosphorous, magnesium, calcium, and potassium contents (Tiwari et al., 2006). Urban vegetation can improve air quality by absorbing some pollution compounds. Vegetation along roadsides has a significantly greater buildup of PAHs in their tissue (Yang et al., 1991; Pathirana et al., 1994), thus reducing the spread of PAHs by absorbing them. Yang et al. (2008) calculated that green roofs in the city of Chicago (19.8 ha) remove 1675 kg of air pollutants per year, mostly ozone (52%), N_2O (27%), and particulate matter with particles smaller than 10 μm in diameter (PM_{10}) (14%).

Urban vegetation can be both a sink and a source of compounds in the troposphere. Leaves of many urban trees are known to remove dust and absorb ozone and other compounds. However, some trees species emit significant amounts of volatile organic compounds (VOCs). Volatile organic compounds are an important precursor for the formation of ozone (Chapter 3, Santosa, 2010, this volume). Many of the gymnosperms are potent emitters of VOCs, but other valued street trees such as sweetgum (*Liquidambar* spp.) and several of the oak species (*Quercus* spp.) are also known to emit large quantities of VOCs (Geron et al., 1995; Benjamin and Winer, 1998). In a comprehensive study in and around New York, where clones of cottonwood (*Populus deltoides*) were planted at urban and rural sites, Gregg et al. (2003) found that urban plant biomass was twice that of rural sites. Using a multiple regression approach, they demonstrated that this was not due to the higher temperatures, higher CO_2 concentrations, and greater N deposition in the city, but that higher mean ozone concentrations in the rural area (as opposed to higher peak ozone concentrations in the city) were responsible for the biomass difference. Paoletti (2009) offered three possible explanations for the phenomenon of higher ozone concentrations in rural areas than in urban areas. First, because ozone is a secondary pollutant, it can be formed at a significant distance away from the precursor (NO_x and VOC) sources. Second, other air pollutants are increased in urban areas, thus facilitating return reactions that break down ozone. Third, rural area have more vegetation that emits biogenic volatile organic compounds (BVOC), which is the more reactive precursor in the process of ozone formation.

Plant phenology starts earlier in urban areas, where the urban heat island induces earlier green-up and flowering, while delaying dormancy (White et al., 2002; Zhang et al., 2004). In addition to the production benefits of increased growing season length, the higher levels of atmospheric CO_2 observed in urban areas (Idso et al., 1998; Day et al., 2002; Gratani and Varone, 2006) generally also increase plant production (Gifford, 2004). Elevated levels of pollen production and increased lengths of the pollen production season can have strong effects on human health. For example, Ziska et al. (2003, 2007a) found that pollen levels of the common ragweed (*Ambrosia artemisiifolia* L.) were consistently higher in urban and urbanizing areas in Maryland than in rural areas. In addition, the pollen season also started earlier in urban areas. On the other hand, the allergen content of the pollen was lower in urban and urbanizing areas than in rural areas, and persistence of ragweed plants with time was reduced in urban areas. A study of another well-known noxious weed, poison ivy [*Toxicodendron radicans* (L.) Kuntze], revealed that its rate of spread, ability to recover from herbivory, and overall production of the irritant urushiol is enhanced by higher CO_2 concentrations (Ziska et al., 2007b). Urban areas can be considered test sites for what may happen to adjacent, nonurban, areas under a climate change scenario where CO_2 concentrations will be higher and air temperatures increased. This may mean that our future world will see an increase in allergenic pollen production and the spread of poison ivy.

Light can be significantly attenuated by urban aerosols, which cause a reduced light intensity and a greater amount of diffuse light. In Athens, Greece, Jacovides et al. (1997) found that photosynthetically active radiation (PAR) could be decreased by urban aerosols by up to 18%, while the increases in diffuse PAR ranged from 7 to 51%. In addition, the urban light environment can be strongly altered by surrounding infrastructure (Grimmond et al., 2001). Shadows cast by surrounding buildings often shorten the period that urban vegetation is exposed to direct sunlight, and can drastically reduce plant photosynthesis and daily carbon gain (Orourke and Terjung, 1981; Kjelgren and Clark, 1992). Light detection and ranging (LIDAR) mapping can be used to predict the light environment in urban areas. These maps can then be used to select the appropriate vegetation for each light environment (Yu et al., 2009).

Insects and Pathogens

Planting of vegetation in urban areas can also lead to enhanced spread of pests and diseases. The emerald ash borer beetle [*Agrilus planipennis* (Fairmaire)], for example, would not be spreading so quickly if ash trees had not been widely planted in urban areas. MacFarlane and Meyer (2005) suggested that the combined high abundance and low genetic diversity of ash trees planted in Michigan cities played a major role in the rapid spread and high impact of the emerald ash borer. Urban areas are particularly susceptible to exotic pest invasions for five main reasons. First, the introduction rate of exotic trees and shrubs is much greater in urban areas. Second, solid wood packing material used for packaging has emerged as a major source of pests (Haack, 2006). Third, the genetic diversity, and associated pest resistance, of most planted ornamental vegetation is low. Fourth, urban vegetation is often under a considerable amount of stress and likely less resistant to invasive pathogens and insect damage. Finally, exotic pests are more likely to find a suitable host due to the large species diversity and widespread use of exotic ornamental plant material in urban areas. Lack of genetic diversity may have played a major role in the rapid spread and enormous damage of Dutch elm disease (*Ophiostoma* spp.), where the widespread preferred planting of several select cultivars that turned out to be very susceptible to the disease had devastating effects on elm inventories in both Europe and North America (Gibbs, 1978).

The degree and type of urbanization not only affects the presence and diversity of vegetation, it also affects the presence of amphibians and reptiles (Vignoli et al., 2009), beetles (Niemela and Kotze, 2009), spiders (Bolger et al., 2008), and a range of arthropods (Bolger et al., 2000). This in turn, may affect how different trophic levels interact with each other. An interesting interaction between location, physiology, and insects was found by Hanks and Denno (1993). In this study on the campus of the University of Maryland they found that the prevalence of an armored scale insect on campus trees was determined by plant water status in an open landscape setting, with insect density decreasing with increasing water stress. This makes sense because scale insects are sap feeders, and sap flow will be decreased in water-stressed plants. However, on the same campus, woodlot trees that were not water stressed were also scale free and should have been covered in scale if plant water status was the only factor determining the density of the scale insect. The difference between the well-watered, scale-covered trees in the open setting and the scale-free trees in the woodlot was explained by the presence of general insect predators in the woodlot that were absent in the open landscape (Hanks and Denno, 1993). Thus, in natural settings, scale population density was inversely related to the pressure of general insect predators, while scale density in the absence of general predators was related to plant water status. This study clearly demonstrated the sometimes complicated interactions among plant responses to the landscape, insect community response to the landscape, and the biological interaction between the plants and the insects. If measurements had been taken only in the open landscape setting, one would have erroneously concluded infestation by armor scale insects is solely a function of plant water status.

A reduced number of chewing insects was likely the cause of the reduced chewing damage that Nuckols and Connor (1995) observed when they compared the amount of chewing damage on seven species of trees in an urban ornamental setting versus those growing in natural forests. They found that trees in urban or ornamental settings had significantly reduced chewing damage (numbers), which is contrary to the view that the supposedly more stressed urban trees are more susceptible to insect attack. They offered three explanations: (i) urban and ornamental trees have higher resistance to insect attack, (ii) insects in urban environments have lower dispersal rates, and (iii) chewing insects have lower survival rates in urban areas. Clearly, the three explanations are not mutually exclusive, but in light of Hanks and Denno's findings in Maryland, a lack of insect pressure may be the more likely cause of the reduced chewing damage. Gardens with less complex vegetation structure, such as single trees or a vegetation structure with several layers missing, were found to support a reduced invertebrate abundance in a survey of gardens in Sheffield, UK (Smith et al., 2006a,c).

The actual size of the park or remnant woodland also affects the degree of insect damage. Christie and Hochuli (2005) showed that trees in larger fragments suffered significantly less leaf damage due to chewing insects than trees in smaller patches. Further analysis showed that this was not due to edge effects, but it was more likely that the size of the treed area itself was the dominant factor determining the amount of damage.

Disturbance and Management

Physical disturbance by humans can play a major role in the success and functioning of urban vegetation, even in larger parks and remnants. Urbanization can significantly affect species survival, as documented by Bagnall (1979). Bagnall's study of a remnant urban forest in New Zealand clearly showed that a lot of damage can be done by excessive, widely scattered, pedestrian traffic and playing children. Similarly, Bhuju and Ohsawa (1998) documented that compaction due to human trampling was obstructing natural succession in a regenerating urban forest near Chiba City, Japan. Nearby construction can strongly affect species composition and survival of individuals. Damage to root systems and compaction during construction and due to trampling after construction strongly affected native species survival in Stockholm, Sweden (Florgard, 2000). Trampling damage in this study was stronger for vegetation growing in shallow, poor, dry soils. Thus, when conserving native vegetation, particular attention should be paid to soil type and, if possible, additional protection of these systems may be warranted. When maintaining native species diversity is not an explicit goal in managing urban vegetation, the solution may be to plant ornamental species that have been specifically selected for their ability to cope with urban environmental conditions (Ware, 1984). Those would generally be species with tolerance of extended periods of drought, as well as a tolerance for flooded soil conditions and the ability to develop a root system in spite of soil compaction and restricted rooting space.

Suppression of the natural disturbance regime can affect vegetation composition and functioning as well. For example, species composition of many natural ecosystems is maintained by the occurrence of occasional low-intensity fires that reduce the dominance of fire sensitive species, such as the invading hardwoods in longleaf pine savannas in Florida. In urban areas, natural fires are actively suppressed, and thus ecosystem functioning is altered. Bringing back some of the natural disturbance regime can potentially restore some of the lost ecosystem functions. Heuberger and Putz (2003) applied growing season fires in urban and suburban neighborhoods in Gainesville, FL and found that burned patches had higher species richness and diversity than control patches. This study showed that, in the Florida Sandhills, regular prescribed low intensity burns could be used to restore species richness and diversity in degraded longleaf pine sandhill communities that are threatened by urbanization.

Plant traits of plants growing in urban riparian zones also shift as a result of altered hydrological cycles due to increasing urbanization. Drought is more common in urban riparian areas, which suffer from flashy stream discharge, than in natural riparian areas that have a more steady stream discharge (Groffman et al., 2003). In a transect study along an urbanization gradient, Groffman et al. (2003) found a marked increase in the proportion of upland species in urban riparian zones of the Gwynns Falls watershed in Baltimore. In a species trait analysis of riparian species growing in 17 watershed sites with differing landscape metrics (ranging from unmanaged forest to urban), Burton et al. (2009) found that as urbanization increased, flood tolerance of the species decreased. Thus, as urbanization changes the hydrology of riparian zones, it also has a strong effect on other ecosystem functions, such as species composition. Realistically, one could not restore the predevelopment species composition without also restoring the predevelopment discharge regimen, something that would be nearly impossible in existing urban areas. In essence, the alteration of the hydrologic cycle has created a novel ecosystem. Both the fire and flooding frequency examples show that management decisions for species richness and diversity should be informed by the effect of changes in disturbance regime as urbanization increases.

A survey of the genetic variation and diversity of the perennial herb *Primula elatior* (L.) Hill (Van Rossum, 2008) demonstrated another issue in urban environments, the high degree of fragmentation, which leads to many small isolated populations that leave species vulnerable to local extinction. A common management response is to increase recruitment within these small populations; however, van Rossum (2008) found that populations with high recruitment rates had lower genetic diversity, presumably due to inbreeding. Instead, van Rossum (2008) recommended that management should be aimed at restoring gene flow between isolated populations to maintain or increase genetic diversity.

Fragmentation of ecosystems through urban development makes some species more vulnerable to extinction. Active management may be necessary to prevent populations of vulnerable species from going extinct, ranging from reducing the activity that threatens the existence of a population to active reintroduction of individuals (Hobbs, 2007). However, there is a fine line between restoring an ecosystem and inserting so much management that preserving a population in the landscape becomes gardening (Hobbs, 2007).

Conclusions

Due to the wide range of habitat types and the highly managed nature of urban vegetation, urban green areas have a large proportion of imported, sometimes invasive, plant species. Urban vegetation provides valuable and critical services to society, but is under considerable stress due to urban environmental conditions such as low soil quality, altered hydrology, reduced light environment, higher air temperatures, reduced nutrient cycling, altered disturbance regime, and increased air pollutants. Management strategies that aim to reduce these stresses, or appropriate selection of plants that can cope with these stresses, will lead to healthier urban plants. Restoration strategies where the aim is to restore native vegetation may not be feasible if presettlement environmental conditions cannot be re-created.

References

Aitkenhead-Peterson, J.A., M.K. Steele, and A. Volder. 2010. Services in natural and human dominated ecosystems. p. 373–390. *In* J. Aitkenhead-Peterson and A. Volder (ed.) Urban ecosystem ecology. Agron. Monogr. 55. ASA, CSSA, and SSSA, Madison, WI.

Akbari, H., M. Pomerantz, and H. Taha. 2001. Cool surfaces and shade trees to reduce energy use and improve air quality in urban areas. Sol. Energy 70:295–310.

Andersson, E. 2006. Urban landscapes and sustainable cities. Ecol. Soc. 11:34.

Bagnall, R.G. 1979. Study of human impact on an urban forest remnant—Redwood Bush, Tawa, near Wellington, New Zealand. N.Z. J. Bot. 17:117–126.

Balakina, J.N., O.V. Makarova, V.V. Bondarenko, L.J. Koudstaal, E.J. Ros, A.J. Koolen, and W.K.P. van Loon. 2005. Simulation of oxygen regime of tree substrates. Urban For. Urban Green. 4:23–35.

Barradas, V.L. 2000. Energy balance and transpiration in an urban tree hedgerow in Mexico City. Urban Ecosyst. 4:55–67.

Beardsell, D. 1981. Soil aeration and trees. Australian Parks and Recreation. November, p. 22–25.

Beckett, K.P., P.H. Freer-Smith, and G. Taylor. 1998. Urban woodlands: Their role in reducing the effects of particulate pollution. Environ. Pollut. 99:347–360.

Bengough, A.G., M.F. Bransby, J. Hans, S.J. McKenna, T.J. Roberts, and T.A. Valentine. 2006. Root responses to soil physical conditions; growth dynamics from field to cell. J. Exp. Bot. 57:437–447.

Benjamin, M.T., and A.M. Winer. 1998. Estimating the ozone-forming potential of urban trees and shrubs. Atmos. Environ. 32:53–68.

Bhuju, D.R., and M. Ohsawa. 1998. Effects of nature trails on ground vegetation and understory colonization of a patchy remnant forest in an urban domain. Biol. Conserv. 85:123–135.

Bolger, D.T., K.H. Beard, A.V. Suarez, and T.J. Case. 2008. Increased abundance of native and non-native spiders with habitat fragmentation. Divers. Distrib. 14:655–665.

Bolger, D.T., A.V. Suarez, K.R. Crooks, S.A. Morrison, and T.J. Case. 2000. Arthropods in urban habitat fragments in southern California: Area, age, and edge effects. Ecol. Appl. 10:1230–1248.

Bolund, P., and S. Hunhammar. 1999. Ecosystem services in urban areas. Ecol. Econ. 29:293–301.

Burian, S.J., and C.A. Pomeroy. 2010. Urban impacts on the water cycle and potential green infrastructure implications. p. 277–296. In J. Aitkenhead-Peterson and A. Volder (ed.) Urban ecosystem ecology. Agron. Monogr. 55. ASA, CSSA, and SSSA, Madison, WI.

Burton, M.L., L.J. Samuelson, and M.D. Mackenzie. 2009. Riparian woody plant traits across an urban–rural land use gradient and implications for watershed function with urbanization. Landscape Urban Plan. 90:42–55.

Burton, M.L., L.J. Samuelson, and S. Pan. 2005. Riparian woody plant diversity and forest structure along an urban–rural gradient. Urban Ecosyst. 8:93–106.

Carter, T., and C.R. Jackson. 2007. Vegetated roofs for stormwater management at multiple spatial scales. Landsc. Urban Plan. 80:84–94.

Carver, A.D., D.R. Unger, and C.L. Parks. 2004. Modeling energy savings from urban shade trees: An assessment of the CITYgreen® energy conservation module. Environ. Manage. 34:650–655.

Chiesura, A. 2004. The role of urban parks for the sustainable city. Landsc. Urban Plan. 68:129–138.

Choi, Y.D., and C. Bury. 2003. Process of floristic degradation in urban, and, suburban wetlands in northwestern Indiana, USA. Nat. Areas J. 23:320–331.

Christie, F.J., and D.F. Hochuli. 2005. Elevated levels of herbivory in urban landscapes: Are declines in tree health more than an edge effect? Ecol. Soc. 10:10. Available at http://www.ecologyandsociety.org/vol10/iss1/art10/. (verified 2 Mar. 2010).

Clergeau, P., J.P.L. Savard, G. Mennechez, and G. Falardeau. 1998. Bird abundance and diversity along an urban–rural gradient: A comparative study between two cities on different continents. Condor 100:413–425.

Cook, F.J., J.H. Knight, and F.M. Kelliher. 2007. Oxygen transport in soil and the vertical distribution of roots. Aust. J. Soil Res. 45:101–110.

Craul, P.J. (ed.) 1992. Urban soil in landscape design. Wiley, Hoboken, NJ.

Currie, B., and B. Bass. 2008. Estimates of air pollution mitigation with green plants and green roofs using the UFORE model. Urban Ecosyst. 11:409–422.

Day, T.A., P. Gober, F.S.S. Xiong, and E.A. Wentz. 2002. Temporal patterns in near-surface CO_2 concentrations over contrasting vegetation types in the Phoenix metropolitan area. Agric. For. Meteorol. 110:229–245.

De Kimpe, C.R. and J.L. Morel. 2000. Urban soil management: A growing concern. Soil Sci. 165:31–40.

Donovan, G.H., and D.T. Butry. 2009. The value of shade: Estimating the effect of urban trees on summertime electricity use. Energy Build. 41:662–668.

Dussaillant, A.R., C.H. Wu, and K.W. Potter. 2004. Richards equation model of a rain garden. J. Hydrol. Eng. 9:219–225.

Effland, W.R., and R.V. Pouyat. 1997. The genesis, classification, and mapping of soils in urban areas. Urban Ecosyst. 1:217–228.

Ehrenfeld, J.G., and E.K. Stander. 2010. Habitat function in urban riparian zones. p. 103–118. In J. Aitkenhead-Peterson and A. Volder (ed.) Urban ecosystem ecology. Agron. Monogr. 55. ASA, CSSA, and SSSA, Madison, WI.

Escobedo, F.J., and D.J. Nowak. 2009. Spatial heterogeneity and air pollution removal by an urban forest. Landsc. Urban Plan. 90:102–110.

Florgard, C. 2000. Long-term changes in indigenous vegetation preserved in urban areas. Landsc. Urban Plan. 52:101–116.

Gaston, K.J., R.A. Fuller, A. Loram, C. MacDonald, S. Power, and N. Dempsey. 2007. Urban domestic gardens (XI): Variation in urban wildlife gardening in the United Kingdom. Biodivers. Conserv. 16:3227–3238.

Gaston, K.J., R.M. Smith, K. Thompson, and P.H. Warren. 2005a. Urban domestic gardens (II): Experimental tests of methods for increasing biodiversity. Biodivers. Conserv. 14:395–413.

Gaston, K.J., P.H. Warren, K. Thompson, and R.M. Smith. 2005b. Urban domestic gardens (IV): The extent of the resource and its associated features. Biodivers. Conserv. 14:3327–3349.

Georgi, N., and K. Zafiriadis. 2006. The impact of park trees on microclimate in urban areas. Urban Ecosyst. 9:195–209.

Geron, C.D., T.E. Pierce, and A.B. Guenther. 1995. Reassessment of biogenic volatile organic compound emissions in the Atlanta area. Atmos. Environ. 29:1569–1578.

Gibbs, J.N. 1978. Intercontinental epidemiology of Dutch Elm Disease. Annu. Rev. Phytopathol. 16:287–307.

Gifford, R.M. 2004. The CO_2 fertilising effect—Does it occur in the real world? The International Free Air CO_2 Enrichment (FACE) Workshop: Short- and long-term effects of elevated atmospheric CO_2 on managed ecosystems, Ascona, Switzerland, March 2004. New Phytol. 163:221–225.

Godefroid, S. 2001. Temporal analysis of the Brussels flora as indicator for changing environmental quality. Landsc. Urban Plan. 52:203–224.

Godefroid, S., and N. Koedam. 2003. How important are large vs. small forest remnants for the conservation of the woodland flora in an urban context? Glob. Ecol. Biogeogr. 12:287–298.

Godefroid, S., and N. Koedam. 2007. Urban plant species patterns are highly driven by density and function of built-up areas. Landscape Ecol. 22:1227–1239.

Golubiewski, N.E. 2006. Urbanization increases grassland carbon pools: Effects of landscaping in Colorado's front range. Ecol. Appl. 16:555–571.

Gratani, L., and L. Varone. 2006. Carbon sequestration by Quercus ilex L. and Quercus pubescens Willd. and their contribution to decreasing air temperature in Rome. Urban Ecosyst. 9:27–37.

Graves, W.R. 1994. Urban soil temperatures and their potential impact on tree growth. J. Arboric. 20:24–27.

Graves, W.R., and M.N. Dana. 1987. Root-zone temperature monitored at urban sites. HortScience 22:613–614.

Graves, W.R., R.J. Joly, and M.N. Dana. 1991. Water-use and growth of honey locust and tree-of-heaven at high root-zone temperature. HortScience 26:1309–1312.

Graves, W.R., and L.C. Wilkins. 1991. Growth of honey locust seedlings during high root-zone temperature and osmotic stress. HortScience 26:1312–1315.

Gregg, J.W., C.G. Jones, and T.E. Dawson. 2003. Urbanization effects on tree growth in the vicinity of New York City. Nature 424:183–187.

Gregg, J.W., C.G. Jones, and T.E. Dawson. 2006. Physiological and developmental effects of O_3 on cottonwood growth in urban and rural sites. Ecol. Appl. 16:2368–2381.

Gregory, J.H., M.D. Dukes, P.H. Jones, and G.L. Miller. 2006. Effect of urban soil compaction on infiltration rate. J. Soil Water Conserv. 61:117–124.

Grime, J.P. 1977. Evidence for existence of 3 primary strategies in plants and its relevance to ecological and evolutionary theory. Am. Nat. 111:1169–1194.

Grimmond, C.S.B., S.K. Potter, H.N. Zutter, and C. Souch. 2001. Rapid methods to estimate sky-view factors applied to urban areas. Int. J. Climatol. 21:903–913.

Grimmond, C.S.B., C. Souch, and M.D. Hubble. 1996. Influence of tree cover on summer-time surface energy balance fluxes, San Gabriel Valley, Los Angeles. Clim. Res. 6:45–57.

Groffman, P.M., D.J. Bain, L.E. Band, K.T. Belt, G.S. Brush, J.M. Grove, R.V. Pouyat, I.C. Yesi-lonis, and W.C. Zipperer. 2003. Down by the riverside: Urban riparian ecology. Front. Ecol. Environ. 1:315–321.

Guevara-Escobar, A., E. Gonzalez-Sosa, C. Veliz-Chavez, E. Ventura-Ramos, and M. Ramos-Salinas. 2007. Rainfall interception and distribution patterns of gross precipitation around an isolated *Ficus benjamina* tree in an urban area. J. Hydrol. 333:532–541.

Guntenspergen, G., and J. Levenson. 1997. Understory plant species composition in remnant stands along an urban-to-rural land-use gradient. Urban Ecosyst. 1:155–169.

Haack, R.A. 2006. Exotic bark- and wood-boring Coleoptera in the United States: Recent establishments and interceptions. Can. J. For. Res. 36:269–288.

Hanks, L.M., and R.F. Denno. 1993. Natural enemies and plant water relations influence the distribution of an armored scale insect. Ecology 74:1081–1091.

Heisler, G.M., and A.J. Brazel. 2010. The urban physical environment: Temperature and urban heat islands. p. 29–56. *In* J. Aitkenhead-Peterson and A. Volder (ed.) Urban ecosystem ecology. Agron. Monogr. 55. ASA, CSSA, and SSSA, Madison, WI.

Herbst, M., J.M. Roberts, P.T.W. Rosier, and D.J. Gowing. 2006. Measuring and modelling the rainfall interception loss by hedgerows in southern England. Agric. For. Meteorol. 141:244–256.

Heuberger, K.A., and F.E. Putz. 2003. Fire in the suburbs: Ecological impacts of prescribed fire in small remnants of longleaf pine (*Pinus palustris*) sandhill. Restor. Ecol. 11:72–81.

Hobbs, R.J. 2007. Managing plant populations in fragmented landscapes: Restoration or gardening? Aust. J. Bot. 55:371–374.

Hope, D., C. Gries, W.X. Zhu, W.F. Fagan, C.L. Redman, N.B. Grimm, A.L. Nelson, C. Martin, and A. Kinzig. 2003. Socioeconomics drive urban plant diversity. Proc. Natl. Acad. Sci. USA 100:8788–8792.

Huinink, J.T.M. 1998. Soil quality requirements for use in urban environments. Soil Tillage Res. 47:157–162.

Idso, C.D., S.B. Idso, and R.C. Balling. 1998. The urban CO2 dome of Phoenix, Arizona. Phys. Geogr. 19:95–108.

Jacovides, C.P., F. Timbios, D.N. Asimakopouolos, and M.D. Steven. 1997. Urban aerosol and clear skies spectra for global and diffuse photosynthetically active radiation. Agric. For. Meteorol. 87:91–104.

Jim, C.Y. 2001. Managing urban trees and their soil envelopes in a contiguously developed city environment. Environ. Manage. 28:819–832.

Jim, C.Y., and W.Y. Chen. 2008. Assessing the ecosystem service of air pollutant removal by urban trees in Guangzhou (China). J. Environ. Manage. 88:665–676.

Kelsey, P., and R. Hootman. 1990. Soil resource evaluation for a group of sidewalk street tree planters. J. Arboric. 16:113–117.

Kjelgren, R.K., and J.R. Clark. 1992. Photosynthesis and leaf morphology of *Liquidambar Styraciflua* L. under variable urban radiant-energy conditions. Int. J. Biometeorol. 36:165–171.

Kjelgren, R.K., and J.R. Clark. 1993. Growth and water relations of *Liquidambar Styraciflua* L. in an urban park and plaza. Trees Struct. Funct. 7:195–201.

Knapp, S., I. Kuhn, O. Schweiger, and S. Klotz. 2008. Challenging urban species diversity: Contrasting phylogenetic patterns across plant functional groups in Germany. Ecol. Lett. 11:1054–1064.

Koh, L.P., and N.S. Sodhi. 2004. Importance of reserves, fragments, and parks for butterfly conservation in a tropical urban landscape. Ecol. Appl. 14:1695–1708.

Kozlowski, T.T. 1999. Soil compaction and growth of woody plants. Scand. J. For. Res. 14:596–619.

Kuttler, W., and A. Strassburger. 1999. Air quality measurements in urban green areas—A case study. Atmos. Environ. 33:4101–4108.

Lemonsu, A., V. Masson, and E. Berthier. 2007. Improvement of the hydrological component of an urban soil-vegetation-atmosphere-transfer model. Hydrol. Processes 21:2100–2111.

Li, F., R.S. Wang, J. Paulussen, and X.S. Liu. 2005. Comprehensive concept planning of urban greening based on ecological principles: A case study in Beijing, China. Landsc. Urban Plan. 72:325–336.

Li, W.F., Z.Y. Ouyang, X.S. Meng, and X.K. Wang. 2006. Plant species composition in relation to green cover configuration and function of urban parks in Beijing, China. Ecol. Res. 21:221–237.

Loeb, R.E. 2006. A comparative flora of large urban parks: Intraurban and interurban similarity in the megalopolis of the northeastern United States. J. Torrey Bot. Soc. 133:601–625.

Long, A.J., and P.K.R. Nair. 1999. Trees outside forests: Agro-, community, and urban forestry. New For. 17:145–174.

Loram, A., K. Thompson, P.H. Warren, and K.J. Gaston. 2008b. Urban domestic gardens (XII): The richness and composition of the flora in five UK cities. J. Veg. Sci. 19:321–330.

Loram, A., J. Tratalos, P.H. Warren, and K.J. Gaston. 2007. Urban domestic gardens (X): The extent and structure of the resource in five major cities. Landscape Ecol. 22:601–615.

Loram, A., P.H. Warren, and K.J. Gaston. 2008a. Urban domestic gardens (XIV): The characteristics of gardens in five cities. Environ. Manage. 42:361–376.

Lorenz, K., and E. Kandeler. 2005. Biochemical characterization of urban soil profiles from Stuttgart, Germany. Soil Biol. Biochem. 37:1373–1385.

MacFarlane, D.W., and S.P. Meyer. 2005. Characteristics and distribution of potential ash tree hosts for emerald ash borer. For. Ecol. Manage. 213:15–24.

Marco, A., T. Dutoit, M. Deschamps-Cottin, J.F. Mauffrey, M. Vennetier, and V. Bertaudiere-Montes. 2008. Gardens in urbanizing rural areas reveal an unexpected floral diversity related to housing density. C. R. Biol. 331:452–465.

Mathieu, R., C. Freeman, and J. Aryal. 2007. Mapping private gardens in urban areas using object-oriented techniques and very high-resolution satellite imagery. Landsc. Urban Plan. 81:179–192.

Matteo, M., T. Randhir, and D. Bloniarz. 2006. Watershed-scale impacts of forest buffers on water quality and runoff in urbanizing environment. J. Water Resour. Plann. Manage. 132:144–152.

McCleery, R. 2010. Urban mammals. p. 87–102. In J. Aitkenhead-Peterson and A. Volder (ed.) Urban ecosystem ecology. Agron. Monogr. 55. ASA, CSSA, and SSSA, Madison, WI.

McDonnell, M.J. 1997. A paradigm shift. Urban Ecosyst. 1:85–86.

McDonnell, M.J., and S.T.A. Pickett. 1990. Ecosystem structure and function along urban-rural gradients—An unexploited opportunity for ecology. Ecology 71:1232–1237.

McDonnell, M.J., S.T.A. Pickett, P. Groffman, P. Bohlen, R.V. Pouyat, W.C. Zipperer, R.W. Parmelee, M.M. Carreiro, and K. Medley. 1997. Ecosystem processes along an urban-to-rural gradient. Urban Ecosyst. 1:21–36.

McKinney, M. 2008. Effects of urbanization on species richness: A review of plants and animals. Urban Ecosyst. 11:161–176.

McKinney, M.L. 2006. Urbanization as a major cause of biotic homogenization. Biol. Conserv. 127:247–260.

Miederer, M., A. Maschka-Selig, and C. Hohl. 1995. Monitoring polycyclic aromatic hydrocarbons (PAHs) and heavy metals in urban soil, compost and vegetation. Environ. Sci. Pollut. Res. Int. 2:83–89.

Moffatt, S.F., S.M. McLachlan, and N.C. Kenkel. 2004. Impacts of land use on riparian forest along an urban–rural gradient in southern Manitoba. Plant Ecol. 174:119–135.

Niemela, J., and D.J. Kotze. 2009. Carabid beetle assemblages along urban to rural gradients: A review. Landsc. Urban Plan. 92:65–71.

Niinemets, U., and J. Penuelas. 2008. Gardening and urban landscaping: Significant players in global change. Trends Plant Sci. 13:60–65.

Nuckols, M.S., and E.F. Connor. 1995. Do trees in urban or ornamental plantings receive more damage by insects than trees in natural forests. Ecol. Entomol. 20:253–260.

Orourke, P.A., and W.H. Terjung. 1981. Relative influence of city structure on canopy photosynthesis. Int. J. Biometeorol. 25:1–19.

Panno, S.V., V.A. Nuzzo, K. Cartwright, B.R. Hensel, and I.G. Krapac. 1999. Impact of urban development on the chemical composition of ground water in a fen–wetland complex. Wetlands 19:236–245.

Paoletti, E. 2009. Ozone and urban forests in Italy. Environ. Pollut. 157:1506–1512.

Pathirana, S., D.W. Connell, and P.D. Vowles. 1994. Distribution of polycyclic aromatic hydrocarbons (PAHs) in an urban roadway system. Ecotoxicol. Environ. Saf. 28:256–269. Pemberton, R.W., and H. Liu. 2009. Marketing time predicts naturalization of horticultural plants. Ecology 90:69–80.

Perata, P., and A. Alpi. 1993. Plant responses to anaerobiosis. Plant Sci. 93:1–17.

Pickett, S.T.A., M.L. Cadenasso, J.M. Grove, C.H. Nilon, R.V. Pouyat, W.C. Zipperer, and R. Costanza. 2001. Urban ecological systems: Linking terrestrial ecological, physical, and socioeconomic components of metropolitan areas. Annu. Rev. Ecol. Syst. 32:127–157.

Pouyat, R., I. Yesilonis, and N. Golubiewski. 2009. A comparison of soil organic carbon stocks between residential turf grass and native soil. Urban Ecosyst. 12:45–62.

Pouyat, R.V., M.J. McDonnell, and S.T.A. Pickett. 1995. Soil characteristics of oak stands along an urban–rural land-use gradient. J. Environ. Qual. 24:516–526.

Pouyat, R.V., K. Szlavecz, I.D. Yesilonis, P.M. Groffman, and K. Schwarz. 2010. Chemical, physical, and biological characteristics of urban soils. p. 119–152. In J. Aitkenhead-Peterson and A. Volder (ed.) Urban ecosystem ecology. Agron. Monogr. 55. ASA, CSSA, and SSSA, Madison, WI.

Pysek, P., Z. Chocholouskova, A. Pysek, V. Jarosik, M. Chytry, and L. Tichy. 2004. Trends in species diversity and composition of urban vegetation over three decades. J. Veg. Sci. 15:781–788.

Ranta, P. 2008. The importance of traffic corridors as urban habitats for plants in Finland. Urban Ecosyst. 11:149–159.

Reichard, S.H. 2010. Inside out: Invasive plants and urban environments. p. 241–252. In J. Aitkenhead-Peterson and A. Volder (ed.) Urban ecosystem ecology. Agron. Monogr. 55. ASA, CSSA, and SSSA, Madison, WI.

Santosa, S.J. 2010. Urban air quality. p. 57–74. In J. Aitkenhead-Peterson and A. Volder (ed.) Urban ecosystem ecology. Agron. Monogr. 55. ASA, CSSA, and SSSA, Madison, WI.

Savard, J.P.L., P. Clergeau, and G. Mennechez. 2000. Biodiversity concepts and urban ecosystems. Landsc. Urban Plan. 48:131–142.

Scharenbroch, B.C., J.E. Lloyd, and J.L. Johnson-Maynard. 2005. Distinguishing urban soils with physical, chemical, and biological properties. Pedobiologia 49:283–296.

Shashua-Bar, L., and M.E. Hoffman. 2000. Vegetation as a climatic component in the design of an urban street- An empirical model for predicting the cooling effect of urban green areas with trees. Energy Build. 31:221–235.

Shochat, E., S. Lerman, and E. Fernández-Juricic. 2010. Birds in urban ecosystems: Population dynamics, community structure, biodiversity, and conservation. p. 75–86. In J. Aitkenhead-Peterson and A. Volder (ed.) Urban ecosystem ecology. Agron. Monogr. 55. ASA, CSSA, and SSSA, Madison, WI.

Simpson, J.R., and E.G. McPherson. 1998. Simulation of tree shade impacts on residential energy use for space conditioning in Sacramento. Atmos. Environ. 32:69–74.

Smith, R.M., K.J. Gaston, P.H. Warren, and K. Thompson. 2005. Urban domestic gardens (V): Relationships between landcover composition, housing and landscape. Landscape Ecol. 20:235–253.

Smith, R.M., K.J. Gaston, P.H. Warren, and K. Thompson. 2006a. Urban domestic gardens (VIII): Environmental correlates of invertebrate abundance. Biodivers. Conserv. 15:2515–2545.

Smith, R.M., K. Thompson, J.G. Hodgson, P.H. Warren, and K.J. Gaston. 2006b. Urban domestic gardens (IX): Composition and richness of the vascular plant flora, and implications for native biodiversity. Biol. Conserv. 129:312–322.

Smith, R.M., P.H. Warren, K. Thompson, and K.J. Gaston. 2006c. Urban domestic gardens (VI): Environmental correlates of invertebrate species richness. Biodivers. Conserv. 15:2415–2438.

Sorace, A. 2001. Value to wildlife of urban-agricultural parks: A case study from Rome urban area. Environ. Manage. 28:547–560.

Stander, E.K., and J.G. Ehrenfeld. 2010. Urban riparian function. p. 253–276. In J. Aitkenhead-Peterson and A. Volder (ed.) Urban ecosystem ecology. Agron. Monogr. 55. ASA, CSSA, and SSSA, Madison, WI.

Taha, H. 1997. Urban climates and heat islands: Albedo, evapotranspiration, and anthropogenic heat. Energy Build. 25:99–103.

Thompson, K., K.C. Austin, R.M. Smith, P.H. Warren, P.G. Angold, and K.J. Gaston. 2003. Urban domestic gardens (I): Putting small-scale plant diversity in context. J. Veg. Sci. 14:71–78.

Thompson, K., S. Colsell, J. Carpenter, R.M. Smith, P.H. Warren, and K.J. Gaston. 2005. Urban domestic gardens (VII): A preliminary survey of soil seed banks. Seed Sci. Res. 15:133–141.

Thompson, K., J.G. Hodgson, R.M. Smith, P.H. Warren, and K.J. Gaston. 2004. Urban domestic gardens (III): Composition and diversity of lawn floras. J. Veg. Sci. 15:373–378.

Thompson, K., and M.A. McCarthy. 2008. Traits of British alien and native urban plants. J. Ecol. 96:853–859.

Tiwari, S., M. Agrawal, and F.M. Marshall. 2006. Evaluation of ambient air pollution impact on carrot plants at a sub urban site using open top chambers. Environ. Monit. Assess. 119:15–30.

Tratalos, J., R.A. Fuller, P.H. Warren, R.G. Davies, and K.J. Gaston. 2007. Urban form, biodiversity potential and ecosystem services. Landsc. Urban Plan. 83:308–317.

Urban, M.C., D.K. Skelly, D. Burchsted, W. Price, and S. Lowry. 2006. Stream communities across a rural–urban landscape gradient. Divers. Distrib. 12:337–350.

van Rensburg, L., G.H.J. Kruger, B. Ubbink, M.C. Scholes, and J. Peacock. 1997a. A phytocentric perspective of Asterolecanium quercicola Bouche infestation on Quercus robur L. trees along an urbanization gradient. S. Afr. J. Bot. 63:25–31.

van Rensburg, L., G.H.J. Kruger, B. Ubbink, J. Stassen, and H. van Hamburg. 1997b. Seasonal performance of Quercus robur L along an urbanization gradient. S. Afr. J. Bot. 63:32–36.

Van Rossum, F. 2008. Conservation of long-lived perennial forest herbs in an urban context: Primula elatior as study case. Conserv. Genet. 9:119–128.

Vignoli, L., I. Mocaer, L. Luiselli, and M.A. Bologna. 2009. Can a large metropolis sustain complex herpetofauna communities? An analysis of the suitability of green space fragments in Rome. Anim. Conserv. 12:456–466.

Volder, A., and W.T. Watson. 2010. Urban forestry. p. 227–240. In J. Aitkenhead-Peterson and A. Volder (ed.) Urban ecosystem ecology. Agron. Monogr. 55. ASA, CSSA, and SSSA, Madison, WI.

Volder, A., T. Watson, and B. Viswanathan. 2009. Potential use of pervious concrete for maintaining existing mature trees during and after urban development. Urban For. Urban Green. 8:249–256.

Von der Lippe, M., and I. Kowarik. 2007. Long-distance dispersal of plants by vehicles as a driver of plant invasions. Conserv. Biol. 21:986–996.

Von der Lippe, M., and I. Kowarik. 2008. Do cities export biodiversity? Traffic as dispersal vector across urban–rural gradients. Divers. Distrib. 14:18–25.

Ware, G. 1990. Constraints to tree growth imposed by urban soil alkalinity. J. Arboric. 16:35–38.

Ware, G.H. 1984. Coping with clay: Trees to suit sites, sites to suit trees. J. Arboric. 10:108–112.

White, M.A., R.R. Nemani, P.E. Thornton, and S.W. Running. 2002. Satellite evidence of phenological differences between urbanized and rural areas of the eastern United States deciduous broadleaf forest. Ecosystems 5:260–273.

Whitlow, T.H., N.L. Bassuk, and D.L. Reichert. 1992. A 3-year study of water relations of urban street trees. J. Appl. Ecol. 29:436–450.

Wilby, R.L., and G.L.W. Perry. 2006. Climate change, biodiversity and the urban environment: A critical review based on London, UK. Prog. Phys. Geogr. 30:73–98.

Williams, N.S.G., J.W. Morgan, M.J. McDonnell, and M.A. McCarthy. 2005. Plant traits and local extinctions in natural grasslands along an urban–rural gradient. J. Ecol. 93:1203–1213.

Williams, N.S.G., M.W. Schwartz, P.A. Vesk, M.A. McCarthy, A.K. Hahs, S.E. Clemants, R.T. Corlett, R.P. Duncan, B.A. Norton, K. Thompson, and M.J. McDonnell. 2009. A conceptual framework for predicting the effects of urban environments on floras. J. Ecol. 97:4–9.

Xiao, Q., and E.G. McPherson. 2002. Rainfall interception by Santa Monica's municipal urban forest. Urban Ecosyst. 6:291–302.

Yang, J., Q. Yu, and P. Gong. 2008. Quantifying air pollution removal by green roofs in Chicago. Atmos. Environ. 42:7266–7273.

Yang, S.Y.N., D.W. Connell, D.W. Hawker, and S.I. Kayal. 1991. Polycyclic aromatic hydrocarbons in air, soil and vegetation in the vicinity of an urban roadway. Sci. Total Environ. 102:229–240.

Yu, B.L., H.X. Liu, J.P. Wu, and W.M. Lin. 2009. Investigating impacts of urban morphology on spatio-temporal variations of solar radiation with airborne LIDAR data and a solar flux model: A case study of downtown Houston. Int. J. Remote Sens. 30:4359–4385.

Yu, C., and W.N. Hien. 2006. Thermal benefits of city parks. Energy Build. 38:105–120.

Zedler, J.B., and M.K. Leach. 1998. Managing urban wetlands for multiple use: Research, restoration, and recreation. Urban Ecosyst. 2:189–204.

Zhang, X.Y., M.A. Friedl, C.B. Schaaf, A.H. Strahler, and A. Schneider. 2004. The footprint of urban climates on vegetation phenology. Geophys. Res. Lett. 31:L12209, doi:10.1029/2004GL020137.

Zipperer, W.C. 2002. Species composition and structure of regenerated and remnant forest patches within an urban landscape. Urban Ecosyst. 6:271–290.

Zipperer, W.C., S.M. Sisinni, R.V. Pouyat, and T.W. Foresman. 1997. Urban tree cover: An ecological perspective. Urban Ecosyst. 1:229–246.

Ziska, L.H., D.E. Gebhard, D.A. Frenz, S. Faulkner, B.D. Singer, and J.G. Straka. 2003. Cities as harbingers of climate change: Common ragweed, urbanization, and public health. J. Allergy Clin. Immunol. 111:290–295.

Ziska, L.H., K. George, and D.A. Frenz. 2007a. Establishment and persistence of common ragweed (*Ambrosia artemisiifolia* L.) in disturbed soil as a function of an urban–rural macro-environment. Glob. Change Biol. 13:266–274.

Ziska, L.H., R.C. Sicher, K. George, and J.E. Mohan. 2007b. Rising atmospheric carbon dioxide and potential impacts on the growth and toxicity of poison ivy (*Toxicodendron radicans*). Weed Sci. 55:288–292.

10

Urban Agriculture:
A Comparative Review
of Allotment and Community Gardens

David L. Iaquinta
Axel W. Drescher

Abstract

In this chapter we explore differences among various forms of urban agriculture, focusing primarily on allotment and community gardens. Starting with clarifying terminology, we introduce connections to various other types of urban gardens in common practice. We examine the historical roots of urban agriculture to demonstrate its long-term persistence, and we consider the relationship of gardening to food security across a sampling of countries, especially during the tumultuous 20th century. The multifunctional role of urban gardens is explored with respect to the ecology of the city. Various contemporary trends are enumerated with selective references. The discussion addresses organic gardening, intercultural gardens, and the transferability of allotment gardens to developing countries in greater detail, focusing on their significance for allotments and community gardens. Gender and tenure issues are addressed throughout the paper. The overall treatment is selective rather than exhaustive and is intended to provide the reader with an integrated understanding of the important roles played by urban gardening. We identify the significant differences that exist between allotments and community gardens as defined herein.

Urban agriculture is a general term used to refer to the large variety of food production forms centered in and around cities, yet food issues are mostly regarded as agricultural and rural issues, a fact that severely limits the visibility of the entire urban food system for policymakers, publics, and even researchers. Other urban systems, such as housing, employment, transportation, and energy, do not share this limitation to the same degree. Pothukuchi and Kaufman (1999) identified three reasons for this low visibility: (i) the historic process by which various issues and policies came to be defined as urban; (ii) the spread of processing, refrigeration, and transportation technology, which along with cheap, abundant energy rendered invisible the loss of farmland around older cities; and

D.L. Iaquinta, Dep. of Sociology, Anthropology and Social Work, Nebraska Wesleyan University, 5000 Saint Paul Ave., Lincoln, NE 68504-2794 (dli@NebrWesleyan.edu); A.W. Drescher, Inst. of Physical Geography, Albert-Ludwigs University, Werthmannstr. 4, D-79085 Freiburg, Germany (Axel.Drescher@sonne.uni-freiburg.de).

doi:10.2134/agronmonogr55.c10

(iii) the continuing institutional separation of urban and rural policy. Despite this, the urban food system connects in important ways to other urban systems such as land use, economic development, and infrastructure and impacts the urban environment in many ways. More directly it contributes significantly to community health and welfare and to metropolitan economies. Urban agriculture has always played a key role in the urban food system and also suffers from low visibility. However, urban agriculture lacks general visibility for intrinsic reasons going beyond the low visibility of the overall urban food system. Urban agriculture has low visibility because it is an activity whose inputs and outputs are distributed across space, time, and stakeholders. Thus, urban agriculture goes largely unseen as an aggregate sector essential to urban vitality, a sector that can benefit greatly from sectoral-level planning. Together with periurban agriculture, urban agriculture represents a growing activity and consequently an emerging arena of research. Urban gardening is the most significant component of urban agriculture from the perspective of the individual practitioner. In terms of land tenure and stakeholder labor, urban gardening refers to three basic types of practices: home gardening, allotment gardening, and community gardening. While there is no universal agreement on the precise meaning of these terms, we adopt the following definitions:

- *Home gardens* are maintained—typically but not always near the home—by individuals or households who have some access to land either customary or legal that they have arranged for themselves.
- *Allotment gardens* are individual parcels of land allocated to individuals or households for personal use; while contiguous, the parcels are worked independently by each household and the land is made available through either government action or private entities (see Flavell, 2003).[1] The participating individual households are organized into a self-governing association.
- *Community gardens* are maintained by a group of individuals or households who produce agricultural goods collectively on a piece of land primarily for their own consumption (Drescher et al., 2006).

This typology of urban gardening addresses the imprecision introduced by the conventional usage of the term *community garden* in the United States, whereby it subsumes both allotments and collectively tended parcels.

Allotments and Community Gardens

Allotment gardens developed earliest in Germany and Britain. In the mid 1800s they were reintroduced as *Schrebergärten* in Germany, where they have flourished for more than a century and a half (Kasch, 2001). Allotment gardens are characterized by a concentration in one place of several small land parcels usually 200 to 400 m^2 each. Individual families are organized into an association, which assigns the land parcels and governs many aspects of the activity. In allotment gardens, parcels are cultivated individually within legal constraints and association rules, as compared with community gardens, where the entire area is tended collectively by a group of people (Holmer et al., 2003). In short, community gardeners share the basic resources of land, water, and sunlight (MacNair, 2002). Community

---

[1] Flavell (2003) defines an allotment as a "small individual parcel of cultivable land rented by an amateur gardener separately from a house, not adjacent to that house, free of any building, and usually in a block with other similar parcels."

gardens are often organized around a particular institution, such as school, workplace, faith organization, or hospital. They may also be organized around social characteristics, such as ethnicity, age, religious orientation, or economic well-being. In such cases they are often referred to as school gardens, children's gardens, or poor gardens, for example. They may also be organized around a specific purpose, as in the case of war gardens during World Wars I and II, relief gardens during the Great Depression, and vacant lot gardens in Canada.

Organic Gardening and Entrepreneurial Urban Agriculture

Some "types" of gardens refer not to their structural aspects but to the methods used, as in the case of organic gardens, or for the nonsubsistence use of the goods produced, as in the case of entrepreneurial urban agriculture. While, such terms identify important aspects of a particular garden, we consider them subcategories since they can apply to any of the three main classes of gardens. Thus, any type of gardening can employ organic methods to a greater or lesser extent; the important point is who makes the decision. In a home garden the decision rests with the individual stakeholder or householder. In a community garden the decision rests with the collective, depending entirely on its governance and organizational structure. In an allotment garden the decision also rests with the individual or household, but is subject to restrictions imposed by the allotment association. Generally speaking, the individual has a voice in the promulgation of allotment governance rules; however, this is not always the case. Some nongovernmental organizations (NGOs) operate what they call community gardens (but which under our typology would be allotment gardens) wherein the rules are established by the NGO with little or no stakeholder input. This configuration seems more common when stakeholders are poor or comprised of immigrant populations.

Entrepreneurial urban agriculture is a new name for an old practice, nonsubsistence production for market exchange. Historically, it has taken a variety of forms and is sometimes akin to truck gardening in the periurban. Kaufman and Bailkey (2000) compiled a substantial inventory of such programs operating in U.S. cities. This focus on nonsubsistence production points to a broader subclassification of our main garden types. Home, allotment, and community gardens can each be used to fulfill three possible key functions: (i) subsistence, (ii) market exchange, and (iii) social exchange, for example, barter or social capital maintenance. The roles of subsistence and market exchange have been long identified in the literature. However, the role of social exchange has been less well explored. Certainly barter has been discussed in its more limited economic exchange capacity, but the use of garden produce to maintain and enhance social capital is much less well developed. In this capacity garden produce is exchanged or given to tacitly strengthen intergenerational, familial, ethnic, or neighborhood bonds between individuals, families, and households.[2] Also, in community gardens with the express goal of community building—sometimes referred to as *donation gardens*—produce is freely available to anyone in the community, regardless of their contribution.[3]

---

[2] Personal observations of the authors in contemporary Roman home gardens and in informal allotments and refugee garden allotments and community gardens throughout the United States.

[3] For example, see http://www.urbanharvest.org/programs/cgardens/types.html or http://www.ecoday. org.nz/16_Gardening/CG_02_Community_Garden_Models.htm (URLs verified 5 Mar. 2010).

Roots of Urban Agriculture

Archaeological evidence of urban agriculture exists around the world. Archaeologists frequently find the remnants of large-scale, often sophisticated, irrigation systems in and around the cities of ancient civilizations (Mougeot, 1994), evidencing the presence of urban agriculture. Such agriculture thrived from the outposts of the Roman Empire in Algeria and Morocco (Greene, 1990; Kehoe, 1988), to the walled gardens of ancient Persia (Subtelny, 1997; Necipoglu, 1997; Bailey, 1997); from Europe's medieval monastery towns (Astill and Langdon, 1997; Hamerow, 2002), to the city states of the Aztecs (Evans, 1990; Smith and Price, 1994) and the terraced farms of Machu Picchu high in the Peruvian Andes (Weatherford, 1993). Indeed, significant variations in urban agricultural ecology likely existed even within a given culture (Dunning et al., 1997). Evidence suggests that such agriculture served many functions beyond the obvious food and fodder production, including the production of fencing, building materials, ornamentals, and medicinal herbs. What is new is the scale of both urbanization (Iaquinta and Drescher, 2002a) on the one hand and the rapid growth in the number of urban poor on the other hand (Iaquinta and Drescher, 2002b). These two trends combine to greatly increase both the opportunity and need for urban agriculture. Estimates are that by 2015 there will be no fewer than 564 cities around the world with one million or more residents. Of these, 425 will be in developing countries (Mwale, 2006).

Early Allotment Gardens in Germany

Not only was urban agriculture a phenomenon of ancient cities, it has remained a part of urban life right through to the present. The full extent of urban agriculture has been underestimated, even during the transition to the modern industrial era (French, 2000). For example, the history of the allotment gardens in Germany is closely connected with the period of industrialization in Europe during the 19th century when a large number of people migrated from rural areas to the cities to find employment and a better life. Often these families were living under extremely poor conditions and suffering from inadequate housing, malnutrition, and other forms of social neglect. To improve their overall condition and to allow them to grow their own food, municipal authorities, churches, and employers often provided open spaces for garden purposes. Originally called *gardens of the poor*, they were later termed *allotment gardens*. One of the first "allotment gardens for the poor" was created around 1806 in the Danish city of Kappeln an der Schlei. Named *Carlsgardens* after the landgrave Carl von Hessen, their main purpose was to reduce hunger and poverty. By 1826 these allotment gardens could be found in 19 cities, spreading to many more cities by the mid 19th century. For example, various gardens were founded in Berlin, including Red Cross gardens, Worker's Gardens, and Railroader gardens along the railway lines (Wille, 1939).

The idea of organized allotment gardening accelerated around 1864, when the so-called *Schreber Movement* started in the city of Leipzig in Saxony. By public initiative areas within the city were leased in order make it possible for children to play in a healthy environment, and in harmony with nature. Later these areas included actual gardens for children, however, adults soon tended to take over the cultivation. This kind of gardening became increasingly popular not only in Germany, but also in many other European countries (Crouch, 2000; Sidblad, 2000).

Fig. 10–1. Children's farm garden with 1008 plots in Thomas Jefferson Park, New York City (source: Library of Congress).

Early Community Gardens in the United States

Community gardens began to appear in the United States as a response to poverty and unemployment resulting from the economic depression between the years of 1893 and 1897 (Goldstein, 1997; Warner, 1987). This period, characterized by capital flight and many failures of banks, railroads, and other industries, resulted in widespread bankruptcies and unemployment (Warner, 1987). Bassett (1979, 1981) termed these gardens created in the period between 1894 and 1917 *potato patches*.

Municipal gardens arose especially in areas of high immigrant population. One such garden was Thomas Jefferson Park in New York City (Fig. 10–1). Opened in 1905, it was

> ...to provide organized play to the children of Little Italy, as the crowded tenement district in East Harlem was then known... A children's farm garden, one of many which flourished in parks in the first half of the 20th century, opened on May 20, 1911 with 1008 plots for children to grow flowers and vegetables. Designed as a place of respite for child laborers, the farm garden later hosted nature study classes and, during the World Wars, provided a lesson in self-sufficiency for local children.[4]

While these early American endeavors were termed *community gardens*, it is hard to determine the degree to which they might have in fact operated more like allotments in the terminology used herein. Juxtaposed to these municipal or community gardens was a style of home gardening among immigrants in particular that resembled the "vacant lot gardening" in Canada, which is discussed more below. Indeed, when economic resources were sufficient, individual households would purchase or lease open land in the city expressly to maintain household gardens for subsistence consumption and preservation (Drescher et al., 2006; Saverino, 1995; MacNair, 2002).

---

[4] New York City Parks and Recreation Department (2010), http://www.nycgovparks.org/parks/thomasjeffersonpark (accessed 23 Mar. 2010). The first such gardens appear to have been initiated in DeWitt Clinton Park as reported in the *Brooklyn Daily Eagle Almanac* (1905, p. 77), "In DeWitt Clinton Park a children's farm garden was maintained during the summer and vegetables were grown to maturity by the young farmers. The results were very gratifying and the experiment has attracted a great deal of attention all over the country, and the department has received many letters commending the work."

Urban Gardening as Food Security

Early British Experience

In a meticulous analysis of evidence from Britain from the 1790s through the late 19th century concerning early allotment extent, production, and impacts, Burchardt (2002) offered "...a critique of the traditional view of the allotment movement as an atavistic phenomenon, rooted in the production relations and social assumptions of a past era." He notes

> [E]vidence provided by the database, is that allotment yields were quite remarkably high—perhaps even double those obtained by farmers. Even more important, these impressive yields translated into correspondingly high profits. Allotments therefore made a much larger contribution to living standards than has hitherto been recognised—far from being a marginal phenomenon; they played a crucial part in the family economies of those of the labouring poor who had them. Furthermore, since the number of allotments rose rapidly after the inception of the second allotment movement... [1830-early 1850s]..., and since these plots were concentrated in those parts of the country where living standards were lowest, allotments significantly affected the lives of many of those who most needed them.

Burchardt (2002) further notes that the root forces giving rise to the allotment movement were fundamentally food security issues:

> The initial layer of explanation... [for allotment provision]... is that both the first and second allotment movements were responses to events which disrupted the order of landed society. In the 1790s this was a severe subsistence crisis in the context of war; in 1830 it was the Captain Swing riots. [T]he second [allotment movement] achieved impressive results... [by]... establishing an active and highly effective national organisation, the... [Labourer's Friend Society (LFS)[5] which]... played a crucial role in bringing allotments to the attention of landowners.

Allotments, he concludes

> ...were a vector for the modernisation of rural society. Not only were the behavioural and attitudinal changes which the allotment movement hoped to achieve forward-looking and liberal in character, but the effects of allotment provision were to bring rural labourers closer to, rather than further from, the mainstream of mid-Victorian urban Britain.

Thus, food security encompasses lack of access to land (i.e., outcome of the enclosure movement), wartime shortages, and civil disorder (i.e., rioting due to food shortages and hunger). Food security has been a motivating force for allotments since the end of the 17th century. In addition to wars and severe depression, demand for allotments has been consistently driven by industrialization, urbanization, and consolidation of rural farm holdings (Flavell, 2003)[6], especially through the mechanism of poverty creation. Nonetheless, urban agriculture is still all too frequently confronted with ignorance and resistance among decision

---

[5] Jones (2003) reports that the LFS was preceded until 1830 by the evangelical Society for Bettering the Conditions and Increasing the Comforts of the Poor (SBCP), which counted William Wilberforce among its prime movers. The LFS in turn gave way to the Society for Improving the Condition of the Laboring Class (SICLC) in 1844.

[6] Flavell focuses his analysis on Sheffield, which he describes as "not merely a major pioneer, but as the first town to possess large numbers of urban allotments, long before the era of statutory provision."

Fig. 10–2. A typical allotment garden in Germany (photograph by Axel Drescher, 2006).

makers and those members of the public who see it as backward and inefficient, views seldom held by practitioners of any stripe.

The Twentieth Century and Two World Wars

During both World Wars I and II, food shortages arose in the urban areas of major contestants on all sides of the conflicts. Added to outright food shortages at home was the need to provide food for deployed troops and allies in the conflict.

Germany

During World War I, conditions in Germany were often desperate, with food in short supply and the urban population suffering from inadequate nutrition. Many cities were cut off from rural sources of agricultural production. When food did reach the cities it was often sold at exorbitant prices on the black market. This situation heightened the need to produce food in and adjacent to the city. In consequence, home and allotment gardens became increasingly important sites for fruit and vegetable production, often essential for survival. In 1919 immediately following World War I, the important food security role played by allotment gardens was formalized in the Small Garden and Small-Rent Land Law. This was the first allotment gardening legislation passed in Germany; it provided security of land tenure and fixed leasing fees. In 1983 the law was amended by the Federal Allotment Gardens Act (Holmer and Drescher, 2005). Today, there are still about 1.4 million allotment gardens in Germany (Fig. 10–2) covering an area of 47,000 ha (Gröning and Wolschke-Bulmahn, 1995).

The situation with respect to food was rather different in World War II. Expansionist militarism, appropriation of food from occupied territories, and the use of prison labor all helped stave off food shortages in Germany until late in the war when allied bombing, the Russian "scorched earth" policy, and labor shortages began to erode food production. Despite this emerging situation and the significant changes in the organization of allotment governance introduced under the National Socialists in the years before WWII[7], allotments in Germany helped to

---

[7] Democratic procedures have a long tradition in many of the allotment gardening associations. However, when the National Socialists took power on 30 January, 1933 they required National Socialist streamlining and enforced it on 29 July, 1933. Consequently, the articles of the associations were changed. This meant blood-and-soil ideology, no more elections, the National Socialist Workers Party (NSDAP) appointed the "leaders" for the community gardening associations, and no Jews as community gardeners—only "Germans of Aryan descent" could operate a garden (Gröning and Wolschke-Bulmahn, 1995; Gröning, 1996).

meet urban food needs during the war. The role played by allotments increased as the war wore on, and ultimately the substantial destruction of infrastructure led to the expansion of allotments into all kinds of available urban space following the war (Fig. 10–3).

Great Britain

In Britain and the United States war gardens were actively promoted by the government in an attempt to augment urban food supplies. Britain employed the concept during WWI with success (Butcher, 1918) and introduced a greatly scaled up version during WWII because of the much greater reliance on food imports by that time (Spudić, 2007).

In part due to the "Dig for Victory" campaign, launched by the government in the autumn of 1939, the British succeeded in meeting roughly one-half the nation's demand for fruit and vegetables from gardening during the Second World War. Great Britain had approx. 1.5 million allotments at the end of WWI and again at the end of WWII (Crouch and Ward, 1997; Howe et al., 2005).

Belgium

Van Molle and Segers (2008) provide a comprehensive review of allotment gardening generally and in Belgium specifically. They report that the influence of the national Ligue du Coin de Terre National Allotment Garden Federation during WWI was substantial: garden complexes increased from 56 to 717 between 1914 and 1918, cultivated land area grew from 800 to 7,300 ha, and the number of members of the Ligue increased from 16,000 to 180,000. By the end of WWI nearly "one-tenth of the Belgian population was using leased allotments to compensate for food shortage and inadequate official food distribution" (Van Molle and Seegers, 2008). While the popularity of allotments was once again quite high during WWII, it waned significantly both between the wars and after WWII in Belgium.

Poland

Poland too has a long history of allotment gardens. In 1997 Poland celebrated 100 years of formal allotment food production. The allotments grew out of the need to absorb displaced rural labor into an urban environment ill equipped in infrastructure and the ability to meet people's basic daily needs. Today allotments still occupy prime city space despite the turbulent economic and political transformations throughout the period from the 19th century to the present. The scale of allotment gardening here is impressive, with gardeners collectively forming the largest class of land managers or users in Poland. Nonetheless, just as elsewhere there is a continuing need to balance food security needs, open space policy, and market demands on land. The current allotment system

> ... combines local and national government policy, administration, and management by an NGO, and deeded private use by individual gardeners. Today this arrangement produces ongoing tension among the stakeholders and a contentious, even healthy, debate about the private and public uses of urban land. (Bellows, 2004)

United States

Urban gardening in the United States reflects many of the trends in Europe. Certainly this was true with urban and community gardens during the two world wars. The story of urban agriculture and community gardens during WWI is well documented—also laden with patriotic fervor—by the National War Garden Commission (Pack, 1919). Although the commission was established by private citizens from across the country as a patriotic response to the nation's needs, with time it took on the legitimacy of a government entity (Fig. 10–4). In 1917 responding to the commission's influence, Americans raised $520 million worth of garden produce in "liberty gardens"—vacant lots, backyards, and unused land.

During WWII, the War Food Administration within the USDA created the National Victory Garden Program Fig. 10–5 under which open spaces and vacant lots across America were converted to "victory gardens" to grow food for everyone, not just the poor (Miller, 2003). In 1942, about 5.5 million gardeners participated in the war garden effort, making seed package sales rise 300%. The USDA estimated more than 20 million garden plots (Fig. 10–6) were planted, with an estimated 9 to 10 million pounds of fruit and vegetables grown annually, 44%

Fig. 10–4. National War Garden Commission promotion for Liberty Gardens in WWI (source: Library of Congress).

Fig. 10–5. World War II victory garden poster (source: Library of Congress).

Fig. 10–6. Victory gardens at Forest Hills, Queens, New York in June 1944 (photograph: Howard R. Hollem, Reproduction Number: LC-USW3-042659-C, source: Library of Congress).

of the fresh vegetables in the United States (Bassett, 1981). Researchers identify six periods of urban agriculture in the United States: potato patches (1890–1930), City Beautiful Movement (1890–1910), WWI liberty gardens (1917–1919), Depression relief gardens (1930–1938), WWII victory gardens (1940–1945), and the Community Garden Movement (1970–present). Notably four of the six periods are associated with episodic geopolitical and economic convulsions.

Canada

In Canada the first allotments were created in the 1890s by the Canadian Pacific Railway (CPR) near its stations and along its developing rail lines. Populating the developing Western region meant attracting pioneers and settlers to isolated areas with few alternatives for recreation. Providing plots for cultivation was seen both as fulfilling this need and as a way to beautify railway property (Bellan, 1958). Ultimately, selling such "wonders of the west" led to the CPR overseeing gardens along "25,749 km of track from coast to coast" (Von Baeyer, 1984).

Von Baeyer (1984) also reports the emergence of so-called "vacant lot gardens" in many Canadian cities by the early 20th century, supported by both municipal governments and the national Department of Agriculture. Mirroring views in the United States, they were seen first "as a form of welfare for the poor, then as a beautifying measure, and finally as patriotic duty during World War I" (Von Baeyer, 1984). While not the same as allotment gardens, vacant lot gardening shares historical roots both with allotment gardening and with a similar pattern of gardening among immigrants in the United States. By 1910 it was a popular movement across Canada, and vacant-lot associations could be found in most Canadian cities by 1916.[8]

Contemporary Shifts in the Motivation for Gardening

The functions and importance of allotment gardening in Germany have changed with time. In times of crisis and widespread poverty from 1850 to 1950 allotment gardening was essentially a part-time job intended to enhance food security and increase the food supply. Presently, the functions of allotment gardening are more varied. In an era of busy working days and fast-paced urban lifestyles, allotment gardens have increasingly become recreational areas and locations for social gatherings. Essentially green oases within oceans of asphalt and cement, they can contribute greatly to the preservation of nature within cities. Admittedly, this "nature" is a highly modified ecology, but it is nonetheless primarily biological rather than built and hosts greater biodiversity than is typically the case for urban green areas. What was previously a part-time job in the allotment is for many considered a hobby today, where a hectic schedule becomes a distant memory and digging in the earth becomes far more than simply growing food and getting a little soil under the fingernails.

Still, in situations of weak economy and high unemployment, gardens once again become increasingly important for food production. Thus, allotment gardens today are on the one hand a hobby and on the other hand a food insurance investment against economic uncertainty. Beyond its value supplementing the household economy and basic diet, for many the garden ensures access to better tasting, healthier, organic food stuffs in a cost-effective manner. In sum, for many gardeners today allotments serve both as a mechanism of insurance risk avoidance and assurance guarantee of access to quality food, while simultaneously providing recreation, fulfilling intrinsically satisfying needs, and building social capital.

The German experience closely echoes Burchardt's (2002, p. 4–5) description of early allotment gardens in Britain during the early 19th century:

> [L]andowners were dependent on labourers being prepared to take it... [i.e., land allotments]. The success of the allotment movement, therefore, was ultimately a function of the strength of demand. Allotments were a major asset to labourers, both in material form and in less tangible ways. But many of the non-material benefits were consequent upon the large increment made by allotments to material living standards. The root cause of this...was the negligible opportunity cost to labourers of cultivating an allotment. This in turn was a consequence of the massive scale of underemployment in rural England between 1815 and the early 1850s, affecting women and children even more than men.

---

[8] Von Baeyer (1984, p. 95) reports the following figures for 1918: Guelph, Ontario: 1600 lots; Montreal, Quebec: 5000 lots; Toronto Ontario: 2060 lots.

Thus, the shift in motivation for allotment gardening in Germany closely reflects the availability of "leisure time," level of economic need, and the overall perceived "marginal cost" to gardeners in economic and social terms.

Addressing community and allotment gardens in the United States (Hynes and Howe, 2005) make much the same point:

> The early history of urban gardens in the United States of America is one of food production in response to war, economic depression, and short-lived civic reform movements. During the past thirty years, a broad-based community garden movement has spawned a wide variety of social, economic, health, and educational benefits in more than 250 cities and towns across the country. A companion food security movement has promoted urban–rural linkages, urban agriculture, and farmers' markets. Studies have shown that community gardens and nearby green space in cities are an important response to needs for nutritious and affordable food, psychological and physiological health, social cohesion, crime prevention, recreation, and life satisfaction, particularly in low-income communities.

Given this shift in the motivation for both allotment and community gardening, it is fair to ask how changing demographics and economic well-being will affect future trends. Certainly increasing affluence produces increased leisure time. Also, the worldwide aging of the population means a much larger proportion of the world's population will have more "leisure" time vis-à-vis retirement, increasing the capacity for intergenerational transfers of information and practice (Iaquinta et al., 1999). These trends combined with changing motivation auger for increasing interest and participation in all forms of gardening. While on the surface this might seem to run counter to the conventional view that increasing affluence means a decreased appetite for agriculture in any form, the key to our view is the expansion of interest due to shifting motivation.

Ecology and Urban Design

An important difference between the way allotment gardens and community gardens have been organized in Germany and the United States, respectively, is in terms of the way they are or are not integrated into the overall urban ecology. In the United States sites for community gardens are entirely dependent on the willingness either of a private land holder, such as a faith-based or civic organization, or of some governmental authority with appropriate jurisdiction to make the land available to gardeners. Therefore, only "surplus" land is used in this capacity, and there is generally little or no concern to maximize the potential benefits beyond the immediate users. This also poses a problem for the gardeners in that their tenure rights are typically not vested in a legally binding fashion. Blomley (2003), citing Schmelzkopf (1995), made this point well:

> Most cities will only grant short-term leases to community gardens on city-owned land (if at all).... Although some community gardens can be fenced (and are thus not an open access commons) they are neither clearly public nor private. If anything, they "transcend the separation... [T]hey are part of the public domain and are the sites of many functions conventionally equated with the private sphere. Domestic activities, nurturing, and a sense of home are explicitly brought outside into the gardens" (Schmelzkopf, 1995). This ambiguity has proven important: the continuing battle over the future of

several hundred community gardens in New York, for example, turns on a "basic problem of classification and geography": that is different appraisals of what constituted public space. City officials regard community gardeners as temporary users of municipal land, whose interests can be superceded (sic) by the "public good." Gardeners, conversely, argue that community gardening produces truly public spaces, predicated on localized community, democratization and interaction.

But there is a larger point to be made here that can only be seen clearly by contrasting the situation with the German allotment system. Through the use of sound urban planning, legally binding tenure rights and municipal coordination, gardens are situated to meet a variety of larger community needs, where problems become opportunities. In Germany, the need to create urban buffers and manage of urban microclimates, solid waste, and nuisance factors such as noise are all considered when constructing new allotments for the city, so too are factors such as access to transportation, open space, green space, water, and sunlight (see also Colding et al., 2006).

For example, some gardens are situated at the fringe of the city so as to facilitate cold air drainage from surrounding mountains, thus reducing urban heat accumulation. Some are positioned between neighborhoods both to create natural buffers and to provide a common ground for intermixing residents. Other allotments are positioned inside the city along extensive rail and highway corridors, thus creating sound and space barriers from noxious elements of the city. Many lie along river courses, allowing natural water flows to be diverted throughout the allotments; combined with simple rainwater catchments erected by individual gardeners, this greatly reduces the demand for city water.

These larger ecosystem services and enhancements to the urban morphology are largely absent in American community gardens. Indeed, as Blomley (2003) has suggested, the side effects of community gardens in the United States can be perverse in terms of reshaping urban morphology: "...their success in stabilizing and beautifying low-income neighborhoods may unintentionally contribute to the process of gentrification." Finally, the scale of allotments in Germany far exceeds the typical community garden complex in the United States. It is the failure to plan for urban agriculture as a sector and to see urban agriculture as a potential solution to many other urban problems that have been identified as the hallmarks of the American system (Iaquinta and Drescher, 2002a).

A recent FAO report (2008) provides both examples of successful urban and periurban agriculture programs (UPA) around the world and a prescription for how to capitalize on ecosystem opportunities while meeting food security needs through urban agriculture:

> The long-term viability of UPA itself depends on how successful farmers and urban officials are at exploiting the potential environmental benefits, minimizing the problems, and finding ways to secure growers access to land and water. Optimal management of urban and periurban resources requires land use planning which views agriculture as an integral component of the urban natural resources system and balances the competitive and synergistic interactions among the users of the natural resources (water, land, air, wastes).
>
> In practice there are several possibilities to safeguard land for... [urban and periurban agriculture].... City councils can provide unused open spaces or brown-fields (case of Germany and many other European countries), unused land under

power lines (case of Lima and Belo Horizonte) or along railways can be reserved using public private partnerships between cities and companies, private land can be provided using tax exemptions (case of Cagayan de Oro), temporarily flooded areas can be reserved for UPA (case of Dar es Salaam), buffer zones for micro-climate conservation provide another option. (FAO, 2008, p. 67–68)

According to the FAO (2008) report, key prerequisites to safeguarding land and achieving synergistic ecosystem gains are (i) political will, (ii) proper legal integration of urban and periurban agriculture into municipal laws, (iii) effective organization of city farmers, and (iv) appropriate institutional designation of urban and periurban agriculture.

These prerequisites apply even in developed countries. For example, Roy (2001) notes that the allotment gardens in Winnipeg did not "contribute to sustainable development to the extent they do in some other cities." They were not located so that they could be used as a coping strategy by the poor. Access was not equitable, and gardeners did not realize the potential economic benefits. Few gardeners used organic gardening techniques, and the city "did not have any regulations or programs in place to promote their use." Roy concludes that "to contribute to sustainable urban development, allotments must be located so that they can be used by those seeking to reduce family food budgets as well as people looking for recreation. In addition, institutional structures need to be in place to ensure their development and vitality."

New Movements and Research

There are many emerging areas of research in the area of urban agriculture, but space prohibits their detailed consideration. While many more could be listed, a partial list includes:

- The relationship of gender to urban agriculture generally and gardening in particular (Hovorka et al., 2009)
- Nutrition related to urban gardens and agriculture (Brown and Jameton, 2000; Pothukuchi, 2005)
- The relationship of urban agriculture to a variety of health issues such as malaria (Dongus, 2009) and outcomes such as "healing gardens" and "gardens of hope" for HIV/AIDS victims, among others (Marcus and Barnes, 1999; Hoekstra, 2008)
- Alternative food networks and their impact on improving access to food for marginalized groups (Abrahams, 2007; Maye et al., 2007)
- Mapping of urban food systems using geographic information systems (GIS), including supermarkets, food outlets, and standard food sources, as well as open spaces and gardens in urban and periurban settings (e.g., Dongus and Drescher, 2006)
- Participatory mapping of open spaces and "informal" food sources in cities, identified by local stakeholders as providing access to food (Drescher and Holmer, 2007)
- Relationship between gardens, their management, and differing periurban types (Iaquinta and Drescher, 2001, 2000)
- Biodiversity conservation through urban agriculture (Colding et al., 2006)
- Rainwater catchments in urban gardens as a means of poverty alleviation (Auerbach, 2005)

The remainder of this section is devoted to an extended consideration of organic agriculture, intercultural gardens, and transferability of allotment gardens. Our goal is twofold. First, we wish to pique the reader's interest regarding new directions for study and application. Second, we want to expand somewhat our integration of social science concepts and make some connections across the information already presented.

Organic Agriculture

Organic agriculture is an imprecise term but generally refers to the production of cultivars without the aid of inorganic fertilizers or chemical herbicides, fungicides, and pesticides. It includes a wide range of techniques, including composting, use of animal and human excreta as fertilizer, biological pest control, crop rotation, interplanting, symbiont planting, and the use of indigenous and heirloom cultivars. In this section we present suggestive rather than exhaustive evidence on the topic; it is based largely on the joint task force report issued by UNCTAD-UNEP (2008). The report examines "...the relationship between organic agriculture and food security in Africa... [and analyzes] ...organic agriculture's impact on food availability as well as natural, social, human, physical, and financial capital in the region". The authors define organic agriculture as: "...a holistic production system based on active agro-ecosystem management rather than on external inputs,... [using] ...both traditional and scientific knowledge". This definition is consistent with both the FAO/WHO Codex Alimentarius guidelines (FAO/WHO Codex Alimentarius Commission, 2001) and the International Federation of Organic Agricultural Movements (IFOAM, 2009) definition. The report authors also argued that organic agriculture must be sustainable such that it not adversely affect natural (Costanza et al., 1997), human (Orr, 1992; Byerlee, 1998; Lieblin et al., 2004; Leeuwis, 2004), or social (Flora and Flora, 1996; Pretty, 2003; Cramb and Culasero, 2003) capital.

Data for the report come chiefly from Kenya, Uganda, and Tanzania since the authors focus on only "certified organic" agriculture, and such data are not widely available in African nations. The focus on certified organic production figures means that the results are probably a conservative estimate of actual production and impacts. The focus on Africa is warranted because if organic urban agriculture can be shown to be realistic and effective in the African context, with its often limited resources, then it is arguably valuable in the context of more affluent nations with a wide variety of stakeholders.

The most important conclusion from the report is that conventional wisdom about organic agriculture and yield is wrong. "Organic agriculture can increase agricultural productivity and can raise incomes with low-cost, locally available and appropriate technologies, without causing environmental damage" (UNCTAD-UNEP, 2008, p. 39). Similar findings were reported by Gibbon and Bolwig (2007).

Further, the report concludes that organic agriculture can improve all five capitals—especially natural, social, and human—by building up natural resources, strengthening communities, and improving human capacity. This strengthening is shown to occur in multiple ways for each of the capital asset classes. Other researchers report similar findings (Ostrom, 2000; Pretty, 2003). Importantly, these findings hold whether the production was primarily for subsistence, domestic market exchange, or foreign exchange. It would be hard to argue that these findings would apply less well to subsistence or small-scale market production in allotments or community gardens specifically, although foreign exchange is not likely significant for allotments or community gardens.

Other key findings of the report include that organic agriculture is:

- Ideally suited for many poor, marginalized smallholder farmers in Africa, because it requires minimal or no external inputs, uses locally and naturally available materials to produce high-quality products, and encourages a systemic approach to farming that is more stress tolerant due to greater diversity
- Less energy and external input dependent
- More likely synonymous with "sustainable" when organic farming principles are adopted as a holistic approach for the whole of an integrated agricultural system, leading to heightened food security for the whole region
- More likely to make a significant contribution to food security and poverty alleviation when combined with policy and institutional support
- Facilitated when obstructionist policies advocating high-input farming management practices are removed
- Inherently management- and knowledge-intensive, which necessitates building the learning and cooperative capacity of individuals and groups
- Shown to increase incomes through cash savings, profits from market exchange, and premium prices for certified organic produce and value added when combined with processing activities

It is true that urban agriculture is not identical to organic agriculture, and it is also true that the evidence cited here is from developing African countries. Nonetheless, the interconnections between the needs and benefits of urban and organic agriculture dovetail closely. One of the strengths of focusing our discussion on the UNCTAD-UNEP task force approach is that the report analyzes the data in the context of the differing dimensions of capital that actually affect people's lives and upon which community well-being rests. If the individual and community benefits derived from recycling organic wastes are factored into the analysis, the gains are magnified exactly to the degree that they reduce municipal costs and challenges that devolve from viewing organic waste as a nuisance problem. Thus, when systems of urban gardening allotment or community are linked to both organic gardening methods and municipal waste recycling programs, the outcomes are significantly more positive for stakeholders of all types, governance structures, and the community as a whole. Indeed, all capital asset classes benefit in multiple ways.

Intercultural Gardening, Women, and Social Integration

Migration is a demographic process as old as the need for primitive populations to find food. The reasons for migration have changed throughout history. However, the 20th century was witness to a scale of change unprecedented in human history in volume, breadth, and direction. In particular, political changes, war, globalization, and demographic development in Europe led to new types of migration and new migrants in many European countries. In consequence social tensions have arisen everywhere between the newcomers and the previous resident population. While migration commonly produces "newcomer–old timer" tensions, they are heightened when superimposed on racial, ethnic, and religious identity differences. The basic problems facing communities are (i) how to integrate the various groups into a constructive social fabric, (ii) what policies need to be developed to accomplish this end, and (iii) how to build the required social support to implement those policies.

Intercultural gardening is the concept recently coined in Germany to focus attention on this process of integration and community building among diverse populations. As with so many aspects of urban gardening, the phenomenon is not a new one. However, the need to focus attention on the phenomenon is new to some places with an already entrenched view of urban gardens and their role in the community. In Germany—and Europe generally—allotments have a long history without any significant role in the social integration of ethnically distinct groups. In this context it makes sense to think about and classify gardens in terms of tenure and labor. Consequently, allotment gardens require a new identity when the focus shifts intentionally to a role outside of their primary and longstanding purpose.

In countries like the United States there is a long history of repeatedly absorbing immigrant groups and a longstanding need to forge a common identity. In such places the development of urban agriculture occurred pari passu with the continual social integration of diverse peoples. Gardening in this context was always seen to have a certain element of community building and likely largely explains the American tradition of preferring to label urban gardens—other than home gardens—as community gardens, whether they are tended collectively or individually. Here we can see why there has been confusion about terminology. The preferred terms on either side of the Atlantic Ocean strongly reflect different historical contexts, social goals, and sociopolitical imperatives. Urban gardening is simply a set of practices situated within these divergent conditions that is consequently perceived in different ways even when the practices are essentially the same.

Nonetheless, social integration is an extremely important component of social relations and is worthy of attention in the context of our study of urban agriculture. In terms of social integration, there are two basic dimensions of intercultural gardening or community gardening in the American lexicon, and they reflect the broader literature on multicultural relations: (i) social integration of the ethnic immigrant population with the dominant national group and (ii) social integration among various ethnic immigrant groups themselves. These processes occur in both types of urban gardens under consideration here, community and allotment. These processes represent two key dimensions of "bridging social capital" (Putnam, 2000), a prerequisite for assimilation and a facilitator of integration.

Gardens in this context are essentially a civil society response to the challenge of integration, and this is why they are often sponsored by faith-based and humanitarian groups. The basic idea goes far beyond just gardening, as the gardens themselves often serve as sites for delivering language courses and training, hosting joint social activities, cooking, and engaging in many other personally valuable and socially integrating activities.[9] In the United States new movements aim to protect gardens in large cities and to convert unused open spaces into new productive areas not simply to enhance the food security of vulnerable populations but also to actively "build community." This civil society response points to

---

[9] Müller (2007) reported that German intercultural gardens embrace a wide-ranging repertoire of activities, especially during the winter months. Activities extend "to language and computer courses, arts and crafts, sport, theatre workshops, intercultural environmental education, neighbourhood networking, music, lectures and counselling, many activities for children, further education in nutrition and gardening, factory tours, and excursions."

the inadequacy of urban planning and the inability of the marketplace alone to meet people's food and social needs.

In the United States, this civil society response manifests in large cities as abandoned and vacant areas are transformed into community gardens where all sorts of people can work in common cause (Wood and Landry, 2007). For example, New York City's urban gardens are rare and threatened because insatiable demand for real estate puts them under constant pressure.

The More Gardens! Coalition works along with community gardeners facing development to preserve threatened gardens. Strategies include physical occupation such as camping out in gardens slated for demolition, coalition building by enlisting the support of local and state politicians, and even legal action such as filing lawsuits against the city. By such methods the coalition has helped save more than 400 gardens from development (Cohen, 2008). Regardless of whether the plots were originally tended collectively or individually, they have also created bridging social capital, built capacity among stakeholders, and created community.

In a 1996 German initiative, refugees, migrants, and German families founded the first "International Garden" in Goettingen. Many Bosnian refugees had settled in Goettingen, and there was a significant need to integrate this new population. This demographic experience was mirrored throughout Germany, and Europe generally, due to the collapse of the Iron Curtain, the ever increasing immigration from non-European—and typically less developed—nations, and the emerging new "supra-identity" of the European Union. Not surprisingly, a new movement emerged and gardens identified as *international, multicultural,* or *intercultural* spread rapidly across Germany and Europe. Founded in 2003, the first intercultural garden in Berlin, for example, is located in an area of 4000 m² in the district Treptow-Köpenick. Here people from Kazakhstan, Vietnam, Russia, Egypt, Hungary, India, Afghanistan, the Ukraine, Bosnia, and Germany garden in allotments and engage in social interaction with one another.

In Germany, the intended purpose of intercultural gardens extends beyond the obvious production of food and aims to increase intercultural communication and integration (Müller, 2007). Gardening offers a chance to exchange recipes and garden produce, but also allows gardeners to exchange experiences and communicate across cultures. The common language is German, which helps immigrants become familiar with the new language. This is particularly important for women, who are more prevalent than men in the gardens.

Women who are unable or not permitted to attend language and culture courses are often able to engage in gardening. In some cases this situation arises due to cultural norms governing acceptable women's roles or traditional male domination that persists in the new environment. In other cases it reflects the more pragmatic need to augment husbands' low incomes with garden production by women. Whatever the cause, it generally reflects the overall low status of women. Yet, at the same time the intercultural garden gives women access to an environment where they can acquire language skills, escape from often cramped housing conditions, develop friendships with other women that strengthen their self confidence and build both the binding and bridging social capitals that help them cope with the family conflicts and rapid social change they may be experiencing as a result of differences in intergenerational socialization. But, while German intercultural gardens are true allotments, they do not look the same as traditional allotments. Müller (2007) describes them this way:

Generally, there is at most one garden hut per project, and trimmed hedges or fenced-in plots are lacking. Discreet boundaries are the rule between the ten to eighty square-metre beds. Sometimes they are demarcated with shoes, sometime with rope, or, often indiscernibly for outsiders, only by different species of plant. Fences are rarely erected, only as protection against rabbits, dogs, and sometimes against young local troublemakers.

Intercultural gardens are designed to include a large commons that includes a fireplace and other optional features such as playground equipment, perhaps a site trailer for the children, sometimes a building, and even an occasional greenhouse. Anyone familiar with a typical German allotment complex will recognize that features often found in individual allotment gardens are moved into the commons area in the intercultural garden. This arrangement facilitates communication, shared socializing over drinks and food, integrated play among children, and opportunities for larger social gatherings to which individual gardeners can invite other friends and family, potentially expanding benefits of social integration.

Transferability of Allotment Gardens to Developing Countries
Given the long success of both allotment and community gardens throughout Europe, the United States, and Canada, it is likely they will continue to flourish. Perhaps a more important set of questions concerns whether they are transferable "technologies" that can succeed in and benefit developing countries. In addressing these questions it is imperative to keep in mind that the value of foreign aid—especially technological aid—is intrinsic not to the technology or aid but to the society in which it was developed (Iaquinta, 1975; Caldwell, 1972; Hughes and Hunter, 1972). This well-documented and long-understood truth underscores the importance of stakeholder participation when planning and implementing such transfers; as research demonstrates, this is certainly the case for allotment gardens (Drescher, 2000; Mata, 2006).

Referring to Drescher's work, Mata (2006) wrote, "Using a plan of action on a pre-established 'blueprint' that emphasizes morphology disengages the model from the functional processes occurring in the surrounding environment, with which users of the model have established a relationship." Drescher (2000) himself has aptly stated, "... urban planning should be more than the preparation of master plans."

Burchardt (2002, p. 4) shed light on the varied motivations by which landowners' largesse translated into allotments in 19th century England. In so doing he foreshadowed the variety of motivations propelling the actions of governments, NGOs, and private land holders today:

> But successful as the LFS[10] was, since it did not let allotments itself it was in the last resort wholly dependent on the willingness of private landowners to provide allotments. The question then becomes: why did landowners decide to let land to labourers?... [T]he most important feature of allotments from a landowner's point of view was precisely the diversity of goals that could be achieved by letting them.

It is this "diversity of goals" that heightens interest in the transferability of allotment gardens to a wide variety of settings in developing countries. In many

---

[10] Laborer's Friend Society; see also Footnote 5.

developing countries the largest landholder is the state or government in one form or another—this is especially true in Africa. Clearly as a landholder, the state has a wide variety of interests in the general welfare of the population at all levels. Thus, the diversity of goals opens a broad spectrum of entry points for considering the role of allotment gardens in addressing a range of national and local challenges.

Africa

Due to the rapid growth of southern African cities, the basic needs of citizens (e.g., shelter, food, education) tend to be undermined, and local authorities often find themselves in a difficult situation, facing ungovernable conditions. Considering cultural norms and traditions, climatic conditions, and natural resource availability, leaders could look to establish allotment gardens as a means to reduce poverty and improve food security in cities. As discussed above, this proved helpful in the European and American context during times of crisis.

The idea of transferring the European model for community food security in urban areas is not new. Similar attempts were made in Eastern Europe (Chatwin, 1998). Experience in Germany and elsewhere (Butcher, 1918) shows that allotment systems are effectively linked to community development programs that introduce democratic rules related to the management of the plots (Gröning and Wolschke-Bulmahn, 1995; Gröning, 1996). Questions remain as to whether allotment gardens inevitably foster these outcomes in all or most situations and under what conditions the outcomes are fostered. Special participatory training programs can further contribute to community building, but such a process needs monitoring and mediation. In southern Africa the group most affected by poverty is female-headed households; they are a likely target group who would be interested in and could be integrated into allotment programs.

Success in transferring the German allotment system to cities in southern Africa depends on effective integration of local institutions, capacity building among interest groups, and the responsiveness of local politicians to community needs. Community development programs based on such partnerships have proved successful in South Africa (Nell et al., 2000; Drescher, 2001). The main problems lie in the institutional context of urban planning and municipal governments. Local governments in many African cities are weak institutions. Municipal councils that started off as colonial institutions were never fundamentally transformed to address growing African urban population (UNCHS, 1998). Local institutions like these are central to development once urban–periurban–rural linkages and the diversity of periurban areas are taken into account (Iaquinta and Drescher, 2000).

Local institutions are highly variable in their form and function, and they are not well understood in the context of urban–periurban interaction. A general presumption is that they are dominated by an overall urban bias and that they largely ignore the legitimacy of rural and periurban interests. While this is often the case for formal, modern local institutions, the situation is often quite different for traditional and informal local institutions. Regarding the situation in periurban areas, local formal and informal institutions, such as farmer groups, water users, and women's collectives, tend to have little influence on decision making in nearby urban centers.

Capacity building for local institutions and the development of interdisciplinary institutional approaches toward the urban–rural continuum are therefore needed if allotments are to be successfully introduced and effective in meeting both end-user needs and larger community goals (Iaquinta and Drescher, 2000). Greater collaboration between research and development capacities in urban planning and those in agricultural development is needed to make urban farming more efficient and sustainable (Mougeot, 1996; Iaquinta, 1975). Finally, community and allotment gardens that have become part of a zoning plan, that is, those given formal legal stature and protection, are harder to relocate (Groening, 1996).

The Philippines

In the Philippines allotment gardens have been introduced using a modified German system. Vegetables are produced in the urban allotment gardens using practices similar to those in rural areas. However, cultivars differ between the two areas and lower levels of agrochemicals are used in the allotments due to their proximity to populated areas (Guanzon and Holmer, 2003). The high diversity of vegetables and production levels improve both food security and income generation.

As already indicated, urban agriculture is compatible with organic gardening, but the two do not always go together. This is certainly the case in Cagayan de Oro (CDO) Mindanao, Philippines, where allotment gardeners are not strongly motivated by environmental concerns and where there exist few government restrictions on the use of agrochemicals:

> Here, market forces often serve in place of an environmental ethic in promoting organic methods. The Cagayan de Oro allotment gardeners are quite market-oriented with 70% of their production marketed directly in the garden, mostly to close neighbors. These customers are well aware of production practices in the allotments and do not accept of produce that has been heavily sprayed. (Holmer et al., 2003)

This situation contrasts with the general food system, where produce comes from distant sources and consumers must make assumptions about production methods and contaminants. In developing countries the contrast between locally produced foodstuffs and those delivered in the formal urban food system is even more dramatic because food safety controls are often lacking and labeling is nonexistent or unreliable.

There are other interesting facets of the CDO gardens. Raised beds, simple rainwater catchments, and linear plantings characterize the layout of the gardens, and the gardens have been supported by local authorities in the barangay city districts (Drescher and Holmer, 2007). Unlike in Europe, where allotment gardens are usually located on public lands owned by government entities, all allotment gardens of Cagayan de Oro are established on private land because of the lack of publicly owned open spaces (Fig. 10–7). Land for the CDO allotment gardens was secured from private landowners by the chairmen of the barangay, who asked if poor residents could use their vacant land for food production only and who supported legislation providing tax relief to landowners who make land available for allotments (Drescher and Holmer, 2007). Poor urban families committed themselves to use the land only for food production and construct no residential structures other than a small shed for tools and other garden implements. Conditions governing land use were formalized and signed by all stakeholders, although no provision was made regarding the time frame for guaranteed access

Fig. 10–7. San Isidro allotment garden at Barangay Kauswagan (photograph: R.J. Holmer).

by the gardeners to ensure that they could recoup longer term investments they might make on the land.

A novel aspect of the CDO work has been the fusion between allotment gardens as a solution to both urban food insecurity and solid waste management (Gerold et al., 2005). The poor benefited by collecting biodegradable waste from households, and the gardens absorbed the free compost thus generated. This collaboration between government and community groups is noteworthy as an example for other cities (Mata, 2006).

However, all is not perfect, as Mata also identified several critical issues regarding the transfer of the German allotment garden model to the Philippines. First, industrial and residential development has consistently encroached on agricultural space. Since 1985, this development has especially threatened local agriculture as opposed to the large-scale corporate agriculture practiced in the area by Del Monte, Tropical Fruits, and Nestle (NEDA, 1997). Absent a comprehensive land use plan, local growers lose space and low-income households lose access to affordable, locally grown vegetables. Second, individual gardeners experienced significant opportunity costs due to unemployment and underemployment that created a mismatch between intermittent income and persistent bills for utilities and housing. Third, gardeners' diets contained insufficient calories since they grew primarily leafy greens, legumes, and other vegetables in the allotments; consequently, they were forced to purchase their staple grain rice on the open market at inflated prices. Thus, the most successful farmers were those that had secure employment, consistent income, and who used the garden to supplement their household diet. Effective urban planning and municipal support are also required.

All this said, the number one requirement for success in transferring allotment garden technology is to recognize "...the importance of integrating all stakeholders' voices into the planning process when adapting models" (Mata, 2006). Arguing for a stronger empirical base in the study of urban cultivation,

Webb (1998) developed an index of crop importance, noted the centrality of individual producers to understanding cultivation practices and indirectly provided strong support for participatory planning and research approaches:

> Two major implications emerge from this index of crop importance. Firstly... it is possible to develop empirical measures linked to household welfare. Secondly, because the index is based on cultivation practice, it is likely to reflect the concerns, experiences, and priorities of the cultivators themselves. For this reason, the findings have a far greater value for policy and planning than those derived hypothetically." (Webb, 1998, p. 212)

So if cultivators' "concerns, experiences, and priorities" are at the heart of the usefulness of the technique and central to our ultimate understanding, what advantage is there in omitting their input during the planning and implementation process? How would their input either lessen the quality of the project or reduce its likely success? In other words, the presumption should be inclusion, and the burden of proof should rest with substantiating the reasons for exclusion rather than the reverse, which is the standard approach.

Conclusions

Urban agriculture in general—and urban gardening in particular—has a long history throughout the world. It is safe to say it is here to stay. A great deal of advocacy exists to promote its utility, and perhaps more claims are made than good research exists to support those claims. Nonetheless, there is an abundant and growing body of research from across the social, biological, and physical sciences addressing many aspects of this phenomenon. We have tried to sample from the breadth of that literature to create a core understanding of just what we do know about the subject, without attempting to be exhaustive. Many other issues are involved, and the reader is encouraged to pursue them using this discussion as a foundation.

One of our primary purposes was to simultaneously clarify some of the terminology surrounding urban gardening while elaborating the significant differences that exist among the constituent types of urban gardening. Certainly, a taxonomy of urban gardening is useful to the degree that it helps the reader understand the phenomena under consideration. However, we have tried to show that while clear terminology can illuminate, the important point is to understand the object of study both as an objective phenomenon operating in the physical world and as a subjective set of perceptions and social interactions. We conclude that significant differences do exist among the various types of urban gardens that must be taken into account when working in the area, whether as a researcher, practitioner, or policymaker.

Fundamentally, urban gardening represents a disturbed natural ecology. It emerges from a complicated interplay of motivations, politics, and individual will. It is hard to clearly identify just why it is such a fascinating subject. Likely it is a combination of many factors: its complexity amid seeming simplicity, its persistence amid its transient nature, its interplay of the physical and the social, its "primitive" appeal to individuals with widely different education and social-class backgrounds, its transformative impact on individual consciousness in the face of seeming mundane activity, or maybe just that the food tastes so good and growing it can be pleasurable. Whatever the reasons, it is an arena of activity that will continue to grow in the future and provide abundant opportunities for research: applied and basic, analytic and integrative. So... see you in the garden.

References

Abrahams, C. 2007. Globally useful conceptions of alternative food networks in the developing south: The case of Johannesburg's urban food supply system. p. 95–114. *In* D. Maye et al. (ed.) Alternative food geographies: Representation and practice. Elsevier, New York.

Astill, G., and J. Langdon (ed.) 1997. Medieval farming and technology: The impact of agricultural change in northwest Europe. E.J. Brill Publ., Leiden, the Netherlands.

Auerbach, R. (ed.) 2005. Rainwater harvesting, organic farming and landcare: A vision for uprooting rural poverty in South Africa. Rainman Landcare Foundation, Durban, South Africa.

Bailey, G. 1997. The sweet smelling notebook: An unpublished Mughal source on garden design. p. 129–139. *In* A. Petruccioli (ed.) Gardens in the time of the Great Muslim empires: Theory and design. E.J. Brill, Leiden, the Netherlands.

Bassett, T.J. 1979. Vacant lot cultivation: Community gardening in America, 1893–1978. M.A. thesis. Dep. of Geography, Univ. of California, Berkeley.

Bassett, T.J. 1981. Reaping on the margins: A century of community gardening in America. Landscape 25:1–8.

Bellan, R.C. 1958. The development of Winnipeg as a metropolitan centre. PBD. diss. Columbia Univ., New York.

Bellows, A.C. 2004. One hundred years of allotment gardens in Poland. Food Foodways 12(4):247–276.

Blomley, N. 2003. Unsettling the city: Urban land and the politics of property. Routledge, New York.

Brooklyn Daily Eagle Almanac. 1905. Vol. XX, no. 1, of the Eagle Library, The Brooklyn Daily Eagle. Jan. 1905. Brooklyn Daily Eagle, New York.

Brown, K.H., and A.L. Jameton. 2000. Public health implications of urban agriculture. J. Public Health Pol. 21:20–39.

Burchardt, J. 2002. The allotment movement in England, 1793–1873. The Boydell Press, The Royal Historical Society, Woodridge, UK.

Butcher, G.W. 1918. Allotments for all: The story of a great movement. Allen and Unwin, London.

Byerlee, D. 1998. Knowledge-intensive crop management technologies: Concepts, impacts and prospects in Asian agriculture. p. 113–133. *In* P. Pingali and M. Hossain (ed.) Impacts of rice research. IRRI, Manila.

Caldwell, L.K. 1972. An ecological approach to international development. p. 927–947. *In* M.T. Farvar and J.P. Milton (ed.) The careless technology. Natural History Press, Garden City, NY.

Chatwin, M.E. 1998. Family allotment gardens in Georgia: Introduction of a European model for community food security in urban areas. Available at http://srdis.ciesin.org/cases/georgia-001.html (verified 4 Mar. 2010). The WBI's CBNRM Initiative, The World Bank.

Cohen, S. 2008. Community gardens: Protecting the planet while feeding it. New York Observer, 12 June 2008.

Colding, J., J. Lundberg, and C. Folke. 2006. Incorporating green-area user groups in urban ecosystem management. AMBIO 35:237–244.

Costanza, R., R. d'Arge, R. de Groot, S. Farber, M. Grasso, B. Hannon, K. Limburg, S. Naeem, R.V. O'Neil, J. Paruelo, R.G. Raskin, P. Sutton, and M. van den Belt. 1997. The value of the world's ecosystem services and natural capital. Nature 387:253–260.

Cramb, R.A., and Z. Culasero. 2003. Landcare and livelihoods: The promotion and adoption of conservation farming systems in the Philippine uplands. Int. J. Agric. Sustainability 1:141–154.

Plate 1–1. A view of the Earth at night from satellite perspective (source: NASA).

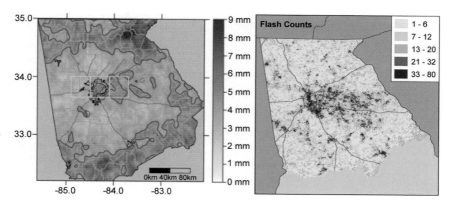

Plate 1–2. Rainfall (left) and lightning flash anomalies (right) for the warm season in Atlanta. Rainfall anomalies were presented in Mote et al. (2007) and lightning flash anomalies were presented in Stallins and Rose (2008). Consult these sources for further information on methods and data.

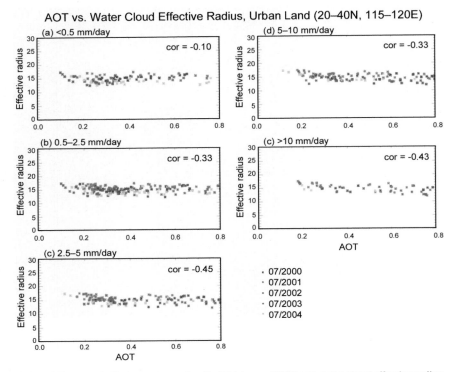

AOT vs. Water Cloud Effective Radius, Urban Land (20–40N, 115–120E)

Plate 1–3. Scatter plot between aerosol optical thickness (AOT) and water cloud effective radius. The data were sampled for the month of July during the interval 2000–2004, over the China Sea (20–40N, 120–150E): (a) for areas with monthly rainfall <0.5 mm d⁻¹; (b) same as (a) except for rainfall between 0.5 and 2.5 mm d⁻¹; (c) for rainfall between 2.5 and 5.0 mm d⁻¹; (d) for rainfall between 5 and 10 mm d⁻¹; and (e) for rainfall events >10 mm d⁻¹ (from Jin and Shepherd, 2008).

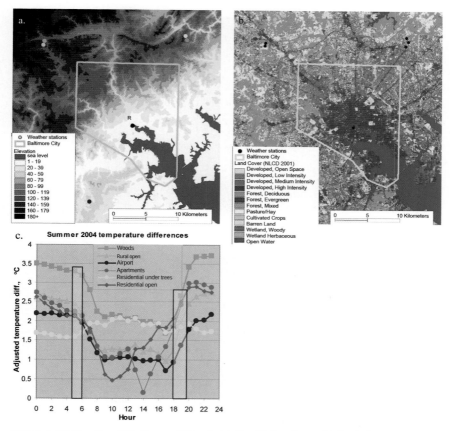

Plate 2–1. (a) Elevation of Baltimore, MD and vicinity with locations of 1.5-m-height temperature measuring sites color-coded to the average temperature differences in panel c. (b) Land use for Baltimore and vicinity, with dark red being most developed, suburban residential mostly medium pink, developed open space such as parks light pink, agriculture yellow and brown, and forest green. (c) Differences in temperature, urban reference (*R*) minus other sites, averaged by hour of the day from May through September in different land-use categories. Temperatures adjusted for elevation difference, assuming a standard atmosphere lapse rate. Range of times of sunrise and sunset indicated by shaded yellow and blue.

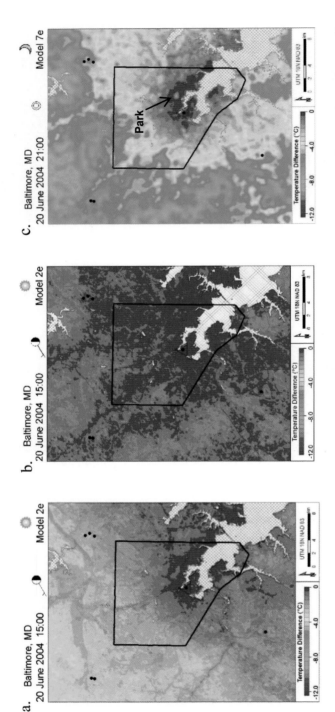

Plate 2–2. (a) Modeled temperature differences (ΔT) at 1.5-m-height air across Baltimore, (black line) and vicinity. Black dots indicate weather stations used in regression modeling to develop prediction equations for ΔT. This map is for 1500 local standard time of a partly cloudy summer day with low wind speeds (<2.6 m s^{-1} = 5 kn), Turner stability Class 2. Water shown by cross-hatched blue. Solid colors indicate ΔT with respect to the warmest temperature on the map (dark red). The coolest (light yellow) is 4.1°C cooler. More than one-half of the predicted temperature difference is due to differences in elevation. (b) With the elevation factor removed from the ΔT equation, the influences of land cover are illustrated for the same time as in 2a; land cover causes a ΔT range of about 1.6°C. (c) With clear sky and low wind speed at night, Turner Class 7, the urban heat island effect is near maximum. A large city park (Patterson) is about 2°C cooler than the dense residential area surrounding it. See Plate 2–1a for elevation map of the Baltimore area and Plate 2–1b for land use. The patterns of elevation and land use are evident in the pattern of predicted ΔT.

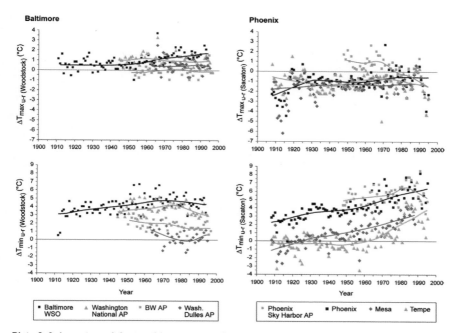

Plate 2–3. Long-term July monthly averages of maximum daily urban temperature minus corre-sponding rural temperatures for stations in and near Baltimore and the same for May temperatures for the Phoenix region (top), and monthly averages of minimum daily urban temperatures minus corresponding rural temperatures (bottom) (Source: Brazel et al., 2000).

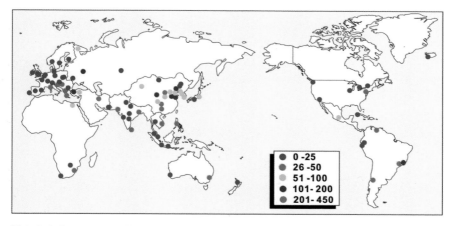

Plate 3–1. Average annual concentrations of SO_2 (μg m^{-3}) in the air of many urban areas in the world in 1995–2001 (adapted from UN DESA, 2005).

Urban Soil Mosaic

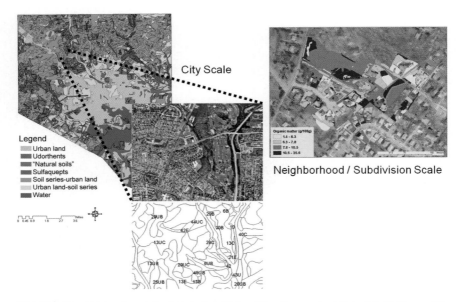

Plate 7–1. The urban soil mosaic shown at multiple scales. Map on the left depicts Baltimore City soil survey units aggregated by native soil series or "natural" soils, urban land, soil series–urban land complex, Udorthents, Sulfaquepts, and urban land–soil series complex (modified from USDA-NRCS, 1998). Center image is an aerial photograph and associated soil map units. Map on the right shows Ecotype Level II polygons for two subdivisions in a suburban neighborhood.

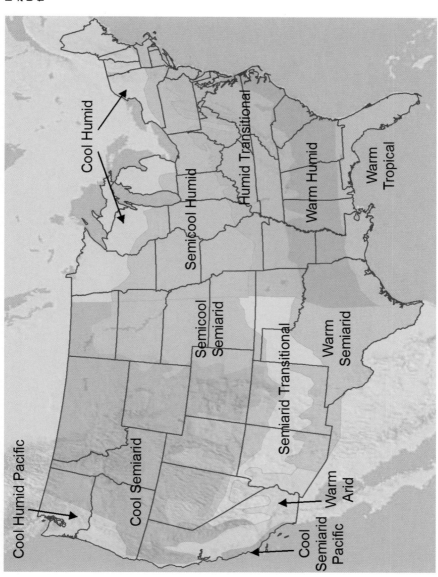

Plate 8–1. Turfgrass climate zones of the continental United States, as adapted from Beard and Beard (2005).

Crouch, D. 2000. Reinventing allotments for the twenty-first century: The UK experience. Acta Hort. ISHS 523:135–142 http://www.actahort.org/books/523/523_18.htm.

Crouch, D., and C. Ward. 1997. The allotment: Its landscape and culture. Five Leaves Publ., Nottingham, UK.

Drescher, A. 2000. Urban and periurban agriculture and planning. FAO-ETC/R AF Discussion paper for electronic conf. Urban and Periurban Agriculture on the Policy Agenda. 21 Aug.–30 Sept. Available at http://www.ruaf.org/node/311 (verified 22 May 2010).

Drescher, A.W. 2001. The German allotment gardens—A model for poverty alleviation and food security in southern African cities? p. 159–167. In Proc. of the Sub-Regional Expert Meeting on Urban Horticulture, Stellenbosch, South Africa. 15–19 Jan. 2001. FAO/Univ. of Stellenbosch. Available at http://foodafrica.nri.org/ (verified 22 May 2010).

Drescher, A.W., and R.J. Holmer (ed.) 2007. Beyond project borderlines—The Asia urbs project. GIS based Urban Environmental Planning and Food Security Project. APT Rep. 16. Univ. of Freiburg, Germany.

Drescher, A.W., R.J. Holmer, and D.L. Iaquinta. 2006. Urban homegardens and allotment gardens for sustainable livelihoods: Management strategies and institutional environments. p. 317–338. In P.K. Nair and B.M. Kumar (ed.) Tropical homegardens: A time tested example of sustainability. Advances in Agroforestry 3. Springer, Dordrecht, the Netherlands.

Dongus, S. 2009. Malaria risk resulting from urban agriculture—Persisting misconception or urgent need for mitigation? Combing physical and human geographic approaches in an operational setting in Dar es Salaam Tanzania. Ph.D. diss. Univ. of Basel, Switzerland.

Dongus, S., and A.W. Drescher. 2006. The use of GIS, GPS and aerial imagery for mapping urban agricultural activities on open space in cities. In GIS Development Weekly, 20 Mar. 2006.

Dunning, N., T. Beach, and D. Rue. 1997. The paleoecology and ancient settlement of the Petexbatun Region, Guatemala. Ancient Mesoamerica 8:255–266.

Evans, S.T. 1990. The productivity of maguey terrace agriculture in central Mexico during the Aztec period. Lat. Am. Antiq. 1:117–132.

FAO. 2008. Urban agriculture for sustainable poverty alleviation and food security. FAO, Rome.

FAO/WHO Codex Alimentarius Commission. 2001. Guidelines for the production, processing, labelling and marketing of organically produced foods. CAC/GL 32-1999. Rev.1-2001. FAO, Rome.

Flavell, N. 2003. Urban allotment gardens in the eighteenth century: The case of Sheffield. Agric. Hist. Rev. 51:95.

Flora, C.B., and J.L. Flora. 1996. Creating social capital. p. 217–225. In W. Vitek and W. Jackson (ed.) Rooted in the land: Essays on community and place. Yale Univ. Press, New Haven, CT.

French, H.R. 2000. Urban agriculture, commons and commoners in the seventeenth and eighteenth centuries: The case of Sudbury, Suffolk. Agric. Hist. Rev. 48(II):171–199.

Gerold, J., A.W. Drescher, and R.J. Holmer. 2005. Kleingärten zur Armutsminderung—Schrebergärten in Cagayan de Oro. Südostasien 214:76–77.

Gibbon, P., and S. Bolwig. 2007. The economics of certified organic farming in tropical Africa: A preliminary analysis. SIDA DIIS Working Paper 2007/3. Subser. Standards and Agro-Food-Exports (SAFE) 7. Available at http://www.diis.dk/graphics/Publications/WP2007/WP2007-3%20til%20web.pdf (verified 5 Mar. 2010).

Goldstein, L.J. 1997. Philadelphia's community garden history. City Farmer, p. 1. Available at http://www.cityfarmer.org/Phillyhistory10.html (verified 5 Mar. 2010).

Greene, K. 1990. The archaeology of the Roman economy. Univ. of California Press, Berkeley.

Groening, G. 1996. Politics of community gardening in Germany. In City Farmer. Available at http://www.cityfarmer.org/german99.html (verified 5 Mar. 2010). Canada's Office of Urban Agriculture, City Farmer's Urban Agriculture Notes, Vancouver, BC, Canada.

Gröning, G. 1996. Politics of community gardening in Germany, presented at the 1996 Annual Conf. of The American Community Gardening Assoc. ACGA Branching Out: Linking Communities Through Gardening, Montréal, Canada. 26–29 Sept. 1996.

Available at http://www.cityfarmer.org/german99.html#introgerman (verified 5 Mar. 2010). City Farmer, Urban Agriculture Notes, Vancouver, BC, Canada.

Gröning, G., and J. Wolschke-Bulmahn. 1995. Von Ackermann bis Ziegelhütte, Ein Jahrhundert Kleingartenkultur in Frankfurt am Main, Klötzer, Wolfgang und Dieter Rebentisch, im Auftrag des Frankfurter Vereins für Geschichte und Landeskunde. Verbindung mit der Frankfurter Historischen Kommission Hg., Studien zur Frankfurter Geschichte, Band 36, Frankfurt am Main.

Guanzon, Y.B., and R.J. Holmer. 2003. Basic cultural management practices for vegetable production in urban areas of the Philippines. Urban Agric. Mag. 10:14–15.

Hamerow, H. 2002. Early medieval settlements: The archaeology of rural communities in north-west Europe 400–900. Oxford Univ. Press, Oxford, UK.

Hoekstra, F. 2008. Gardens of hope—Urban micro-farming and HIV/AIDS. Available at http://www.ruaf.org/node/1558 (accessed 29 June 2009, verified 5 Mar. 2010). Resource Centres on Urban Agriculture and Food Security (RUAF).

Holmer, R.J., M.T. Clavejo, S. Dongus, and A.W. Drescher. 2003. Allotment gardens for Philippine cities. Urban Agric. Mag. 11:29–31.

Holmer, R.J., and A.W. Drescher. 2005. Building food secure neighbourhoods: The role of allotment gardens. Urban Agric. Mag. 15:19–20.

Hovorka, A., H. De Zeeuw, and M. Njenga (ed.) 2009. Women feeding cities: Mainstreaming gender in urban agriculture and food security. Practical Action Publ., Rugby, UK.

Howe, J., K. Bohn, and A. Viljoen. 2005. Food in time: The history of English open urban space as a European example. p. 95–107. In A. Viljoen et al. (ed.) CPULs: Continuous productive urban landscapes. Designing urban agriculture for sustainable cities. Architectural Press, Oxford, UK.

Hughes, C.C., and J.M. Hunter. 1972. The role of technological development in promoting disease in Africa. p. 69, 101. In M.T. Farvar and J.P. Milton (ed.) The careless technology. Natural History Press, Garden City, NY.

Hynes, H.P., and G. Howe. 2005. Urban horticulture in the contemporary United States: Personal and community benefits. In R. Junge-Berberovic et al. (ed.) International Conference on Urban Horticulture, Waedenswil, Switzerland. 31 Jan. 2004. Available at http://www.actahort.org/books/643/643_21.htm (verified 5 Mar. 2010).

Iaquinta, D.L. 1975. Food production: Systems in conflict? M.S. thesis. Univ. of Wisconsin, Madison.

Iaquinta, D.L. 1999. The impact of aging populations on food security: The role of intergenerational transfers and land tenure dynamics. Sustainable Development Working Paper, FAO Library Call number 39 Ia6E. FAO, Rome.

Iaquinta, D.L., and A.W. Drescher. 2000. Defining peri-urban: Rural–urban linkages and institutional connections. In FAO Corporate Document Repository, Land Reform: Land Settlement and Cooperatives. Available at ftp://ftp.fao.org/docrep/fao/003/X8050T/X8050T03.pdf or http://www.fao.org/DOCREP/003/X8050T/x8050t02.htm#P13_235 (verified 1 July 2009). FAO, Rome.

Iaquinta, D.L., and A.W. Drescher. 2001. More than the spatial fringe: An application of the periurban typology to planning and management of natural resources in the peri-urban, Workshop at the DPU International Conference, Rural–Urban Encounters: Managing the Environment of the Peri-Urban Interface, Royal Holloway University of London. 9–10 Nov. 2001. Available at http://www.ucl.ac.uk/dpu/pui/events/EPM_conf_abstracts.htm#DefiningthePeri-Urban (verified 1 July 2009).

Iaquinta, D.L., and A.W. Drescher. 2002a. Food security in cities—A new challenge to development. p. 983–994. In C.A. Brebbia et al. (ed.) The sustainable city II—Urban regeneration and sustainability. Advances in architecture. WIT Press, Ashurst, Southampton, UK.

Iaquinta, D.L., and A.W. Drescher. 2002b. Urbanization—Linking development across the changing landscape, Special Chapter for State of Food and Agriculture. Available at http://www.fao.org/fileadmin/templates/FCIT/PDF/sofa.pdf (verified 1 July 2009). FAO, Rome.

Iaquinta, D.L., J. Du Guerny, and L. Stloukal. 1999. Linkages between rural population ageing, intergenerational transfers of land and agricultural production: Are they important? SD Dimensions. Available at http://www.fao.org/sd/WPdirect/WPan0039. htm (verified 5 Mar. 2010). FAO, Rome.

IFOAM. 2009. Definition of organic agriculture. Available at http://www.ifoam.org/grow-ing_organic/definitions/doa/index.html (accessed 6 June 2009, verified 1 July 2009). International Foundation of Organic Agriculture Movement, Bonn, Germany.

Jones, P. 2003. Digging up the past–Jeremy Burchardt the allotment movement in England, 1793–1873. Hist. Workshop J. 56:272–277.

Kasch, G. 2001. Deutsches Kleingärtnermuseum in Leipzig: Deutschlands Kleingärtner vom 19. zum 21. Jahrhundert. Band 4. Sächsische Landesstelle für Museumswesen, Chemnitz, Germany.

Kaufman, J., and M. Bailkey. 2000. Farming inside cities: Entrepreneurial urban agriculture in the United States. Working Paper WP00JK1. Available at http://queencityfarm.org/FarmingInsideCities.pdf (verified 1 July 2009). Lincoln Institute of Land Policy, Cambridge, MA.

Kehoe, D.P. 1988. The economics of agriculture on Roman imperial estates in North Africa. Vandenhoeck and Ruprecht, Göttingen, Germany.

Leeuwis, C. 2004. Communication for rural innovation. Blackwell Publishing, Oxford, UK.

Lieblin, G., E. Østergaard, and C. Francis. 2004. Becoming an agroecologist through action education. Int. J. Agric. Sustain. 23:147–153.

MacNair, E. 2002. The garden city handbook. Polis Project on Ecological Governance, Univ. of Victoria, BC, Canada.

Marcus, C.C., and M. Barnes. 1999. Healing gardens: Therapeutic benefits and design recommendations. John Wiley and Sons, New York.

Mata, C.T. 2006. Examining problems with implementing the German allotment garden model: Learning from Bugo Barangay. Planning Forum 12:81–101.

Maye, D., L. Holloway, and M. Kneafsey. 2007. Alternative food geographies: Representation and practice. Elsevier Ltd., Amsterdam.

Miller, C. 2003. In the sweat of our brow: Citizenship in American domestic practice during WWII—Victory gardens. J. Am. Cult. 26(3):395–409.

Mougeot, L. 1994. The rise of city farming: Research must catch up with reality. ILEIA Newsl. 104:4–5.

Mougeot, L. 1996. Introduction: An improving domestic and international environment for African urban agriculture. Special Issue: Urban Agriculture in Africa. Afric. Urban Q. 112/3:137–152.

Müller, C. 2007. Intercultural gardens—Urban places for subsistence production and diversity. German J. Urban Stud. 46(1). Available at http://www.difu.de/publikationen/dfk/en/welcome.shtml (verified 5 Mar. 2010).

Mwale, F.P.. 2006. Growing better cities. Part 1. The Issue. Available at http://www.idrc.ca/in_focus_cities/ev-95353-201-1-DO_TOPIC.html (added 24 Mar. 2006, modified 2 May 2006, verified 5 Mar. 2010). IDRC, Ottawa, ON, Canada.

NEDA. 1997. Socoeconomic profile for Cagayan de Oro, region X. National Economic and Development Authority, Manila.

Necipoglu, G. 1997. The suburban landscape of sixteenth-century Istanbul as a mirror of classical Ottoman garden culture. p. 32–71. In A. Petruccioli (ed.) Gardens in the time of the great Muslim empires: Theory and design. Brill Academic Publishers, Leiden, the Netherlands.

Nell, W.T., S.J. Wessels, J. Mokoka, and S. Machedi. 2000. A creative multidisciplinary approach towards the development of food gardening. Development Southern Africa, 175(5):807–819.

New York City Parks and Recreation Department. 2010. Thomas Jefferson Park. Available at http://www.nycgovparks.org/parks/thomasjeffersonpark (verified 22 May 2010).

Orr, D. 1992. Ecological literacy. SUNY Press, Albany, NY.

Ostrom, E. 2000. Social capital: A fad or fundamental concept? p. 172–214. In P. Dasgupta and S. Serageldin (ed.) Social capital: A multifaceted perspective. The World Bank, Washington, DC.

Pack, C.L. 1919. The war garden victorious. Available at http://www.earthlypursuits.com/WarGarV/WarGardTitle.htm (verified 1 July 2009). National War Garden Commission, J.B. Lippincott Company, Philadelphia, PA.

Pothukuchi, K. 2005. Building community infrastructure for healthy communities: Evaluating action research components of an urban health research programme. Plann. Pract. Res. 202:127–146.

Pothukuchi, K., and J.L. Kaufman. 1999. Placing the food system on the urban agenda: The role of municipal institutions in food systems planning. Agric. Human Values 162:213–224.

Pretty, J. 2003. Social capital and the collective management of resources. Science 302:1912–1915.

Putnam, R.D. 2000. Bowling alone: The collapse and revival of American community. Simon and Schuster, New York.

Roy, M. 2001. The rhetoric and reality of allotment gardens and sustainable development: The case of allotment gardens in Winnipeg, Manitoba. M.A. thesis. Univ. of Manitoba, Winnipeg, MB, Canada.

Saverino, J.L. 1995. 'Domani ci zappa': Italian immigration and ethnicity in Pennsylvania. Pa. Folklife 45(Autumn):2–22.

Sidblad, S. 2000. Swedish perspectives of allotment and community gardening. Acta Hort. ISHS 523:151–160.

Schmelzkopf, K. 1995. Urban community gardens as contested space. Geogr. Rev. 853:364–381.

Smith, M.E., and T.J. Price. 1994. Aztec-period agricultural terraces in Morelos, Mexico: Evidence for household-level agricultural intensification. J. Field Archaeol. 21:169–179.

Spudić, S. 2007. The new victory garden. Royal Horticultural Society diss., Wisley Diploma in Practical Horticulture. Available at http://homepage.mac.com/cityfarmer/AllotmentReportweb.pdf (verified 1 Mar. 2010).

Subtelny, E.M. 1997. Agriculture and the Timerud Chahargagh: The evidence from a medieval Persian agricultural manual. p. 110–128. In A. Petruccioli (ed.) Gardens in the time of the great Muslim empires: Theory and design. Brill Academic Publishers, Leiden, the Netherlands.

UNCTAD-UNEP. 2008. Organic agriculture and food security in Africa. Prepared for the UNEP-UNCTAD Capacity-Building Task Force on Trade, Environment and Development, UNCTAD/DITC/TED/2007/15. Available at http://www.foodfirst.org/files/pdf/UN_Organic%20Agriculture_Africa_2008.pdf (verified 5 Mar. 2010). United Nations, New York.

UNCHS. 1998. Privatization of municipal services in East Africa: A governance approach to human settlements management. United Nations Centre for Human Settlements, UN-Habitat, Nairobi, Kenya.

Van Molle, L., and Y. Segers. 2008. Micro-farming on other men's land. Allotments from the 19th to the 21st century: Belgian history in a global perspective. HUB Research paper 2008/23. Available at http://www.hubrussel.be/Documenten/Internet/PDF/HUB%20research%20paper%20reeks/HRP2008/HRP23.pdf (verified 5 Mar. 2010). Hogeschool, Univ. Brussels, Belgium.

Von Baeyer, E. 1984. Rhetoric and roses: A history of Canadian gardening 1900–1930. Fitzhenry and Whiteside, Markham, ON.

Warner, S.B., Jr., and H. Durlach. 1987. To dwell is to garden: A history of Boston's community gardens. Northeastern Univ. Press, Boston.

Weatherford, J. 1993. Early Andean experimental agriculture p. 64–77. In S. Harding (ed.) The "racial" economy of science: Toward a democratic future. Indiana Univ. Press.

Webb, N. 1998. Urban cultivation: Food crops and their importance. Dev. South. Afr. 15(2):201–213.

Wille, K. 1939. Entwicklung und wirtschaftliche Bedeutung des Kleingartenwesens. Frankf.

Wood, P., and C. Landry. 2007. The intercultural city: Planning for diversity advantage. Earthscan Publ., London, UK.

11

Urban Forestry

Astrid Volder
W. Todd Watson

Abstract

Urban forestry is defined as "the art, science, and technology of managing trees and forest resources in and around urban community ecosystems for the physiological, sociological, economic, and aesthetic benefits trees provide to society" (Miller, 1997, cf. Konijnendijk et al., 2006). Spatial scales of urban forestry range from a single tree to large forested stands. Urban forestry employs an ecosystem-based approach that includes non-tree factors such as soil, climate, surrounding vegetation, and the built landscape. Trees in urban environment provide many ecosystem services and are generally considered to be valuable. Examples of services provided by urban trees include air filtration, shade, evaporative cooling, recreational opportunities, wildlife habitat, and carbon sequestration. This chapter discusses recent literature on the benefits that urban trees provide to society, some of the potentially negative aspects of urban trees, and issues related to urban forest management, valuation of urban trees, and the general environment in which urban trees are grow.

Although the term *urban forestry* was not commonly used until the late 20th century (Konijnendijk et al., 2006), urban forestry basically started when human beings began to tend to and plant trees in the towns and villages or near the camps where they lived. The trees were used for fruit and nut production, fuel, timber, shade, and as a central meeting place. In the United States, local public tree plantings were organized and protection laws were issued as early as the mid 1600s (Ricard, 2005). The first expansive garden and arboretum in America, Magnolia Plantation and Gardens, was planted in South Carolina in 1671 (Campana, 1999). However, the development of large city parks and the use of street trees for beautification became more common in the 19th century when some of the famous American city parks and park systems were developed by Frederick Olmsted (e.g., Central Park in New York City and the park systems designed for Buffalo, NY and Milwaukee, WI) and contemporaries (e.g., the Forest Preserve district in Chicago). Smaller towns and cities developed community forests as well, places that initially provided the towns with fuel and timber, but today play

A. Volder, Dep. of Horticultural Sciences, Texas A&M University, TAMU 2133, College Station, TX 77843-2133 (a-volder@tamu.edu); W.T. Watson, Dep. of Ecosystem Science and Management, Texas A&M University, TAMU 2138, College Station, TX 77843-2138 (t-watson@tamu.edu).

doi:10.2134/agronmonogr55.c11

a central role in providing recreational opportunities and clean water. With the onset of urban forestry came specialist caretakers, such as arborists and city foresters. These specialists manage trees from the scale of individual street trees to large-scale urban parks. Managing urban trees involves tree planting, general maintenance, insect diagnosis and control, disease diagnosis and control, fertilization, irrigation, and tree removal. Other duties also include assessing the potential for the tree to become a public hazard (e.g., its vulnerability to wind or its ability to crack sidewalks, walls, and foundations), value assessment, assessing damages to the tree, choosing the right tree species to plant, and keeping a general inventory of the city trees.

The urban forest includes all woody plants in and around the city, such as street trees, garden trees, park trees, and planted or remnant forest stands. Because of their relatively large sizes, trees provide more benefits and present greater liabilities than other urban vegetation (Table 11–1). The maintenance costs per plant are also higher. Trees in the urban forest provide a range of ecosystem services, such as removal of dust and air pollution, carbon sequestration, reducing energy use through shading (in hot climates) or reduced wind velocity (in cold climates), cooling the urban climate through evapotranspiration, improved neighborhood and landscape aesthetics, and enhancing the psychological well-being of people living in urban areas (Crow et al., 2006).

The urban forest is very spatially fragmented and consists of a range of sizes, from individual trees, to trees aligned along a grid pattern (e.g., trees along streets or in parking lots), to small groups of trees (e.g., around buildings and in small parks), to larger stands (e.g., larger parks, remnant forest).

Trees in the Urban Environment

Parks and Gardens

Species composition depends on the function of the park. For example, Taipei urban parks with multiple uses—from nature preservation to human recreation—had greater tree species diversity than riverside parks and street verges (Jim and Chen, 2008b). Jim and Chen (2008a) also found very little similarity in tree species diversity between their study sites. Of the 164 tree species recorded, no single tree species was present in all 30 communities surveyed, and there was little overlap between the three landscape types surveyed (urban park, riverside park, and street verges). Trees in parks are often subjected to very compacted soils due to high foot traffic pressure, which causes reduced soil porosity, increased resistance to root growth, reduced transmission of air and water, which can exacerbate the negative effects of both flooding and drought, and altered nutrient and carbon cycling. Soil chemistry can also be drastically altered due to the aforementioned changes in soil structure, as well as due to altered rates of litter input and removal of topsoil during the development of the park. For example, Jim (1998a) found that soil bulk density in Victoria Park in Hong Kong's exceeded 1.75 Mg m^{-3}. Soils in this park also had a very alkaline pH, a lack of organic matter, limited cation exchange capacity, and low exchangeable cation, N, and P concentrations (Jim, 1998a).

Urban private gardens are generally smaller than those in rural areas due to space restrictions. Still, urban gardens contain a vast reservoir of trees. Even in a city as densely built as Hong Kong residents had an average of five trees

Table 11–1. Some examples of studies referencing positive and negative effects of trees in urban environments.

Environmental benefits	References
Improve air quality	Beckett et al., 1998; Beckett et al., 2000; Givoni, 1991; Jim and Chen, 2008a
Reduce urban heat island	Akbari et al., 2001; Avissar, 1996; Georgi and Zafiriadis, 2006; Shashua-Bar et al., 2009; Spronken-Smith and Oke, 1998
Reduce noise pollution	Bolund and Hunhammar, 1999; Van Renterghem and Botteldooren, 2008
Provide shade in parking areas and other high use areas	Akbari et al., 2001; Georgi and Zafiriadis, 2006; Robitu et al., 2006
Reduce energy use through shade or wind reduction	Akbari, 2002; Carver et al., 2004; Donovan and Butry, 2009; Heisler, 1986; Simpson, 2002
Carbon sequestration	Golubiewski, 2006; Gratani and Varone, 2006; Nowak and Crane, 2002
Intercept rainfall and reduce the amount, velocity, and peak time of runoff	Cook, 2007; Guevara-Escobar et al., 2007; Xiao and McPherson, 2002; Xiao et al., 2000
Other benefits	
Increased property values	Jim and Chen, 2006; Payton et al., 2008; Tyrvainen, 1997; Tyrvainen and Miettinen, 2000
Recreational opportunities	Li et al., 2005; Roovers et al., 2002
Wildlife habitat	Clergeau et al., 1998; Pickett et al., 2001; Sorace, 2001
Source of timber and food	Bjorheden et al., 2003; Long and Nair, 1999
Aesthetic beauty	Crow et al., 2006; Lafortezza et al., 2009; Yang et al., 2009; Zhang et al., 2007
Reduce human stress, improve human health	Crow et al., 2006; Ferris et al., 2001; Groenewegen et al., 2006; Maas et al., 2006; Takano et al., 2002; Tzoulas et al., 2007
Reduce crime	Kuo and Sullivan, 2001
Potential negatives	
Damage to or interference with urban infrastructure	Doherty et al., 2000; Grabosky et al., 2001; Hamilton, 1984b; McPherson, 2000; Randrup et al., 2001
Contribute to emission of volatile organic compounds	Diem, 2000; Donovan et al., 2005; Geron et al., 1995; Llusia et al., 2002; Manning, 2008; Paoletti, 2009; Rennenberg et al., 2006
Dangerous situations due to tree failure	James et al., 2006; Niklas, 2000; Schmidlin, 2009

in their backyards (Jim, 1993). Tree species diversity in these gardens increased with a decreasing plot size, suggesting that people preferred variety even in very confined spaces. Trees were also important in Tunesian gardens, where people planted citrus trees in urban and suburban areas (Khelifa et al., 2008). The use of trees in gardens is governed by both environmental and nonenvironmental factors. In a comprehensive study surveying the factors that determine garden type in Hobart, Australia, Kirkpatrick et al. (2007) found that the percentage of trees in the front yard increased with increasing household income. Increased tree canopy cover has been associated with improved suburban sustainability (Ghosh and Head, 2009). For example, the value of the 400,000 trees in the urban forest in Canberra, Australia has been estimated at US$20 to 67 million for a 4-yr period, in terms of energy reduction, pollution mitigation, and carbon sequestration (Brack, 2002).

Streets

Surveying municipal urban foresters in the United States, Kielbasso et al. (1988) found that municipal urban forestry programs spend the majority of their budgets on street trees. Street trees suffer from a range of adverse environmental conditions, including exposure to car exhaust, high air temperatures and vapor pressure deficit, reduced litter input and nutrient availability, compacted soils, altered hydrology, and, in northern climates, exposure to deicing salts (Kelsey and Hootman, 1990). Street trees in heavy traffic areas had greater Pb levels (Pal et al., 2000) and exposure to exhaust fumes led to reduced chlorophyll concentrations (Durrani et al., 2004). The effects of some urban conditions are reversible. For example, Oleksyn et al. (2007) found that a combination of mulch and fertilizer addition can be used to restore urban tree tissue nitrogen concentrations and photosynthesis rates. Lack of space for roots to proliferate can cause significant problems belowground. Girdled roots that cut into the tree trunk are a common problem in urban areas. On the other hand, tree roots often cause pavement to crack at significant cost. Allowing ample space above and below ground for tree expansion is critical, but often overlooked when landscapes are designed or redeveloped (Jim, 1998b).

Riparian Zones

Riparian forests play an important role in maintaining high water quality input in streams (Chapter 13, Stander and Ehrenfeld, 2010, this volume). With increasing urbanization, it is likely that the species composition of riparian forests will change, and this may ultimately affect the functioning of riparian zones. A study of riparian zone species composition along an urbanization gradient in western Georgia revealed that with increasing urbanization the importance of the invasive nonnative shrub *Ligustrum sinense* Lour. increased (Burton et al., 2005). The same study also showed that species diversity in riparian forests decreased with increasing urbanization. Urban riparian areas also tend to suffer from flashy hydrology that increases stream flow during precipitation events, but also increases drought between precipitation events. This can lead to a gradual shift toward a more drought-tolerant, upland, species composition in urban riparian forests (Groffman et al., 2003).

Forests

Urban forests can be remnants of forests that were there before European settlement, but generally are forests that have developed, either spontaneously regenerated or deliberately planted, after initial clearing by settlers. Forest age and disturbance regime will have strong effects on species composition (Vallet et al., 2008). For example, on the basis of pollen records and aerial photographs, Loeb (1992) was able to reconstruct the history of the Turtle Pond watershed in northeastern Queens, which was abandoned from agricultural use in the late 1800s. Early development of this urban forest was mainly influenced by a combination of human influence (potential deliberate planting of eastern redcedar [*Juniperus virginiana* L.]), natural succession, and the impact of disease and insect outbreaks (chestnut blight [*Cryphonectria parasitica* (Murrill) M.E. Barr], hickory bark beetle [*Scolytus quadripinosus* (Say.)], dogwood anthracnose [*Discula destructiva* Redlin]). However, after 1973, relative importance of birch (*Betula* spp.), sweetgum (*Liquidambar styraciflua* L.), and maple (*Acer* spp.) start to decline, most likely due to increased stress from recreational use that may be affecting reproductive output in these species (Loeb, 1992).

Benefits of Trees in Urban Environments

Urban trees provide significant energy benefits when planted strategically. In hot climates they can help cool buildings and other structures by providing shade, while in cold climates they can help reduce heat loss from buildings by reducing wind speed. Individual oak (*Quercus* spp.) trees in Rome have been shown to remove up to 185 kg of CO_2 per tree per year, depending on tree size (Gratani and Varone, 2006). Urban forests in Chile have been shown to be effective at reducing air pollutants, with the effectiveness varying with air pollutant concentrations and regionally and temporally (Escobedo and Nowak, 2009). Urban forests were more effective per square meter of leaf area in areas of lower socioeconomic status and when pollutant levels were high. Forests in urban watersheds also play a crucial role in maintaining a clean water source (Baron et al., 2002; Beck, 2005; Cadenasso et al., 2008).

Shading of buildings has been shown to reduce cooling energy costs by 7.1% in Sacramento, CA (Simpson and McPherson, 1998) and by 6.1 to 15.4% in Carbondale, IL (Carver et al., 2004). Maximum air temperature reduction due to the presence of trees in Tessaloniki, Greece was 24% (Georgi and Zafiriadis, 2006), a substantial reduction in a city where summer air temperatures regularly spike over 40°C. Location of the tree with regards to providing maximum shade benefits during northern hemisphere summer (i.e., west and south side of a building) reduced summertime electricity use by 5.2% in Sacramento, while trees planted on the north side provided no benefits in terms of shade (Donovan and Butry, 2009). In a detailed analysis of the structure, function, and value of trees along streets and in parks in Fort Collins, CO; Cheyenne, WY; Bismarck, ND; Berkeley, CA; and Glenndale, AZ, Greg McPherson and coworkers estimated that each tree provided services valued between $31 and $89 per year (McPherson et al., 2005).

Some benefits of the presence of urban trees are more difficult to quantify than environmental benefits. Benefits such as improving the emotional state of people who look at greenery, providing an opportunity for exercise, and providing habitat for wildlife, among others, are more difficult to express in monetary terms. People generally feel better when presented with a tree-lined street than a treeless street, and this may alter human behavioral patterns. For example, the presence of trees can encourage people to seek out treed locations for their shopping, and it may also induce people to spend more time in these areas. Wolf (2005) found that consumers in general were willing to drive longer to visit a shopping district with trees rather than one without trees. These consumers also indicated that they would visit more frequently, and for large cities, would spend more time in the district for each visit.

Urban forest habitat is valuable for a range of wildlife, particularly birds (Mortberg and Wallentinus, 2000; Mortberg, 2001; Sandstrom et al., 2006), but also mammals (Gillies and Clout, 2003), marsupials (FitzGibbon et al., 2007), amphibians and reptiles (Drinnan, 2005; Gillies and Clout, 2003; Gonzalez-Garcia et al., 2009), and invertebrate species (Gillies and Clout, 2003). Urban forests also provide valuable shaded and protected habitat for herbaceous plant species, including rare species (Godefroid and Koedam, 2003), and have been reported to reduce crime and increase a sense of safety in residents (Kuo and Sullivan, 2001; Kuo et al., 1998).

The value of an urban tree that is not used for timber production can be based on emotional attachment, aesthetic value, or strictly the functional benefits of having that tree at that location. Thus, value can be very subjective. Assigning a monetary value to trees that are not easily replaced is very difficult and a topic of much debate. Most valuation formulas include factors such as tree size, location of the tree, tree species, and tree physical condition, with trained appraisers weighing each factor. Watson (2002) mentioned two types of methods to establish tree values, one that uses tree size to set an initial value that is then adjusted for species, location, health, and other attributes of the tree, and a method where a point system is used that converts to a monetary value. When five commonly used appraisal methods were compared for six individual trees of different species, appraised values differed dramatically (Watson, 2002). For example, median appraisal (among nine experienced appraisers) of an individual European beech tree ranged from $13,900 to $167,212 among the five methods. Estimates by individual appraisers using the same method for the same tree also varied dramatically; the most consistent method was the point-based standard tree evaluation method (STEM) with a coefficient of variation of 18 (Watson, 2002). This study underscores how difficult it is to assign a monetary value to a tree, even for experienced professionals. Within the United States the most accepted methods are described in the *Guide for Plant Appraisal* (CTLA, 2000), which provides guidelines for market value, income, and replacement cost approaches to appraising tree values.

Management

Keeping a detailed tree inventory is essential for good urban forest management, and detailed management plans are needed to keep trees healthy and provide early intervention when trees may be involved in potentially hazardous situations such as interception of power lines and wood decay. Collection of data beyond an initial inventory and tree sizes is essential (Baker, 1993; McPherson, 1993). A survey of municipal urban foresters in the southern United States revealed that more than 60% of communities had tree inventories (Watson, 2003). Government funding and increased availability of computers were major factors in the doubling of tree inventories described in earlier surveys (Giedraitis and Kielbaso, 1982). Modern techniques integrate accurate physical location details using Global Positioning System (GPS) data, image data, and additional information from Geographical Information Systems (GIS) all in one package (Tait et al., 2009). Ideally, urban forest management plans should improve the benefits provided by trees while limiting tree liabilities and maintenance costs. Ninety percent of the respondents to a survey of municipal urban foresters said that the benefits derived from having a tree inventory were well worth the costs associated with conducting and managing the inventory (Watson, 2003).

Planting of trees is often an afterthought during urban development, and often there is little consideration of potential undesirable traits such as excessive fruit drop, potential for falling limbs and wind damage, loss of sap causing sticky residue on cars and other fixtures, and emission of biogenic volatile organic compounds. Surprisingly, mature tree size is often overlooked as an important criterion when selecting trees for planting (Pauleit, 2003; Pauleit et al., 2002), and trees change from an ornamental asset to a liability due to excessive shade, litter-fall, falling limbs, and roots that destroy infrastructure. Too many times, monetary considerations are the overriding factor, and unsuitable tree species are planted in too small a space. Planting trees in areas where there is no space for canopy or root system expansion leads to problems where restricted growth space often leads to stunted tree growth or premature death. Damage done by tree roots to pavements or belowground infrastructure often results in tree removal (Hamilton, 1984a,b; McPherson, 2000). A lot of these problems can be prevented if initial site design would include allocating proper space for tree growth and would take into account the desired (e.g., fast growing, shade canopy) and undesired traits of a tree growing at that specific location. Liability for damage done by hazardous trees resides with the landowner, and this places responsibility for tree maintenance with the landowner (Mortimer and Kane, 2004). A tree's legal owner has a duty to prevent the tree from doing damage, and if that duty was not met and the tree damages another person or their property, the tree's owner can be held legally responsible (Mortimer and Kane, 2004).

Tree failure is an important issue in urban areas where the higher population density increases the risk of injury of people and damage to property. Between 1995 and 2007 there were 407 deaths from wind-related tree failures in the United States (Schmidlin, 2009). Tree failure is mostly caused by exposure to windstorms and heavy snow and ice loading, but some species (e.g., *Eucalyptus* spp.) also drop branches in response to drought. In general, as the tree crown captures more wind, larger bending moments are created at the tree base (James et al., 2006; Niklas, 2000). However, tree shape and the ratio of branch mass to trunk mass also determine the risk of tree failure (James et al., 2006; Niklas, 2000). During wind storms, wind direction and speed are rarely constant, and leaves, branches, and trunk move in different directions as the trunk sways. All this movement within the canopy creates forces in different directions and has a dampening effect. The independent movement of tree branches reduces the amplitude of trunk sway and increases the stability of the tree (James et al., 2006). Pruning leaves and twigs to reduce drag can also reduce the risk of tree failure; however, the effect of overpruning on overall tree health should be kept in mind (Kane and Harris, 2008). Trees become vulnerable to toppling when root systems are compromised, which reduces the ability of trees to resist pulling forces (Smiley, 2008). Other causes of tree failure include chronic exposure to windy conditions, which causes cracking of the stem, eventually leading to stem failure (Mattheck and Bethge, 1990).

Trees also become more vulnerable to failure as stems and branches are exposed to wood-decaying fungi. As trees get older they become more susceptible to decay and can become hazardous during even mild windstorms (Terho and Hallaksela, 2008; Terho et al., 2007). Hollowing out of the stem reduces the ability of the tree to withstand external forces. Because much of the decay may not be directly visible to the naked eye, these defects often remain undiscovered until after tree failure occurs. Recent technical advances now allow arborists to create noninvasive high resolution images of the degree of decay in individual living tree trunks (Habermehl and Ridder, 1993; Koizumi and Hirai, 2006; Lin et al., 2008), which allows trees in hazardous locations (near roadways, buildings, and busy pedestrian areas) to be inspected regularly for structural defects.

A large proportion of urban tree management is related to the trimming of trees to keep branches away from electrical wires. Many electric companies employ full-time utility arborists and crews of certified line trimmers that share management and maintenance of street trees with municipal arborists (Doherty et al., 2000). Surprisingly, a survey of utility arborists and tree wardens in Massachusetts revealed that 72% of communities with more than 10,000 inhabitants did not have a street tree inventory and 83% of communities with more than 10,000 inhabitants did not have a street tree management plan. Lack of knowledge about inventory and lack of planning drastically reduce the effectiveness of street tree maintenance (Doherty et al., 2000). Positive findings in this survey include a high level of cooperation between municipal foresters and utility arborists, as well as a change in pruning strategy since the 1970s. Traditionally trees that interfered with electrical wires were headed back a specified distance from the wires or rounded over, resulting in deformed, ugly shaped trees that are more vulnerable to decay (Dahle et al., 2006). A move to directional lateral pruning that directs tree growth away from the lines, combined with a cyclical approach (2–5 yr pruning rotation) to tree trimming that prevents crisis situations has drastically reduced the number of misshapen trees in Massachusetts (Doherty et al., 2000). Improving choices made at the time of tree planting toward smaller trees and planting sites further away from power lines could reduce the cost of street tree maintenance.

Once trees or shrubs have been removed or trimmed, a sizeable amount of woody plant debris is left. A 1995 survey of the total amount of yard waste generated in the United States shows that 13,561,612 m³ (17,737,919 cubic yards) of urban tree and landscape residue are generated each year, 98% of which are woody plant related (Whittier et al., 1995). Of this woody debris, 42% was given away, 17% went to the landfill, 15% was left on site or used on site, and only 6% was recycled. Leaving waste on site can work if there is a large leaf litter component (Cothrel et al., 1997). However, the woody debris can also be used to generate energy. Murphey et al. (1980) performed an analysis of three possible waste management strategies by an urban tree maintenance firm in Houston that needed to dispose of 30,000 tons of wood waste annually. The simplest alternative was to truck the waste to a landfill 46 km (28 miles) away. Another option was to team up with the local power company, chip the wood, and burn it to produce electricity. The third strategy was to chip the wood and sell it to a third party for cash. For the second option, the energy gains were converted to a dollar value to be returned to the urban tree maintenance firm. The cost–benefit analysis showed that the second option would be most cost effective and illustrated that urban wood waste can be used as a viable energy source (Murphey et al., 1980). Ironically, the company chose to go with the third alternative and sell the wood chips for cash, most likely to ensure a steady cash flow rather than become involved in the energy business.

Conclusions

Urban forestry is the holistic management of trees in urban areas, ranging from individual street trees to forested stands. Urban forests are an integral part of the urban ecosystem and provide many valuable ecosystem services. The monetary value of these services is difficult to estimate, but general estimates in the literature range from $70 to $100 per tree per year. Management of urban forests should be aimed at optimizing these services through selecting the appropriate species and providing the required soil and environmental conditions to attain maximum tree health.

References

Akbari, H. 2002. Shade trees reduce building energy use and CO_2 emissions from power plants. Environ. Pollut. 116:S119–S126.

Akbari, H., M. Pomerantz, and H. Taha. 2001. Cool surfaces and shade trees to reduce energy use and improve air quality in urban areas. Sol. Energy 70:295–310.

Avissar, R. 1996. Potential effects of vegetation on the urban thermal environment. Atmos. Environ. 30:437–448.

Baker, F.A. 1993. Monitoring the urban forest- case studies and evaluations. Environ. Monit. Assess. 26:153–163.

Baron, J.S., N.L. Poff, P.L. Angermeier, C.N. Dahm, P.H. Gleick, N.G. Hairston, R.B. Jackson, C.A. Johnston, B.D. Richter, and A.D. Steinman. 2002. Meeting ecological and societal needs for freshwater. Ecol. Appl. 12:1247–1260.

Beck, M.B. 2005. Vulnerability of water quality in intensively developing urban watersheds. Environ. Model. Softw. 20:381–400.

Beckett, K.P., P.H. Freer-Smith, and G. Taylor. 1998. Urban woodlands: Their role in reducing the effects of particulate pollution. Environ. Pollut. 99:347–360.

Beckett, K.P., P.H. Freer-Smith, and G. Taylor. 2000. The capture of particulate pollution by trees at five contrasting urban sites. Arboricult. J. 24:209–230.

Bjorheden, R., T. Gullberg, and J. Johansson. 2003. Systems analyses for harvesting small trees for forest fuel in urban forestry. Biomass Bioenerg. 24:389–400.

Bolund, P., and S. Hunhammar. 1999. Ecosystem services in urban areas. Ecol. Econ. 29:293–301.

Brack, C.L. 2002. Pollution mitigation and carbon sequestration by an urban forest. Environ. Pollut. 116:S195–S200.

Burton, M.L., L.J. Samuelson, and S. Pan. 2005. Riparian woody plant diversity and forest structure along an urban-rural gradient. Urban Ecosyst. 8:93–106.

Cadenasso, M.L., S.T.A. Pickett, P.M. Groffman, L.E. Band, G.S. Brush, M.E. Galvin, J.M. Grove, G. Hagar, V. Marshall, B.P. McGrath, J.P.M. Oneil-Dunne, W.P. Stack, and A.R. Troy. 2008. Exchanges across land-water-scape boundaries in urban systems: Strategies for reducing nitrate pollution. Ann. NY Acad. Sci. 1134:213–232.

Campana, R.J. 1999. Arboriculture: History and development in North America. Michigan State Univ. Press, East Lansing.

Carver, A.D., D.R. Unger, and C.L. Parks. 2004. Modeling energy savings from urban shade trees: An assessment of the CITYgreen® energy conservation module. Environ. Manage. 34:650–655.

Clergeau, P., J.P.L. Savard, G. Mennechez, and G. Falardeau. 1998. Bird abundance and diversity along an urban–rural gradient: A comparative study between two cities on different continents. Condor 100:413–425.

Cook, E.A. 2007. Green site design: Strategies for stormwater management. J. Green Build. 2:46–56.

Cothrel, S.R., J.P. Vimmerstedt, and D.A. Kost. 1997. In situ recycling of urban deciduous litter. Soil Biol. Biochem. 29:295–298.

Crow, T., T. Brown, and R. De Young. 2006. The Riverside and Berwyn experience: Contrasts in landscape structure, perceptions of the urban landscape, and their effects on people. Landscape Urban Plan. 75:282–299.

CTLA. 2000. Guide for plant appraisal. 9th ed. Intl. Soc. for Arboriculture, Champaign, IL.

Dahle, G.A., H.H. Holt, W.R. Chaney, T.M. Whalen, J. Grabosky, D.L. Cassens, R. Gazo, R.L. McKenzie, and J.B. Santini. 2006. Decay patterns in silver maple (*Acer saccharinum*) trees converted from roundovers to V-trims. Arboricult. Urban For. 32:260–264.

Diem, J.E. 2000. Comparisons of weekday–weekend ozone: Importance of biogenic volatile organic compound emissions in the semi-arid southwest USA. Atmos. Environ.
34:3445–3451.

Doherty, K.D., H.D.P. Ryan, and D.V. Bloniarz. 2000. Tree wardens and utility arborists: A
management team working for street trees in Massachusetts. J. Arboric. 26:38–46.

Donovan, G.H., and D.T. Butry. 2009. The value of shade: Estimating the effect of urban
trees on summertime electricity use. Energy Build. 41:662–668.

Donovan, R.G., H.E. Stewart, S.M. Owen, A.R. Mackenzie, and C.N. Hewitt. 2005. Development and application of an urban tree air quality score for photochemical pollution
episodes using the Birmingham, United Kingdom, area as a case study. Environ. Sci.
Technol. 39:6730–6738.

Drinnan, I.N. 2005. The search for fragmentation thresholds in a Southern Sydney Suburb.
Biol. Conserv. 124:339–349.

Durrani, G.F., M. Hassan, M.K. Baloch, and G. Hameed. 2004. Effect of traffic pollution on
plant photosynthesis. J. Chem. Soc. Pak. 26:176–179.

Escobedo, F.J., and D.J. Nowak. 2009. Spatial heterogeneity and air pollution removal by an
urban forest. Landscape Urban Plan. 90:102–110.

Ferris, J., C. Norman, and J. Sempik. 2001. People, land and sustainability: Community
gardens and the social dimension of sustainable development. Soc. Policy Admin.
35:559–568.

FitzGibbon, S.I., D.A. Putland, and A.W. Goldizen. 2007. The importance of functional connectivity in the conservation of a ground-dwelling mammal in an urban Australian
landscape. Landscape Ecol. 22:1513–1525.

Georgi, N., and K. Zafiriadis. 2006. The impact of park trees on microclimate in urban areas.
Urban Ecosyst. 9:195–209.

Geron, C.D., T.E. Pierce, and A.B. Guenther. 1995. Reassessment of biogenic volatile organic
compound emissions in the Atlanta area. Atmos. Environ. 29:1569–1578 .

Ghosh, S., and L. Head. 2009. Retrofitting the suburban garden: Morphologies and some
elements of sustainability potential of two Australian residential suburbs compared.
Aust. Geogr. 40:319–346.

Giedraitis, J.P., and J.J. Kielbaso. 1982. Municipal tree management. Rep. 14 (1). Intl. City
Manage. Assoc., Washington, DC.

Gillies, C., and M. Clout. 2003. The prey of domestic cats (Felis catus) in two suburbs of
Auckland City. N.Z. J. Zool. 259:309–315.

Givoni, B. 1991. Impact of planted areas on urban environmental quality—A review. Atmos.
Environ. Part B. Urban Atmos. 25:289–299.

Godefroid, S., and N. Koedam. 2003. How important are large vs. small forest remnants
for the conservation of the woodland flora in an urban context? Glob. Ecol. Biogeogr.
12:287–298.

Golubiewski, N.E. 2006. Urbanization increases grassland carbon pools: Effects of landscaping in Colorado's front range. Ecol. Appl. 16:555–571.

Gonzalez-Garcia, A., J. Belliure, A. Gomez-Sal, and P. Davilla. 2009. The role of urban
greenspaces in fauna conservation: The case of the iguana Ctenosaura similis in the
'patios' of Leon city, Nicaragua. Biodivers. Conserv. 18:1909–1920.

Grabosky, J., N. Bassuk, L. Irwin, and H. van Es. 2001. Shoot and root growth of three tree
species in sidewalks. J. Environ. Hortic. 19:206–211.

Gratani, L., and L. Varone. 2006. Carbon sequestration by Quercus ilex L. and Quercus pubescens Willd. and their contribution to decreasing air temperature in Rome. Urban
Ecosyst. 9:27–37.

Groenewegen, P.P., A.E. den Berg, S. de Vries, and R.A. Verheij. 2006. Vitamin G: Effects of
green space on health, well-being, and social safety. BMC Public Health 6:149.

Groffman, P.M., D.J. Bain, L.E. Band, K.T. Belt, G.S. Brush, J.M. Grove, R.V. Pouyat, I.C. Yesilonis, and W.C. Zipperer. 2003. Down by the riverside: Urban riparian ecology. Front.
Ecol. Environ. 1:315–321.

Guevara-Escobar, A., E. Gonzalez-Sosa, C. Veliz-Chavez, E. Ventura-Ramos, and M. Ramos-Salinas. 2007. Rainfall interception and distribution patterns of gross precipitation around an isolated *Ficus benjamina* tree in an urban area. J. Hydrol. 333:532–541.

Habermehl, A., and H.W. Ridder. 1993. Application of computed tomography for non-destroying investigations of wood in living trees—Studies on park and avenue trees. Holz Als Roh-Und Werkstoff 51:101–106.

Hamilton, W.D. 1984a. Sidewalk/curb-breaking tree roots. 2. Management to minimise existing pavement problems by tree roots. Arboricult. J. 8:223–234.

Hamilton, W.D. 1984b. Sidewalk/curb-breaking tree roots. 1. Why tree roots cause pavement problems. Arboricult. J. 8:37–44.

Heisler, G.M. 1986. Effects of individual trees on the solar radiation climate of small buildings. Urban Ecol. 9:337–359.

James, K.R., N. Haritos, and P.K. Ades. 2006. Mechanical stability of trees under dynamic loads. Am. J. Bot. 93:1522–1530.

Jim, C.Y. 1993. Trees and landscape of a suburban residential neighborhood in Hong Kong. Landscape Urban Plan. 23:119–143.

Jim, C.Y. 1998a. Soil characteristics and management in an urban park in Hong Kong. Environ. Manage. 22:683–695.

Jim, C.Y. 1998b. Impacts of intensive urbanization on trees in Hong Kong. Environ. Conserv. 25:146–159.

Jim, C.Y., and W.Y. Chen. 2006. Impacts of urban environmental elements on residential housing prices in Guangzhou (China). Landscape Urban Plan. 78:422–434.

Jim, C.Y., and W.Y. Chen. 2008a. Assessing the ecosystem service of air pollutant removal by urban trees in Guangzhou (China). J. Environ. Manage. 88:665–676.

Jim, C.Y., and W.Y. Chen. 2008b. Pattern and divergence of tree communities in Taipei's main urban green spaces. Landscape Urban Plan. 84:312–323.

Kane, B., and R. Harris. 2008. Does pruning reduce the risk of tree failure? Arborist News 17:46–48.

Kelsey, P., and R. Hootman. 1990. Soil resource evaluation for a group of sidewalk street tree planters. J. Arboric. 16:113–117.

Khelifa, S.B., H. Rejeb, and N. Souayah. 2008. Observations of *Citrus aurantium* exposed to urban stress in Tunisian cities: Behaviour analysis of in vitro-plants faced to salt stress. J. Food Agric. Environ. 6(2):418–421.

Kielbasso, J., B. Beauchamp, K. Larison, and C. Randall. 1988. Trends in urban forest management. Rep. 20(1). Intl. City Manage. Assoc., Washington, DC.

Kirkpatrick, J.B., G.D. Daniels, and T. Zagorski. 2007. Explaining variation in front gardens between suburbs of Hobart, Tasmania, Australia. Landscape Urban Plan. 79:314–322.

Koizumi, A., and T. Hirai. 2006. Evaluation of the section modulus for tree-stem cross sections of irregular shape. J. Wood Sci. 52:213–219.

Konijnendijk, C.C., R.M. Ricard, A. Kenney, and T.B. Randrup. 2006. Defining urban forestry—A comparative perspective of North America and Europe. Urban For. Urban Green. 4:93–103.

Kuo, F.E., and W.C. Sullivan. 2001. Environment and crime in the inner city—Does vegetation reduce crime? Environ. Behav. 33:343–367.

Kuo, F.E., M. Bacaicoa, and W.C. Sullivan. 1998. Transforming inner-city landscapes—Trees, sense of safety, and preference. Environ. Behav. 30:28–59.

Lafortezza, R., G. Carrus, G. Sanesi, and C. Davies. 2009. Benefits and well-being perceived by people visiting green spaces in periods of heat stress. Urban For. Urban Green. 8:97–108.

Li, F., R.S. Wang, J. Paulussen, and X.S. Liu. 2005. Comprehensive concept planning of urban greening based on ecological principles: A case study in Beijing, China. Landscape Urban Plan. 72:325–336.

Lin, C.J., Y.C. Kao, T.T. Lin, M.J. Tsai, S.Y. Wang, L.D. Lin, Y.N. Wang, and M.H. Chan. 2008. Application of an ultrasonic tomographic technique for detecting defects in standing trees. Int. Biodeterior. Biodegrad. 62:434–441.

Llusia, J., J. Penuelas, and B.S. Gimeno. 2002. Seasonal and species-specific response of VOC emissions by Mediterranean woody plant to elevated ozone concentrations. Atmos. Environ. 36:3931–3938.

Loeb, R.E. 1992. Long-term human disturbance of an urban park forest, New York City. For. Ecol. Manage. 49:293–309.

Long, A.J., and P.K.R. Nair. 1999. Trees outside forests: Agro-, community, and urban forestry. New For. 17:145–174.

Maas, J., R.A. Verheij, P.P. Groenewegen, S. de Vries, and P. Spreeuwenberg. 2006. Green space, urbanity, and health: How strong is the relation? J. Epidemiol. Community Health 60:587–592.

Manning, W.J. 2008. Plants in urban ecosystems: Essential role of urban forests in urban metabolism and succession toward sustainability. Int. J. Sust. Dev. World Ecol. 15:362–370.

Mattheck, C., and K. Bethge. 1990. Wind breakage of trees initiated by root delamination. Trees-Struct. Funct. 4:225–227.

McPherson, E.G. 1993. Monitoring urban forest health. Environ. Monit. Assess. 26:165–174.

McPherson, E.G. 2000. Expenditures associated with conflicts between street tree root growth and hardscape in California, United States. J. Arboricult. 26(6):289–297.

McPherson, G., J.R. Simpson, P.J. Peper, S.E. Maco, and Q.F. Xiao. 2005. Municipal forest benefits and costs in five U.S. cities. J. For. 103:411–416.

Miller, R.W. 1997. Urban forestry: Planning and managing green spaces. 2nd ed. Prentice Hall, Englewood Cliffs, NJ.

Mortberg, U., and H.G. Wallentinus. 2000. Red-listed forest bird species in an urban environment—Assessment of green space corridors. Landscape Urban Plan. 50:215–226.

Mortberg, U.M. 2001. Resident bird species in urban forest remnants; landscape and habitat perspectives. Landscape Ecol. 16:193–203.

Mortimer, M.J., and B. Kane. 2004. Hazard tree liability in the United States: Uncertain risks for owners and professionals. Urban For. Urban Green. 2:159–165.

Murphey, W.K., J.G. Massey, and A. Sumrall. 1980. Converting urban tree maintenance residue to energy. J. Arboric. 6:85–88.

Niklas, K.J. 2000. Computing factors of safety against wind-induced tree stem damage. J. Exp. Bot. 51:797–806.

Nowak, D.J., and D.E. Crane. 2002. Carbon storage and sequestration by urban trees in the USA. Environ. Pollut. 116:381–389.

Oleksyn, J., B.D. Kloeppel, S. Lukasiewicz, P. Karolewski, and P.B. Reich. 2007. Ecophysiology of horse chestnut (*Aesculus hippocastanum* L.) in degraded and restored urban sites. Pol. J. Ecol. 55:245–260.

Pal, A., K. Kulshreshtha, K.J. Ahmad, and M. Yunus. 2000. Changes in leaf surface structures of two avenue tree species caused by auto-exhaust pollution. J. Environ. Biol. 21:15–21.

Paoletti, E. 2009. Ozone and urban forests in Italy. Environ. Pollut. 157:1506–1512.

Pauleit, S. 2003. Urban street tree plantings: Indentifying the key requirements. Proc. Inst. Civil Eng.-Municip. Eng. 156:43–50.

Pauleit, S., N. Jones, G. Garcia-Martin, J.L. Garcia-Valdecantos, L.M. Riviere, L. Vidal-Beaudet, M. Bodson, and T.B. Randrup. 2002. Tree establishment practice in towns and cities—Results from a European survey. Urban For. Urban Green. 1:83–96.

Payton, S., G. Lindsey, J. Wilson, J.R. Ottensmann, and J. Man. 2008. Valuing the benefits of the urban forest: A spatial hedonic approach. J. Environ. Plan. Manage. 51:717–736.

Pickett, S.T.A., M.L. Cadenasso, J.M. Grove, C.H. Nilon, R.V. Pouyat, W.C. Zipperer, and R. Costanza. 2001. Urban ecological systems: Linking terrestrial ecological, physical, and socioeconomic components of metropolitan areas. Annu. Rev. Ecol. Syst. 32:127–157.

Randrup, T.B., E.G. McPherson, and L.R. Costello. 2001. A review of tree root conflicts with sidewalks, curbs, and roads. Urban Ecosyst. 5:209–225.

Rennenberg, H., F. Loreto, A. Polle, F. Brilli, S. Fares, R.S. Beniwal, and A. Gessler. 2006. Physiological responses of forest trees to heat and drought. Plant Biol. 8:556–571.

Ricard, R.M. 2005. Shade trees and tree wardens: Revising the history of urban forestry. J. For. 103:230–233.

Robitu, M., M. Musy, C. Inard, and D. Groleau. 2006. Modeling the influence of vegetation and water pond on urban microclimate. Sol. Energ. 80:435–447.

Roovers, P., M. Hermy, and H. Gulinck. 2002. Visitor profile, perceptions and expectations in forests from a gradient of increasing urbanisation in central Belgium. Landscape Urban Plan. 59:129–145.

Sandstrom, U.G., P. Angelstam, and G. Mikusinski. 2006. Ecological diversity of birds in relation to the structure of urban green space. Landscape Urban Plan. 77:39–53.

Schmidlin, T.W. 2009. Human fatalities from wind-related tree failures in the United States, 1995–2007. Nat. Hazards 50:13–25.

Shashua-Bar, L., D. Pearlmutter, and E. Erell. 2009. The cooling efficiency of urban landscape strategies in a hot dry climate. Landscape Urban Plan. 92:179–186.

Simpson, J.R. 2002. Improved estimates of tree-shade effects on residential energy use. Energ. Build. 34:1067–1076.

Simpson, J.R., and E.G. McPherson. 1998. Simulation of tree shade impacts on residential energy use for space conditioning in Sacramento. Atmos. Environ. 32:69–74.

Smiley, E.T. 2008. Root pruning and stability of young willow oak. Arboricult. Urban For. 34:123–128.

Sorace, A. 2001. Value to wildlife of urban-agricultural parks: A case study from Rome urban area. Environ. Manage. 28:547–560.

Spronken-Smith, R.A., and T.R. Oke. 1998. The thermal regime of urban parks in two cities with different summer climates. Int. J. Remote Sens. 19:2085–2104.

Stander, E.K., and J.G. Ehrenfeld. 2010. Urban riparian function. p. 253–276. In J. Aitkenhead-Peterson and A. Volder (ed.) Urban ecosystem ecology. Agron. Monogr. 55. ASA, CSSA, and SSSA, Madison, WI.

Tait, R.J., T.J. Allen, N. Sherkat, and M.D. Bellett-Travers. 2009. An electronic tree inventory for arboriculture management. Knowl. Base. Syst. 22:552–556.

Takano, T., K. Nakamura, and M. Watanabe. 2002. Urban residential environments and senior citizens' longevity in megacity areas: The importance of walkable green spaces. J. Epidemiol. Community Health 56:913–918.

Terho, M., and A.M. Hallaksela. 2008. Decay characteristics of hazardous Tilia, Betula, and Acer trees felled by municipal urban tree managers in the Helsinki City Area. Forestry 81:151–159.

Terho, M., J. Hantula, and A.M. Hallaksela. 2007. Occurrence and decay patterns of common wood-decay fungi in hazardous trees felled in the Helsinki City. For. Pathol. 37:420–432.

Tyrvainen, L. 1997. The amenity value of the urban forest: An application of the hedonic pricing method. Landscape Urban Plan. 37:211–222.

Tyrvainen, L., and A. Miettinen. 2000. Property prices and urban forest amenities. J. Environ. Econ. Manage. 39:205–223.

Tzoulas, K., K. Korpela, S. Venn, V. Yli-Pelkonen, A. Kazmierczak, J. Niemela, and P. James. 2007. Promoting ecosystem and human health in urban areas using green infrastructure: A literature review. Landscape Urban Plan. 81:167–178.

Vallet, J., H. Daniel, V. Beaujouan, and F. Roze. 2008. Plant species response to urbanization: Comparison of isolated woodland patches in two cities of north-western France. Landscape Ecol. 23:1205–1217.

Van Renterghem, T., and D. Botteldooren. 2008. Numerical evaluation of tree canopy shape near noise barriers to improve downwind shielding. J. Acoust. Soc. Am. 123:648–657.

Watson, G. 2002. Comparing formula methods of tree appraisal. J. Arboric. 28:11–18.

Watson, W.T. 2003. Evolution of municipal urban forestry in the South. p. 14–17. In C. Kollin (ed.) 2003 National Urban Forestry Conference Proceedings. American Forests, Washington, DC.

Whittier, J., D. Rue, and S. Haase. 1995. Urban tree residues: Results of the first national inventory. J. Arboric. 21:57–62.

Wolf, K.L. 2005. Business district streetscapes, trees, and consumer response. J. For. 103:396–400.

Xiao, Q., and E.G. McPherson. 2002. Rainfall interception by Santa Monica's municipal urban forest. Urban Ecosyst. 6:291–302.

Xiao, Q.F., E.G. McPherson, S.L. Ustin, and M.E. Grismer. 2000. A new approach to modeling tree rainfall interception. J. Geophys. Res. Atmos. 105:29173–29188.

Yang, J., L. Zhao, J. McBride, and P. Gong. 2009. Can you see green? Assessing the visibility of urban forests in cities. Landscape Urban Plan. 91:97–104.

Zhang, Y., A. Hussain, J. Deng, and N. Letson. 2007. Public attitudes toward urban trees and supporting urban tree programs. Environ. Behav. 39:797–814.

12

Inside Out: Invasive Plants and Urban Environments

Sarah Hayden Reichard

Abstract

Introduced plants used in urban locations have many beneficial uses, including reducing erosion, sequestering carbon, and providing a colorful and beautiful landscape. A small portion of them become aggressively invasive, however, and have a number of impacts, including competition for resources, changes in ecosystem properties, and an increase in wildfire frequency. Urban areas are sources for wildland invaders for several reasons. The large number of landscape species used, and the large number of individuals of each species, increases the probability of invasion through propagule pressure. Seeds travel outside of urban areas to wildlands by several vectors, including wind, animals (especially birds), and motor vehicles. Fortunately, not all introduced species are invasive and using "Codes of Conduct," or best management practices, may help reduce the number of invasive species used in urban areas, preventing escape to surrounding areas.

Scientists divide the epochs that make up the geologic timescales using terms like the *Holocene* and the *Pleistocene*. Some have suggested that future humans will term our current epoch the *Homogocene* because of the blending of not only cultures and economies, but also plant and other species. Our world is beginning to look the same everywhere—the same plants dot the roadsides and wild landscapes. These "invasive" species are impacting native plants and animals in ways we are only beginning to understand. Urban systems are the source of many problems in wildlands, and the urban systems themselves are also impacted.

Invasive plants are nonnative species that are capable of colonizing everything from your backyard to wildlands, develop self-sustaining populations, and become disruptive and even dominant. Other terms used include *aliens* or *exotics*, two terms meaning nonnative. Not all countries have the same concerns about invasive species. Those that were settled more recently by people of European heritage, such as the United States and Australia, tend to be more worried about their impact. The native indigenous people rarely moved species across long distances, so regional floras were mostly intact when the Europeans arrived, bringing familiar species.

S.H. Reichard, School of Forest Resources, University of Washington, Box 354115, Seattle, WA 98195 (reichard@u.washington.edu).

doi:10.2134/agronmonogr55.c12

Places such as Europe, however, have had centuries of people moving long distances across the continent. In those areas, biologists tend to make a distinction between *archeophytes*, or those plants that are not native but were present before explorers brought plants back from the "new world," and *neophytes*, or those introduced after Columbus and subsequent explorers (Pyšek, 1998).

Why Is There Concern about Invasive Plants?

Introduced invasive plants are now recognized as a major environmental issue, responsible at least in part for nearly 60% of the imperiled plant species in the United States (Wilcove et al., 1998) and costing several billion dollars annually in damages and control costs (Pimentel et al., 2005). The ecological impacts are wide ranging (Mack et al., 2000) and include everything from simple competition for resources such as water and light to complex changes in the food web. Vines are effective competitors for light when they cover trees during the summer growing months and prevent sunlight from reaching either the trees or the forest floor. Some invaders, such as American southwest invader saltcedar (*Tamarix ramosissima* Ledeb.), also have higher water use than native species (Sala et al., 1996). Their uptake of water can lower the water table below the root zone of native species, increasing competition for this valuable resource.

Replacement of a native species by an aggressive nonnative species can lead to a number of changes in plant and animal communities and ecosystems. For instance, a study of giant knotweed [*Fallopia sachalinensis* (F. Schmidt) Ronse Decr., syn. *Polygonum sachalinense* F. Schmidt] in Washington State found that it dominated the banks of rivers and streams, replacing native trees (Urgenson et al., 2009). It produced 70% less leaf litter than the trees, and the amount of nitrogen in the leaves was very different. The knotweed reabsorbed 76% of the nitrogen into the rhizomes and roots, while the native species—willows (*Salix* spp.), alders [*Alnus glutinosa* (L.) Gaertn.], and cottonwoods (*Populus* spp.)—only resorbed 5 to 33%, depending on the species. Fewer nitrogen-rich leaves falling onto the ground and into the water resulted in less food for insects, which in turn meant less food for fish and other animals.

Invasive plants may alter natural disturbance regimes, especially the frequency of fire in ecosystems maintained by fire. For instance, in the shrub-steppe communities in the Great Basin of the United States, fires have historically occurred every 60 to 100 yr, but following the invasive of cheatgrass (*Bromus japonicus* Thunb.), fires are occurring every 3 to 5 yr (Whisenant, 1990). Native shrubs cannot regenerate in that timeframe, and the invasion has resulted in millions of hectares that are essentially a monoculture.

Some of the effects can be long-lasting, well after the plants are removed. While a lack of nitrogen in the Urgenson et al. (2009) example was a problem, some plants can add nitrogen to the soil, which is not always good. Some soils, such as sandy dunes, igneous soils derived from volcanic activity, and prairie soils, are naturally low in nitrogen and the plants native to them are adapted to this. When plants that "fix," nitrogen (convert nitrogen to the forms useable by the plants) invade these types of areas they may increase the soil nitrogen. Vitousek and Walker (1989) documented a number of effects of *Morella faya* (Aiton) Wilbur, a shrub invading Hawaii, and found that it substantially altered nitrogen when it invaded relatively fresh lava flows. Such a change can push the ecosystem beyond the tolerances of native species and facilitate invasion of other nonnatives. Nitrogen may remain elevated for an extended period of time, making restoration after removal more difficult. For instance, the grassland soils around nitrogen-fixing Scotch broom [*Cytisus scoparius* (L.) Link] remained high in total nitrogen and nitrate and lower in pH than nearby soils several months after the plants were removed (Dougherty and Reichard, 2004). Plants other than native species are therefore better able to establish, and the Scotch broom facilitates a community of invasive plants. Restoration of these areas will therefore require waiting for the nitrogen levels to slowly drop naturally or actively trying to change soil chemistry.

There is some debate about whether invasive species are "passengers" or "drivers" of environmental change (MacDougall and Turkington, 2005). Many people think that invasive species are causing, or driving, the alterations of systems, while others think that they are the result of changes in hydrology, disturbance, or other differences in how systems function, acting as passengers to the change. In the examples above, the invasions are clearly driving change, but they may be present because of other changes, acting as passengers as well. For instance, changes such as increased impermeable surfaces in urban areas may increase flooding in rivers, which may then in turn increase invasion along the banks. Unfortunately, determining whether species are responding to or causing environmental change can be very difficult for many reasons. It would be best to know if they were passengers before beginning control work, because if underlying problems remain, the species will likely return, but it is usually unrealistic. It is also possible that the same species may be a driver or a passenger under different circumstances. In urban areas where there is a highly altered landscape, many invasive plants are likely passengers, but the same species in a wildland may be affecting change as a driver.

Why Are There so Many Invasive Species in Urban Locations?

Urban areas tend to have very high levels of invasive plants. A study of the urban vegetation of Berlin found the outskirts of the city were about 30% nonnative invasives, but the urban center was closer to 50% (Kowarik, 1995). This may occur for many reasons. First, urban areas are highly disturbed—soil is removed or compacted, streets are paved, curbs alter surface water movement, and so on. Disturbance and invasive species are often closely linked (D'Antonio et al., 2000; Hobbs 2000). Invaders have traits that allow them to exploit disturbance better than most native species, including a short time from seed germination to fruit production, vegetative reproduction, and high seed production. These traits allow the plants to grow and reproduce quickly following a reduction in competition from other species because of the disturbance.

Urban ecosystems have perhaps the optimal disturbance conditions for invasion. Continually disturbed areas, such as agricultural fields that are tilled regularly, may have invasion from herbaceous species, while wildlands that see minimal disturbance may not have enough reduction in competition to allow establishment. Plants and animals already established may provide a form of biological resistance to invasion (Simberloff, 1986). Urban locations, however, have moderate levels of disturbance along roadways, railroads, greenbelts, and parks. The *intermediate disturbance theory* (Connell, 1978) suggests some removal of competitors because of disturbed conditions, but not continual disturbance, are ideal for invasion. Depending on location, native species may be less equipped to quickly adapt to disturbance, and nearby introduced species can capitalize on the reduction in competition.

There has been the suggestion that forests are less invaded than open areas, including urban zones, because the multilayered canopies typical of forests capture most sunlight before it filters to the forest floor, impairing seed germination of most species, and providing a buffering against disturbances (Corlett, 1992). Communities with high species richness (i.e., the number of species found there) may also be resistant to invasions (Elton, 1958), but terrestrial plants do not conform to this theory as well as other organisms (Mack et al., 2000).

Another reason there are so many invasive species in urban locations is the large amount of source material. Many introduced plants are used to landscape homes, businesses, and parks, and some may become invasive (Chapter 9, Volder, 2010, this volume). Probably about 60% of all invasive plants are introduced for landscape purposes, and the total is higher for woody plants, perhaps as much as 82% (Reichard, 1997). If the plants have some of the types of traits already discussed, they may be able to invade outside gardens.

There is often an assumption that federal governments screen newly introduced plants for their invasiveness before they allow introduction. In general, this is not true. In recent years Australia and New Zealand have required these assessments, but other countries have not. The United States and the European Union are investigating ways to do this, but are not currently assessing the risk of invasiveness of the plants, only whether they have insects or pathogens. In the future we may see more regulation. Some importing nurseries are concerned about these environmental issues and are selective in their inventory, but many are not.

Spread Following Introduction

Invasions tend to have several phases (Fig. 12–1). Species first arrive, either through accidental methods, such as seed contamination, or intentionally, for food, forage, landscape, or other use. Increasingly, most species are intentional introductions, especially for use in urban gardens. Accidentally introduced species may not be

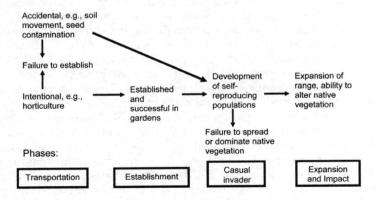

Fig. 12–1. A species goes through several phases before it can be considered invasive. Some casual invaders may be in early stages of a lag phase and become more serious, but many remain in this status indefinitely and do not cause environmental problems.

climatically suitable for the introduction site, but intentionally introduced species are usually matched for the climate and may receive supplemental water and nutrition to allow them to establish. A small percentage of these species may become casual invasives, self-sowing in the garden or establishing small populations, often in disturbed areas near the growing sites. Of these, a small percentage may become a problem either in the urban area, or by dispersing out of it through various methods into surrounding wildlands.

There is a correlation between the number of plants in a species found in an area, either in a location or across time, and the probability of invasion. This is often termed *propagule* or *inoculation pressure*, and the concept is simple: the more plants that are used, the greater the likelihood that the seeds will land in places that are suitable for germination and growth. For instance, a British analysis of catalogs from several nurseries found that the frequency of sale in the 19th century predicted naturalization today (Dehnen-Schmutz et al., 2007a). Additional study also found that species whose seeds were sold then at less expensive prices were more likely to be invasive in current years, suggesting they may have been purchased more because of their reasonable price, which meant more propagules for dispersal (Dehnen-Schmutz et al., 2007b). Mack (2000) also suggested that the link between cultivation for horticulture and agriculture and invasion may also be due to the nature of cultivation. By reducing environmental stochasticity through protection from predators, parasites, drought, frost, and so on, a seed bank develops in the soil from which invasions can occur.

The *minimum resident time*, the time from introduction until recorded invasion, is also correlated with invasion. In an analysis of a longtime nursery in Florida, the probability of invasion increased the longer species were sold, with 70% of the species sold more than 30 yr becoming naturalized, if not fully invasive (Pemberton and Liu, 2009). In Chile, species with a shorter minimum resident time had a more limited geographical spread than those that had been there longer (Castro et al., 2005). Similarly, species that were recorded as naturalized early in New Zealand were more likely to be widespread in the country now (Gatehouse, 2008), and vines that had been documented earlier in Australia were more widely distributed (Harris et al., 2007). This phenomenon may be correlated with propagule pressure because the longer a species is available, the more individuals are likely to be found in gardens. It may also allow for greater opportunities for dispersal along roads or other vectors.

Many plants may be present in gardens for a substantial period of time before they begin to move outside of cultivation. These "sleeper weeds" may be used in gardens for some time before they begin to spread. The time between when they are introduced and when they begin to invade is the *lag phase*. There are many reasons this may occur. Some are extrinsic to the biology of the species. For instance, it may just be a matter of a species increasing in popularity or some other factor related to propagule pressure. In other instances, it may be a change in conditions, such as an episodic disturbance such as a hurricane, or a change in nutrient enrichment such as from nitrogen-fixing species that alter the soil chemistry to facilitate additional invasions. There may also be intrinsic, biological reasons for a lag phase. When a species is introduced many times independently, the initial genetic types might have been less invasive, but subsequent introductions are more aggressive. In some cases, however, the lag phase may be a matter of perception. Knowledgeable people may simply fail to observe that a species is moving from gardens, especially if it does not have colorful flowers or fruits. For instance, it is believed that serrated tussock grass [*Nassella trichotoma* (Nees) Hack. ex Arechav.], a serious problem in pastures, was introduced into Australia in the early 1900s, but it was not reported until 1935. By then, it was already widespread.

Species that are not initially invasive may become so after repeated introductions of the same species causes hybrid genotypes to develop. For instance, the ornamental pear (*Pyrus calleryana* Decne.), a native of China, has long been a popular ornamental in commercial and residential areas, especially in the eastern United States. It was considered noninvasive due to self-incompatibility, but repeated introductions of new cultivated varieties from different regions of China introduced sufficiently different genotypes for reproduction to occur. Fruit production is now possible, and it is in the early stages of invasion (Culley and Hardiman, 2009).

What Difference Does It Make if They Invade Urban Areas?

Having more plants in cities is generally a good thing—they help slow falling rainwater and reduce erosion, they provide cooling for buildings by shading, and aid in reducing global climate change by absorbing carbon (Chapter 18, Aitkenhead-Peterson et al., 2010, this volume). So what is wrong with having a few more of them spreading along roadsides and into urban greenbelts and parks? Many, if not most, urbanites derive their "sense of place" or impression of the natural surroundings of their regions, by exposure to small patches of nature in the city. A relatively small number of people find their way to wildlands. Environmental psychologists think the recognition of place helps to develop concern and connect us with the history of our surroundings. The philosopher and writer Wendell Berry summarized this idea: "You can't know who you are until you know where you are" (Berry, 1994). He identified 17 rules for developing a sustainable community, the second of which is that nature, including air, water, land, and native organisms, should be considered members of the community. Therefore, by keeping some patches of nature in cities similar to wild areas, we are helping people achieve that sense of place they need to move us toward greater sustainability.

Urban dwellers often do not see the connection between the cities and the areas that surround them. They think of the geography as distinct, rather than continuous. Plants in urban areas spread out into wildland areas, and wild areas surrounding cities may be the most invaded. For instance, in New Zealand, invasive species in wildlands were more common in regions with higher population density (Gatehouse, 2008).

Invasive species often move from cities by natural mechanisms such as bird consumption. Many plant species have carbohydrate- and protein-rich fruits to attract avian dispersers. The seeds survive passage through the digestive tract and are defecated or regurgitated in a new location. While bird dispersal of seeds is found mostly among woody invasive plants and only about 50% of them overall, it is common in forest invaders (Reichard and Hamilton, 1997). Most frugivorous birds are generalists and will forage on a number of species, increasing the likelihood of visitation and dispersal of horticultural introductions (Reichard et al., 2001). The distribution of perch sites along the urban to wildland gradient also may factor into spread from urban areas. A study in Belgium found that the spread of an invasive trees species from urban areas depended on the presence of trees where birds could rest and defecate (Deckers et al., 2005). The length of time seeds can remain in the bird's digestive tract is variable, but can be as high as 100 h (Proctor, 1968), and some species may travel considerable distances between foraging and roosting sites (Moulton, 1993; Nakamura and Miyazawa, 1997; Waterhouse, 1997) and while migrating (White and Stiles, 1992). Models have shown that bird dispersal is theoretically an effective vector for plant invasion (Moody and Mack, 1988), leading to numerous small populations that escape detection because of the scattered distribution. Seeds moving through the digestive system of birds often have the hard seed coat ruptured, facilitating germination.

Natural waterways may also be responsible for the movement of invasive plants. The previously discussed *Fallopia sachalinense* is known to spread downstream by fractured rhizomes during floods. Seeds may also be transported. One study documented that all populations of giant hogweed (*Heracleum mantegazzianum* Sommier & Levier) in the drainage of the Auschnippe River in Germany were descended from a single individual plant that was probably cultivated in 1982 in a garden next to the river (Schepker, 1998, as related in Kowarik, 2003). Those living near rivers and streams have a particular need to carefully manage their land.

Of greater concern for spreading invasive plants out of cities is movement by vehicles and recreation. A number of studies have documented that vehicles may carry a number of seeds (Clifford, 1959; Wace, 1977; Schmidt, 1989). Being on a vehicle, however, does not mean that it would necessarily be dislodged from it. A study in Berlin documented this using a motorway tunnel with a high wall dividing the lanes going out of the city with those going in (von der Lippe and Kowarik, 2007). Seed traps were put at ground level 150 m into the tunnels and periodically emptied. The contents were put on pots of soil and germinated in a greenhouse. The study found that 11,818 plants in 204 species were dislodged from the cars and landed in the traps during 1 yr. About one-half were not native to Berlin and 39 were known to be problematic somewhere in the world. Most of the species were not found near the tunnel and seed traps placed near the mouths of the tunnels at slightly higher levels did not have these species, so they were not likely entering by wind. This study nicely documented that cars moving from urban to wild areas are major vectors of invasive plants. As the seeds fall off vehicles along roadsides, they may establish populations if required growing conditions are met. Densities of invasive species are usually highest adjacent to the roads, but in a study in Indiana, individuals were found even 30 m into adjacent forests (Flory and Clay, 2006). Paved roads may have higher levels of invasive species, perhaps because of increased vehicle use leading to increased seed numbers or because water shed from the pavement increased soil moisture levels and improved germination (Gelbard and Belnap, 2003). Roadside vegetation management may control adjacent invasive plants, but unless it is done on a regular basis, plants will likely establish farther from the road and increase invasion into wildlands. Vehicles are not the only such vector moving urban plants out of cities. We often use recreational equipment such as mountain bikes and even ice chests in urban areas and use them again outside cities. Even small amounts of soil may contain many seeds. For instance, prairie grasslands may contain 300 to 800 seeds per square meter of soil and coniferous forests up to 1000 seeds (Silvertown et al., 1987).

What Can People in Urban Areas Do?

Urban locations can be considered "staging areas" for invasion into wildlands. The abundant landscaping, nurturing of plants in gardens, disturbance of soils, and movement of wind, birds, and vehicles from urban to wilder areas all increase the likelihood of invasion. As propagule pressure grows, the likelihood of seed dispersal into surrounding areas by cars or natural dispersal agents such as birds increases. It is important that people living in urban areas take responsibility for the plants they grow.

The problem may seem overwhelming, but there are several simple steps that people in cities can do to prevent or control invasions in their gardens, city parks and greenbelts, and surrounding areas. First, because so many invasive plants are introduced as garden plants, gardeners can make careful choices in selecting garden additions. Plant selection is a value-laden activity—we select our choices based on how the plants look and what they need to grow. Including potential for invasion is one more factor to consider when making the selection. Increasingly, government agencies, nonprofits, and even the garden centers are making lists of regional invasive species available. Because, as previously stated, few governments assess risk of invasion before a new species is introduced, it is especially important that urban gardeners consider it.

In 2001 a group of people got together in St. Louis to establish "codes of conduct," or best management practices, about invasive species for gardeners (Baskin, 2002). Known as the St. Louis Declaration, the codes of conduct for gardeners are summarized below.

- Plant only environmentally safe species in your gardens. Work toward and promote new landscape design that is friendly to regional ecosystems.
- Seek information on which species are invasive in your area. Sources could include botanical gardens, horticulturists, conservationists, and government agencies. Remove invasive species from your land and replace them with non-invasive species suited to your site and needs.
- Ask for only noninvasive species when you acquire plants. Do not trade plants with other gardeners if you know they are species with invasive characteristics.
- Request that botanical gardens and nurseries promote, display, and sell only noninvasive species. Volunteer at botanical gardens and natural areas to assist ongoing efforts to diminish the threat of invasive plants.
- Ask garden writers and other media to emphasize the problem of invasive species and provide information. Request that garden writers promote only noninvasive species.
- Help educate your community and other gardeners in your area through personal contact and in such settings as garden clubs and other civic groups. Invite speakers knowledgeable on the invasive species issue to speak to garden clubs, master gardeners, schools, and other community groups.
- Seek the best information on control of invasive plant species and organize neighborhood work groups to remove invasive plant species under the guidance of knowledgeable professionals.
- Participate in early warning systems by reporting invasive species you observe in your area. Determine which group or agency should be responsible for reports emanating from your area.

It should also be noted that many invasive garden plants have a number of cultivated varieties, or cultivars. *Cultivars* are selected genotypes that may have a unique growth habit, leaf shape, or color or flower or fruit characteristics from others in the species. These are given special names at the end of the species names, usually set apart with single quotation marks. Cultivars may be less invasive than the wild-type species, but this is usually not assured. In general, cultivars with variegated leaves grow more slowly and may be safer.

Because most invasions start in the urban staging areas, informed urbanites should note if they see a species growing wild that they have not previously detected and bring it to the attention of knowledgeable people. These can be agencies involved in noxious weed control, parks employees, botanical garden staff, university employees, or others. Several studies have shown that when control work begins when the invasions are small, it is possible to eradicate the invasion completely. If, however, the populations become large and numerous, the species will likely continue to spread. Municipal agencies, park departments, and nonprofits all assist with removal of invasive plants in urban parks and greenbelts. They usually welcome citizen volunteers. It can be very rewarding to return a park to greater native species composition and help to achieve Wendell Berry's rule of considering native organisms as members of our larger community.

We may be inevitably entering the Homogocene, but awareness about invasive plants and a few actions taken now can help us maintain our "sense of place." Through careful management, urban areas can maintain pockets of native vegetation in parks and greenbelts. Through careful plant selection, urban landscaping need not become a danger to the preservation of wildland plant and animal communities.

References

Aitkenhead-Peterson, J.A., M.K. Steele, and A. Volder. 2010. Services in natural and human dominated ecosystems. p. 373–390. *In* J. Aitkenhead-Peterson and A. Volder (ed.) Urban ecosystem ecology. Agron. Monogr. 55. ASA, CSSA, and SSSA, Madison, WI.

Baskin, Y. 2002. The greening of horticulture: New codes of conduct aim to curb plant invasions. Bioscience 52:464–471.

Berry, W. 1994. Sex, community freedom and economy: Eight essays. Pantheon, New York.

Castro, S.A., J.A. Figueroa, M. Muñoz-Schick, and F.M. Jaksic. 2005. Minimum resident time, biogeographical origin, and life cycle as determinants of the geographical extent of naturalized plants in continental Chile. Divers. Distrib. 11:183–191.

Clifford, H.T. 1959. Seed dispersal by motor vehicles. J. Ecol. 47:311–315.

Connell, J.H. 1978. Diversity in tropical rainforests and coral reefs. Science 199:1302–1310.

Corlett, R.T. 1992. The ecological transformation of Singapore, 1819–1990. J. Biogeogr. 19:411–420.

Culley, T., and N. Hardiman. 2009. The role of intraspecific hybridization in the evolution of invasiveness: A case study of the ornamental pear tree *Pyrus calleryana*. Biol. Invasions 11:1107–1119.

D'Antonio, C., T. Dudley, and M. Mack. 2000. Disturbance and biological invasions: Direct effects and feedbacks. p. 429–468. *In* L. Walker (ed.) Ecosystems of disturbed ground. Vol. 16. Elsevier Science, New York.

Deckers, B., K. Verheyen, M. Hermy, and B. Muys. 2005. Effects of landscape structure on the invasive spread of black cherry *Prunus serotina* in an agricultural landscape in Flanders. Belgium Ecogeogr. 28:99–109.

Dehnen-Schmutz, K., J. Touza, C. Perrings, and M. Williamson. 2007a. The horticultural trade and ornamental plant invasions in Britain. Conserv. Biol. 21(1):224–231.

Dehnen-Schmutz, K., J. Touza, C. Perrings, and M. Williamson. 2007b. A century of the ornamental plant trade and its impact on invasion success. Divers. Distrib. 13:527–534.

Dougherty, D., and S. Reichard. 2004. Factors affecting the control of *Cytisus scoparius* and restoration of invaded sites. Plant Prot. Q. 19:137–142.

Elton, C.S. 1958. The ecology of invasions by animals and plants. Methuen, London.

Flory, S.L., and K. Clay. 2006. Invasive shrub distribution varies with distance to roads and stand age in eastern deciduous forests in Indiana, USA. Plant Ecol. 184:131–141.

Gatehouse, H. 2008. Ecology of the naturalisation and geographic distribution of the non-indigenouse seed plant species of New Zealand. Ph.D. diss. Lincoln University, New Zealand.

Gelbard, J., and J. Belnap. 2003. Roads as conduits for exotic plant invasions in a semiarid landscape. Conserv. Biol. 17:420–432.

Harris, C., B. Murray, G. Hose, and M. Hamilton. 2007. Introduction history and invasive success in exotic vines introduced to Australia. Divers. Distrib. 13:467–475.

Hobbs, R. 2000. Land-use changes and invasions. p. 55–64. *In* H. Mooney and R. Hobbs (ed.) Invasive species in a changing world. Island Press, Washington, DC.

Kowarik, I. 1995. On the role of alien species in urban flora and vegetation. p. 85–103. *In* P. Pyšek et al. (ed.) Plant invasions—General aspects and special problems. Academic Publishing, Amsterdam, the Netherlands.

Kowarik, I. 2003. Human agency in biological invasions: Secondary releases foster naturalization and population expansion of alien plant species. Biol. Invasions 5:293–312.

MacDougall, A.S., and R. Turkington. 2005. Are invasive species the drivers or passengers of change in degraded ecosystems? Ecology 86:42–55.

Mack, R.N. 2000. Cultivation fosters plant naturalization by reducing environmental stochasticity. Biol. Invasions 2:111–122.

Mack, R.N., D. Simberloff, W.M. Lonsdale, H. Evans, M. Clout, and F.A. Bazzaz. 2000. Biotic invasions: Causes, epidemiology, global consequences and control. Ecol. Appl. 10:689–710.

Moody, M.E., and R.N. Mack. 1988. Controlling the spread of plant invasions: The importance of nascent foci. J. Appl. Ecol. 25:1009–1021.

Moulton, M.P. 1993. The all-or-none pattern in introduced Hawaiian passeriforms: The role of competition sustained. Am. Nat. 141:105–119.

Nakamura, H., and Y. Miyazawa. 1999. Movements, space use, and social organization of radio-tracked common cuckoos during the breeding season in Japan. Jpn. J. Ornithol. 46:23–54.

Pemberton, R.W., and H. Liu. 2009. Marketing time predicts naturalization of horticultural plants. Ecology 90:69–80.

Pimentel, D., R. Zuniga, and D. Morrison. 2005. Update on the environmental and economic costs associated with alien-invasive species in the United States. Ecol. Econ. 52:273–288.

Proctor, V.W. 1968. Long-distance dispersal of seeds by retention in the digestive tract of birds. Science 160:321–322.

Pyšek, P. 1998. Alien and native species in central European urban floras: A quantitative comparison. J. Biogeogr. 25:155–163.

Reichard, S. 1997. Preventing the introduction of invasive plants. p. 215–227. *In* J. Luken and J. Thieret (ed.) Assessment and management of plant invasions. Springer-Verlag, New York.

Reichard, S., and C.W. Hamilton. 1997. Predicting invasions of woody plants introduced into North America. Conserv. Biol. 11(1):193–203.

Reichard, S.H., L. Chalker-Scott, and S. Buchanan. 2001. Interactions among non-native plants and birds. p. 179–224. *In* J. Marzluff et al. (ed.) Avian ecology and conservation in an urbanizing world. Kluwer Academic Publ., Dordrecht, the Netherlands.

Sala, A., S. Smith, and D. Devitt. 1996. Water use by *Tamarix ramosissima* and associated phreatophytes in a Mojave Desert floodplain. Ecol. Appl. 6:888–898.

Schepker, H. 1998. Wahrnehmung, Ausbreitung und Bewertung von Neophyten. Ein Analyse der problematischen nichteinheimischen Pflanzenarten in Niedersachen. ibidem, Stuttgart, Germany.

Schmidt, W. 1989. Plant dispersal by motor car. Plant Ecol. 80:147–152.

Silvertown, H., M. Franco, and J. Harper. 1987. Plant life histories: Ecology, phylogeny, and evolution. Cambridge Univ. Press, Cambridge, UK.

Simberloff, D. 1986. Introduced insects: A biogeographic and systematic perspective. p. 3–24. In H.A. Mooney and J.A. Drake (ed.) Ecology of biological invasions in North America and Hawaii. Springer-Verlag, New York.

Urgenson, L.S., S.H. Reichard, and C.B. Halpern. 2009. Community and ecosystem consequences of giant knotweed (Polygonum sachalinense) invasion into riparian forests of western Washington, USA. Biol. Conserv. 142:1536–1541.

Vitousek, P.M., and L. Walker. 1989. Biological invasion by Myrica faya in Hawai'i: Plant demography, nitrogen-fizzation, ecosystem effects. Ecol. Monogr. 59:247–265.

Volder, A. 2010. Urban plant ecology. p. 179–198. In J. Aitkenhead-Peterson and A. Volder (ed.) Urban ecosystem ecology. Agron. Monogr. 55. ASA, CSSA, and SSSA, Madison, WI.

Waterhouse, R.D. 1997. Some observations on the ecology of the rainbow lorikeet Trichoglossus haematodus in Oatley, South Sydney. Corella 21:17–24.

Wace, N. 1977. Assessment of dispersal of plant species. The carborne flora in Canberra. Proc. Ecol. Soc. Aust. 10:167–186.

Whisenant, S.G. 1990. Postfire population dynamics of Bromus japonicus. Am. Midl. Nat. 123:301–308.

White, D.W., and E.W. Stiles. 1992. Bird dispersal of fruits of species introduced into eastern North America. Can. J. Bot. 70:1689–1696.

Wilcove, D.S., D. Rothstein, J. Dubrow, A. Phillips, and E. Losos. 1998. Quantifying threats to imperiled species in the United States. Bioscience 48:607–615.

von der Lippe, M., and I. Kowarik. 2007. Long distance dispersal of plants by vehicles as a driver of plant invasions. Conserv. Biol. 21:986–996.

13 🏛

Urban Riparian Function

Emilie K. Stander
Joan G. Ehrenfeld

Abstract

Urban riparian zones perform hydrological and biogeochemical functions that represent highly valued ecosystem services. In some cases human drivers unique to urban landscapes have been shown to alter functional capacity. However, the scientific literature on urban riparian function is currently too limited in functional and geographical scope to make broad generalizations. Functions related to nitrogen removal, exotic species invasions, and hydrologic alterations due to impervious surface cover have received the most research attention. Greenway planning, stream restoration, and new approaches to stormwater management hold promise for increasing urban riparian functional capacity.

*R*iparian zones, defined as the biotic ecosystems located on the banks of rivers and lakes (Naiman and Decamps, 1997), have been the subject of much research in the past four decades due to their unique location at the interface between terrestrial and aquatic landscapes. In accordance with their geomorphic position along river corridors, they tend to be linear, narrow landscape features. They are ecosystems defined by disturbance events such as floods, droughts, erosion, abrasion, and freezing (Naiman and Decamps, 1997) that reroute stream and drainage channels, accumulate new streambanks, and create heterogeneous floodplains with soils of mixed types (Flite et al., 2001). Plant and animal species have adapted to the dynamic conditions in riparian zones, thriving on the subsidies received from both neighboring upland and aquatic systems and surviving stressful conditions better than nonadapted species (Odum et al., 1979). These various factors result in a high degree of physical, chemical, and biotic heterogeneity, which in turn generates very high species densities and makes them among the most biodiverse of terrestrial ecosystems (Naiman et al., 1993).

Riparian zones are open ecosystems that both receive energy and materials from their adjacent uplands and receiving streams and export energy and materials to downstream ecosystems (Naiman and Decamps, 1997). Their position at the interface between upland terrestrial and aquatic systems means they function as

E.K. Stander, USEPA, Urban Watershed Management Branch, 2890 Woodbridge Ave., MS-104, Edison, NJ 08837 (stander.emilie@epa.gov); J.G. Ehrenfeld, Dep. of Ecology, Evolution, and Natural Resources, Rutgers University, 14 College Farm Rd., New Brunswick, NJ 08901 (ehrenfel@rci.rutgers.edu).

doi:10.2134/agronmonogr55.c13

boundary ecosystems (Cadenasso et al., 2008), serving as nexus or control points for the transfer of energy and materials from terrestrial uplands to receiving streams as moderated by hydrologic flows (Hedin et al., 1998). The combination of dynamic redox conditions and high inputs of organic matter (Norton and Fisher, 2000) results in complex and variable biogeochemical processes occurring within small areas (Hedin et al., 1998). Many functions performed by riparian wetlands represent societally important ecosystem services (Mitsch and Gosselink, 2000), including water quality maintenance through biogeochemical transformations, flood storage, carbon storage and sequestration, maintenance of biodiversity and wildlife habitat, maintenance of trophic structure and food webs, streambank stabilization, flood water storage and moderation of stream hydrographs, nutrient cycling processes, and the provision of recreational and aesthetic resources (Costanza et al., 1997; Naiman and Decamps, 1997; Bolund and Hunhammar, 1999; Norton and Fisher, 2000).

However, there are disturbances, unique to the urban setting, that alter structure in urban riparian zones (Alberti, 2005; Marzluff et al., 2008), raising the question of whether structural changes result in changes in functional capacity. Ehrenfeld (2000) found that different metrics for functional assessment and restoration were needed in urban wetlands because of their unique attributes. Some of these structural changes are caused by disturbances in the actual riparian zone, including ditches, berms, vegetation removal and management, creation of formal and informal trail systems, and dumping of trash, and some are caused indirectly by dimensions of the upland urban land cover, including impervious surfaces connected directly to streams and altered atmospheric chemistry (e.g., ozone formation and increased NO_x from vehicle emissions [Ehrenfeld et al., 2003; Stander, 2007]). Because of these unique urban disturbances, it is likely that urban riparian zones will not perform the same functions as nonurban riparian zones or perform those functions at the same level. Collins et al. (2000) warned that traditional ecological models cannot always be transferred directly to urban ecosystems because these models do not account for human drivers that may affect ecological patterns and processes.

In this chapter we review the existing literature on urban riparian function with an emphasis on drawing comparisons between function in urban versus nonurban riparian zones. We were surprised to find very few studies of specifically urban riparian function in the literature, a finding noted by others (Allan et al., 2008). Of the small number of studies we did find, most were related to NO_3^- removal function and altered biodiversity maintenance function (i.e., loss of native species, presence of exotic species). Many of the studies were conducted in the United States and for some functions were disproportionately located in the eastern United States. There is a clear need to expand the geographic and functional scope of the work that is being undertaken in urban riparian systems, and we will highlight areas where there are research gaps and opportunities to make important contributions to the field. We will review functions that urban riparian zones are likely to be performing, as well as those that may be lacking in urban systems compared with nonurban systems, and also functions that are less well represented in the literature, including the removal of nutrients and pollutants other than nitrate (NO_3^-). The chapter will cover functions related to hydrological and biogeochemical functions. We also highlight some management concerns related to urban riparian function.

Hydrological Functions

Maintenance of Hydrologic Regimes

Riparian ecosystems are highly prized for their role in maintaining the hydrologic regimes that shape freshwater aquatic systems. Surface and groundwater flows that pass through riparian zones influence the timing, frequency, and intensity of storm flows and maintain base flow levels in receiving streams, with cascading effects on aquatic habitat quality and water quality of surface water (Paul and Meyer, 2001). Urbanization has both direct and indirect impacts on the role of riparian zones in maintaining hydrologic regimes. In some cases these changes have been well documented in the scientific literature; in others, more research is needed to determine functional levels provided by riparian systems in urban landscapes.

A number of studies have documented the impacts of impervious surface cover and stormwater management practices on hydrology in riparian systems and their receiving streams. Impervious surfaces, such as roads and roofs, prevent rainfall from infiltrating into the groundwater, as would occur in natural ecosystems (Arnold and Gibbons, 1996; Brabec et al., 2002). In natural systems stormwater that infiltrates into the groundwater discharges to streams slowly over a period of weeks or months. In urban areas rainfall is instead collected in storm drains and channeled immediately to streams, where large stormwater volumes, or peak flows, erode stream banks (Chapters 14 [Burian and Pomeroy, 2010] and 15 [Steele et al., 2010], this volume; Booth 1990; Booth and Jackson, 1997). This set of changes, collectively termed *urban stream syndrome* (Paul and Meyer, 2001; Meyer et al., 2005; Walsh et al., 2005a,b), draws down water tables in adjacent riparian zones, causing groundwater-borne nutrients and pollutants from uplands to flow beneath the biologically active surface soils, where many removal processes occur, and discharge to streams (Fig. 13–1; Groffman et al., 2002; Ehrenfeld et al., 2003; Stander and Ehrenfeld, 2009a). The riparian zone is thus disconnected from the uplands and the stream (Groffman et al., 2003). Other direct hydrological alterations common in the urban setting, such as ditches and berms, can also result in hydrological disconnection (Fig. 13–2; Ehrenfeld and Schneider, 1991; Booth and Jackson, 1997; Ehrenfeld, 2000; Ehrenfeld et al., 2003).

Riparian zones are typically valued for their maintenance of groundwater flows and discharges to streams, contributing to characteristic stream baseflow

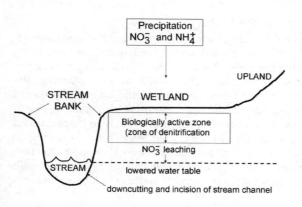

Fig. 13–1. Conceptual diagram demonstrating the effects of urban altered hydrology on water tables and NO$_3^-$ removal in urban riparian wetlands.

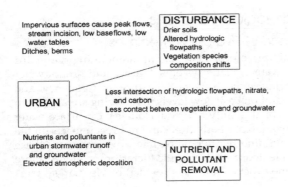

Impervious surfaces cause peak flows, stream incision, low baseflows, low water tables
Ditches, berms

DISTURBANCE
Drier soils
Altered hydrologic flowpaths
Vegetation species composition shifts

Fig. 13–2. Relationships between urbanization and natural riparian characteristics that explain potentially lower capacity for nutrient and pollutant removal in urban riparian wetlands.

URBAN

Less intersection of hydrologic flowpaths, nitrate, and carbon
Less contact between vegetation and groundwater

Nutrients and polluntants in urban stormwater runoff and groundwater
Elevated atmospheric deposition

NUTRIENT AND POLLUTANT REMOVAL

conditions (Matteo et al., 2006). In urban landscapes less infiltration of precipitation as a result of paving of groundwater recharge areas and deforestation at the watershed and riparian scales results in a regional lowering of the groundwater, which may also be responsible for lower water tables in urban riparian zones and lower baseflows in streams (Arnold and Gibbons, 1996; Groffman et al., 2003). Most of the research on urban baseflows has been focused on urbanization at the watershed scale (Rose and Peters, 2001) rather than the role of urban riparian zones in particular.

Streambank and channel stability are another important function of riparian systems that contributes to the maintenance of hydrologic regimes in receiving streams. Severe streambank erosion and resulting loss of property has often been ascribed in part to riparian deforestation (Simon and Steinemann, 2000; Bledsoe and Watson, 2001). Riparian vegetation, especially tree roots, stabilize streambanks, thus minimizing erosion and channel enlargement, to a degree dependent on vegetation type and density (Henshaw and Booth, 2000; Florsheim et al., 2008). Riparian forest and grass vegetation also provides roughness that slows surface flows, thus reducing peak flows and bank erosion (Bledsoe and Watson, 2001; McBride and Booth, 2005). Streambank revegetation is often included in stream restoration practices, with some documented success for reducing bank erosion and increasing slope stability (Simon and Steinemann, 2000), although this has not proven to have cascading positive effects on the biota of receiving streams (Sudduth and Meyer, 2006).

Water Storage

The value of riparian forests for reducing flood magnitude and damage has been recognized for a long time. In China, laws required the planting of trees along the moats around cities during the Zhou Dynasty (1100–770 BCE; Yu et al., 2006), and the law was supported by observations of the value of these plantings (by the thinker Guan Zhi, 770 BCE); there followed a long history of explicit recognition of the value of riparian vegetation and revegetation for flood management (Yu et al., 2006). Water storage is another function that is likely to be performed by urban riparian zones. Municipalities in the United States often protect riparian zones and other wetlands in urban and suburban areas for flood protection. This is a particularly important function in watersheds that are urbanizing, with more land being converted to impervious surfaces, as well as in older urban areas that already contain a high percentage of impervious surfaces and thus are more vulnerable to flooding during intense storms. In areas predicted to have more intense storms as a result of climate change, this will be an increasingly important function. The percentage of municipal lands performing this function that are specifically urban riparian zones has not been quantified, nor has the level of function that they are performing. The water storage capacity of urban riparian zones is therefore another important knowledge gap that deserves research effort.

Biogeochemical Functions

Nutrient and Pollutant Removal for Water Quality Protection

Riparian zones have received much attention for their ability to remove nutrients and pollutants emanating from agricultural land uses (Lowrance et al., 1984; Cooper, 1990; Haycock and Pinay, 1993; Hill, 1996; Norton and Fisher, 2000; Hey, 2002; Hey et al., 2005; Mayer et al., 2007). More recently there has been increased interest in pursuing riparian buffers as a management tool for reducing nonpoint-source pollution from urban land uses (Groffman and Crawford, 2003; Hoffman and Baattrup-Pedersen, 2007). This pollution reaches urban riparian zones in the form of nutrient and pollutant loads in stormwater runoff that washes off impervious surfaces (USEPA, 1983; Faulkner, 2004). Urban stormwater and groundwater has been shown to contain varying loads of nutrients and pollutants, including heavy metals, sediments, pathogens, pesticides, and pharmaceuticals (USEPA, 1983; USGS, 1999). If urban riparian zones are able to perform nutrient and pollutant removal functions as successfully as agricultural riparian buffers reportedly have, they could be used to manage water quality in urban streams.

Nitrogen

The vast majority of research on the nutrient and pollutant removal capability of urban riparian zones has focused on NO_3^- removal. The focus on NO_3^- is a direct outgrowth of the high mobility of NO_3^- in soils, which has fostered a rich body of research on NO_3^- removal in riparian buffers between agricultural uplands and streams. Nitrate transported in streams is considered a primary cause of coastal eutrophication (Conley et al., 2009). Diaz and Rosenberg (2008) reported 405 hypoxic zones globally, including large estuaries like the Chesapeake Bay and the Gulf of Mexico in the United States and the Baltic, Black, and northern Adriatic Seas in Europe. Increasingly, growing urban land cover is emerging as an important source of nutrients, particularly NO_3^-, to receiving waters (USGS, 1999). If riparian zones in urban watersheds can perform the same NO_3^- removal functions as their analogs downgradient of agricultural land uses, urban riparian ecosystems could be put forward as a management strategy for decreasing NO_3^- pollution in urban streams. The specific mechanisms for NO_3^- removal are denitrification (the microbial transformation of NO_3^- to nitrogen gases that are released to the atmosphere), plant uptake, microbial immobilization (NO_3^- assimilation into microbial biomass), storage in soil organic matter, and groundwater mixing (Lowrance et al., 1997). Of these mechanisms, denitrification is the only one that represents a permanent removal of NO_3^- from the system. In all the other mechanisms, NO_3^- is stored within the system and can become remobilized under changing conditions (Groffman and Crawford, 2003).

To legitimately use urban riparian zones as a management tool for reducing NO_3^- pollution in receiving streams, it is essential to confirm that urban riparian zones are able to carry out NO_3^- removal functions and prevent the release of NO_3^- to receiving streams. Very few studies have attempted to do this, with varying results (Hanson et al., 1994a; Addy et al., 1999; Groffman et al., 2002; Groffman and Crawford, 2003; Stander, 2007; Kaushal et al., 2008; Stander and Ehrenfeld, 2009a,b). Some studies have found that, contrary to the NO_3^- sink characteristics of agricultural riparian buffers, in some cases urban riparian zones may be releasing NO_3^- to receiving streams, and thus acting as sources of NO_3^- in urban landscapes (Groffman et al., 2002; Stander, 2007). The aforementioned hydrologic changes that occur in urban landscapes are the likely cause of NO_3^- export from some urban riparian zones. In other cases, urban riparian zones display high levels of NO_3^- retention (Hanson et al., 1994a; Addy et al., 1999; Stander, 2007). In some urban riparian zones in northern New Jersey, water tables were located close to the soil surface in the shallow, biologically active surface soils, similar to water tables in less disturbed, forested riparian zones (Stander and Ehrenfeld, 2009a), and the explanation for the lack of hydrologic disconnection at these sites is not clear. Simplified input–output budgets, based on NO_3^- inputs in throughfall and outputs through leaching to the shallow groundwater, suggested that two urban riparian wetlands in New Jersey were acting as NO_3^- sources to receiving streams, but two others were acting as sinks (Stander, 2007). Here denitrification was considered to be a mechanism for retention and was not counted as an output. In a grassy riparian wetland located in a residential backyard in Rhode Island, NO_3^- bearing groundwater moved through deep soils containing patches of organic matter that served as hotspots for denitrification (Addy et al., 1999). In another riparian zone in Rhode Island, located downslope of a dense residential housing development, rates of net N mineralization, nitrification, litter N content, and soil inorganic N pools were higher than in a paired site located downgradient of an undeveloped, forested upland, and were thus displaying symptoms of N saturation. However, a simplified input–output budget based on estimated groundwater NO_3^- as inputs and estimated outputs from denitrification suggested that the suburban riparian wetland was still functioning as a NO_3^- sink (Hanson et al., 1994a).

None of these studies conducted a complete mass balance analysis of NO_3^- inputs and outputs, but rather they measured processes related to NO_3^- retention, such as rates of denitrification (actual and potential), net nitrification, net N mineralization, microbial biomass N, and NO_3^- leaching. Table 13–1 provides a summary of nitrogen cycling data from all reviewed studies of urban riparian wetlands and a selection of studies from nonurban riparian wetlands, as well as a range of urban and nonurban ecosystem types, for the purposes of comparison. Actual rates of denitrification have been measured in urban riparian surface soils and near-stream sediments in five studies located in the Piedmont physiographic province of the northeastern United States (Hanson et al., 1994b; Addy et al., 1999; Groffman et al., 2002; Kaushal et al., 2008; Stander and Ehrenfeld, 2009a; Table 13–1). Denitrification rates measured in urban riparian wetlands were often lower than rates reported for nonurban riparian wetlands (Ettema et al., 1991; Addy et al., 2002; Clement et al., 2003; Kellogg et al., 2005), although in some cases they were in the same range (Lowrance et al., 1984; Lowrance, 1992; Groffman et al., 1996; Jacinthe et al., 1998; Clement et al., 2002). In cases where denitrification rates

Table 13–1. Summary of N cycling function rates in a range of urban and nonurban ecosystem types.

Ecosystem type/location	Land cover type[†]	Denitrification[‡] Actual $\mu g\ N\ kg^{-1}\ soil\ d^{-1}$	Denitrification[‡] Potential $\mu g\ N\ kg^{-1}\ soil\ h^{-1}$	Nitrification $mg\ N\ kg^{-1}\ soil\ d^{-1}$	N mineralization $mg\ N\ kg^{-1}\ soil\ d^{-1}$	Reference
Riparian wetlands/Rhode Island	SU	7–43[b]	0.0–0.13 C horizon soil matrix; 0.0–2.5 patch material			Addy et al. (1999)
Riparian wetlands/Rhode Island	SU		157–1064	<0.0–1.1	0.03–0.94	Groffman et al. (1992)
Riparian wetlands/Maryland	U/SU	0.7–7[a]	29–2196		0.35	Groffman et al. (2002)
Riparian wetlands/Maryland	U/SU		230–7590	0.15		Groffman and Crawford (2003)
Riparian wetlands/Rhode Island	SU	0.15–1.2[a,e,f]		0.4–1.0	0.75–0.9	Hanson et al. (1994a)
Riparian wetlands/Rhode Island	SU	0.1–262[d]				Hanson et al. (1994b)
Riparian wetlands/Maryland	U/SU					Kaushal et al. (2008)
Riparian wetlands/New Jersey	U/SU	0–27[a]		<0.1–0.9	0.1–0.8	Stander and Ehrenfeld (2009a)
Non-riparian wetlands/New Jersey	U/SU	0–60[e]		<0.1–0.6	<0.1–1.8	Stander and Ehrenfeld (2009a)
Non-riparian wetlands/New Jersey	U/SU			0.494–10.514	10.00–12.31	Zhu and Ehrenfeld (1999)
Retention basins/Arizona	U	0.3–5.2[a,e,g]	390–1151	0.186–0.642	0.199–0.668	Zhu et al. (2004)
Upland forest/Maryland	U			<0.1–0.5	0.1–0.5	Groffman et al. (2006)
Upland forest/New York	U			0–1.4	0–2.3	Zhu and Carreiro (2004)
Riparian wetlands/Rhode Island	NU	60–185[d]				Addy et al. (2002)
Riparian wetlands/northwestern France	NU	12–20[a,e,f]	4.2–270	–0.1–1.4	0.1–1.4	Clement et al. (2002)
Riparian wetlands/northwestern France	NU	186–689[a]				Clement et al. (2003)
Riparian wetlands/Georgia	NU	8.8–83.5[a]				Ettema et al. (1991)
Riparian wetlands/Pennsylvania	NU		0–210			Flite et al. (2001)
Riparian wetlands/Rhode Island	NU		52–1084	–0.03–2.66		Groffman et al. (1992)
Riparian wetlands/Rhode Island	NU	0.00–4.10[e]				Groffman et al. (1996)
Riparian wetlands/Rhode Island	NU			–0.1–0.1	0.1–0.4	Hanson et al. (1994a)
Riparian wetlands/Rhode Island	NU	<0.1–3[b]	0.079 C horizon soil matrix; 0.41–7.3 patch material			Jacinthe et al. (1998)

Table continued.

Table 13–1. Continued.

Ecosystem type/location	Land cover type†	Denitrification‡		Nitrification	N mineralization	Reference
		Actual	Potential			
		μg N kg⁻¹ soil d⁻¹	μg N kg⁻¹ soil h⁻¹	mg N kg⁻¹ soil d⁻¹	mg N kg⁻¹ soil d⁻¹	
Riparian wetlands/Rhode Island	NU	<1–330[d]				Kellogg et al. (2005)
Riparian wetlandsGeorgia	NU	0.66[a,e,f]				Lowrance et al. (1984)
Riparian wetlands/Georgia	NU	28.95[a]	0–4900			Lowrance (1992)
Non-riparian wetlands/Minnesota	NU				0.73–6.6	Bridgham et al. (1998)
Non-riparian wetlands/New Jersey	NU			0.000–0.041	3.61–4.94	Zhu and Ehrenfeld (1999)
Upland forest/Michigan	NU	<0.1–1.3[a,e,f]				Groffman et al. (1993)
Upland forest/New York and Connecticut	NU			0–0.268	0–2.1	Zhu and Carreiro (2004)

† U, urban; SU, suburban; NU, nonurban.

‡ Potential denitrification rates measured by denitrification enzyme activity. Actual denitrification rates were measured using the following methods: (a)C_2H_2 block, (b) mesocosm N gas production, (c) Br⁻ plume, (d) push–pull; when converting areal-based denitrification rates to volume-based rates, assumed: (e) 10-cm depth, (f) 1.1 g cm⁻³ bulk density, (g) 1.1 g cm⁻³ bulk density, (f) 1.3 g cm⁻³ bulk density.

were lower in urban than in nonurban wetlands, the likely culprit is hydrological disconnection. Denitrification is an anaerobic process that requires the saturated conditions typically found in highly functioning, hydrologically connected wetlands. The surface soils, usually defined as the top 30 cm, are where most of the biological activity occurs because this is the location of the root zone and the most active microbial populations (Chapter 7, Pouyat et al., 2010, this volume). When lower water tables cause these soils to become drier than they were before urbanization, denitrification is inhibited (Fig. 13–1, 13–2), as demonstrated in the Maryland and New Jersey studies (Table 13–1; Groffman et al., 2002; Stander and Ehrenfeld, 2009a), as well as in an agricultural riparian zone with an incised stream (Bohlke et al., 2007). Kaushal et al. (2008) found that hydrologic reconnection of urban riparian zones through streambank restoration resulted in higher denitrification rates in near-stream sediments in restored sites compared with unrestored sites. Restored sites had denitrification rates comparable to or higher than nonurban riparian zones (Ettema et al., 1991; Groffman et al., 1996; Addy et al., 2002; Clement et al., 2003; Kaushal et al., 2008). Denitrification rates in urban riparian wetlands were similar to other nonriparian urban ecosystems (Table 13–1), including stormwater retention basins (Zhu et al., 2004) and nonriparian wetlands (Stander and Ehrenfeld, 2009a), which is not surprising given that other urban ecosystems are subjected to similar types of hydrologic disturbances that may impact denitrification rates. Rates were comparable with nonurban upland forests (Groffman et al., 1993), which is also expected because the aerobic soils of upland forests support lower rates of denitrification than the naturally saturated soils of undisturbed wetlands.

Several studies measured rates of potential denitrification using denitrification enzyme activity (DEA) methodology. In this technique soil samples collected in the field are incubated in anaerobic conditions in the laboratory, and a source of C and a supply of NO_3^- are added to produce ideal conditions for denitrifying bacteria in the soil. These measurements thus represent the maximum denitrification rates that could be expected in that particular soil given its texture and other defining characteristics, when the main drivers of denitrification—anaerobicity, C, and NO_3^-—are not limiting. Potential denitrification rates in urban and suburban riparian wetlands were well within the range of nonurban riparian wetlands and urban stormwater retention basins (Table 13–1; Groffman et al., 1992; Lowrance, 1992; Zhu et al., 2004). In some cases urban rates were considerably higher than nonurban rates (Flite et al., 2001; Clement et al., 2002; Jacinthe et al., 1998). Groffman and Crawford (2003) found that potential denitrification rates were more variable in urban than in rural soils, but this is likely due to the wider range of soil moisture and organic matter conditions in the urban sites (Groffman and Crawford, 2003). Mean DEA in patches of organic matter (i.e., hotspots) in Rhode Island riparian wetlands were much higher than in the soil matrix (Addy et al., 1999), demonstrating the importance of organic matter, in addition to anaerobicity, as a control on denitrification rates. The fact that potential denitrification rates in urban riparian wetlands are similar to nonurban wetlands suggests that urban soils retain the potential for high levels of denitrification function. If the proper conditions, particularly anoxia and high organic matter content, could be restored to urban riparian wetlands, they could resume their role as NO_3^- sinks in urban landscapes.

There are a number of other site- and landscape-level factors known to control denitrification function in undisturbed riparian systems that may or may not play an important role in urban systems. Soil characteristics are an important site-level driver of anaerobic conditions in riparian wetlands, and they also regulate the availability of C, which serves as an electron donor to drive the denitrification reaction. The presence of hydric soils, characterized by fine texture (i.e., clayey soils), fine pore structure, low hydraulic conductivity, and low redox potential, is typically considered a useful indicator of high denitrification capacity (Kellogg et al., 2005). In urban riparian zones soils are often coarser textured than in nonurban riparian systems (Ehrenfeld et al., 2003), and coarse-textured soils may allow N inputs to be leached to the groundwater and discharged to the receiving stream because of the fast infiltration and low contact time between N in infiltrating water and soil. Several studies in nonurban riparian zones found lower rates of NO_3^- removal in riparian wetlands with coarse-textured soils (Cooper, 1990; Hanson et al., 1994b; Clausen et al., 2000). However, coarser soils at the watershed scale allow for subsurface delivery of NO_3^- from uplands to riparian zones, and NO_3^- removal is more effective in subsurface than surface flows, according to studies in nonurban riparian systems (Lowrance et al., 1984; Peterjohn and Correll 1984; Osborne and Kovacic, 1993; Norton and Fisher, 2000; Mayer et al., 2007). Although subsurface soils typically have lower denitrification potential than surface soils as a result of low C availability and less biological activity (Lowrance, 1992; Kellogg et al., 2005), buried soil horizons and patches of organic matter deposits can serve as hotspots for denitrification even in subsurface soils, coarse-textured soils, or soils with low denitrification rates in the general soil matrix (Gold et al., 1998; Jacinthe et al., 1998; Addy et al., 1999; Hill et al., 2000; McClain et al., 2003; Kellogg et al., 2005). The availability of NO_3^- is another important site-level controlling factor on denitrification rates. Elevated rates of atmospheric N deposition in more intensely urban landscapes (Lovett et al., 2000; Stander, 2007) combined with high loads and/or concentrations of NO_3^- in stormwater runoff, urban groundwater, and surface waters that flood riparian zones (USEPA, 1983; USGS, 1999) suggest that NO_3^- should be available in urban riparian systems. However, hydrologic flowpaths must transport that NO_3^- to soils with enough labile C and the proper redox conditions to allow denitrification to occur. The width of the riparian buffer zone has been hypothesized as a factor regulating the degree to which NO_3^-, soil C, and hydrologic flowpaths can intersect. However, Mayer et al. (2007) found that buffer width explained only a small amount (9%) of variation in NO_3^- removal effectiveness in nonurban riparian systems (Table 13–2). These relationships have not been tested in urban riparian systems.

A number of landscape-level characteristics and properties also affect denitrification. These include hydrogeomorphology, topography, and seasonal and climatic effects (Cirmo and McDonnell, 1997; Naiman and Decamps, 1997; Mayer et al., 2007). Vidon and Hill (2004) suggested using maps of hydrogeomorphic classes to predict NO_3^- retention capacity because hydrogeomorphology is a predictor of hydrological flow paths and retention times. However, the predictive power of hydrogeomorphology may be diminished in urban settings where dry riparian soils and lowered water tables caused by hydrological disconnection may override the importance of hydrogeomorphology in controlling NO_3^- removal processes like denitrification (Stander and Ehrenfeld, 2009a,b). Steeper topography affects NO_3^- removal capacity by delivering more NO_3^- to riparian wetlands

Table 13–2. Patterns in NO$_3^-$ removal effectiveness of riparian buffers of varying width, vegetation type, and hydrologic inputs (modified from Mayer et al., 2007).

Buffer variable	N	Mean removal effectiveness	R^2	p
		% ± 1 SE		
All studies	88	67.5 ± 4.0	0.09	0.005
Width category				
0–25 m	45	57.9 ± 6.0	0.01	0.5
26–50 m	24	71.4 ± 7.8	0.00	0.8
>50 m	19	85.2 ± 4.8	0.03	0.5
Flowpath				
Surface	23	41.6 ± 7.1	0.21	0.03
Subsurface	65	76.7 ± 4.3	0.02	0.3
Vegetation type				
Forest	31	72.2 ± 6.9	0.04	0.3
Forested wetland	7	85.0 ± 5.2	0.00	1.0
Herbaceous	32	54.0 ± 7.5	0.21	0.009
Herbaceous/forest	11	79.5 ± 7.3	0.39	0.04
Wetland	7	72.3 ± 11.9	0.01	0.9

via surface runoff, resulting in less opportunity for denitrification (Lowrance et al., 1997). Denitrification can also be influenced by climatic and seasonal effects. Soil temperatures and precipitation patterns, which historically follow climatic and seasonal patterns, are altered in urban landscapes by the urban heat island effect (Chapter 2, Heisler and Brazel, 2010, this volume) and climate change. The consequent affects on biogeochemical processes and thus NO$_3^-$ retention dynamics have not been studied in urban riparian systems. Seasonal human behaviors can also affect NO$_3^-$ removal dynamics. In agricultural settings pulses of NO$_3^-$ exports often coincide with fertilizer applications (Peterjohn and Correll, 1984; Verchot et al., 1997). The effects of seasonal turfgrass fertilizer applications in urban landscapes on denitrification and NO$_3^-$ removal in urban riparian zones have not been studied to our knowledge.

Rates of nitrification, the microbial production of NO$_3^-$ from NH$_4^+$, have also been cataloged in urban riparian wetlands in Maryland, New Jersey, and Rhode Island (Table 13–1). In disconnected wetlands, nitrification rates would be expected to be higher than in less disturbed wetlands with saturated soils (Groffman et al., 2002; Stander and Ehrenfeld, 2009a,b). Nitrification is an aerobic microbial process that is facilitated in the drier, oxygenated soils often found in disconnected urban riparian wetlands. For the most part, ranges of nitrification rates measured in urban riparian soils are comparable to nonurban riparian soils (Table 13–1), although individual studies found significantly higher rates in urban riparian versus urban nonriparian wetlands (Stander and Ehrenfeld, 2009a) and urban riparian versus nonurban riparian wetlands (Hanson et al., 1994a), as expected. Ranges of nitrification rates in urban, nonwetland soils were also similar to urban riparian soils (Table 13–1; Zhu and Carreiro, 2004; Zhu et al., 2004; Groffman et al., 2006). The highest rates were found in nonriparian urban wetland soils (Table 13–1; Zhu and Ehrenfeld, 1999). Similar patterns were shown in N mineralization rates (Table 13–1), with comparable ranges of rates in urban riparian wetlands and nonurban riparian wetlands (Zhu and Ehrenfeld, 1999) and urban retention basins and upland forests. Zhu and Ehrenfeld (1999) also recorded higher N mineralization rates in nonriparian versus riparian wetlands in urban landscapes (Table 13–1).

Other indicators of N saturation and enrichment that could result in NO_3^- releases from urban riparian wetlands have been documented in the same set of urban riparian studies. Soil inorganic N levels are higher in some urban sites than in nonurban sites (Hanson et al., 1994a; Groffman et al., 2002), but not in other urban sites (Groffman and Crawford, 2003). Hanson et al. (1994a) found higher N concentrations in herbaceous and red maple (*Acer rubrum* L.) leaf litter at the enriched suburban riparian site compared with the control, undeveloped site. Microbial biomass N was also higher at the enriched site (Hanson et al., 1994a), but Groffman and Crawford (2003) found no differences in microbial biomass C or N between urban and rural riparian wetlands in Baltimore, MD. Nitrate concentrations in soil leachate in New Jersey urban riparian wetlands ranged from below detection to 5.5 mg L^{-1}, while leachate from forested upland urban soils in Maryland ranged from 0.1 to 12 mg L^{-1} NO_3^- (Groffman et al., 2006; Stander, 2007).

Plant uptake is another important NO_3^- removal mechanism. Nitrate uptake by canopy trees may explain higher NO_3^- concentrations in soil and groundwater, as well as high potential rates of N mineralization and nitrification in the N-enriched riparian zone downslope of a suburban residential development in Rhode Island (Hanson et al., 1994a). Nitrate uptake increased foliar N content, resulting in an increase in litter N content. Higher litter N content stimulated higher rates of N mineralization and nitrification. Groundwater NO_3^- removal by plant uptake depends on shallow groundwater flow, perhaps above impermeable soil layers, intersecting the root zone, and a strong root–soil connection (Lowrance et al., 1997; Yeakley et al., 2003). Vegetation type also impacts NO_3^- removal capacity. Kansiime et al. (2007) found that changes in plant community composition in Ugandan riparian wetlands near Kampala City and other urban areas potentially affected N removal because native vegetation (papyrus [*Cyperus giganteus* Vahl]) could take up and store more N than other plant community types. A popular question in riparian research is the relative effectiveness of forested versus grassy riparian buffers for NO_3^- removal. Most researchers hypothesize that while grass buffers are able to remove sediment and particulate N through trapping, they are less effective at removing dissolved N than forest buffers (Daniels and Gilliam, 1989; Verchot et al., 1997). In a meta-analysis of the largely nonurban riparian literature on NO_3^- removal, Mayer et al. (2007) found that the effectiveness of grass buffers increases with buffer width in a nonlinear fashion, while NO_3^- removal in forested buffers was not correlated with buffer width (Table 13–2). Some studies have found poor removal in nonurban grass buffers (Osborne and Kovacic, 1993; Hubbard and Lowrance, 1997), while others have found them to be effective (Haycock and Pinay, 1993; Schnabel et al., 1996; Clausen et al., 2000; Lee et al., 2003; Schoonover and Williard, 2003), depending on soil type and depth of confining layers (Vidon and Hill, 2004). Other studies, including urban riparian zones in Baltimore (Groffman and Crawford, 2003), have found no appreciable difference between forest and grass buffers for NO_3^- removal function, although in one case this may have been due to the presence of tree roots in the subsoil of the grassed sites (Addy et al., 1999).

The vast majority of studies on urban riparian functions related to NO_3^- removal have been conducted in the Piedmont physiographic province of the northeastern United States. In this small group of studies, some urban riparian wetlands have shown symptoms of N saturation and may be acting as sources of NO_3^- to receiving streams, while others still display seasonally high water tables and rates of nitrification and denitrification similar to riparian wetlands located in less developed landscapes. There is clearly a need, however, to expand these studies to other physiographic regions, nontemperate zones, arid and semiarid regions, and indeed other parts of the world, to develop more detailed and complete N budgets for urban riparian systems and to specifically relate N processes, including denitrification, to hydrologic changes resulting from urban land use. Other mechanisms for NO_3^- retention, including plant and microbial uptake, also deserve more research attention in urban riparian zones.

Phosphorus

Phosphorus is probably the best studied pollutant after N, as it is an important export from agricultural land uses and increasingly from urban landscapes as well (Chapter 15, Steele et al., 2010, this volume; USGS, 1999). It is critical to note that on a per unit basis, P has a greater impact on freshwater aquatic systems than N (Lowrance et al., 1997) and is increasingly associated with coastal eutrophication (Conley et al., 2009). Both natural and anthropogenic sources of P contribute to P loads in receiving streams. Phosphorus is naturally found in shallow groundwater and in soils; both can serve as P sources to receiving streams (Sonoda and Yeakley, 2007). At the site scale, soil properties, particularly pH, anaerobicity, and Fe and Al content, exert the most control on P availability to plants and microbes, as well as P retention and release to groundwater and streams (Hogan and Walbridge, 2007). In the Willamette River basin, Sonoda and Yeakley (2007) compared P retention ability in urban and nonurban riparian soils and found that urban soils had higher pH and lower concentrations of Fe and Al, causing the urban soils to be less retentive of P than nonurban soils in this river basin. Moving from the nonurban headwaters to lower, more urban locations in the watershed, soluble reactive P (SRP) concentrations and SRP and total P loadings were higher in streams and beneath-stream groundwater. Their results also suggest that both groundwater and P-enriched surface runoff from upland urban impervious surfaces contribute to these higher stream values. Deforestation of riparian zones in urban catchments and a resulting decrease in biological P uptake could also be contributing to higher P concentrations and loadings in urban streams (Sonoda and Yeakley, 2007). Hogan and Walbridge (2007) found the highest P concentrations in surface soil and plant tissues and soil P saturation in urban riparian wetlands located in watersheds with intermediate levels of urban intensity in northern Virginia. Total P concentrations were also highest in urban wetland soils at intermediate urban intensities in the Puget Sound Basin (Horner et al., 2001). This pattern is likely explained by reduced P inputs at higher urban intensities due to hydrologic disconnection and the routing of wastewaters through sewage systems that export water directly to coastal waters (Hogan and Walbridge, 2007).

Phosphorus in stormwater runoff from impervious surfaces is often bound to sediment and suspended particles. In urban watersheds sediment export is common due to soil erosion; this is especially common in urbanizing watersheds because of erosion at construction sites, in contrast to older urban watersheds with more stable land uses (Trimble, 1997; Groffman et al., 2003; Faulkner, 2004). Erosion rates in urbanizing watersheds are remarkably higher than in agricultural or undisturbed watersheds (50,000 Mg km^{-2} yr^{-1} in urban, 4000 Mg km^{-2} yr^{-1} in agricultural, <100 in undisturbed) (Faulkner, 2004). A high proportion of annual export can occur during very high flow events (Royer et al., 2006). In agricultural landscapes high P discharges often occur in the spring and summer in association with fertilizer inputs and intense storms (Peterjohn and Correll, 1984). Other urban sources of P to receiving streams include fertilizers, septic fields, leaking sanitary sewers, and illicit connections between sanitary sewers and stormwater drainage systems (Paul and Meyer, 2001). Osborne and Wiley (1988) found that even though urban land covered only 5% of the mostly agricultural Salt Fork watershed in Illinois, it was the controlling factor on concentrations of dissolved P in receiving streams year round. Phosphorus adsorbed in soils in upland land uses, either from fertilizer use or from septic fields, can be remobilized following a change in conditions, such as a change in pH, anoxia, vegetation structure, or soil erosion, and be released to streams (Paul and Meyer, 2001).

Phosphorus management is more challenging than N. The main mechanisms for P removal include particulate P trapping by vegetation, sediment removal, infiltration, and consequent adsorption to soil and clay particles, and vegetative and microbial uptake (Lowrance et al., 1997; Faulkner, 2004). None of these removal mechanisms are permanent, as the P is stored within the system and is vulnerable to remobilization. There is no microbial transformation analogous to denitrification that is capable of permanent P removal, and this presents a major challenge to P management through the use of riparian zones, urban and otherwise (Lowrance et al., 1997). The best approach is to maximize sediment control and infiltration and promote vegetative uptake. Often grassy riparian zones are targeted for P removal due to their increased ability to trap sediments compared to forested systems (Lowrance et al., 1997; Liu et al., 2008). However, forested riparian zones in agricultural settings have demonstrated success in sediment and P retention (Cooper et al., 1987; Gilliam, 1994; Craft and Casey, 2000). Because urban riparian zones receive higher amounts of runoff and higher concentrations of P in runoff, sediment trapping is likely to be a more important means of P removal in urban riparian zones.

Heavy Metals

Urban wetlands and detention basins are known to contain higher concentrations of heavy metals, including Pb, Zn, Cu, Cr, Cd, and Ni, in their soils and sediments than their nonurban analogs (Parker et al., 1978; Horner et al., 2001; Faulkner, 2004), and within urban watersheds, wetland soils in watersheds dominated by industrial land uses have higher metals concentrations than watersheds dominated by residential land uses (Murray et al., 2004). Like P, heavy metals are often transported in urban stormwater runoff via adsorption to sediment particles. As a result, sediment trapping and removal and infiltration are important removal mechanisms for heavy metals (Faulkner, 2004). The dynamics of heavy metal occurrence and removal in urban riparian zones is underrepresented in the scientific literature and warrants a more intensive research effort.

Other Pollutants

Very little is known about the capacity of urban riparian zones to remove other pollutants, such as pesticides, polyaromatic hydrocarbons, polychlorinated biphenols, and pharmaceuticals, which are transported into riparian zones in urban stormwater. Concentrations of polyaromatic hydrocarbons are higher in wetland sediments associated with commercial land uses than residential, industrial, or undeveloped land cover (Kimbrough and Dickhut, 2006) as a result of nonpoint-source pollution from automobiles. One possible concern is that these pollutants, either by themselves or in combination with other pollutants, could potentially inhibit denitrification. High chloride levels have been shown to inhibit denitrification (Hale and Groffman, 2006), and the effects of metals vary (Bernhardt et al., 2008). The effects of pharmaceuticals are not yet known, but it appears that antibiotics may inhibit denitrification (Bernhardt et al., 2008). Pesticides, of which a third of U.S. use is in urban areas at double the application rate of agricultural systems (Paul and Meyer, 2001), negatively impact denitrifier growth rates and activity (Bernhardt et al., 2008). Pathogens, particularly *Escherichia coli* and fecal coliforms, are routinely conveyed through the subsurface of urban riparian zones as a result of leaky combined sewer overflow (CSO) pipes (Rose, 2007). It is not clear what effect pathogens could have on soil processes or whether urban riparian zones are able to remove pathogens through infiltration. This area is particularly deserving of additional research.

Given the importance of many pollutants in urban waters (Allan, 2004), it is indeed surprising that so little research has been devoted to examining pollutant removal and sequestration in urban riparian systems. Urban riparian systems appear to have high potential for removing, sequestering, and transforming many kinds of pollutants in urban ecosystems, given the right hydrological conditions. Aside from a recommendation to reconnect urban riparian systems more extensively to surface waters, and to increase contact times between polluted waters and riparian ecosystems, there is insufficient research available to make more detailed recommendations for riparian management for water quality protection.

The Future of Urban Riparian Zones

Urban riparian zones carry out, or have the potential to carry out, many functions of value to society, yet these functions are often compromised by the conditions of urban environments. Furthermore, although some aspects of urban riparian ecology are reasonably well documented, particularly issues of plant community composition and avian ecology, for many functions and processes, there is a surprising lack of data to evaluate either the level of functioning or the factors affecting function that are unique to the urban environment.

From our review, two primary factors emerge that limit the functioning of urban riparian zones. First, altered hydrology clearly affects all aspects of both biogeochemical function and also plant community structure and dynamics (Chapter 6, Ehrenfeld and Stander, 2010, this volume). A number of actions can be taken to ameliorate hydrologic disconnection. Old mosquito ditching and ditching from former agricultural use can be filled, which would promote higher water tables within inner flood plain areas. Stream restoration, including restoration and revegetation of streambanks, protection of banks from erosion, and restoration of channel complexity may also be useful. Although Kaushal et al. (2008) found that streambank restoration resulted in higher near-stream rates of NO_3^- removal, the overall mixed success of stream restoration projects (Selvakumar et al., 2008) suggests that stream restoration may have limited effectiveness in protecting riparian zones. In particular, restored streambanks remain susceptible to future erosion caused by high urban peak flows. The effects of transforming the land surface from pervious to impervious cover throughout a watershed is one major cause of hydrologic disturbance, but addressing these effects is beyond the scope of restoration projects aimed at the riparian site scale. Basin-wide improvements in stormwater management are likely to be the most effective methods of improving the components of urban hydrology, particularly flood frequency and magnitude, which affect riparian systems (Walsh et al., 2005b). Low impact development (LID) approaches like rain gardens, pervious pavement, and green roofs (Chapters 19 [Rowe and Getter, 2010] and 20 [Li et al., 2010], this volume), and best management practices like stormwater detention and retention basins (Chapter 20, Li et al., 2010, this volume), can significantly reduce peak flows and thus prevent streambank erosion (Walsh et al., 2005b). Ideally, LID practices should be implemented first to achieve the desired reduction in peak flows; then streambank stabilization should move forward to restore hydrologic connection between uplands, riparian zones, and surface waters. Because LID structures have demonstrated the ability to perform some water quality functions (Davis et al., 2001; Kim et al., 2003; Dietz and Clausen, 2005; Hsieh and Davis, 2005; Hunt et al., 2006; Davis, 2007), some researchers have suggested they can replace riparian zones in urban landscapes (Bernhardt and Palmer, 2007; Bernhardt et al., 2008; Cadenasso et al., 2008). Thus, the land–water interface would become spatially dispersed throughout watersheds and perform water quantity and quality functions at the points where stormwater is generated (Walsh et al., 2005b; Cadenasso et al., 2008). However, more research is needed to quantify and improve the pollutant removal capacity of individual LID structures, as well as the overall effectiveness of dispersed LID structures at the watershed scale (Davis, 2007; Shuster et al., 2008; Thurston et al., 2008; Stander and Borst, 2010).

Second, the fragmentation of riparian ecosystems is pervasive throughout urban areas (Alberti, 2005; Nilsson et al., 2005; Hamer and McDonnell, 2008; Schlesinger et al., 2008). As noted above, the loss of connectivity along rivers is a primary effect of urban development, with impacts on both the biota and on ecosystem functions. Conservation and land management planning in urban areas often includes the development of greenways, or park and open space systems along rivers (Chapter 21, Beer, 2010, this volume). Discussion of greenway planning frequently centers on issues of accessibility, connectivity among urban land uses (e.g., residential and commercial areas), and implementation (Lindsey, 2003; Floress et al., 2009) and is traced back to Frederick Law Olmstead's influence on city planning. However, although greenway planners cite "balance with nature" as an organizing principle of greenway development, there is apparently little interchange between greenway planners and riparian ecologists about integrating riparian habitat protection and conservation into open space planning. Even greenway planning based on river systems does not explicitly consider the management of the riparian system as an ecosystem, separately from human and economic considerations (Toccolini et al., 2006). Conversely, discussion of urban planning and open space issues is notably lacking from summaries of riparian ecology (Naiman et al., 2005). One recent study of planning in Shellharbor, Australia (Parker et al., 2008) provides a good example of the kind of integration that is needed. In this study, Parker et al. (2008) used principles of ecological connectivity to design a greenway corridor system for a 154-km^2 urban region that would link two forest reserves on opposite sides of the site. Riparian habitat emerged as an optimal, and logical, basis for such an open space system that would also benefit human residents. We suggest that for both the enhancement of riparian ecosystems and the development of open space systems in urban areas there is an urgent need for planners and ecologists to collaborate on developing greenway systems that simultaneously provide values for urban residents and support the ecological functions of riparian ecosystems.

Notice

The USEPA, through its Office of Research and Development, funded and managed, or partially funded and collaborated in, the research described herein. It has been subjected to the Agency's peer and administrative review and has been approved for external publication. Any opinions expressed in this paper are those of the author(s) and do not necessarily reflect the views of the Agency; therefore, no official endorsement should be inferred. Any mention of trade names or commercial products does not constitute endorsement or recommendation for use.

References

Addy, K.L., A.J. Gold, P.M. Groffman, and P.A. Jacinthe. 1999. Ground water nitrate removal in subsoil of forested and mowed riparian buffer zones. J. Environ. Qual. 28:962–970.

Addy, K.L., D.Q. Kellogg, A.J. Gold, P.M. Groffman, G. Ferendo, and C. Sawyer. 2002. In situ push–pull method to determine ground water denitrification in riparian zones. J. Environ. Qual. 31:1017–1024.

Alberti, M. 2005. The effects of urban patterns on ecosystem function. Int. Reg. Sci. Rev. 28:168–192.

Allan, C.J., P. Vidon, and R. Lowrance. 2008. Frontiers in riparian zone research in the 21st century. Hydrol. Processes 22:3221–3222.

Allan, J.D. 2004. Landscapes and riverscapes: The influence of land use on stream ecosystems. Annu. Rev. Ecol. Evol. Syst. 35:257–284.

Arnold, L.C., Jr., and C.J. Gibbons. 1996. Impervious surface coverage: The emergence of a key environmental indicator. J. Am. Plann. Assoc. 6:243–259.

Beer, A.R. 2010. Greenspaces, green structure, and green infrastructure planning. p. 431–448. In J. Aitkenhead-Peterson and A. Volder (ed.) Urban ecosystem ecology. Agron. Monogr. 55. ASA, CSSA, and SSSA, Madison, WI.

Bernhardt, E.S., L.E. Band, C.J. Walsh, and P.E. Berke. 2008. Understanding, managing, and minimizing urban impacts on surface water nitrogen loading. Ann. N.Y. Acad. Sci. 1134:61–96.

Bernhardt, E.S., and M.A. Palmer. 2007. Restoring streams in an urbanizing world. Freshwater Biol. 52:738–751.

Bledsoe, B.P., and C.C. Watson. 2001. Effects of urbanization on channel instability. J. Am. Water Resour. Assoc. 37(2):255–270.

Bohlke, J.K., M.E. O'Connell, and K.L. Prestegaard. 2007. Ground water stratification and delivery of nitrate to an incised stream under varying flow conditions. J. Environ. Qual. 36:664–680.

Bolund, P., and S. Hunhammar. 1999. Ecosystem services in urban areas. Ecol. Econ. 29:293–301.

Booth, D.B. 1990. Stream-channel incision following drainage-basin urbanization. Water Resour. Bull. 26(3):407–417.

Booth, D.B., and C.R. Jackson. 1997. Urbanization of aquatic systems: Degradation thresholds, stormwater detection, and the limits of mitigation. J. Am. Water Resour. Assoc. 33(5):1077–1090.

Brabec, E., S. Schulte, and P.L. Richards. 2002. Impervious surfaces and water quality: A review of current literature and its implications for watershed planning. J. Plann. Lit. 16(4):499–514.

Bridgham, S.D., K. Updegraff, and J. Pastor. 1998. Carbon, nitrogen, and phosphorus mineralization in northern wetlands. Ecology 79(5):1545–1561.

Burian, S.J., and C.A. Pomeroy. 2010. Urban impacts on the water cycle and potential green infrastructure implications. p. 277–296. *In* J. Aitkenhead-Peterson and A. Volder (ed.) Urban ecosystem ecology. Agron. Monogr. 55. ASA, CSSA, and SSSA, Madison, WI.

Cadenasso, M.L., S.T.A. Pickett, P.M. Groffman, L.E. Band, G.S. Brush, M.F. Galvin, J.M. Grove, G. Hagar, V. Marshall, B.P. McGrath, J.P.M. O'Neil-Dunne, W.P. Stack, and A.R. Troy. 2008. Ann. N. Y. Acad. Sci. 1134:231–232.

Cirmo, C.P., and J.J. McDonnell. 1997. Linking the hydrologic and biogeochemical controls of nitrogen transport in near-stream zones of temperate-forested catchments: A review. J. Hydrol. 199:88–120.

Clausen, J.C., K. Guillard, C.M. Sigmund, and K.M. Dors. 2000. Water quality changes from riparian buffer restoration in Connecticut. J. Environ. Qual. 29:1751–1761.

Clement, J.C., G. Pinay, and P. Marmonier. 2002. Seasonal dynamics of denitrification along topohydrosequences in three different riparian wetlands. J. Environ. Qual. 31:1025–1037.

Clement, J.C., R.M. Holmes, B.J. Peterson, and G. Pinay. 2003. Isotopic investigation of denitrification in a riparian ecosystem in western France. J. Appl. Ecol. 40(6):1035–1048.

Collins, J.P., A. Kinzig, N.B. Grimm, W.F. Fagan, D. Hope, J.G. Wu, and E.T. Borer. 2000. A new urban ecology. Am. Sci. 88(5):416–425.

Conley, D.J., H.W. Paerl, R.W. Howarth, D.F. Boesch, S.P. Seitzinger, K.E. Havens, C. Lancelot, and G.E. Likens. 2009. Controlling eutrophication: Nitrogen and phosphorus. Science 323:1014–1015.

Cooper, A.B. 1990. Nitrate depletion in the riparian zone and stream channel of a small headwater catchment. Hydrobiologia 202:13–26.

Cooper, J.R., J.W. Gilliam, R.B. Daniels, and W.P. Robarge. 1987. Riparian areas as filters for agricultural sediment. Soil Sci. Soc. Am. J. 51:416–420.

Costanza, R., R. d'Arge, R. de Groot, S. Farber, M. Grasso, B. Hannon, K. Limburg, S. Naeem, R. O'Neill, J. Paruelo, R. Raskin, P. Sutton, and M. van den Belt. 1997. The value of the world's ecosystem services and natural capital. Nature 387(15):253–260.

Craft, C., and W.P. Casey. 2000. Sediment and nutrient accumulation in floodplain and depressional freshwater wetlands of Georgia, USA. Wetlands 20:323–332.

Daniels, R.B., and J.W. Gilliam. 1989. Sediment and chemical load reduction by grass and riparian filters. Soil Sci. Soc. Am. J. 60:246–251.

Davis, A.P. 2007. Field performance of bioretention: Water quality. Environ. Eng. Sci. 24(8):1048–1064.

Davis, A.P., M. Shokouhian, H. Sharma, and C. Minami. 2001. Laboratory study of biological retention for urban storm water management. Water Environ. Res. 73(1):5–14.

Diaz, R.J., and R. Rosenberg. 2008. Spreading dead zones and consequences for marine systems. Science 321(5891):926–929.

Dietz, M.E., and J.C. Clausen. 2005. A field evaluation of rain garden flow and pollutant treatments. Water Air Soil Pollut. 167:123–138.

Ehrenfeld, J.G. 2000. Evaluating wetlands within an urban context. Urban Ecosyst. 4:69–85.

Ehrenfeld, J.G., H.B. Cutway, R.H. Hamilton IV, and E. Stander. 2003. Hydrologic description of forested wetlands in northeastern New Jersey, USA—An urban/suburban region. Wetlands 23:685–700.

Ehrenfeld, J.G., and J.P. Schneider. 1991. *Chamaecyparis thyoides* wetlands and suburbanization: Effects on hydrology, water quality and plant community composition. J. Appl. Ecol. 28(2):467–490.

Ehrenfeld, J.G., and E.K. Stander. 2010. Habitat function in urban riparian zones. p. 103–118. *In* J. Aitkenhead-Peterson and A. Volder (ed.) Urban ecosystem ecology. Agron. Monogr. 55. ASA, CSSA, and SSSA, Madison, WI.

Ettema, C.H., R. Lowrance, and D.C. Coleman. 1991. Riparian soil response to surface nitrogen input: Temporal changes in denitrification, labile and microbial C and N pools, and bacterial and fungal respiration. Soil Biol. Biochem. 31:1609–1624.

Faulkner, S.P. 2004. Urbanization impacts on the structure and function of forested wetlands. Urban Ecosyst. 7:89–106.

Flite, O.P., R.D. Shannon, R.R. Schnabel, and R.R. Parizek. 2001. Nitrate removal in a riparian wetland of the Appalachian Valley and Ridge physiographic province. J. Environ. Qual. 30:254–261.

Floress, K., A. Baumgart-Getz, L.S. Prokopy, and J. Janota. 2009. The quality of greenways planning in northwest Indiana: A focus on sustainability principles. J. Environ. Plann. Manage. 52:61–78.

Florsheim, J.L., J.F. Mount, and A. Chin. 2008. Bank erosion as a desirable attribute of rivers. Bioscience 58(6):519–529.

Gilliam, J.W. 1994. Riparian wetlands and water quality. J. Environ. Qual. 23:896–900.

Gold, A.J., P.A. Jacinthe, P.M. Groffman, W.R. Wright, and R.H. Puffer. 1998. Patchiness in groundwater nitrate removal in a riparian forest. J. Environ. Qual. 27:146–155.

Groffman, P.M., D.J. Bain, L.E. Band, K.T. Belt, G.S. Brush, J.M. Grove, R.V. Pouyat, I.C. Yesilonis, and W.C. Zipperer. 2003. Down by the riverside: Urban riparian ecology. Front. Ecol. Environ. 1:315–321.

Groffman, P.M., N.J. Boulware, W.C. Zipperer, R.V. Pouyat, L.E. Band, and M.F. Colosimo. 2002. Soil nitrogen cycle processes in urban riparian zones. Environ. Sci. Technol. 36:4547–4552.

Groffman, P.M., and M.K. Crawford. 2003. Denitrification potential in urban riparian zones. J. Environ. Qual. 32:1144–1149.

Groffman, P.M., A.J. Gold, and R.C. Simmons. 1992. Nitrate dynamics in riparian forests: Microbial studies. J. Environ. Qual. 21:666–671.

Groffman, P.M., G. Howard, A.J. Gold, and W.M. Nelson. 1996. Microbial nitrate processing in shallow groundwater in a riparian forest. J. Environ. Qual. 25:1309–1316.

Groffman, P.M., R.V. Pouyat, M.L. Cadenasso, W.C. Zipperer, K. Szlavecz, I.D. Yesilonis, L.E. Band, and G.S. Brush. 2006. Land use context and natural soil controls on plant community composition and soil nitrogen and carbon dynamics in urban and rural forests. For. Ecol. Manage. 236:177–192.

Groffman, P.M., D.R. Zak, S. Christensen, A. Mosier, and J.M. Tiedje. 1993. Early spring nitrogen dynamics in a temperate forest landscape. Ecology 74(5):1579–1585.

Hale, R.L., and P.M. Groffman. 2006. Chloride effects on nitrogen dynamics in forested and suburban stream debris dams. J. Environ. Qual. 35:2425–2432.

Hamer, A.J., and M.J. McDonnell. 2008. Amphibian ecology and conservation in the urbanising world: A review. Biol. Conserv. 141:2432–2449.

Hanson, G.C., P.M. Groffman, and A.J. Gold. 1994a. Symptoms of nitrogen saturation in a riparian wetland. Ecol. Appl. 4(4):750–756.

Hanson, G.C., P.M. Groffman, and A.J. Gold. 1994b. Denitrification in riparian wetlands receiving high and low groundwater nitrate inputs. J. Environ. Qual. 23:917–922.

Haycock, N.E., and G. Pinay. 1993. Groundwater nitrate dynamics in grass and poplar vegetated riparian buffer strips during the winter. J. Environ. Qual. 22:273–278.

Hedin, L.O., J.C. von Fischer, N.E. Ostrom, B.P. Kennedy, M.G. Brown, and G.P. Robertson. 1998. Thermodynamic constraints on nitrogen transformations and other biogeochemical processes at soil–stream interfaces. Ecology 79(2):684–703.

Heisler, G.M., and A.J. Brazel. 2010. The urban physical environment: Temperature and urban heat islands. p. 29–56. In J. Aitkenhead-Peterson and A. Volder (ed.) Urban ecosystem ecology. Agron. Monogr. 55. ASA, CSSA, and SSSA, Madison, WI.

Henshaw, P.C., and D.B. Booth. 2000. Natural restabilization of stream channels in urban watersheds. J. Am. Water Resour. Assoc. 36(6):1219–1236.

Hey, D.L. 2002. Nitrogen farming: Harvesting a different crop. Restor. Ecol. 10:1–10.

Hey, D.L., L.S. Urban, and J.A. Kostel. 2005. Nutrient farming: The business of environmental management. Ecol. Eng. 24:279–287.

Hill, A.R. 1996. Nitrate removal in stream riparian zones. J. Environ. Qual. 25:743–755.

Hill, A.R., K.J. Devito, S. Campagnolo, and K. Sanmugadas. 2000. Subsurface denitrification in a forest riparian zone: Interactions between hydrology and supplies of nitrate and organic carbon. Biogeochemistry 51(2):193–223.

Hoffmann, C.C., and A. Baattrup-Pedersen. 2007. Re-establishing freshwater wetlands in Denmark. Ecol. Eng. 30:157–166.

Hogan, D.M., and M.R. Walbridge. 2007. Urbanization and nutrient retention in freshwater riparian wetlands. Ecol. Appl. 17(4):1142–1155.

Horner, R.R., S.S. Cooke, L.E. Reinelt, K.A. Ludwa, and N.T. Chin. 2001. Water quality and soils. p. 237–254. In A.L. Azous and R.R. Horner (ed.) Wetlands and urbanization: Implications for the future. Lewis Publ., Boca Raton, FL.

Hsieh, C.H., and A.P. Davis. 2005. Evaluation and optimization of bioretention media for treatment of urban storm water runoff. J. Environ. Eng. 131(11):1521–1531.

Hubbard, R.K., and R. Lowrance. 1997. Assessment of forest management effects on nitrate removal by riparian buffer systems. Trans. ASAE 40:383–391.

Hunt, W.F., A.R. Jarrett, J.T. Smith, and L.J. Sharkey. 2006. Evaluating bioretention hydrology and nutrient removal at three field sites in North Carolina. J. Irrig. Drain. Eng. 132(6):600–608.

Jacinthe, P.A., P.M. Groffman, A.J. Gold, and A. Mosier. 1998. Patchiness in microbial nitrogen transformations in groundwater in a riparian forest. J. Environ. Qual. 27:156–164.

Kansiime, F., E. Kateyo, H. Oryem-Origa, and P. Mucunguzi. 2007. Nutrient status and retention in pristine and disturbed wetlands in Uganda: Management implications. Wetlands Ecol. Manage. 15:453–467.

Kaushal, S.S., P.M. Groffman, P.M. Mayer, E. Striz, and A.J. Gold. 2008. Effects of stream restoration on denitrification in an urbanizing watershed. Ecol. Appl. 18(3):789–804.

Kellogg, D.Q., A.J. Gold, P.M. Groffman, K. Addy, M.H. Stolt, and G. Blazejewski. 2005. In situ groundwater denitrification in stratified, permeable soils underlying riparian wetlands. J. Environ. Qual. 34:524–533.

Kim, H., E.A. Seagren, and A.P. Davis. 2003. Engineered bioretention for removal of nitrate from stormwater runoff. Water Environ. Res. 75(4):355–367.

Kimbrough, K.L., and R.M. Dickhut. 2006. Assessment of polycyclic aromatic hydrocarbon input to urban wetlands in relation to adjacent land use. Mar. Pollut. Bull. 52:1355–1363.

Lee, K.-H., T.M. Isenhart, and R.C. Schultz. 2003. Sediment and nutrient removal in an established multi-species riparian buffer. J. Soil Water Conserv. 58:1–7.

Li, M.-H., B. Dvorak, and C.Y. Sung. 2010. Bioretention, low impact development, and stormwater management. p. 413–430. In J. Aitkenhead-Peterson and A. Volder (ed.) Urban ecosystem ecology. Agron. Monogr. 55. ASA, CSSA, and SSSA, Madison, WI.

Lindsey, G. 2003. Sustainability and urban greenways—Indicators in Indianapolis. J. Am. Plann. Assoc. 69:165–180.

Liu, H., Q.H. Weng, and D. Gaines. 2008. Spatio-temporal analysis of the relationship between WNV dissemination and environmental variables in Indianapolis, USA. Int. J. Health Geogr. 7:66.

Lovett, G.M., M.M. Traynor, R.V. Pouyat, M.M. Carreiro, W.X. Zhu, and J.W. Baxter. 2000. Atmospheric deposition to oak forests along an urban–rural gradient. Environ. Sci. Technol. 34:4294–4300.

Lowrance, R. 1992. Groundwater nitrate and denitrification in a coastal plain riparian forest. J. Environ. Qual. 21:401–405.

Lowrance, R., L.S. Altier, J.D. Newbold, R.R. Schnabel, P.M. Groffman, J.M. Denver, D.L. Correll, J.W. Gilliam, J.L. Robinson, R.B. Brinsfield, K.W. Staver, W. Lucas, and A.H. Todd. 1997. Water quality functions of riparian forest buffers in Chesapeake Bay watersheds. Environ. Manage. 21(5):687–712.

Lowrance, R., R. Todd, J. Fail, Jr., O. Hendrickson, Jr., R. Leonard, and L. Asmussen. 1984. Riparian forests as nutrient filters in agricultural watersheds. Bioscience 34(6):374–377.

Marzluff, J.M., E. Shulenberger, W. Endlicher, M. Alberti, G. Bradley, C. Ryan, C. ZumBrunnen, and U. Simon (ed.) 2008. Urban ecology: An international perspective on the interaction between humans and nature. Springer, New York.

Matteo, M., T. Randhir, and D. Bloniarz. 2006. Watershed-scale impacts of forest buffers on water quality and runoff in urbanizing environment. J. Water Resour. Plann. Manage. 132(3):144–152.

Mayer, P.M., S.K. Reynolds, M.D. McCutchen, and T.J. Canfield. 2007. Meta-analysis of nitrogen removal in riparian buffers. J. Environ. Qual. 36:1172–1180.

McBride, M., and D.B. Booth. 2005. Urban impacts on physical stream condition: Effects of spatial scale, connectivity, and longitudinal trends. J. Am. Water Resour. Assoc. 41(3):565–580.

McClain, M.E., E.W. Boyer, C.L. Dent, S.E. Gergel, N.B. Grimm, P.M. Groffman, S.C. Hart, J.W. Harvey, C.A. Johnston, E. Mayorga, W.H. McDowell, and G. Pinay. 2003. Biogeochemical hot spots and hot moments at the interface of terrestrial and aquatic ecosystems. Ecosystems 6:301–312.

Meyer, J.L., M.J. Paul, and W.K. Taulbee. 2005. Stream ecosystem function in urbanizing landscapes. J. North Am. Benthol. Soc. 24(3):602–612.

Mitsch, W., and J.G. Gosselink. 2000. Wetlands. 3rd ed. Van Nostrand Reinhold, New York.

Murray, K.S., D.T. Rogers, and M.M. Kaufman. 2004. Heavy metals in an urban watershed in southeastern Michigan. J. Environ. Qual. 33:163–172.

Naiman, R.J., and H. Decamps. 1997. The ecology of interfaces: Riparian zones. Annu. Rev. Ecol. Syst. 28:621–658.

Naiman, R., H. Décamps, and M.E. McClain. 2005. Riparia. Ecology, conservation and management of streamside communities. Elsevier Academic Press, San Diego, CA.

Naiman, R.J., H. Decamps, and M. Pollock. 1993. The role of riparian corridors in maintaining regional biodiversity. Ecol. Appl. 3:209–212.

Nilsson, C., C.A. Reidy, M. Dynesius, and C. Revenga. 2005. Fragmentation and flow regulation of the world's large river systems. Science 308:405–408.

Norton, M.M., and T.R. Fisher. 2000. The effects of forest on stream water quality in two coastal plain watersheds of the Chesapeake Bay. Ecol. Eng. 14:337–362.

Odum, E.P., J.T. Finn, and E.H. Franz. 1979. Perturbation theory and the subsidy-stress gradient. Bioscience 29:349–352.

Osborne, L.L., and D.A. Kovacic. 1993. Riparian vegetated buffer strips in water-quality restoration and stream management. Freshwater Biol. 29:243–258.

Osborne, L.L., and M.J. Wiley. 1988. Empirical relationships between land use/cover and stream water quality in an agricultural wetland. J. Environ. Manage. 26(1):9–27.

Parker, G.R., W.W. McFee, and J.M. Kelly. 1978. Metal distribution in forested ecosystems in urban and rural northwestern Indiana. J. Environ. Qual. 7:337–342.

Parker, K., L. Head, L.A. Chisholm, and N. Feneley. 2008. A conceptual model of ecological connectivity in the Shellharbour Local Government Area, New South Wales, Australia. Landsc. Urban Plan. 86:47–59.

Paul, M.J., and J.L. Meyer. 2001. Streams in the urban landscape. Annu. Rev. Ecol. Syst. 32:333–365.

Peterjohn, W.T., and D.L. Correll. 1984. Nutrient dynamics in an agricultural watershed: Observations on the role of a riparian forest. Ecology 65(5):1466–1475.

Pouyat, R.V., K. Szlavecz, I.D. Yesilonis, P.M. Groffman, and K. Schwarz. 2010. Chemical, physical, and biological characteristics of urban soils. p. 119–152. In J. Aitkenhead-Peterson and A. Volder (ed.) Urban ecosystem ecology. Agron. Monogr. 55. ASA, CSSA, and SSSA, Madison, WI.

Rose, S. 2007. The effects of urbanization on the hydrochemistry of base flow within the Chatahoochee River Basin (Georgia, USA). J. Hydrol. 341:42–54.

Rose, S., and N.E. Peters. 2001. Effects of urbanization on streamflow in the Atlanta area (Georgia, USA): A comparative hydrological approach. Hydrol. Processes 15:1441–1457.

Rowe, D.B., and K.L. Getter. 2010. Green roofs and garden roofs. p. 391–412. In J. Aitkenhead-Peterson and A. Volder (ed.) Urban ecosystem ecology. Agron. Monogr. 55. ASA, CSSA, and SSSA, Madison, WI.

Royer, T.V., M.B. David, and L.E. Gentry. 2006. Timing of riverine export of nitrate and phosphorus from agricultural watersheds in Illinois: Implications for reducing nutrient loading to the Mississippi River. Environ. Sci. Technol. 40(13):4126–4131.

Schlesinger, M.D., P.N. Manley, and M. Holyoak. 2008. Distinguishing stressors acting on land bird communities in an urbanizing environment. Ecology 89:2302–2314.

Schnabel, R.R., L.F. Cornish, W.L. Stout, and J.A. Shaffer. 1996. Denitrification in a grassed and a wooded, valley and ridge, riparian ecotone. J. Environ. Qual. 25:1230–1235.

Schoonover, J.E., and K.W.J. Williard. 2003. Ground water nitrate reduction in giant cane and forest riparian buffer zones. J. Am. Water Resour. Assoc. 39:347–354.

Selvakumar, A., T.P. O'Connor, and S. Struck. 2008. Evaluation of receiving water improvements from stream restoration (Accotink Creek, Fairfax City, VA). EPA/600/R-08/110, Cincinnati, OH.

Shuster, W.D., M.A. Morrison, and R. Webb. 2008. Front-loading urban stormwater management for success—A perspective incorporating current studies on the implementation of retrofit low-impact development. Cities Environ. 1(2):article 8.

Simon, K., and A. Steinemann. 2000. Soil bioengineering: Challenges for planning and engineering. J. Urban Plann. Dev. 126(2):89–102.

Sonoda, K., and J.A. Yeakley. 2007. Relative effects of land use and near-stream chemistry on phosphorus in an urban stream. J. Environ. Qual. 36:144–154.

Stander, E.K. 2007. The effects of urban hydrology and elevated atmospheric deposition on nitrate retention and loss in urban wetlands. Ph.D. diss. Graduate Program in Ecology and Evolution, Graduate School. Rutgers Univ., New Brunswick, NJ.

Stander, E.K., and M. Borst. 2010. Enhancing rain garden design to promote nitrate removal: A hydraulic test of a media carbon amendment. J. Hydrol. Eng. (in press).

Stander, E.K., and J.G. Ehrenfeld. 2009a. Rapid assessment of urban wetlands: Do hydrogeomorphic classification and reference criteria work? Environ. Manage. 43(4):725–742.

Stander, E.K., and J.G. Ehrenfeld. 2009b. Rapid assessment of urban wetlands: Functional assessment model development and evaluation. Wetlands 29(1):261–276.

Steele, M.K., W.H. McDowell, and J.A. Aitkenhead-Peterson. 2010. Chemistry of urban, suburban, and rural surface waters. p. 297–340. In J. Aitkenhead-Peterson and A. Volder (ed.) Urban ecosystem ecology. Agron. Monogr. 55. ASA, CSSA, and SSSA, Madison, WI.

Sudduth, E.B., and J.L. Meyer. 2006. Effects of bioengineered streambank stabilization on bank habitat and macroinvertebrates in urban streams. Environ. Manage. 38(2):218–226.

Thurston, H.W., A.H. Roy, W.D. Shuster, M.A. Morrison, M.A. Taylor, and H. Cabezas. 2008. Using economic incentives to manage stormwater runoff in the Shepherd Creek watershed. Part I. EPA/600/R-08-129. USEPA, Cincinnati, OH.

Toccolini, A., N. Fumagalli, and G. Senes. 2006. Greenways planning in Italy: The Lambro River Valley greenways system. Landsc. Urban Plan. 76:98–111.

Trimble, S.W. 1997. Contribution of stream channel erosion to sediment yield from an urbanizing watershed. Science 278:1442–1444.

USEPA. 1983. Results of the nationwide urban runoff program: Volume 1. Final report. Water Planning Division, Washington, DC.

USGS. 1999. The quality of our nation's waters—Nutrients and pesticides: USGS Circular 1225. USGS, Denver, CO.

Verchot, L.V., E.C. Franklin, and J.W. Gilliam. 1997. Nitrogen cycling in piedmont vegetated filter zones: I. Surface soil processes. J. Environ. Qual. 26:327–336.

Vidon, P.G.F., and A.R. Hill. 2004. Landscape controls on nitrate removal in stream riparian zones. Water Resour. Res. 40:W03201.

Walsh, C.J., T.D. Fletcher, and A.R. Ladson. 2005a. Stream restoration in urban catchments through redesigning stormwater systems: Looking to the catchment to save the stream. J. North Am. Benthol. Soc. 24:690–705.

Walsh, C.J., A.H. Roy, J.W. Feminella, P.D. Cottingham, P.M. Groffman, and R.P. Morgan. 2005b. The urban stream syndrome: Current knowledge and the search for a cure. J. North Am. Benthol. Soc. 24(3):706–723.

Yeakley, J.A., D.C. Coleman, B.L. Haines, B.D. Kloeppel, J.L. Meyer, W.T. Swank, B.W. Argo, J.M. Deal, and S.F. Taylor. 2003. Hillslope nutrient dynamics following upland riparian vegetation disturbance. Ecosystems 6:154–167.

Yu, K.J., D.H. Li, and N.Y. Li. 2006. The evolution of Greenways in China. Landsc. Urban Plan. 76:223–239.

Zhu, W.-X., and M.M. Carreiro. 2004. Temporal and spatial variations in nitrogen transformations in deciduous forest ecosystems along an urban-rural gradient. Soil Biol.
Biochem. 36:267–278.

Zhu, W.-X., N.D. Dillard, and N.B. Grimm. 2004. Urban nitrogen biogeochemistry: Status
and processes in green retention basins. Biogeochemistry 71:177–196.

Zhu, W.-X., and J.G. Ehrenfeld. 1999. Nitrogen mineralization and nitrification in suburban
and undeveloped Atlantic white cedar wetlands. J. Environ. Qual. 28(2):523–529.

14

Urban Impacts on the Water Cycle and Potential Green Infrastructure Implications

Steven J. Burian

Christine A. Pomeroy

Abstract

The urban water cycle is mediated by a coupled built–natural system where the hydrologic stores and fluxes are highly modified through a combination of geologic, terrain, climate, development, and human behavior factors. Studies for more than 50 yr have sought to characterize the urban effects on the water cycle. Yet, predicting urbanization impacts at the necessary spatial and temporal resolution remains a challenge. Moreover, quantifying the cascading impacts on associated cycles and contaminant stores and fluxes becomes more difficult as the layers of urban system complexity are expanded and interconnected. To contribute to the advancement of the understanding of urban impacts on the water cycle, our objective in this chapter is to establish known effects of traditional urban development on the water cycle, describe emerging water cycle modifications associated with low-impact development, and identify current uncertainties in need of further research.

Urban growth is recognized as one of the most pressing environmental concerns facing the current generation. In the 20th century, urbanization increased the world's metropolitan population from 220 million to 2.8 billion (United Nations Population Fund, 2009), contributing to significant changes in land surface characteristics. In the conterminous United States, for example, the amount of impervious surfaces associated with cities grew to approximately the area of the state of Ohio (Elvidge et al., 2004). Current projections suggest the problem will intensify in the near future. By 2050 the global urban population is expected to increase from the current 3.5 billion (6.6 billion total) to 6.5 billion (9.1 billion total) (United Nations Population Division, 2009). During this same period the urban population in China, India, and the United States is estimated to increase by 67, 300, and 44%, respectively. The projected urban growth will challenge regional carrying capacities and have profound impacts on the water cycle at the local, regional, and global scales.

S.J. Burian (burian@eng.utah.edu) and C.A. Pomeroy (christine.pomeroy@utah.edu), Dep. of Civil and Environmental Engineering, The University of Utah, 122 S. Central Campus Dr., Suite 104, Salt Lake City, UT 84112.

doi:10.2134/agronmonogr55.c14

A city casts a direct footprint that is easily ascertained by a view from above. However, there are few (if any) cities whose water needs can be supported based on the direct footprint and whose water impacts are contained within or immediately adjacent to the direct footprint. Water demand is met by appropriation of supply from nearby and possibly distant watersheds and underlying aquifers. Hydrologic modification, contaminated discharges (e.g., stormwater, combined-sewer overflows), and treated effluent directly impact waterways and water quality in and adjacent to the urban footprint, and the effects cascade to downstream water bodies.

The water system supporting an urban area is composed of an intricate, highly regulated network of rivers, streams, lakes, and reservoirs and so-called "gray infrastructure" elements, including pipes, channels, junctions, diversions, bypasses, pumps, regulators, storage and treatment facilities, and other appurtenances that collect, convey, store, and treat water, wastewater, and stormwater. Streams, rivers, and other natural water bodies in an urban environment have been highly modified from their original state by channelization, flow diversions, discharges, and other hydraulic modifications. Furthermore, traditional riparian corridors and buffer zones are often replaced by housing, parking lots, roadways, and other urban structures.

The imported water, supplied from surface water or groundwater sources, is typically treated and distributed for a range of municipal and industrial uses. The finished water is conveyed by an intricate distribution network using pumps and gravity. Complementing centralized treatment and distribution networks are the use of secondary water distribution systems and decentralized alternatives implementing water reuse and rainwater harvesting. Wastewater generated by municipal and industrial water users is collected in a complex system of underground conduits (i.e., sewers). In general, sanitary sewers are built to follow surface topography, taking advantage of gravity to facilitate transport of the wastewater to treatment facilities where physical, chemical, and biological processes are employed to clean the water before discharge to receiving water systems (Metcalf & Eddy, Inc., 2003).

Stormwater runoff generated in urban areas during rainfall events may be conveyed in combination with sanitary wastewater to a treatment facility (a so-called combined-sewer system). Combined-sewer systems were commonly constructed during the 19th century in Europe and the United States when connections from failing cesspools and privies were made to existing stormwater drainage pipes (Burian et al., 1999, 2000; Burian, 2001; Melosi, 2000). Alternatively, the system may be designed with separate conduits to convey wastewater and stormwater (i.e., separate-sewer system). In the United States, most drainage systems constructed recently have been separate, but approximately 800 cities and 40 million people continue to be served by legacy combined-sewer systems. Both combined- and separate-sewer systems contribute discharges to the urban hydrologic system and impact the receiving water body hydrology and ecosystems.

The urban water budget describes the volumes of water in the surface, subsurface, and atmospheric compartments of the urban environment and the rate of movement of water between compartments (Welty, 2009). The water budget can be described mathematically for a control volume that extends some distance below the surface up into the atmosphere as:

$$P + F + I = D + ET + \Delta r + \Delta S + \Delta A + f \qquad [1]$$

Here, all terms have units of depth of water per unit time. The left side of Eq. [1] represents water mass added to the urban control volume, while the right side represents mechanisms of transport out. P is precipitation, F is water released into the atmosphere by combustion, I is the net rate of urban water supply, D is the rate of discharge via sanitary sewers, ET is the rate of evapotranspiration, Δr is the net runoff by overland flow and in stormwater drainage pipes, ΔS is the rate of change of water storage in the volume, ΔA is the net advection of moisture into or out of the urban volume, and f is the infiltration of water into deeper layers of soil (Oke, 1987). For urban sites that have been carefully selected, horizontal advection can be neglected (Eq. [2]). If the rate of water supply is separated into the indoor (I_{in}) and outdoor components (I_{out}), the balances simplify to:

$$P + F + I_{in} + I_{out} = D + ET + \Delta r + \Delta S + f \qquad [2]$$

Urbanization and the associated water supply, wastewater collection and treatment, and stormwater infrastructure have a wide range of impacts on the water budget components and the related movement of environmental contaminants. The degree of urbanization has been suggested as the most dominant factor altering the water budget of an area (Claessens et al., 2006). The removal of indigenous vegetation, grading of land surfaces, compaction of pervious surfaces, construction of impervious surfaces and landscaped areas, and importation of water change the fluxes and stores of water and associated pollutants, as illustrated for a hypothetical case in Fig. 14–1. The

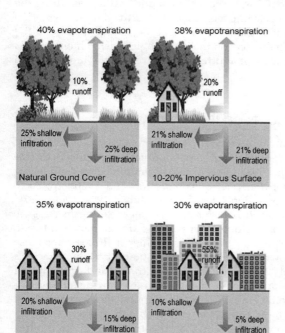

40% evapotranspiration

10% runoff

25% shallow infiltration

25% deep infiltration

Natural Ground Cover

38% evapotranspiration

20% runoff

21% shallow infiltration

21% deep infiltration

10-20% Impervious Surface

35% evapotranspiration

30% runoff

20% shallow infiltration

15% deep infiltration

35-50% Impervious Surface

30% evapotranspiration

55% runoff

10% shallow infiltration

5% deep infiltration

75-100% Impervious Surface

Fig. 14–1. Changes to hydrologic cycle with urbanization (Federal Interagency Stream Restoration Working Group, 1998).

characteristics of urban configuration and the accompanying water, wastewater, and stormwater infrastructure control the type and magnitude of water cycle modification in cities.

A plethora of studies have been performed to quantify water cycle impacts of urban areas implementing gray infrastructure elements (i.e., pipes and channels), and many of these are summarized in this chapter. In addition to impacts from traditional gray infrastructure approaches, new water cycle impacts are emerging as different ways to configure urban areas, manage water, and design water infrastructure systems are being introduced under the auspices of sustainable development. One of the more widespread concepts being implemented is green infrastructure associated with low-impact development (LID) (Prince George's County-Maryland, 1999; Chapter 20, Li et al., 2010, this volume). Green infrastructure (GI) is comprised of the interconnected networks of natural and constructed ecological systems within, around, and between urban areas (Tzoulas et al., 2007). It includes components such as hydrologically functional landscapes, bioretention, infiltration, bioswales, greenways, rainwater harvesting, pervious pavement, and neighborhood-scale constructed wetlands, stormwater detention, retention, and infiltration systems. The intent of GI is to provide additional ecosystem, human health, and water-energy efficiency benefits to the core stormwater management function. From a planning perspective there is the objective to integrate green space into the urban fabric (Sandström, 2002; Chapter 21, Beer, 2010, this volume). Urban areas implementing GI and decentralized approaches are expected to impact the water cycle much differently from those that have implemented centralized gray infrastructure approaches. There remains, however, great uncertainty regarding the form and magnitude of water cycle impacts associated with GI implementation at the site, local, municipal, and regional scales. In this chapter we extend the review of the traditional urban impacts on the hydrologic cycle to include a discussion of the expected impacts of GI and recent research on the impacts of GI on water cycle processes.

Urban Impacts on the Water Cycle

Uncontrolled urbanization contributes to many environmental problems, including deforestation (Masek et al., 2000), landscape modification, reduced surface energy fluxes (Cleugh and Oke, 1986), urban climate modification (Landsberg, 1981), precipitation modification (Shepherd, 2005; Burian and Shepherd, 2005), flooding (Hollis, 1975), water pollution (Heaney and Huber, 1984; Klein, 1979), and air pollution (Romero et al., 1999). In terms of the water cycle, urban areas are known to increase surface runoff if stormwater controls are not installed (Ando et al., 1984; DeWalle et al., 2000), and to create a flashier hydrologic response (Leopold, 1968). Less well known and less certain are the impacts on precipitation (Chapter 1, Shepherd et al., 2010, this volume), infiltration and groundwater recharge, soil moisture, and evapotranspiration (ET). The following subsections describe the impacts of traditional gray infrastructure approaches to urban development on streamflow and runoff, infiltration and recharge, and soil moisture and ET. Within each subsection we describe the potential impacts of GI implementation, including highlights of recent research.

Runoff, Streamflow, and Receiving Water Quality

Traditional Urban Impacts

The nature of impacts to the water cycle caused by channel modification, diversions, and contaminant inputs depends on the characteristics of the urban area; the size, intensity, and frequency of precipitation events; the size and characteristics of the receiving water (e.g., degree of mixing and dilution and assimilative capacity); and the beneficial uses of the receiving water (e.g., water supply, recreation, ecological habitat). As the physical structure of the hydrologic system is modified by urbanization, so are the magnitudes and qualities of associated fluxes and reservoirs. Before urbanization, dry weather streamflow is sustained by surrounding groundwater aquifers. However, as urbanization progresses, anthropogenic inputs to and diversions from streamflow increase, and in ultraurban environments the dry-weather surface flow may be reduced significantly and/or dominated by point-source discharges, including effluent from municipal and industrial treatment facilities and nuisance flows from the stormwater drainage system (e.g., excess irrigation water, inappropriate discharges, crossconnection flows). In some cases dry weather or low flows may be important sources of annual pollutant loads (McPherson et al., 2005) and are known to be a major governor of the structure of aquatic communities (Nilsson et al., 2003).

Increased flashiness of runoff response is characteristic of urban impacts on the water cycle (Leopold, 1968). Runoff volumes and peak discharges are well-known effects of urban development. The removal of trees and vegetation during urbanization reduces interception capacity, thereby increasing the amount of rainfall reaching the ground surface. The compaction of soil during construction (Pitt et al., 2002) and the introduction of impervious surfaces (Schueler, 1994) reduce depression storage, infiltration, and groundwater recharge. These altered surface characteristics, combined with the introduction of efficient drainage systems, increase the drainage density of urban catchments, which decreases the travel time of overland flow to the drainage system and thus augments the frequency of higher magnitudes of stormwater runoff rates and streamflow (e.g., Booth and Jackson, 1997).

The concern with stormwater discharges has traditionally been downstream flooding, channel erosion, and physical damages to drainage infrastructure. Flood control strategies historically involved implementation of storage reservoirs and channelization of streams and rivers in urbanizing environments. Elevated high flows during wet weather are a concern not only from a flood control and stream geomorphology perspective, but also for habitat alterations and aquatic life impacts. Increased frequency and duration of elevated discharge rates in urban streams (Hollis, 1975) contribute to accelerated stream bank and bed erosion (Hammer, 1972), resulting in physical habitat destruction (USEPA, 1997). It has been concluded that the physical modification and hydrologic changes of the water body from stormwater discharges may be more important than chemical contaminant inputs for determining ecological impacts to water bodies (Pitt and Bissonnette, 1984; Booth and Jackson, 1997; Roesner et al., 2001).

Increased erosion increases turbidity in the water column, thereby reducing light penetration and visibility and increasing fish gill clogging and burial of bottom-dwelling flora and fauna, which contribute to a shift of macroinvertebrate and fish species and density (USEPA, 1977; Chapter 16; Roy et al., 2010, this volume). In addition, sediment deposition fills impoundments, complicates water treatment, and reduces aesthetic quality of water bodies, all of which have tremendous environmental and economic ramifications. Removal of the riparian corridor also eliminates tree and vegetation canopy coverage over waterways, resulting in increased in-stream temperatures (Brown and Krygier, 1970) and detrimental effects on aquatic life. Klein (1979) documented the reduction in macroinvertebrate diversity in urban streams in Maryland and noted the impact to be present when watershed imperviousness exceeded 10 to 15%. Numerous studies have since represented urbanization by the percentage of imperviousness and corroborated the 10% threshold of imperviousness as the critical condition for stream impacts (e.g., Booth and Jackson, 1997).

Increased stress is also placed on urban water resources by increased bacterial and toxic pollutant loading. The transport of wastes carrying disease-causing pathogens during dry and wet weather to receiving water bodies has the potential to create severe nuisance conditions with dramatic public health ramifications. For example, a study of stormwater discharges to Santa Monica Bay found strong evidence of an increased risk of a broad range of adverse health effects (e.g., fever, ear discharge, gastrointestinal illness) caused by swimming in ocean water close to stormwater outfalls when indicator coliform densities were high (Haile et al., 1996). A range of urban sources of toxic pollutants contribute to receiving waters, including those emanating from urban landscape maintenance (e.g., yards, golf courses, and parks), which has been highlighted as a significant cause of pesticide contamination of urban waterways and shallow aquifers (USGS, 1999). Also, numerous emerging contaminants have recently been detected at trace levels in urban-influenced waterways (Chapter 15, Steele et al., 2010, this volume). Drugs used in human medical care may enter the environment via discharges of municipal sewage, hospital effluents, sewage sludges, landfill leachates, domestic septic tanks, and production residues (Verstraeten et al., 2002). Over-the-counter sales and hospital use of common pharmaceuticals such as ibuprofen and aspirin have been estimated to yield consumption rates of up to 1000 t yr^{-1} in Germany, a figure that rivals the annual use of the most important herbicides that are widely found in surface waters (Verstraeten et al., 2002).

Input of oxygen-demanding substances and nutrients associated with sanitary and stormwater discharges can lower the dissolved oxygen (DO) content in the receiving water to levels harmful to aquatic life, resulting in limited biological diversity, degraded aesthetic quality, and toxicity impacts. There have been numerous studies of the effect of dry- and wet-weather discharges on DO levels in receiving water bodies (e.g., Keefer et al., 1979), which have found that the response is complicated by downstream processes, and the effects of the modified DO on aquatic life is difficult to isolate from the effects of other waste discharges (Field and Pitt, 1990). Excessive inputs of nutrients such as bioavailable nitrogen and phosphorus can accelerate eutrophication, causing increased algae and aquatic plant growth and potentially harmful algal blooms (De Jonge et al., 2002). This may in turn lead to reduced DO levels, toxic conditions, and fish kills. Nutrient pollution has been identified as a likely cause of hypoxia (i.e., reduced dissolved oxygen levels) in coastal areas (e.g., dead zone in Gulf of Mexico, Chesapeake Bay), and the dry- and wet-weather contribution of nutrients from urban areas through surface and atmospheric pathways may be significant (Boesch et al., 2001).

Historical remedies of urban water pollution relied on redirection of the waste stream. For example, as a remedy to direct sanitary discharges, interceptor sewers were constructed to divert the dry- and wet-weather flows to a strategically located discharge point or a treatment facility. Redirection solutions merely transferred the problem downstream. In the case of interceptor sewers, they lacked the design capacity to transport the combined sewage during large rainfall events; hence, relief structures called combined-sewer overflows were required (Moffa, 1997). The combined-sewer overflows have been a major cause of water quality and habitat problems for decades due to elevated temperatures of discharges, high velocities, and input of bacteria, floatable material, solids, toxic organic compounds and trace metals, oxygen-demanding substances, and nutrients (Field and Turkeltaub, 1981; USEPA, 1996). Most modern sewer systems in the United States are designed to convey sanitary flow separately from stormwater in a separate-sewer system. However, groundwater infiltration and rainfall-derived infiltration and inflow to the sanitary line during wet weather may overwhelm the system and require overflow controls called sanitary-sewer overflows (SSOs). In the United States, point-source dry-weather discharges were the first to be targeted by federal regulations with the passage of the Clean Water Act in 1972. However, as control of point-source discharges increased, nonpoint-source discharges emerged as a significant source of contaminants to water bodies. Recently, watershed-wide planning strategies, the Total Maximum Daily Load (TMDL) program, and the National Pollutant Discharge Elimination System (NPDES) program have been introduced to protect urban waterways from potentially harmful dry- and wet-weather and point and nonpoint-source discharges. Currently, nonpoint sources of pollution, especially those associated with wet-weather flows, are identified as one of the primary causes of degraded water quality and ecosystems in urban areas of the United States (USEPA, 1983, 2004).

The USGS National Water Quality Assessment (NAWQA) Program is the primary source for long-term, nationwide monitoring data characterizing the quality of groundwater, surface water, and aquatic ecosystems, with more than 35 urban sites included in the study (USGS, 1999). Results from a decade of NAWQA observations have shown concentrations of phosphorus exceeded the desired USEPA goal to control plant growth in more than 70% of urban streams sampled, pesticide concentrations exceeded at least one USEPA guideline for protecting aquatic life in 93% of urban streams sampled, and fecal coliform bacteria commonly exceeded recommended standards for water recreation (Hamilton et al., 2004; Coles et al., 2004). Decades of research have shown that the quality of urban stormwater discharges even without sanitary input is sufficient to degrade receiving water quality and ecosystems (e.g., Burton and Pitt, 2002). A recent national water quality report to Congress (USEPA, 2004) identified urban runoff storm sewers as a leading source of impairment of lakes, ponds, reservoirs, and estuaries. This attention to stormwater has led to the proliferation of regulatory actions and best management practices as remedies. Realization of the impact of channelization has initiated efforts to naturalize fluvial systems in urban areas to enable them to once again support healthy, biologically diverse aquatic ecosystems (Wade et al., 2001). Construction sites have also come under scrutiny recently as a significant source of sediments discharged to water bodies.

Traditional urban development substantially modifies the water cycle, requiring a range of mitigation measures and controls to be implemented to meet flood control, water quality, and habitat protection objectives. In some cases gray infrastructure approaches have been effective at controlling runoff quantity and quality and minimizing the urban impact on the water cycle. However, recently a new approach is emerging implementing distributed or decentralized controls and infrastructure approaches. The impact of this philosophy on runoff and streamflow is described next.

Green Infrastructure Impacts

A recent change in stormwater management philosophy in the United States accompanied the emergence of LID. Decentralized stormwater management approaches were identified as a way to recreate the natural hydrologic cycle promoting sustainable water resources and healthy ecosystems, while at the same time mitigating downstream flood and water quality impacts. Low-impact development implements a range of decentralized, and often vegetated, practices (i.e., green infrastructure) to control stormwater close to its source and to promote infiltration. These and other GI practices retain stormwater close to its point of generation and release it at a slower rate through infiltration into the soil, evaporation and transpiration into the atmosphere, and controlled release into the traditional gray infrastructure system. Intuitively, the impacts of GI on the water cycle are expected to be significantly different from those described above for traditional gray infrastructure.

In general, GI reduces runoff volume, peak discharge, and pollutant loading. Permeable pavement may have the longest record of documented research among the variety of GI practices. The configuration of a permeable pavement system can be monolithic or modular, operating as a detention or retention system. When operating as a detention system, permeable pavement has in general been found to decrease peak discharge and pollutant concentrations in field, laboratory, and numerical modeling studies (Jackson and Ragan, 1974; Field et al., 1982; Brattebo and Booth, 2003; Bean et al., 2007; Collins et al., 2008). When operating as a retention system, permeable pavement reduces not only peak discharge and pollutant concentrations, but also reduces runoff volume. Similar to permeable pavement, other GI practices may operate as detention or retention systems, and similar impacts on the water cycle have been observed. In particular, the reduction of peak discharge, runoff volume, and pollutant loading by bioretention systems (Davis et al., 2009; Chapter 20, Li et al., 2010, this volume) and green roofs (Carter and Rasmussen, 2007; Chapter 19, Rowe and Getter, 2010, this volume) has been well documented. Geometrical properties, slope, and vegetation and soil characteristics have been identified as important factors influencing water quantity and quality control (VanWoert et al., 2005; Carter and Jackson, 2007; Getter et al., 2007).

Although the performance of individual GI practices has been fairly well studied, many design questions remain (Davis et al., 2009). In addition, the effectiveness of individual GI components for controlling stormwater runoff is clear, but the impact of upscaling to the municipal level is uncertain. For example, the City of Philadelphia recently proposed a plan to perform widespread implementation of GI in the next 20 yr (Bauer, 2009). This is an ambitious effort, but the impact on runoff is uncertain. Rational considerations (i.e., the effectiveness of individual components should sum) and modeling studies (Holman-Dodds et al., 2003) have shown upscaling to be effective, but a comprehensive study of benefits and impacts, especially those that may be unforeseen due to hidden linkages and cascading effects and thus not represented in past modeling studies, has not been performed.

An interesting potential opportunity for GI approaches to enhance ecosystem and human health in urban environments has recently emerged as an additional benefit beyond the water cycle. Tzoulas et al. (2007) provided an interdisciplinary review of the social science, ecosystem health, and public health literature in the context of GI. Their study led to a conceptual framework highlighting the dynamic factors and complex interactions of GI and the built environment affecting ecosystem and human health. Relying on ecological indicators has also emerged as a new approach to design of stormwater best management practices (Pomeroy, 2007).

Although GI impacts on the water cycle seem beneficial for runoff, streamflow, and water quality, the impacts on other components of the water cycle (e.g., infiltration, recharge, and ET) carry additional benefits that are not commonly factored into urban stormwater management decisions. The next two subsections describe the impacts of traditional gray infrastructure on these water cycle components and highlight the implications of introducing GI.

Infiltration, Recharge, and Groundwater Quality

Traditional Urban Impacts

Intuitively, the traditional urban impact of increasing runoff generation should cause a decrease in infiltration; however, stormwater management practices and overirrigation of landscapes may cause a net increase of infiltration and groundwater recharge with urbanization. As described above, traditional urban designs introduce substantial amounts of directly connected impervious surfaces. Impervious surfaces by definition reduce infiltration capacity (Pauleit and Duhme, 2000) and have been found to reduce infiltration, recharge, and baseflow (Klein, 1979; Simmons and Reynolds, 1982). The unexpected hydrologic alteration of pervious areas compacted during development has been documented in numerous studies to decrease infiltration at a developed site (Boyer and Burian, 2002; Pitt et al., 2008). Jones et al. (1987) described the effect of soil compaction on many factors, including increased runoff and erosion. Schueler (1996a) discussed, in general, the compaction of urban soils and examined how soil compaction increases the bulk density of urban soils to 1.5 to 1.9 g mL^{-1} due to development, which is approaching the density of concrete. Schueler (1996b) also described practices to minimize compaction in urban soils during construction or to reverse it after it occurs, including soil tilling, soil loosening, selective grading, soil amendments, compost amendments, time, and reforestation. Compost amendment and reforestation allowed for the most decrease in soil bulk density (both methods allowed a decrease of 0.25–0.35 g mL^{-1}). The conclusion, however, was that it is difficult, if not impossible, to restore a compacted soil's original structure.

In general, pervious areas in cities are likely to have highly variable infiltration rates (Pitt et al., 2001). Studies have found excavation procedures and turf placement to control infiltration rates as much as soil characteristics (Hamilton and Waddington, 1999). Sandy soils are not affected by water content as much as by compaction, while clayey soils are not as influenced by compaction as water content (Pitt and Lantrip, 2000). Temporal patterns in observed infiltration rates in urban areas led Raimbault et al. (2002) to conclude the design of retention and infiltration facilities should be based on actual rainfall data and must be able to incorporate both in-soil flows and the effects of the drainage systems. Most associate infiltration of stormwater with pervious areas, but water may pass through cracks in pavement and act as significant sources of recharge to the subsurface (Ragab et al., 2003; Sharp et al., 2006).

Expansion of the urban footprint often precedes expansion of the municipal sewage collection system. Expanding subdivisions are often initially constructed with septic systems to reduce costs, leading to enhanced downward flux of anthropogenic pollutants, and resulting in water quality degradation in underlying shallow aquifers. In the intervening decades before expansion of sewage collection systems, many cities have discovered elevated nitrates and pharmaceuticals in shallow groundwater (e.g., Seiler et al., 1999). In semiarid and arid areas, septic systems may provide the vast majority of groundwater recharge (Seiler, 1999). The surface reuse of treated wastewater (gray water recycling and use of reclaimed water) for irrigation of residential lawns, golf courses, and parks in urban areas also may change recharge rates and slowly increase the concentrations of natural constituents in soils and groundwater.

Urban areas in general may impact groundwater resources by extraction or alteration of the spatial distribution of quantity and quality from recharge. For example, water (e.g., stormwater, irrigation) infiltrated to the subsurface is retained and may be released by evaporation or transpiration, or may recharge groundwater. Unfortunately, the pathways and flux rates between the surface and subsurface in urban areas are not well defined. Underneath the urban hydrologic footprint, the effects of urbanization are similar to the effects of karstification (Sharp and Garcia-Fresca, 2003). Water mains and sewer systems may develop leaks after construction and provide pathways for enhanced recharge (Yang et al., 1999), and irrigation in urban areas may contribute significantly to recharge to the subsurface (Al-Rashed and Sherif, 2001; Meyer, 2005). Lerner (2002) reported that in Nottingham, UK about 50 km of water-carrying piping exist per square kilometer of urban area. If the average width of trenches constructed to accommodate this piping were 1 m, then 5% of the city area would have a highly modified permeability structure. Determination of these pathways is a prerequisite to determination of groundwater utilization rates and forecasting of low surface flows as a result of land use change (Nilsson et al., 2003). However, quantifying recharge in the urban footprint is problematic because such complexity makes the application of traditional techniques (e.g., the application of Darcy's Law via a groundwater flow model) difficult and prone to errors (Lerner, 2002). Given this uncertainty, it is generally assumed in areas with substantial water importation that the amount of recharge will increase, but more research is needed to accurately define the water budget to project impacts to infiltration, recharge, and groundwater quality from urban development.

Green Infrastructure Impacts

As described above, GI components are typically designed for the infiltration of captured runoff. The implications of this on the water budget are expected to be highly variable from city to city and within a city. Roof runoff to soakaways and infiltration boreholes or basins designed to dispose of storm runoff are unique to the urban environment (Price, 1994). Although only documented in a limited number of studies, these sources may be 30 to 50% of total recharge to a basin, with leaking water mains typically being the largest of these sources (Lerner, 2002). Kronaveter et al. (2001) used a hydrologic model to (i) analyze effects of urban development on infiltration and runoff and (ii) evaluate GI and other practices designed to increase infiltration. Their model showed that allowing runoff from rooftops to drain to the yard or garden and sending runoff through infiltration strips increased infiltration in a residential lot by 18% of the annual rainfall.

If the GI component is designed to infiltrate the captured stormwater runoff, the effects on infiltration and recharge are intuitive. However, the in-place performance may not meet the expectations of the design or may change with time. In general, the measured infiltration rates of the native subsoil beneath the structure provide a reasonable estimate of performance. Relatively little change to infiltration capacity has been observed over periods of several years for bioretention systems (Emerson and Traver, 2008), but more long-term performance data are needed. Some seasonal variation in infiltration rates is expected, and reduced infiltration during colder months has been observed for infiltration best management practices (Emerson and Traver, 2008), but more research is needed for a range of climate, topographic, and geological conditions. In addition, studies are needed to better quantify how the impacts of GI on infiltration will scale to the municipal level.

Although GI practices are promoted as a means to control stormwater pollution, infiltration of stormwater presents a concern with the uncertain fate and transport of contaminants (Mikkelsen et al., 1994). Long-term case studies of stormwater infiltration systems have found higher concentrations of stormwater pollutants (e.g., heavy metals) in the surface and subsurface soil systems compared with background concentrations (Mikkelsen et al., 1997; Pitt et al., 1999). In areas where the soil is sandy or a shallow water table exists, stormwater infiltration may increase the risk of groundwater contamination (Fischer et al., 2003). However, absorbable contaminants prevalent in stormwater runoff are readily removed in the soil matrix of stormwater infiltration systems constructed with sufficient depth of soil above the groundwater table (Pitt et al., 1999).

Soil Moisture and Evapotranspiration
Traditional Urban Impacts
The expeditious removal of stormwater for flood control purposes would be expected to decrease soil moisture and in turn ET magnitudes. However, stormwater management practices concentrate water in detention and retention ponds, infiltration basins, and other practices seeking to limit downstream flood impacts and treat polluted runoff. These may lead to concentrations of elevated soil moisture and increased evaporation from open water sources or transpiration from vegetated stormwater management practices. This is an area of significant uncertainty associated with dry-weather hydrologic processes that do not normally fall under the purview of drainage design and thus have been neglected in terms of stormwater management. However, the implications for ecology and urban climate and subsequent cascading effects are significant (Burian et al., 2009).

Urbanization is known to have a profound effect on the surface radiation budget and exchange of heat, moisture, and momentum between the land surface and the atmosphere (Oke, 1987; Chapters 1 [Shepherd et al., 2010] and 2 [Heisler and Brazel, 2010], this volume). Altered surface cover properties (e.g., albedo), turbulent kinetic energy, and mechanical mixing by buildings act to modify surface fluxes. Depending on changes in vegetation structure and irrigation patterns the impact on evaporative fluxes can be positive or negative.

In general, urbanization has been shown to decrease ET rates in the coterminous United States (Peterson et al., 1995), and a similar trend has been observed in the eastern United States. Urban development in eastern U.S. watersheds generally involves loss of vegetation and replacement of deciduous forests with built landscapes and impermeable surfaces. Removal of deciduous forests results in a net decrease of transpiration, causing decreased latent heat flux, increased sensible heat flux, and increased surface flows. Studies in Columbia, MD, for example, have shown urban areas have ET rates lower than rural areas (Landsberg, 1981), and a larger study of 51 watersheds in the eastern United States has confirmed a decrease in ET rates (Dow and DeWalle, 2000).

In the semiarid western United States ET rates may actually be greater in residential locations because of irrigation of landscapes introduced during urbanization (Grimmond and Oke, 1986; Burian et al., 2007). Urban and especially suburban development in much of the western United States has replaced vegetation in semiarid and arid regions with grass and tree species that are native to humid temperate environments with annual rainfall above 1000 mm. Due to the large evaporative demand in arid regions, and since mesic plant species in arid climates necessarily require large amounts of irrigation water, urbanization in the West has the potential to substantially increase ET fluxes relative to native vegetation. Typical ET rates of well-watered grasses and deciduous forests are more than 3 to 5 mm d^{-1} (Wilson and Baldocchi, 2000; Meyers, 2001), while those of semiarid and arid ecosystems are an order of magnitude smaller. Latent heat fluxes (i.e., ET) in irrigated arid situations can exceed incoming radiation as energy is advected horizontally from surrounding hot dry lands (Rosenberg and Verma, 1978). The same pattern has been observed in urban parks (Spronken-Smith et al., 2000), and it is reasonable to assume that ET rates from irrigated lawns in hot climates will also exceed available energy.

The implication of altered soil moisture and ET in urban areas is potentially acute. Changes in ET, combined with the modified urban surfaces act to alter temperatures in urban areas compared with adjacent nonurban areas. In the extreme cases, reduced ET in urban areas may increase surface and air temperatures (Landsberg, 1981; Akbari et al., 2001; Chapter 2, Heisler and Brazel, 2010, this volume), contributing to the formation of the urban heat island (UHI) (Carlson et al., 1981; Oke, 1987; Grimmond and Oke, 1995; Chapter 2, Heisler and Brazel, 2010, this volume). The hydrologic cycle, microclimate, energy use, and urban form/ landscape are interrelated through a complex system of connections and feedbacks (Burian et al., 2009) that remain uncertain and difficult to incorporate into urban design.

Green Infrastructure Impacts
There have been several studies of the impact of vegetation in urban areas on ET fluxes and microclimate (Bonan, 2000; Dimoudi and Nikolopoulou, 2003), but relatively few studies have been performed to quantify the impacts of individual or municipal-scale GI components on soil moisture and ET. The amount of water captured by GI elements that is released to the atmosphere versus that recharging

groundwater is unknown. In some cases GI might be designed to not infiltrate. In those cases, evaporation and transpiration are the only mechanisms with which runoff volume can be reduced. Runoff reduction through evaporation has been quantified for permeable pavement systems constructed with impermeable liners (Pratt et al., 1989). Plot-scale outdoor studies have further determined the dependence of evaporation from permeable pavement systems on meteorological variables with smaller evaporation magnitudes found in humid environments (Newton, 2005). Although the evaporation from the permeable pavement unit is possible, there is some question regarding the impact of permeable pavement (and other GI components) on evaporation from the underlying soils due to a reducing effect similar to having mulch on top of soil. The relative balance of increased evaporation from the storage water and the decreased evaporation from the covered soil is unknown.

Evapotranspiration processes are critical for the proper functioning of bioretention systems. However, few observations have been made of ET from bioretention systems, and there is not enough data to characterize the impact on this hydrologic process for a given set of design criteria. A study in Louisburg, NC was able to show ET accounting for the fate of 15 to 20% of all inflow water on an annual basis (Sharkey, 2006). The lack of studies quantifying ET has limited the application of ET calculations in the design of stormwater management structures (Pitt et al., 2008). Evapotranspiration impacts from implementation of greens roofs have been observed and simulated because of their important link to the urban heat island (Bass et al., 2002). However, there has not been a comprehensive study of diverse municipal-scale GI implementation on the ET, microclimate, and meteorology.

Similar to the other impacts discussed earlier, the major area of uncertainty associated with GI implementation is the impact to the water cycle at the municipal scale. Introducing vegetation for green infrastructure, UHI mitigation, and aesthetics may reduce temperatures (Rosenfeld et al., 1998; Akbari et al., 2001; Kottmeier et al., 2007; Jeyachandran et al., 2010), save energy, and provide stormwater management benefits, but the required irrigation affects the hydrologic cycle by increasing water demand (Grimmond and Oke, 1986; Guhathakurta and Gober, 2007), especially in semiarid and arid areas (Crutzen, 2004; Guhathakurta and Gober, 2007). Also, removing irrigated landscapes can have the opposite effect of increasing UHI and energy demands (Jeyachandran, 2009). This is a complex tradeoff important in urban water management decision making at many scales, and it extends from energy and water use to influences on air quality (Cardelino and Chameides, 1990) and human health (Harlan et al., 2006).

Evapotranspiration modification combined with changes in surface properties, water use, runoff, soil moisture, and vegetation in urban landscapes has a significant impact on the water cycle (Grimmond and Oke, 1986; Claessens et al., 2006). The effect of introducing GI on microclimate, energy use, and water use is difficult to determine and is usually not incorporated in design. The impact is difficult to predict because long-term observations of the urban hydrologic system in a range of climate–geographic–urban form are not yet available to develop the necessary modeling tools. Further, the effect of modified hydrologic cycle observations and microclimate (Akbari et al., 2001) by modifying surface fluxes of heat (sensible heating) and water vapor (latent heating) is unknown.

Summary and Research Needs

This chapter reviewed the impacts of traditionally developed urban areas on the water cycle and presented water cycle impacts of emerging GI practices. The literature is replete with studies of the impact of gray infrastructure approaches on the water cycle, but drawing general conclusions about the impacts to the water cycle is challenging because of the wide array of urban configurations and climate regimes. Rather, one must rely on studies fitting specified conditions or make inferences from similar studies. There are few studies of GI impacts on the water cycle, and most are focused at the component or site scale, which makes general conclusions even more difficult to define. Finding studies related to particular conditions or the ability to draw inferences is much less likely than with traditional urban impacts. To address the knowledge gap of GI impacts on the water cycle, the critical research needs are to:

- Quantify GI impacts on runoff, infiltration, groundwater recharge, soil moisture, and ET at the site scale for a range of design and climate characteristics
- Characterize mechanisms responsible for pollutant removal, especially as they relate to design variables
- Define impacts of GI on carbon and nutrient cycles
- Evaluate upscale effects of GI implementation—if it is to be applied at the municipal scale what are the implications for the hydrologic cycle?
- Collect long-term observations of new GI urbanization and areas with significant retrofit implementing GI practices
- Ascertain impact of climate change on GI applicability and performance and define possible design modifications
- Coordinate research across climate and ecoregions—much of the GI research is focused on humid climates, but the most rapid urbanization is taking place in the western United States

As this chapter and monograph demonstrate, a multidisciplinary perspective is necessary to achieve an integrated balance of hydrologic systems and the built environment by use of GI. The multiple objectives of urban water management require an integrated approach incorporating stakeholder input. In particular contexts, many of the public health, aquatic life, aesthetic, and socioeconomic impacts caused by dry- and wet-weather surface discharges are fairly well understood scientifically, and further improvements to correct these problems are dependent on overcoming institutional challenges by linking science and policy more effectively.

Urban hydrologic impacts in the past have resulted from decisions made at the large scale associated with centralized infrastructure (e.g., combined-sewer overflows, sanitary-sewer overflows, stormwater discharges). However, as decentralized approaches are introduced, the importance of individual decisions in urban areas increases as individual responsibility for operation and maintenance of the systems increases. How do traditional public works departments evolve to meet the challenges with decentralized systems or hybrid centralized-decentralized systems, and how does public education play a role?

Policy changes for GI must understand the potential impacts on urban hydrology. Management policies set at the municipal, state, and federal levels are developed and passed by elected officials who rely on technical staff and lobbyists for advice, but the officials often lack fundamental understanding of the principles of hydrologic science, especially in cases such as GI, where studies are lacking. Such policies establish the legal and regulatory context that directly, or indirectly, affects various aspects of the hydrologic system (i.e., water rights laws, water quality regulations, water supply system requirements, land management policies, regional metropolitan planning, land development regulations, property taxes, construction and maintenance of water infrastructure, and water use fees). In some places policy changes to encourage GI must take into account the water laws governing human interaction with the water cycle. Concerns for possible legality of GI approaches and decentralized approaches must be considered.

References

Akbari, H., M. Pomerantz, and H. Taha. 2001. Cool surfaces and shade trees to reduce energy use and improve air quality in urban areas. Sol. Energ. 70(3):295–310.

Al-Rashed, M.F., and M.M. Sherif. 2001. Hydrogeological aspects of groundwater drainage or the urban areas in Kuwait City. Hydrol. Processes 15:777–795.

Ando, Y., K. Musiake, and Y. Takahasi. 1984. Modeling of hydrologic processes in a small urbanized hillslope basin with comments on the effects of urbanization. J. Hydrol. 68(1–4):61–83.

Bass, B., R. Stull, S. Krayenjoff, and A. Martilli. 2002. Modelling the impact of green roof infrastructure on the urban heat island in Toronto. Green Roof Infrastruct. Monit. 4(1):2–3.

Bauer, S. 2009. Breaking ground with a $1.6 billion plan to tame water. Available at www.philly.com (accessed 28 Sept. 2009, verified 16 Mar. 2010).

Bean, E.Z., W.F. Hunt, and D.A. Bidelspach. 2007. Evaluation of four permeable pavement sites in Eastern North Carolina for runoff reduction and water quality impacts. J. Irrig. Drain. Eng. 133(6):583–592.

Beer, A.R. 2010. Greenspaces, green structure, and green infrastructure planning. p. 431–448. In J. Aitkenhead-Peterson and A. Volder (ed.) Urban ecosystem ecology. Agron. Monogr. 55. ASA, CSSA, and SSSA, Madison, WI.

Boesch, D.F., R.B. Brinsfield, and R.E. Magnien. 2001. Chesapeake Bay eutrophication: Scientific understanding, ecosystem restoration, and challenges for agriculture. J. Environ. Qual. 30(2):303–320.

Bonan, G.B. 2000. The microclimates of a suburban Colorado (USA) landscape and implications for planning and design. Landsc. Urban Plan. 49:97–114.

Booth, D.B., and C.R. Jackson. 1997. Urbanization of aquatic systems—Degradation thresholds, stormwater detention, and the limits of mitigation. Water Resour. Bull. 33:1077–1090.

Boyer, M., and S.J. Burian. 2002. The effects of construction activities and the preservation of indigenous vegetation on stormwater runoff rates in urbanizing landscapes. Proc. Indigenous Vegetation within Urban Development, Uppsala, Sweden. 14–16 Aug. 2002.

Brattebo, B.O., and D.B. Booth. 2003. Long-term stormwater quantity and quality performance of permeable pavement systems. Water Res. 37:4369–4376.

Brown, G.W., and J.T. Krygier. 1970. Effects of clearcutting on stream temperature. Water Resour. Res. 6(4):1133–1139.

Burian, S.J. 2001. Developments in water supply and wastewater management in the United States during the nineteenth century. Water Resour. Impact 3(5):14–18.

Burian, S.J., S.J. Nix, S.R. Durrans, R. Pitt, C.Y. Fan, and R. Field. 1999. The historical development of wet weather flow management. J. Water Resour. Plann. Manage. 125(1):3–13.

Burian, S.J., S.J. Nix, R.E. Pitt, and S.R. Durrans. 2000. Urban wastewater management in the United States: Past, present, and future. J. Urban Technol. 7(3):33–62.

Burian, S.J., E. Pardyjak, C. Forster, S. Bush, I. Jeyachandran, P. Ramamurthy, and N. Augustus. 2007. Interdisciplinary study of coupled water-biophysical-climate systems in a semi-arid urban environment. *In* AMS 7th Symposium on the Urban Environment, San Diego, CA. 10–13 Sept. 2007. AMS, Washington, DC.

Burian, S.J., E. Pardyjak, I. Jeyachandran, N. Augustus, P. Ramamurthy, C. Forster, and B. Skousen. 2009. Climate and urban form effects on the water and energy budgets in a residential area of Salt Lake City, Utah. *In* 89th AMS Annual Meeting, Eighth Symposium on the Urban Environment, Phoenix, AZ. 11–15 Jan. 2009. AMS, Washington, DC.

Burian, S.J., and J.M. Shepherd. 2005. Effect of urbanization on the diurnal rainfall pattern in Houston. Hydrol. Processes 19(5):1089–1103.

Burton, G.A., Jr., and R.E. Pitt. 2002. Stormwater effects handbook, Lewis Publ., Boca Raton, FL.

Cardelino, C.A., and W.L. Chameides. 1990. Natural hydrocarbons, urbanization, and urban ozone. J. Geophys. Res. 95(D9):13971–13979.

Carlson, T.N., J.K. Dodd, S.G. Benjamin, and J.N. Cooper. 1981. Satellite estimation of the surface energy balance, moisture availability and thermal inertia. J. Appl. Meteorol. 20(1):67–87.

Carter, T.L., and C.R. Jackson. 2007. Vegetated roofs for stormwater management at multiple spatial scales. Landsc. Urban Plan. 80(1–2):84–94.

Carter, T.L., and T.C. Rasmussen. 2006. Hydrologic behavior of vegetated roofs. J. Am. Water Resour. Assoc. 42(5):1261–1274.

Claessens, L., C. Hopkinson, E. Rastetter, and J. Vallino. 2006. Effect of historical changes in land use and climate on the water budget of an urbanizing watershed. Water Resour. Res. 42:W03426, doi:10.1029/2005WR004131.

Cleugh, H.A., and T.R. Oke. 1986. Suburban–rural energy balance comparisons in summer for Vancouver, B.C. Boundary-Layer Meteorol. 36(4):351–369.

Coles, J.F., T.F. Cuffney, G. McMahon, and K.M. Beaulieu. 2004. The effects of urbanization on the biological, physical, and chemical characteristics of coastal New England streams. USGS Professional Paper 1695. USGS, Reston, VA.

Collins, K.A., W.F. Hunt, and J.M. Hathaway. 2008. Hydrologic comparison of four types of permeable pavement and standard asphalt in eastern North Carolina. J. Hydrol. Eng. 13(12):1146–1157.

Crutzen, P.J. 2004. New directions: The growing urban heat and pollution "island" effect—Impact on chemistry and climate. Atmos. Environ. 38(21):3539–3540.

Davis, A.P., W.F. Hunt, R.G. Traver, and M. Clar. 2009. Bioretention technology: Overview of current practice and future needs. J. Environ. Eng. 135(3):109–117.

De Jonge, V.N., M. Elliott, and E. Orive. 2002. Causes, historical development, effects and future challenges of a common environmental problem: Eutrophication. Hydrobiologia 475–476:1–19.

DeWalle, D.R., B.R. Swistock, T.E. Johnson, and K.J. McGuire. 2000. Potential effects of climate change and urbanization on mean annual streamflow in the United States. Water Resour. Res. 36(9):2655–2664.

Dimoudi, A., and M. Nikolopoulou. 2003. Vegetation in the urban environment: Microclimatic analysis and benefits. Energy Build. 35:69–76.

Dow, C.L., and D.R. DeWalle. 2000. Trends in evaporation and Bowen ratio on urbanizing watersheds in eastern United States. Water Resour. Res. 36(7):1835–1843.

Elvidge, C.D., C. Milesi, J.B. Dietz, B.T. Tuttle, P.C. Sutton, R. Nemani, and J.E. Vogelmann. 2004. U.S. constructed area approaches the size of Ohio. EOS 85(24), doi:10.1029/2004EO240001.

Emerson, C.H., and R.G. Traver. 2008. Multi-year and seasonal variation of infiltration from stormwater best management practices. J. Irrig. Drain. Eng. 134(5):598–605.

Federal Interagency Stream Restoration Working Group. 1998. Stream corridor restoration: Principles, processes, and practices.

Field, R., H. Masters, and M. Singer. 1982. Porous pavement: Research, development and demonstration. Transp. Eng. J. 108(TE3):244–258.

Field, R., and R.E. Pitt. 1990. Urban storm-induced discharge impacts: US Environmental Protection Agency Research Program Review. Water Sci. Technol. 22(10/11):1–7.

Field, R., and R. Turkeltaub. 1981. Urban runoff receiving water impacts: Program overview. J. Environ. Eng. Div. 107(1):83–100.

Fischer, D., E.G. Charles, and A.L. Baehr. 2003. Effects of stormwater infiltration on quality of groundwater beneath retention and detention basins. J. Environ. Eng. 129(5):464–471.

Getter, K.L., D.B. Rowe, and J.A. Andresen. 2007. Quantifying the effect of slope on extensive green roof stormwater retention. Ecol. Eng. 31(4):225–231.

Grimmond, C.S.B., and T.R. Oke. 1986. Urban water balance. 2: Results from a suburb of Vancouver, British Columbia. Water Resour. Res. 22(10):1404–1412.

Grimmond, C.S.B., and T.R. Oke. 1995. Comparison of heat fluxes from summertime observations in the suburbs of four North American cities. J. Appl. Meteorol. 34(4):873–889.

Guhathakurta, S., and P. Gober. 2007. The impact of the Phoenix urban heat island on residential water use. J. Am. Plann. Assoc. 73(3):317–329.

Hamilton, G.W., and D.V. Waddington. 1999. Infiltration rates on residential lawns in central Pennsylvania. J. Soil Water Conserv. 54:564–568.

Hammer, T.R. 1972. Stream channel enlargement due to urbanization. Water Resour. Res. 8:139–167.

Haile, R.W., J. Alamillo, K. Barrett, R. Cressey, J. Dermond, C. Ervin, A. Glasser, N. Harawa, P. Harmon, J. Harper, C. McGee, R.C. Millikan, M. Nides, and J.S. Witte. 1996. An epidemiological study of possible adverse health effects of swimming in Santa Monica Bay. Rep. Santa Monica Bay Restoration Project, Santa Monica, CA.

Hamilton, P.A., T.L. Miller, and D.N. Myers. 2004. Water quality in the Nation's streams and aquifers—Overview of selected findings, 1991–2000. USGS Circ. 1265. USGS, Reston, VA.

Harlan, S.L., A.J. Brazel, L. Prashad, W.L. Stefanov, and L. Larsen. 2006. Neighborhood microclimates and vulnerability to heat stress. Soc. Sci. Med. 63:2847–2863.

Heaney, J.P., and W.C. Huber. 1984. Nationwide assessment of urban runoff impact on receiving water quality. Water Resour. Bull. 20(1):35–42.

Heisler, G.M., and A.J. Brazel. 2010. The urban physical environment: Temperature and urban heat islands. p. 29–56. In J. Aitkenhead-Peterson and A. Volder (ed.) Urban ecosystem ecology. Agron. Monogr. 55. ASA, CSSA, and SSSA, Madison, WI.

Hollis, G.E. 1975. The effect of urbanization on floods of different recurrence interval. Water Resour. Res. 11(3):431–435.

Holman-Dodds, J., A. Bradley, and K. Potter. 2003. Evaluation of hydrologic benefits of infiltration based urban storm water management. J. Am. Water Resour. Assoc. 39:205–215.

Jackson, T.J., and M. Ragan. 1974. Hydrology of porous pavement parking lots. J. Hydrol. Div. 100:1739–1752.

Jeyachandran, I. 2009. Effects of landscape modification on evapotranspiration, microclimate, energy use, and water use in urban environments. Ph.D. diss. Univ. of Utah, Salt Lake City.

Jeyachandran, I., S.J. Burian, E. Pardyjak, C.A. Pomeroy, and B. Skousen. 2010. Impact of green infrastructure on urban energy fluxes and microclimate. EWRI India 2010 Conference Proc. ASCE, Reston, VA.

Jones, A., E. Dickey, D. Eisenhauer, and R. Wiese. 1987. Identification of soil compaction and its limitations to root growth. Inst. of Agriculture and Nat. Res., University of Nebraska, Lincoln.

Keefer, T.N., R.K. Simons, and R.S. McQuivey. 1979. Dissolved oxygen impact from urban storm runoff. EPA-600-2-79-150. USEPA, Cincinnati, OH.

Klein, R.D. 1979. Urbanization and stream quality impairment. Water Resour. Bull. 15(4):948–963.

Kottmeier, C., C. Biegert, and U. Corsmeier. 2007. Effects of urban land use on surface temperature in Berlin: Case study. J. Urban Plann. Dev. 133(2):128–137.

Kronaveter, L., U. Shamir, and A. Kessler. 2001. Water-sensitive urban planning: Modeling on-site infiltration. J. Water Resour. Plann. Manage. 127(2):78–88.

Landsberg, H. 1981. The urban climate. Academic Press, New York.

Leopold, L.B. 1968. Hydrology for urban planning—A guidebook on the hydrologic effects of urban land use. USGS Circ. 554. USGS, Washington, DC.

Lerner, D. 2002. Identifying and quantifying urban recharge; a review. Hydrogeol. J. 10:143–152.

Li, M.-H., B. Dvorak, and C.Y. Sung. 2010. Bioretention, low impact development, and stormwater management. p. 413–430. In J. Aitkenhead-Peterson and A. Volder (ed.) Urban ecosystem ecology. Agron. Monogr. 55. ASA, CSSA, and SSSA, Madison, WI.

Masek, J.G., F.E. Lindsay, and S.N. Goward. 2000. Dynamics of urban growth in the Washington, DC metropolitan area, 1973–1996, from Landsat observations. Int. J. Remote Sens. 21(18):3473–3486.

McPherson, T.N., S.J. Burian, M.K. Stenstrom, H.J. Turin, M.J. Brown, and I.H. Suffet. 2005. Comparison of dry and wet weather flow nutrient loads from a Los Angeles watershed. J. Am. Water Resour. Assoc. 41(4):959–969.

Melosi, M.V. 2000. The sanitary city: Urban infrastructure in America from colonial times to the present. The Johns Hopkins Univ. Press, Baltimore, MD.

Metcalf & Eddy, Inc. 2003. Wastewater engineering: Treatment and reuse. 4th ed. McGraw-Hill, New York.

Meyer, S.C. 2005. Analysis of base flow trends in urban streams, northeastern Illinois, USA. Hydrogeol. J. 13(5–6):871–885.

Meyers, T.P. 2001. A comparison of summertime water and CO_2 fluxes over rangeland for well watered and drought conditions. Agric. For. Meteorol. 106:205–214.

Mikkelsen, P.S., M. Häfliger, M. Ochs, P. Jacobsen, J.C. Tjell, and M. Boller. 1997. Pollution of soil and groundwater from infiltration of highly contaminated stormwater—A case study. Water Sci. Technol. 36(8–9):325–330.

Mikkelsen, P.S., G. Weyer, C. Berry, Y. Walden, V. Colandini, S. Poulsen, D. Grotehusmann, and R. Rohlfing. 1994. Pollution from urban stormwater infiltration. Water Sci. Technol. 29(1–2):293–302.

Moffa, P.E. 1997. The combined sewer overflow problem: An overview. p. 1–26. In P.E. Moffa (ed.) Control and treatment of combined sewer overflows. 2nd ed. Van Nostrand Reinhold, New York.

Newton, D.B. 2005. The effectiveness of modular porous pavement as a stormwater treatment device. Ph.D diss. Griffith Univ., Brisbane.

Nilsson, C., J.E. Pizzuto, G.E. Moglen, M.A. Palmer, E.H. Stanley, N.E. Bockstael, and L.C. Thompson. 2003. Ecological forecasting and the urbanization of stream ecosystems: Challenges for economists, hydrologists, geomorphologists, and ecologists. Ecosystems 6:659–674.

Oke, T.R. 1987. Boundary layer climates. Routledge, London.

Pauleit, S., and F. Duhme. 2000. Assessing the environmental performance of land cover types for urban planning. Landsc. Urban Plan. 52(1):1–20.

Peterson, T.C., V.S. Golubev, and P.Y. Groisman. 1995. Evaporation losing its strength. Nature 377(6551):687–688.

Pitt, R., and P. Bissonnette. 1984. Bellevue Urban Runoff Program, Summary Report. USEPA and the Storm and Surface Water Utility, Bellevue, WA.

Pitt, R., S.-E. Chen, and S. Clark. 2002. Compacted urban soils effects on infiltration and bioretention stormwater control designs. Proc. of Ninth International Conf. on Urban Drainage, Portland, OR.

Pitt, R., S.-E. Chen, S.E. Clark, J. Swenson, and C.K. Ong. 2008. Compaction's impacts on urban storm-water infiltration. J. Irrig. Drain. Eng. 134(5):652–658.

Pitt, R., S.-E. Chen, and C.K. Ong. 2001. Measurements of infiltration rates in compacted urban soils. p. 534–538. In Proc. of an Engineering Foundation Conf. ASCE, Snowmass Village, CO.

Pitt, R., S. Clark, and R. Field. 1999. Groundwater contamination potential from stormwater infiltration practices. Urban Water 1(3):217–236.

Pitt, R., and J. Lantrip. 2000. Infiltration through disturbed urban soils. p. 1–22. In W. James (ed.) Advances in modeling the management of stormwater impacts. Vol. 8. Computational Hydraulics International, Guelph, ON, Canada.

Pitt, R., J. Voorhees, and S. Clark. 2008. Evapotranspiration and related calculations for stormwater biofiltration devices: Proposed calculation scenario and data. p. 309– 340. *In* W. James et al. (ed.) Stormwater and urban water systems modeling. Monogr. 16. Computational Hydraulics International, Guelph, ON, Canada.

Pomeroy, C.A. 2007. Evaluating the impacts of urbanization and stormwater management practices on stream response. Ph.D. diss. Dep. of Civil Engineering, Colorado State Univ., Fort Collins.

Pratt, C.J., J.D. Mantle, and P.A. Schofield. 1989. Urban stormwater reduction and quality improvement through use of permeable pavements. Water Sci. Technol. 21(8/9):769–778.

Price, M. 1994. Drainage from roads and airfields to soakaways: Groundwater pollutant or valuable recharge? J. Inst. Water Environ. Manage. 8(5):468–479.

Prince George's County-Maryland. 1999. Low impact development design strategies. Prince George's County, Dep. of Environmental Resources, Programs and Planning Div., Largo, MD.

Ragab, R., P. Rosier, A. Dixon, J. Bromley, and J.D. Cooper. 2003. Experimental study of water fluxes in residential area: 2. Road infiltration, runoff and evaporation. Hydrol. Processes 17:2423–2437.

Raimbault, G., E. Berthier, M. Mosini, and C. Joannis. 2002. Urban stormwater infiltration and soil drainage. Proc. of Ninth International Conf. on Urban Drainage, Portland, OR.

Roesner, L.A., B.P. Bledsoe, and R.W. Brashear. 2001. Are best management-practice criteria really environmentally friendly? J. Water Resour. Plann. Manage. 5/6:150–154.

Romero, H., M. Ihl, A. Rivera, P. Zalazar, and P. Azocar. 1999. Rapid urban growth, land-use changes and air pollution in Santiago, Chile. Atmos. Environ. 33(24–25):4039–4047.

Rosenberg, N.J., and S.B. Verma. 1978. Extreme evapotranspiration by irrigated alfalfa: A consequence of the 1976 Midwestern drought. J. Appl. Meteorol. 17:934–941.

Rosenfeld, A.H., H. Akbari, J.J. Romm, and M. Pomerantz. 1998. Cool communities: Strategies for heat island mitigation and smog reduction. Energy Build. 28(1):51–62.

Rowe, D.B., and K.L. Getter. 2010. Green roofs and garden roofs. p. 391–412. *In* J. Aitkenhead-Peterson and A. Volder (ed.) Urban ecosystem ecology. Agron. Monogr. 55. ASA, CSSA, and SSSA, Madison, WI.

Roy, A.H., M.J. Paul, and S.J. Wenger. 2010. Urban stream ecology. 2010. p. 341–352. *In* J. Aitkenhead-Peterson and A. Volder (ed.) Urban ecosystem ecology. Agron. Monogr. 55. ASA, CSSA, and SSSA, Madison, WI.

Sandström, U.F. 2002. Green infrastructure planning in urban Sweden. Plann. Pract. Res. 17(4):373–385.

Schueler, T.S. 1994. The importance of imperviousness. Watershed Protect. Tech. 1(3):100–111.

Schueler, T.S. 1996a. The compaction of urban soils. Watershed Protect. Tech. 3(2):661–665.

Schueler, T.S. 1996b. Can urban soil compaction be reversed? Watershed Protect. Tech. 3(2):666–669.

Seiler, R. 1999. A chemical signature for ground water contaminated by domestic wastewater. Ph.D. diss. Univ. of Nevada, Reno.

Seiler, R., S. Zaugg, J. Thomas, and D. Howcroft. 1999. Caffeine and pharmaceuticals as indicators of waste water contamination in wells. Ground Water 37(3):405–410.

Sharkey, L.J. 2006. The performance of bioretention areas in North Carolina: A study of water quality, water quantity, and soil media. M.S. thesis. North Carolina State Univ., Raleigh.

Sharp, J.M., L.N. Christian, B. Garcia-Fresca, and C.A. Stewart. 2006. Changing recharge and hydrogeology in an urbanizing area—Example of Austin, Texas, USA. Geol. Soc. America, Abst. with Programs (Ann. Mtg.) 38(7):289.

Sharp, J.M., and B. Garcia-Fresca. 2003. Effects of urbanization of groundwater resources, recharge rates, and flow patterns. Geol. Soc. America, Abst. with Programs (Ann. Mtg.) 35:158.

Shepherd, J.M. 2005. A review of current investigations of urban-induced rainfall and recommendations for the future. Earth Interact. 9(12):1–27.

Shepherd, J.M., J.A. Stallins, M.L. Jin, and T.L. Mote. 2010. Urbanization: Impacts on clouds, precipitation, and lightning. p. 1–28. *In* J. Aitkenhead-Peterson and A. Volder (ed.) Urban ecosystem ecology. Agron. Monogr. 55. ASA, CSSA, and SSSA, Madison, WI.

Simmons, D.L., and R.S. Reynolds. 1982. Effect of urbanization on base flow of selected south-shore streams, Long Island, New York. Water Resour. Bull. 18(5):797–805.

Spronken-Smith, R.A., T.R. Oke, and W.P. Lowry. 2000. Advection and the surface energy balance across an irrigated urban park. Int. J. Climatol. 20:1033–1047.

Steele, M.K., W.H. McDowell, and J.A. Aitkenhead-Peterson. 2010. Chemistry of urban, sub-urban, and rural surface waters. p. 297–340. *In* J. Aitkenhead-Peterson and A. Volder (ed.) Urban ecosystem ecology. Agron. Monogr. 55. ASA, CSSA, and SSSA, Madison, WI.

Tzoulas, K., K. Korpela, S. Venn, V. Yli-Pelkonen, A. Kaźmierczak, J. Niemela, and P. James. 2007. Promoting ecosystem and human health in urban areas using Green Infrastructure: A literature review. Landsc. Urban Plan. 81:167–178.

United Nations Population Division. 2009. World urbanization prospects. Available at http://esa.un.org/unpd/wup/index.htm (accessed 31 Mar. 2009, verified 22 May 2010).

United Nations Population Fund. 2009. Introduction: Peering into the dawn of an urban millennium. Available at http://www.unfpa.org/swp/2007/english/introduction.html (accessed 31 Mar. 2010, verified 22 May 2010).

USEPA. 1977. Suspended and dissolved solids effects on freshwater biota: A review. EPA-600-3-77-042. USEPA, Corvallis, OR.

USEPA. 1983. Final report of the Nationwide Urban Runoff Program, Vol. I. USEPA, Washington, DC.

USEPA. 1996. Combined sewer overflows and the multimetric evaluation of their biological effects: Cases studies in Ohio and New York. EPA-823-R-96-002, USEPA, Office of Water, Washington, DC.

USEPA. 1997. Urbanization and streams: Studies of hydrologic impacts. EPA-841-R-97-009. USEPA, Office of Water, Washington, DC.

USEPA. 2004. National Water Quality Inventory: Report to Congress. EPA 841-R-08-001. USEPA, Washington, DC.

USGS. 1999. The quality of our nation's waters—Nutrients and pesticides. USGS Circ. 1225. USGS, Reston, VA.

VanWoert, N.D., D.B. Rowe, J.A. Andresen, C.L. Rugh, R.T. Fernandez, and X. Lan. 2005. Green roof stormwater retention: Effects of roof surface, slope, and media depth. J. Environ. Qual. 34(3):1036–1044.

Verstraeten, I.M., T. Heberer, and T. Scheytt. 2002. Occurence, characteristics, transport, and fate of pesticides, pharmaceuticals, industrial products, and personal care products at riverbank filtration sites. p. 175–228. *In* C. Ray et al. (ed.) Riverbank filtration: Improving source water quality. Kluwer Academic Publ., Dordrecht, the Netherlands.

Wade, R.J., B.L. Rhoads, J. Rodríguez, M. Daniels, D. Wilson, E.E. Herricks, F. Bombardelli, M. Garcia, and J. Schwartz. 2001. Integrating science and technology to support stream naturalization near Chicago, Illinois. J. Am. Water Resour. Assoc. 38(4):931–944.

Welty, C. 2009. The urban water budget. p. 17–28. *In* L.A. Baker (ed.) The water environment of cities. Springer, New York.

Wilson, K.B., and D.D. Baldocchi. 2000. Seasonal and interannual variability of energy fluxes over a broadleaved temperate deciduous forest in North America. Agric. For. Meteorol. 100:1–18.

Yang, Y., D.N. Lerner, M.H. Barrett, and J.H. Tellham. 1999. Quantification of groundwater recharge in the City of Nottingham, UK. Environ. Geol. 38:183–198.

15

Chemistry of Urban, Suburban, and Rural Surface Waters

Meredith K. Steele
William H. McDowell
Jacqueline A. Aitkenhead-Peterson

Abstract

The chemistry of urban surface waters is influenced by natural biogeochemical processes that interact with the physical structures as well as the many activities and material inputs associated with dense human habitation. Concentrations of sediment, nitrogen, phosphorus, natural and novel organic compounds, metals, chloride, sulfate, calcium, magnesium, sodium, and potassium all change as the percentage of urbanized land increases. Unique urban chemistries result from many factors in the urban environment. They include high rates of watershed loading that increase with human activity and population density; more direct hydrologic pathways that route runoff and waste streams to surface waters via drainage systems and wastewater treatment plants; changes in the capacity of the landscape to retain and transform nutrient and organic matter inputs due to loss of vegetative cover and effective riparian zones; and alteration of streambed sediments, streambed form, and rates of bank erosion by increased peak flows. The conditions that create the uniquely urban biogeochemical cycles often lead to a degradation of water quality, but the extent of degradation can vary dramatically. Improvements in waste management, drainage systems, and stream restoration have been shown to reduce degradation or rehabilitate water quality in urban areas. The following is a review of available knowledge on global urban water chemistry and the chemical cycling in urban watersheds, which results in distinct urban biogeochemistry.

Throughout the world, the number of people living in cities is growing. By 2050, 70% of the people on the planet will live in urban areas, with 85% of the population in developed nations and 67% of less developed nations living in urban areas (United Nations Population Division, 2008). Rapid population growth in urban areas causes both more extensive and more intensive development. This development and the activities that occur following development have a significant impact on the chemistry of surface water within and downstream of cities.

M.K. Steele (mkbilek@gmail.com) and J.A. Aitkenhead-Peterson (jpeterson@ag.tamu.edu), Dep. of Soil and Crop Sciences, Texas A&M University, 2474 TAMU, College Station, TX 77843; W.H. McDowell, Dep. of Natural Resources, James Hall, University of New Hampshire, Durham, NH 03824 (bill.mcdowell@unh.edu).

doi:10.2134/agronmonogr55.c15

The *National Water Quality Inventory: 2000 Report* identified runoff within urban ecosystems as a leading source of water quality impairment to surface waters, ranking it fourth for rivers and streams, third for lakes, and second for estuaries (USEPA, 2002). Pathogens, sediments, toxic chemicals, heavy metals, nutrients, salinity, and carbon are potential impairments caused by urbanization. Bernhardt and Palmer (2007) described the state of urban streams as "gutters," simplified channels that carry high loads of water and other elements away from urban centers. Both intentionally and unintentionally, people have used the natural process of collection and removal provided by streams and rivers to simply carry away wastes and residues. This collection system integrates human activities and natural processes on the land and in the water, resulting in what has been called a "distinct urban biogeochemistry" (Kaye et al., 2006). Almost every object and organism in a city interacts with water at some point in its existence, from cradle or manufacturing to its grave or disposal. Water has been called the "universal solvent," and thus every interaction has the potential to alter water chemistry and quality. However, the system is robust, and generally the most common activities and objects in cities have the largest impact.

Cities are characterized by a large fraction of impervious surfaces, and almost all of the urban impacts that manifest themselves as water quality changes have some connection to increases in impervious surfaces. Impervious surfaces are relatively smooth, reduce water infiltration into the soil, and are designed to shed water as quickly as possible. These conditions create an ideal situation for water to transport anything on the surface. The other common element of urban systems that impacts stream chemistry is the use of wastewater treatment facilities (WWTF). Changes in impervious surface are typically considered to be nonpoint or diffuse sources of pollution, while WWTF are technically considered to be point sources because they provide a single, easily identified input to surface waters. The actual impacts of WWTF on water quality are more complicated, however, because leaks in the sewer lines can discharge raw sewage into surface waters as an effective nonpoint source, and for combined sewer systems, the runoff from impervious systems can be treated and thus can act to reduce nonpoint source inputs to surface waters.

This chapter describes terrestrial and aquatic sources and sinks, the resulting concentrations and/or loads in surface runoff, and the in-stream processes that regulate the retention or loss in surface waters of sediment, nutrients (nitrogen and phosphorus), carbon (natural and novel), metals, and major anions and cations in urban ecosystems.

Sediment

Human activities associated with urbanization generally have higher erosion rates per area than many agricultural and rural activities (Brandt et al., 1972; Novonty and Olem, 1994). The reported erosion rates both during and after urban construction range from 5 to 50,000 Mg km^{-2} yr^{-1} (Table 15–1). In contrast, estimates of erosion rates in agriculture range from 100 to 4000 Mg km^{-2} yr^{-1} (Brandt et al., 1972). The highest reported erosion rates for urban areas are generally associated with construction (Wolman and Schick, 1967; Maniquiz et al., 2009), and the majority of erosion estimates in established urban areas are less than 1000 Mg km^{-2} yr^{-1} (Fig. 15–1). A limited number of studies have investigated stream

Table 15–1. Erosion rates for various urban activities and land uses reported in the literature.

	Erosion rate	Location	Reference
	Mg km^{-2} yr^{-1}		
		Active construction	
Industrial complex	6,290	Korea	Maniquiz et al., 2009
Road	290	Korea	Maniquiz et al., 2009
Recreational area	10,560	Korea	Maniquiz et al., 2009
Athletic facility	21,050	Korea	Maniquiz et al., 2009
Construction site	96	Washington, USA	Reinelt, 1996
Urban development	2,630	Korea	Maniquiz et al., 2009
Building sites	49,000	Illinois, USA	Illinois EPA, 1995
Road cuts	50,000	Georgia, USA	Georgia Department of Transportation, 1993
Construction site	35,056	Maryland, USA	Wolman and Schick, 1967
		Post construction	
Industrial complex	469	Korea	Maniquiz et al., 2009
Recreational area	1,820	Korea	Maniquiz et al., 2009
Urban Development	152	Korea	Maniquiz et al., 2009
Athletic facility	1,600	Korea	Maniquiz et al., 2009
Commercial	80	Washington, USA	Horner, 1992
Commercial	28,032	Maryland, USA	Wolman and Schick, 1967
Industrial park	25,228	Maryland, USA	Wolman and Schick, 1967
Urban (developing)	813	Maryland, USA	Wark and Keller, 1963
Urban (developing)	648	Maryland, USA	Wark and Keller, 1963
Urban (developing)	560	District of Columbia, USA	Wark and Keller, 1963
Urban (developing)	371	Maryland, USA	Wark and Keller, 1963
		Residential	
Low density	5	Washington, USA	Reinelt, 1996
Moderate density	32	Washington, USA	Horner, 1992
High density	35	Washington, USA	Reinelt, 1996
Housing	8,409	Maryland, USA	Guy, 1963
Housing	3,959	Maryland, USA	Wolman and Schick, 1967
Housing	1,962	Maryland, USA	Guy and Ferguson, 1962
Housing	1,138	Virginia, USA	Holeman and Geiger, 1959
		Channel erosion	
San Diego Creek ~50% Urbanized	46.5	California, USA	Trimble, 1997
Issaquah Creek ~19% Urbanized	3.5	Washington, USA	Nelson and Booth, 2002

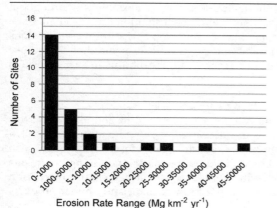

Fig. 15–1. Frequency of measured erosion rate estimates in urban ecosystems found in the literature. Data synthesized from references in Table 15–1.

Erosion Rate Range (Mg km^{-2} yr^{-1})

channel erosion rates in urban areas. These studies suggest that a significant portion of the urban contribution to sediment loads comes from high peak flows and decreases in vegetation that result in increased stream channel erosion (Trimble, 1997; Nelson and Booth, 2002). Trimble (1997) suggests that construction of armoring and stabilization of urban channels should be implemented to prevent the degradation and loss of sediments. Maniquiz et al. (2009) found active construction contributed the majority of the sediment eroded from several types of urban development sites; only 20 to 40% of sediment loss occurred before and after construction, and in some cases post-construction erosion was lower than its undeveloped state. Their study also demonstrated that losses of sediments during and after construction were not equal for all types of facilities; athletic and recreational facilities had higher post-construction erosion rates than other types of urban facilities (Maniquiz et al., 2009). The texture of stream sediment may also be affected by urbanization with increasing proportions of coarse-textured sands and reduction in fine-textured silts and clays (Finkenbine et al., 2000; Pizzuto et al., 2000). This may explain why correlations between urban land cover and total suspended solids have not always been found (Brett et al., 2005). Sediment eroded from an upland catchment area suspended in the water column and settled on the bed plays an important role in the release and adsorption of many chemical constituents (e.g., heavy metals and phosphorus), physical properties (e.g., turbidity and temperature), as well as biotic processes.

Nitrogen and Phosphorus

The concentrations and loads of nitrogen and phosphorus in surface waters have received enormous attention in the past several decades (e.g., Grimm et al., 2000; Paul and Meyer, 2001; Bernhardt et al., 2008). Interest in these two elements results from their key roles as limiting nutrients for aquatic biota and the eutrophication of lakes, estuaries, rivers, and streams (Huisman et al., 2005; Diaz et al., 2008). Eutrophication of surface water is considered one of the leading causes of dead zones in estuaries, where the high nutrient concentrations increase algal primary productivity and result in a depletion of oxygen when algal biomass is subsequently decomposed. Previously nitrogen was thought to control dead zones in marine environments, while phosphorus was more important in controlling freshwater eutrophication, but new evidence suggests that a dual approach of considering nitrogen and phosphorus together may reduce impacts throughout a drainage network and provide a more permanent solution (Conely et al., 2009). While much research has been aimed at quantifying agricultural contributions to nutrient loads, urbanization has been recognized for some time as a cause of high loads and concentrations of nitrogen and phosphorus in surface waters (USEPA, 2002).

Urban Nitrogen Cycling

Reactive nitrogen is an important component of life and its abundance controls the rate of many biological processes. *Reactive nitrogen* refers to inorganic nitrogen, such as ammonium and nitrate, as well as organic forms of nitrogen, such as amino acids. Both forms can be utilized by the majority of living organisms. Henceforth, the term *nitrogen* (N) will refer to reactive nitrogen in this discussion. Humans require N and consume it in the form of plant and animal products.

Human alteration of the landscape by agriculture has increased the delivery of N to surface waters (Boyer et al., 2002), both by increases in coverage of N-fixing crops (legumes and rice [*Oryza sativa* L.] in association with cyanobacteria) and by the synthesis of inorganic nitrogen fertilizers (Galloway et al., 2004). In combination with fossil fuel combustion, the amount of reactive nitrogen produced by humans is expected to exceed natural production by the year 2050 (Galloway et al., 2004; IPCC, 2007). Nitrogen fertilizers revolutionized agriculture during the 20th century and altered the N cycle in large areas of the earth. Urban centers indirectly concentrate much of this N fertilizer in the form of food shipped in from agricultural areas, dispersed throughout the population, and reconcentrated at wastewater treatment facilities (Groffman et al., 2004). Food is not the only source of N inputs to urban ecosystems and is only one of the pathways N takes through the urban environment. Like N cycling in pristine regions, urban nitrogen cycling is complex. Several components make up the nitrogen cycle for a given area: inputs, transformations, storage, and outputs (or losses).

Inputs of Reactive Nitrogen

In a natural terrestrial ecosystem, the primary inputs of reactive N are from atmospheric deposition of N_2 fixed by lightning, and fixation by naturally occurring leguminous plants. The amount of reactive nitrogen produced by lightning globally ranges from 3 to 10 Tg N yr^{-1} (Prather et al., 2001; Galloway et al., 2004). Nitrogen inputs into urban watersheds are very complex, and few studies have attempted N budgets for urban watersheds as almost every commodity has some N component (Van Breeman et al., 2002; Bernhardt et al., 2008). The importation of water into urban areas is an important N input. Large volumes of water imported from outside the watershed can act as an N loading mechanism. The concentration of N in this water is generally low, but the volumes are very high, resulting in high input loads to the watershed (Kennedy et al., 2007; Wolf et al., 2007). For example, Warren-Rhodes and Koenig (2001) found that 8 tons of N, which represented 3% of all imported N annually, was imported within water to Hong Kong in 1997.

Food is another N import to an urban watershed. Groffman et al. (2004) used an assumed human N consumption (and excretion) rate of 12 g of N per capita per day based on the findings by Bleken and Bakken (1998). Based on five urban watershed mass balances of nitrogen (Baker et al., 2001; Faerge et al., 2001; Warren-Rhodes and Koenig, 2001; Groffman et al., 2004; Wollheim et al., 2005), Bernhardt et al. (2008) estimated that food imported into urban watersheds contribute between 13 and 90% of imported N. Pet food may be a particularly important part of the total food load because N in food is released as waste from the animal and is more likely to be deposited onto lawns and other green spaces and subject to leaching and runoff, rather than entering the wastewater stream with human wastes (Baker et al., 2001). For the Baltimore Long Term Ecological Research (LTER), pet waste–derived N can be a larger annual input (17 kg N ha^{-1} yr^{-1}) than either fertilizer application or atmospheric deposition (Baker et al., 2001; Groffman et al., 2004).

Fertilizer applications and fossil fuel combustions may also be a significant input. Groffman et al. (2004) estimated that lawn fertilization in Baltimore (14.4 kg N ha^{-1} yr^{-1}) was greater than atmospheric deposition. In Hong Kong the estimate for lawn fertilization was about 8 kg N ha^{-1} yr^{-1} (Warren-Rhodes and Koenig, 2001). A survey of home owners found application rates were highly variable;

however, rates (in mass per area) were similar for areas with low population density area, with typically larger lawn areas, and high density areas, with less total lawn area (Law et al., 2004). Localized deposition of N from fossil fuel combustion may be a significant contributor of NH_3 to surface runoff when NH_3 is given off as a byproduct of catalytic conversion of NO_x compounds (Maestre and Pitt, 2005; Bernhardt et al., 2008). Baker et al. (2001) estimated that dry deposition in Phoenix, AZ is 15 kg NO_x–N ha^{-1} and 3.5 kg NH_3–N ha^{-1} annually in urban areas, compared with the 1.9 kg N ha^{-1} yr^{-1} received via wet deposition in Phoenix, AZ. Many other sources of N get deposited onto impervious surfaces (e.g., excrement dropped by migrating birds) or funneled into the wastewater treatment system (e.g., industrial byproducts) and contribute to loading, but our knowledge of these is still limited.

Movement, Retention, and Storage

The majority of N inputs in undisturbed or agricultural ecosystems first make contact with vegetation or soil. This is an important difference between "natural" N cycles and urban N cycles, because contact with soil and vegetation provides an increased opportunity for N to be immobilized by plants and microbes. N_2 converted to ammonia can either be taken up directly by plants or transformed in the soil to nitrate and then immobilized in the form of organic nitrogen compounds by plants. If nitrogen is limiting, then most of the N will be scavenged by plants and microbes. The majority of organic nitrogen compounds returned to the soil will be mineralized to ammonia, transformed back to nitrate, and readsorbed by plants. Nitrate not adsorbed has several possible fates—it can leach to ground and surface waters in aerobic conditions, and in anaerobic conditions it can be reduced to volatile N_2O and N_2 compounds.

Nitrogen loads in urban ecosystem are less likely to make contact with the soil and vegetation. Urban ecosystems have a greater chance of N being deposited on impervious surfaces, thus reducing the chance that the N is immobilized. Gobel et al. (2007) estimated average concentration ranges from rainwater and roofs and trafficked impervious surfaces to contain 0.10 to 3.39 mg L^{-1} of ammonium and 1.54 to 5.00 mg L^{-1} nitrate. Nitrogen on impervious surfaces is a result of atmospheric deposition, NO_x release from fossil fuel combustion, excreta from urban animals, dust settling, and fertilizer overspray. Furthermore, the N brought into an urban ecosystem in the form of food bypasses the outside environment almost entirely. The majority of N in food is consumed and concentrated at WWTF, and waste food is delivered to landfills. Food N thus does not come into contact with the environment or soil until it is discharged as wastewater effluent. Where on-site waste disposal is practiced (typically in low-density urban areas), however, a portion of the N in food will be filtered through the soil, taken up by plants, or may end up percolating through to groundwater. Hatt et al. (2004) found that river N concentrations were directly related to septic tank density in Melbourne, Australia; however, the total loads of N in urban streams were more directly related to the amount of effective impervious surfaces. Nitrogen delivered to WWTF undergoes treatment before its release to surface waters. In 2004, approximately 70% of wastewater treatment facilities treated waste by primary and secondary treatment, which entails settling out solids and aerobic biological digestion; few use tertiary treatment, which enables the removal of the majority of N from the wastewater stream (USEPA, 2004). Nitrogen released in effluent after secondary treatment is generally in inorganic form (primarily nitrate), and the effect on surface waters in urban ecosystems is a reduction in the ratio of dissolved organic nitrogen to total dissolved nitrogen (DON/TDN) relative to surface waters in urban ecosystems that do not contain a WWTF (Aitkenhead-Peterson et al., 2009).

Little information is available on the sinks of N in urban ecosystems. Depending on the soil type, climate, and hydrology, lawns and green spaces have less runoff than impervious surfaces and are predicted to be one of the N sinks (Bernhardt et al., 2008). Fertilized green spaces such as lawns, athletic fields, and golf courses may not function as sinks because of the high N inputs from fertilizer. King et al. (2007) investigated N loss from a golf course in Austin, TX and found that 3% of applied N fertilizer was lost in stormwater. While not a large proportion, it does indicate that these spaces are less likely to function as sinks. Landfills are also predicted to be sinks, as they accumulate the majority of solid waste from municipalities (Bernhardt et al., 2008). The infrastructure itself may become a sink for N as it becomes enriched in N-containing consumer products (Kennedy et al., 2007). In a Baltimore LTER study, the overall N retention was found to be 75% of all inputs for suburban areas, only slightly less than agricultural ecosystems with 77% retention, but much lower than a forested catchment that retained 95% of N inputs (Groffman et al., 2004). Baker et al. (2001) suggested that the majority of N input to Phoenix, AZ is lost through denitrification to the atmosphere. Losses other than those to surface water include aerosolized and volatilized N, dust loss, and N transported out of the watershed by humans in various products or wastes.

Nitrogen Loss to Surface Waters

Bernhardt et al. (2008) compared five nitrogen budgets for urban watersheds in Phoenix, AZ (Baker et al., 2001), Baltimore (Groffman et al., 2004), Boston (Wollheim et al., 2005), Bangkok (Faerge et al., 2001), and Hong Kong (Warren-Rhodes and Koenig, 2001) and suggested that despite the greater likelihood that N deposited on the landscape will end up in surface waters, the percentage of total nitrogen inputs lost to surface water is relatively small, with only 3% of N inputs (Phoenix, AZ), 10% of N inputs (Baltimore, MD), and 15% of N inputs to a small catchment in the suburbs of Boston exported to stream waters (Baker et al., 2001; Groffman et al., 2004; Wollheim et al., 2005). These values are lower than might be expected because most N in the waste stream is exported out of these basins and thus does not enter surface waters. Impacts to surface waters are greater in developing urban areas with less comprehensive wastewater treatment and more localized or limited treatment. For example, 90% of N inputs to Bangkok province and approximately 50% of N inputs to the city of Hong Kong were lost as outputs to surface waters (Faerge et al., 2001; Warren-Rhodes and Koenig, 2001).

While the percentage of inputs lost to surface waters by urban watersheds is relatively low in industrialized cities, the total export of N from urban and suburban watersheds through surface waters is higher than exports from less populated watersheds. Groffman et al. (2004) found that total N exports from three suburban and one urban watershed ranged from 4.9 to 11.2 kg N ha^{-1} yr^{-1} from 1999 to 2001, substantially larger exports than the range of 0.51 to 0.48 kg N ha^{-1} yr^{-1} for a forested reference watershed for the same period. In a study of 42 watersheds around the world, Peierls et al. (1991) found that watershed nitrate

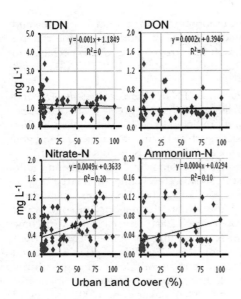

Fig. 15–2. Relationship between surface water concentrations of total dissolved nitrogen (TDN), dissolved organic nitrogen (DON), nitrate nitrogen (Nitrate-N), and ammonium nitrogen (ammonium N) and the percentage of urban land cover in watersheds reported in the literature. Source data are obtained from the following references: Aitkenhead-Peterson et al., 2009; Bahar and Yamamaro, 2008; Bedore et al., 2008; Bhatt and McDowell, 2007; Brett et al., 2005; Chang and Carlson, 2005; Chea et al., 2004; Cunningham et al., 2009; Daniels et al., 2002; Fitzpatrick et al., 2007; Lewis et al., 2007; Liu et al., 2000; Rose and Peters, 2001; Schoonover et al., 2005; Smart et al., 1985; Von Schiller, 2008; Zampella et al., 2007.

concentrations and exports were strongly correlated with population density in the watershed. However, agriculture has a significant effect on the N exports in large watersheds that include both the production and consumption of food. Investigations on the Mississippi watershed showed that the largest contributor to N enrichment of the watershed and the Gulf of Mexico is agriculture (Mitsch et al., 2001; Robertson et al., 2009).

A range of N species concentrations have been reported in the literature. Overall, total dissolved nitrogen (TDN) and dissolved organic nitrogen (DON) concentrations show no trend with increasing percentage of urban land use in watersheds without wastewater treatment effluent affecting stream N. However, both nitrate N and ammonium N are weakly positively correlated with increasing urban land use (Fig. 15–2). Treated and untreated effluent had a significant impact on the concentrations of all N species regardless of percentage urban land use (Table 15–2). Surface water impacted by untreated sewage has significantly higher ammonium than nitrate concentrations (Bhatt and McDowell, 2007).

Riparian and In-Stream Nitrogen Processing

Nitrogen cycling in streams is altered by urbanization due to changes in stream hydrology, biology, riparian zone function, and rates of N delivery that accompany urbanization. These changes result in alteration of the uptake, transformation, and release of nitrogen. Increases in urban impervious surface area and stormwater drainage infrastructure increase the surface runoff loading, peak discharges, and annual water export of receiving streams (McMahon and Cuffney, 2000; Rose and Peters, 2001; Walsh et al., 2005; Chapter 14, Burian and Pomeroy, 2010, this volume).

Most of the N entering urban streams bypasses riparian zones, since it is delivered directly to the stream through stormwater drainage systems and wastewater treatment facilities (Chapter 14, Burian and Pomeroy, 2010, this volume). Natural riparian zones are the interface between the surface water and terrestrial

Table 15–2. Chemistry of urban surface waters directly receiving wastewater treatment effluent and those not directly receiving effluent.†

	No effluent	Effluent
pH	7.66	7.67
Electrical conductivity, μS cm^{-1}	444	790
Sodium, mg L^{-1}	36.3	104.1
Potassium, mg L^{-1}	3.46	7.99
Magnesium, mg L^{-1}	10.6	11.5
Calcium, mg L^{-1}	29.2	32.8
Chloride, mg L^{-1}	26.8	84.2
Sulfate, mg L^{-1}	27.4	28.0
Total dissolved N, mg L^{-1}	1.08	12.60
Ammonium N, mg L^{-1}	0.04	5.30
Nitrate-N, mg L^{-1}	0.51	5.38
Bicarbonate, mg L^{-1}	128	193
Phosphate-P, mg L^{-1}	0.07	1.43
Dissolved organic N, mg L^{-1}	0.42	1.88
Dissolved organic C, mg L^{-1}	10.3	19.0

† Data sources: Aitkenhead-Peterson et al., 2009; Bahar and Yamamaro, 2008; Bedore et al., 2008; Bhatt and McDowell, 2007; Brett et al., 2005; Chang and Carlson, 2005; Chea et al., 2004; Cunningham et al., 2009; Daniels et al., 2002; Fitzpatrick et al., 2007; Lewis et al., 2007; Liu et al., 2000; Rose and Peters, 2001; Schoonover et al., 2005; Smart et al., 1985; Von Schiller, 2008; Zampella et al., 2007.

systems. Vegetation along riparian zones has the capacity to take up and immobilize nutrients before they enter the stream channel. Roughness from vegetation and its litter slows down overland flow, encouraging infiltration. Riparian zones, with water tables close to the soil surface, tend to be more saturated with water for longer time periods, inducing reducing conditions. Saturation reduces oxygen availability, and nitrate inputs to riparian zones are likely to undergo denitrification resulting in production of volatile N_2O and N_2 compounds. Bypassing these natural filters, where they would otherwise be present in the urban landscape, results in more N being added directly to the stream. Even when water does pass through riparian zones in urban ecosystems, they may no longer be effective filters for N. Groffman et al. (2002) found that urban streams were highly incised and had lower water tables. These lower water tables created more aerobic soils and reduced the riparian zone capacity for denitrification, possibly contributing to the increased concentrations of nitrate in urban streams (Groffman et al., 2002).

Changes to ecosystem processes within the stream channel may alter N cycling and increase N concentrations. Industrial and municipal waste and high chloride concentrations have been shown to decrease the denitrification potential in streams (Richards and Knowles, 1995; Hale and Groffman, 2006). A study of denitrification across a large land cover gradient found no significant relationship between land cover and denitrification, possibly due to the wide range of conditions that occur within land covers (Mulholland et al., 2009). However, the efficiency of denitrification decreased as nitrate levels increased (Mulholland et al., 2009). This result suggests that surface waters receiving effluent high in nitrate will export a higher fraction than those that receive lower-concentration wastewater treatment effluents (Table 15–2).

Urban streams often have altered courses and hardened banks to prevent the stream from meandering. This can homogenize the streambed and reduce the ability of streams to store carbon, decreasing the potential for N to be metabolized (Meyer et al., 2005). A study by Hall et al. (2009) found that a structural equation model based on ecosystem metabolism, hydraulic parameters, and N concentrations was able to explain 79% of the variability in log uptake length of nitrate. Uptake length increased with discharge and increasing nitrate concentrations, where land use indirectly correlated with uptake length via gross primary productivity (Hall et al., 2009). Increases in dissolved organic carbon (DOC) concentration in the Sacramento River in California have also been attributed to upstream WWTF discharge into the river (Sickman et al., 2007). The few studies on urban stream metabolism have variable results from very low to high metabolisms (Meyer et al., 2005; Grimm et al., 2005). A study by Von Schiller et al. (2008) found that ammonium demand (measured as uptake velocity) decreased along the forested to urban gradient in response to increases in ammonium, DON, and DOC. One of the primary concerns with high N loading in urban ecosystems is the potential for eutrophication of fresh and estuarine waters.

In addition to the complex factors that are already known to affect N cycling in urban aquatic ecosystems, additional effects on N cycling may also be occurring due to the host of contaminants entering urban systems. Polynuclear aromatic hydrocarbons (PAHs), pharmaceuticals, personal care chemicals, and metals all have potential effects on N metabolism in streams, but they have barely begun to be investigated (Bernhardt et al., 2008). Nitrogen concentrations and loads transported from urban areas are also dependent on levels of available phosphorus in the water (Paerl et al., 2004). If phosphorus is limiting biotic growth then the excess N will not be utilized and transported downstream (Paerl, 2009). Problems with eutrophication and low oxygen can be displaced further downstream of significant sources where the two nutrients are both found in excess quantities (Paerl, 2009)

Urban Phosphorus

Surface water in urban catchments around the world generally has higher phosphorus concentrations than surface water in rural catchments (Meybeck, 1998; Winter and Duthie, 2000; Bhatt and McDowell, 2007). For example, a 10-yr record of catchments in the greater Seattle area found stream water phosphorus concentrations were correlated with urban land cover, and most urban streams had on average 95% higher total phosphorus and 122% higher soluble reactive phosphorus than the most forested streams (Brett et al., 2005). Even where the percentage of urbanized land in a catchment is relatively low (<10%), urban areas can still have significant impacts on P concentrations and loads (Osborne and Wiley, 1988). In a U.S. Midwest catchment, urbanization was a dominant factor even though urban areas constituted only 5% of the area, yet urban land use controlled dissolved phosphorus concentration throughout the year (Osborne and Wiley, 1988).

High concentrations generally lead to higher loads and exports from the watershed, and phosphorus loads generated in urban areas can influence downstream reaches and discharge points. Inputs of P to an urban watershed include fertilizers, human and pet food, atmospheric deposition, and P-containing consumer and industrial products (Davis and Gentry, 2000). Fewer detailed budgets of P loading are currently available compared with N budgets. Davis and Gentry (2000) found that 35% of the annual P input into the urbanized Illinois River watershed was lost through river exports to the greater Mississippi River watershed. Part of this exported P was generated from sewage discharged from WWTFs in the Chicago area that contributed an estimated 70% of P load in the Illinois River and ~5% of the P load to the Gulf of Mexico from the Mississippi River watershed (Davis and Gentry, 2000). The proportion of inputs of P into an urbanized region lost through surface water is variable. In another P budget of the upper Potomac River Basin, Jaworski et al. (1992) found that 34% of the imported P was lost from the watershed.

Sources

Wastewater Treatment Facilities and Septic Systems

Discharge of wastewater to urban surface waters has been found to be a major contributor to the total P loads (La Valle, 1975; Davis and Gentry, 2000; Bowes et al., 2005). In Illinois, WWTF effluent contributed 47% of the total P load to the state's rivers (Davis and Gentry, 2000). A study of WWTF P inputs found that introducing an 80% reduction in P load from the seven largest WWTFs in the English Warwickshire Avon catchment resulted in an estimated decrease in total P export of 378 t yr^{-1} (52% of the total load) (Bowes et al., 2005). However, because of the extremely high nutrient loadings that exist in many UK rivers, nutrient removal from these large WWTFs alone are unlikely to reduce P concentrations to a desired concentration of 0.2 mg L^{-1}. This would require tertiary treatment at not only large, but also medium and smaller facilities (Hilton et al., 2002; Bowes et al., 2005). In a multiple regression analysis of 24 catchments in the Windsor, Ontario area, 76% of the variation in stream orthophosphate concentrations was explained by the percentage of watershed households connected to city sewers, while garden fertilizer use and precipitation phosphate content accounted for 4 and 2% of the variation, respectively (La Valle, 1975). Jarvie et al. (2006) argued that point discharge of effluent is a greater threat to urban water quality than diffuse sources from urban and agricultural runoff. They reasoned that WWTF effluent contains the primary soluble reactive P (SRP; <0.45 μm, molybdate reactive), and because SRP is more easily utilized by aquatic organisms, streams receiving effluent are at a larger risk for eutrophication than those with P loads primarily from particulate sources (Jarvie et al., 2006). Bans on P-based detergents, which reduced the SRP in WWTF effluent, were successful in decreasing algal blooms in the freshwater portions of the Neuse River, North Carolina (Paerl et al., 2004).

In a properly functioning septic system, soils also adsorb P from the system's leach field and therefore reduce the potential for septic systems to contribute significantly to P loading in urban streams. Several studies have concluded that septic systems do not contribute to P loading in urban surface waters (Hoare, 1984; Hatt et al., 2004; Jarvie et al., 2006). Where P was filtered out by the soil, nitrate has been connected to surface water concentrations (Hoare, 1984; Hatt et al., 2004). Gerritse et al. (1995) also found that stream concentrations were unaffected by septic systems; however, they found a minimum travel time of P was between 1 and 8 yr based on column leaching, which suggests that with time there is a potential for P from septic systems to reach groundwater or surface water in older communities. Studies have found that the potential for leaching P from soil increases in soil with greater than 10% degree of soil P saturation (Heckrath et al., 1995; Hooda et al., 2000). In Ireland, evidence suggests that septic systems, primarily poorly maintained systems, contributed to high P loading in high-density rural populations (Arnscheidt et al., 2007). Similarly, Jarvie et al. (2006) found a correlation between SRP during high flow in catchments where no

WWTF exists and concluded that the SRP was most likely from septic systems and small plants that discharge onto the floodplain being washed in as the water table rises.

Impervious Surfaces

Impervious surfaces have been implicated as a source of many contaminants in urban environments, and are among the main sources for dissolved P in many northern climates (Bannerman et al., 1993; Hatt et al., 2004). Hatt et al. (2004) found strong correlations between P loads and impervious surface and connectivity of impervious surfaces and streams via stormwater drainage systems. Phosphorus from natural sources such as pollen deposition from trees, leaching of P from plant tissue, and airborne particulate deposition, as well as anthropogenic sources such as road sand or misapplied fertilizer can accumulate on impervious surfaces, making their impact on urban water resources critical to assess (Sharpley, 1981; Dorney, 1985; Banks and Nighswander, 1999; Hu et al., 2001; Ahn and James, 2001; Burian et al., 2002; Easton et al., 2007).

Fertilization

The resulting buildup of P in agricultural soil from overfertilization with either chemical fertilizers or manures has been a recognized problem for water quality and the prevention of cultural eutrophication (Sharpley et al., 1994). Similar problems occur in urban ecosystems where overfertilization with chemical fertilizers, biosolids used in sod production, or manures from pets can lead to nutrient buildups in soil compared with less human-impacted systems (Baker et al., 2001; Pouyat et al., 2007). Fertilizer use in urban catchments has been found to contribute to the elevated P concentrations in streams (LaValle, 1975; Waschbusch et al., 1999). For example, lawns and streets were found to be the primary source of phosphorus to urban streams in Madison, WI. As a result of fertilizer application, lawns were estimated to contribute between 49 and 61% of the total P load in urban streams (Waschbusch et al., 1999). Influencing the amount and timing of P fertilization in urban ecosystem through education and extension may be more difficult than in agriculture because of the greater number of residents in any given urban watershed. Working with lawn care companies may provide a more effective avenue to influence a greater area of fertilized turf and horticultural areas in a city.

Groundwater

The exchange of water between ground and surface waters has a strong influence on stream chemistry and nutrient fluxes (Pionke et al., 1988; Fiebig et al., 1990; Triska et al., 1993; Sonoda and Yeakley, 2007). However, less is known about the nature of this relationship in urban ecosystems. Sonoda and Yeakley (2007) studied the urbanizing Johnson Creek in Portland, OR and concluded that in both less disturbed and urban streams groundwater contributed to the total P load; however, in urban systems the relationship between stream and groundwater P concentrations can be masked by other inputs, and overall P concentrations were more closely related to land-use patterns than groundwater. Meross (2000), studying the same watershed, found that sections of Johnson Creek were altered through channelization and rerouting due to urban development, but that only 6% of the precipitation was hydrologically disconnected from the creek and therefore could not explain the lack of influence groundwater had on urban

stream chemistry in this watershed (Sonoda and Yeakley, 2007). Stream alterations such as channeling, armoring, and the connectivity between ground and
surface water are highly variable within and among urban centers, and therefore
further research is needed on the influence that these alterations have on concentrations of groundwater P and on stream P loads. Because riparian zones function
as interfaces between groundwater and surface waters in urban ecosystems as
well as other more natural environments, they likely play a key role in regulating
the relationship between ground- and surface-water concentrations and load of P
(Chapter 13, Stander and Ehrenfeld, 2010, this volume).

Erosion

Urban activities and stream hydrology increase erosion of sediments from
upland soils and stream channels (Wolman and Schick, 1967; Trimble, 1997; Nelson and Booth, 2002; Maniquiz et al., 2009). Sediments washed into urban streams,
and streambeds are more likely to be coarser textured and therefore more likely
to release adsorbed P to the water column (Finkenbine et al., 2000; Pizzuto et
al., 2000; McDowell and Sharpley, 2001). Stream banks and other fine soil materials can have increased adsorption capacity; as a result, their erosion has less
potential to contribute to high P in the water column, and they may act as sinks
for P in rivers and streams (McDowell and Sharpley, 2001). However, erosion of
these materials may increase eutrophication downstream in lakes and reservoirs,
and as oxygen decreases, eroded sediments may release P to the water column
(McDowell and Sharpley, 2001).

"Chemical Time Bombs"

Several scenarios in urban ecosystems may result in high levels of P being released
in short periods of time. Phosphorus is often stored within the soil matrix due to
overfertilization, adsorption of P from septic systems, or saturation of riparian
soils. Phosphorus can be mobilized by changes in land-management practices,
particularly those that increase water and wind erosion, and release P to surface
waters in large quantities (Bennett et al., 1999; Conley et al., 2002). This effect has
been called the "chemical time bomb" (Stigliani et al., 1991) and is of particular concern when previously agricultural land is cleared for urban growth and
large amounts of sediments have been mobilized (Bennett et al., 1999; Maniquiz
et al., 2009). Riparian zones are an example of another type of "time bomb" that
exists when soils with high P concentration become anaerobic and chemically
reduced. These soil conditions have the potential to release large amounts of P
when they become saturated and reducing conditions occur (Patrick and Khalid,
1974). Remobilization of P under hypoxic conditions in lakes and the annual variation in sediment has long been known to release P in quantities up to an order of
magnitude greater than other, more controllable P sources (Mortimer, 1941; Ingall
et al., 1993; Conley et al., 2002).

Characterization and Phosphorus in Stream Processing

Phosphorus in streams is present in several metabolically available forms and
can be immobilized by chemical, adsorptive, and biotic processes. On the basis of
measurements in streams of Missouri's Ozark Plateau, Smart et al. (1985) found
urbanization increased total P as a result of increased particle-associated P and
dissolved P levels. Removal and sequestration of P within an urban stream may
be affected by the concentration of other dissolved ions; however this has not

always been shown to be true in urban catchments. In Chicago area streams, for example, 79% of the total P was found in the dissolved form despite the high levels of Ca present (Bedore et al., 2008). Bedore et al. (2008) suggested that removal of P from the water column of these streams was inhibited by the high concentrations of dissolved P in the presence of Ca^{+2} and Mg^{+2}. Although the presence of Ca with P is generally considered to aid removal of P from the water column, dissolved P concentrations greater than 0.61 mg P L^{-1} inhibit calcite nucleation and disrupt crystal growth, resulting in coprecipitation of Ca and P only at low to moderate initial concentrations of dissolved P (House and Donaldson, 1986). The majority of dissolved P removal from the water column may occur through adsorption to sediments during both high and low flows (Gibson and Meyer, 2007; House and Denison, 2002). Antecedent weather conditions are also important in regulating P adsorption to sediments, because they control the pool of fine sediment and associated P available for remobilization during storm pulses (House and Denison, 2002). McDowell and Sharpley (2001) compared bank sediments and reported that sandier bed sediments released P more readily and supported a higher P concentration in the water. Several studies have found adsorption of P was correlated with Fe and organic matter concentrations, and P in sediments was adsorbed on Fe and organic complexes (McDowell and Sharpley, 2001; Bedore et al., 2008).

Phosphorus is also removed from the water column and temporarily stored by biotic activity. McDaniel and David (2009) found biotic activity contributed to a mean range of 26 to 40% of total P uptake in Illinois rivers, and was correlated with sediment organic matter content. Haggard et al. (1999) reported a similar average of 38% uptake by the biotic community in Oklahoma streams. Dissolved P more easily supports microbial metabolic activity than organic P; therefore, the presence of high dissolved P has also been suggested to inhibit the mineralization of organic P in the water column, increasing the potential for organic P to accumulate in the sediment and be transported downstream (Bedore et al., 2008). The persistence and cycling within a stream reach is an important indicator of utilization by aquatic organisms and the amount of P that will be transported downstream (Peterson et al., 2001). Small streams across biomes are efficient at retaining and cycling nutrients; however, small streams receiving point-source inputs are typically less efficient than less disturbed streams (Haggard et al., 2001, 2005; Marti et al., 2004; Peterson et al., 2001; Pollock and Meyer, 2001). In Vermont, Meals et al. (1999) used a P associated with a dye tracer to spike the LaPlatte River, a eutrophic river fed by a WWTF in the fall and winter seasons. In the fall, 39% of the added P was retained over 12 h, and at the end of 48 h 38% of added P (4 mg P m^{-2}) was still retained in the stream reach. However, during the winter all the P added was exported from the reach within 24 h (Meals et al., 1999). Phosphorus retention and uptake lengths can be influenced by many different factors, including flow, temperature, concentration gradient, total suspended solids, and biological activity (Meals et al., 1999; Gibson and Meyer, 2007). In the large, urbanized Chattahoochee River in Atlanta, GA, P uptake lengths (i.e., the distance traveled by a P molecule before being removed from the water column) were highly variable, with some measured dates having no uptake and others indicating a release of soluble reactive phosphorus within the river reach, but on average the uptake lengths for P were many kilometers (Gibson and Meyer, 2007). Several studies suggest that short- and medium-term P uptake and release is a function of biotic activity and bioavailable P, rather than sediment sorption (McDowell et al., 2003; McDaniel and David, 2009).

Carbon

Total carbon in surface and ground waters consists of natural organic compounds derived primarily from the decomposition of plants and animals, inorganic carbon that can be in the form of CO_2, HCO_3^- or CO_3^{2-}, and novel, human-made organic carbon. This section will discuss natural and novel organic carbon.

Natural Organic Carbon

The biogeochemical cycling of organic carbon is arguably the most important process in aquatic ecosystems because of its central role in regulating many other elemental cycles and providing food for consumer organisms (Chrost, 1989; Findley and Sinsabaugh, 1999; Harbott and Grace, 2005). Important ecosystem functions of organic carbon (OC) include providing energy to microbial consumers and subsequently to higher trophic levels, as well as regulating the availability of dissolved nutrients and metals (Chrost, 1989; Findlay and Sinsabaugh 1999). Despite its important role, however, the amounts, sources, quality, and functions of carbon in urban aquatic ecosystems are poorly understood compared with less disturbed aquatic ecosystems (Paul and Meyer, 2001).

The sources of organic carbon in urban surface waters include both point-source discharges and nonpoint sources. Sickman et al. (2007) found point-source urban wastewater discharges made up about 60% of DOC inputs to the Sacramento River; the remaining 40% was contributed through nonpoint sources. Other studies have found that DOC was not significantly greater in urban streams receiving WWTF effluent (Daniels et al., 2002; Aitkenhead-Peterson et al., 2009). Several studies have shown increasing concentrations of DOC with high flow and storm events, indicating that carbon sources in the surface water were not the result of high carbon levels in the groundwater that contribute to baseflow (Chang and Carlson, 2005; Hook and Yeakley, 2005). Nonpoint sources of OC in urban regions are most often related to the amount of remaining vegetation within cities. Positive relationships between DOC concentration and the percentage of remaining forested cover within urban areas have suggested that leaf litter may contribute to DOC concentrations in urban streams (Chang and Carlson, 2005). During storm events, Hook and Yeakley (2005) found that 70 to 74% of DOC export was contributed by remnant riparian areas. Carbon dating analysis suggests that DOC in nonpoint source C in the Sacramento River is derived primarily from leaching of older soil organic matter (Sickman et al., 2007). However, in an urban to rural gradient in south-central Texas, DOC concentrations were not related to remaining forest, but were more closely associated with urban open areas such as golf courses, sports parks, and neighborhood lawns (Aitkenhead-Peterson et al., 2009). Novel organic compounds washed from impervious surfaces during storm events also contribute to the total organic carbon loads in urban watersheds (Eganhouse et al., 1981; Xian et al., 2007). In a single storm event sampling of the Los Angeles River, 60% of the total extractable organics was estimated to be hydrocarbons of anthropogenic origin (Eganhouse et al., 1981). Streams without a WWTF are often depleted of OC in the benthic material. In a comparison of two forested and four urban catchments, average organic matter standing stocks were significantly lower in urban streams near Atlanta, GA caused by scouring of the highly mobile sandy substrates in urban channels as a result of more severe high flow events (Paul and Meyer, 2001). More research is needed on DOC sources in urban areas with a variety of climates, urban land uses, and water chemistries.

Despite the critical role of DOC in aquatic ecosystems, limited work has been done on DOC quality in urban stream ecosystems (Eganhouse et al., 1981; Paul and Meyer, 2001). Storm events can increase both dissolved and particulate OC concentrations, but less is known about the baseflow proportions of particulate and dissolved organic carbon (McConnell, 1980; Paul and Meyer, 2001). In an urbanized creek, carbohydrate concentrations in particulate organic matter (POM) were higher than that in POM of a nearby forested reference stream, suggesting that urbanization affects the nature of transported organic matter (Sloane-Richey et al., 1981). The carbon associated with sewage effluent is generally more labile than DOC from natural sources, causing high biological oxygen demand and oxygen deficits associated with storms in urban environments (McConnell, 1980; Faulkner et al., 2000; Ometo et al., 2000).

Bioavailability and metabolism of different types of DOC in urban surface water affect both the concentrations of DOC and the microbial communities supported by them. Results from analysis of relative extracellular enzyme activity (EEA) rates by Harbott and Grace (2005) in streams with roughly similar DOC concentrations show that there was a shift in DOC bioavailability depending on the origin of organic substrates in each stream. Large variations in EEA suggest diverse sources of DOC from urban areas (Harbott and Grace, 2005). A study by Paul (1999) also found that leaf litter from different tree species decayed at different rates in an urban Atlanta, GA stream compared with a rural one. Compared with forested streams, all coarse and fine particles released in urban Atlanta streams traveled much further before leaving the water column (Paul, 1999). These findings, combined with decrease in organic matter storage, indicate that urban streams retain less organic matter and suggest that secondary production could be limited (Paul, 1999). A comparison of three rivers in Michigan by Ball et al. (1973) found the urban river had higher gross primary production and community respiration than the forested river.

Novel Organic Carbon

The presence of novel organic compounds in urban streams and rivers due to human activity is a pervasive and well-studied result of urbanization. Concentrations of these compounds depend on the source and the intensity of human activity, with new novel C compounds constantly emerging.

Hydrocarbons

Concentrations and loads of novel hydrocarbons in urban streams and rivers can be very high and have been found at levels stressful to sensitive stream organisms (Latimer and Quinn, 1998). Particularly disturbing is the total amounts of these compounds entering bays, estuaries and the ocean. A substantial amount of fuel oil (485,000 L) enters the Narragansett Bay, RI, USA each year (Hoffman et al., 1982, Latimer and Quinn, 1998). Similarly, the Los Angeles River alone contributes about 1% of the annual world petroleum hydrocarbon input to the ocean (Eganhouse et al., 1981). In the U.S. Mid-Atlantic, fluxes of polycyclic aromatic hydrocarbons (PAH) from the primarily urban Anacostia

River were exceptionally high, comparable with the much larger Potomac River, and are thought to be an important source of PAH to the Chesapeake Bay (Foster et al., 2000).

Further north in the Chesapeake Bay's largest tributary, fluvial transport of PAHs in the Susquehanna River were correlated with basin precipitation, snow-melt runoff, and river discharge, factors that are consistent with flow-driven, nonpoint-source inputs (Preston et al., 1989; Foster et al., 2000). Impervious surfaces, particularly roads and parking lots, are probably the most well-known contributor of novel organic compounds to urban streams and rivers. Paul and Meyer (2001) noted, "It is difficult to find automobile parking spaces without oil stains in any city." A literature synthesis by Gobel et al. (2007) reported that average concentrations of PAHs and mineral organic hydrocarbons (MOH) in runoff from impervious surfaces associated with vehicle traffic are much higher than urban green spaces or roofing runoff (Table 15–3). Relationships have been found between river oil and grease, population density, and the percentage of impervious surface in an urban Tampa Bay, FL watershed (Xian et al., 2007). Although natural aliphatic hydrocarbons are present in streams, the overwhelming majority are due to street runoff carrying petroleum-based compounds (Hunter et al., 1979; Mackenzie and Hunter, 1979; Eganhouse et al., 1981; Whipple and Hunter, 1979). A study on tributaries of the Chesapeake Bay watershed (Mid-Atlantic, USA) found that the PAHs present were characteristic of weathered or combusted petroleum products (Foster et al., 2000). Eganhouse et al. (1981) suggested the following nonpoint sources of hydrocarbons in the Los Angeles River basin: vehicular exhaust particles, lubricating oils, atmospheric fallout (e.g., from forest fires, combustion of fossil fuels, and eolian transport of bio-organics), fuel oils, spillage of crude and refined petroleum products (during production, processing, or transportation), leached and eroded pavement, and natural biogenic sources on land, among others. Mahler et al. (2005) reported that runoff from parking lots in Austin, TX constructed with coal-tar emulsion sealcoat had mean concentrations of PAHs 65 times higher than the mean concentration from unsealed asphalt and cement lots and suggested that seal coat may be responsible for the majority of the PAH in surface waters. Other potential sources of novel PAHs are industrial effluents, particularly those that use organic solvents (Yamamoto et al., 1997).

Less is known about the in-stream processing and cycling of these types of compounds compared with other novel organic compounds (Beasley and Kneale, 2002). Daniels et al. (2000) found that hydrophobic novel organic compounds

Table 15–3. Average concentrations of polycyclic aromatic hydrocarbons (PAH) and mineral organic hydrocarbons (MOH) reported in the literature.†

Source	Type	PAH	MOH
		$\mu g\ L^{-1}$	$mg\ L^{-1}$
Rainwater	n/a	0.39	0.38
Roof runoff	n/a	0.44	0.70
Trafficked areas runoff	pedestrian/cycle	1.00	0.16
	car park	3.50	0.16
	service road	4.50	0.16
	main road	1.65	4.17
	motorway	2.61	4.76

† Data adapted from Gobel et al. (2007).

declined in concentration and were present to a depth of 1 m, although the more soluble compounds showed no clear distribution. Lack of a pattern with depth suggests that the more soluble compounds were traveling through the sediment. Estimates of the fraction of hydrocarbons associated with particles range from 86 to 93% of the total hydrocarbon load (MacKenzie and Hunter, 1979; Hoffman et al., 1982). Several PAHs are known to have carcinogenic properties, but other effects are not well understood (Christensen et al., 1975; Eisler, 1987; Mastran et al., 1994). Their effects are varied, but they are known to affect mammals (including humans), birds, invertebrates, plants, amphibians, and fish (USEPA, 2003). In aquatic systems the toxicity of PAH increases with increasing molecular weight (Eisler, 1987). Species shifts and loss of populations of macroinvertebrates have been shown in several studies (Whipple, 1981; Shutes, 1984; Clements et al., 1994; Maltby et al., 1995a,b).

Pesticides

Beautiful lawns and green spaces are generally a valued part of urban landscapes and balance the large areas of impervious surfaces, but from the perspective of urban water quality, urban green spaces and lawn care may also contribute to detectable concentrations of pesticides (Struger and Fletcher, 2007). The contribution of pesticides to total organic carbon concentrations is very small; however, these compounds may impact aquatic ecosystem function and quality of water for drinking at very low concentrations.

The frequency of pesticides detected in urban streams is high, and concentrations often exceed those set for the protection of aquatic biota (USGS, 1999; Hoffman et al., 2000). A wide range of pesticides has been found (Table 15–4), including insecticides, fungicides, and herbicides. Of particular concern is the frequent detection of organochlorine pesticides, including DDT [1,1,1-trichloro-2,2-bis(4-chlorophenyl)ethane] and DDT metabolites in water, sediment, and fish tissue samples in watersheds dominated by urban land uses (Yamamoto et al., 1997; USGS, 1999; Black et al., 2000).

Golf courses are often thought of as large contributors to pesticide concentrations; however, Struger and Fletcher (2007) found that golf courses in the Don and Humber watersheds, tributaries of Lake Ontario near Toronto, Canada were not significant contributors of pesticides, possibly due to the integrated pest management plans required by the city. They suggested that detection of pesticides is more likely due to use in home lawn care (Struger and Fletcher, 2007). Compared with agriculture, the contribution of pesticides to watersheds by urbanized areas may be greater than expected based on acreage. A study comparing biocide concentrations in agricultural and urban streams in seven locations around the United States found disproportionately higher concentrations of insecticides than herbicides in urban streams relative to agricultural streams (Hoffman et al., 2000). Although the herbicide contribution of all urban areas is likely very small compared with the contribution from agriculture, the contribution of insecticides may be of similar magnitude in urban and agricultural areas and should not be overlooked (Hoffman et al., 2000).

Pharmaceuticals and Personal Care Products

Wastewater effluent carries a wide range of carbon compounds, and while some studies have found WWTFs contribute a significant portion of the bulk DOC loads in urban streams and rivers, others have found no difference between

Table 15–4. Novel organic compounds found in urban surface waters (W), sediments (S), potable drinking water (P), and effluent (E) reported in the literature.

Type	Compound	Source	Reference
Herbicides	prometon	W,S	Hoffman et al., 2000
			Daniels et al., 2000
	simazine	W	Hoffman et al., 2000
	atrazine	W	Hoffman et al., 2000
			Struger and Fletcher, 2007
	tebuthiuron	W	Hoffman et al., 2000
	metalor	W	Hoffman et al., 2000
	linuron	S	Daniels et al., 2000
	2,4-D	W	Struger and Fletcher, 2007
	bromacil	W	Struger and Fletcher, 2007
	dicamba	W	Struger and Fletcher, 2007
	metachlor	W	Struger and Fletcher, 2007
	metribuzin	W	Struger and Fletcher, 2007
	mecoprop	W	Struger and Fletcher, 2007
	benfluralin	W	USGS, 2007
	trifluralin	W	USGS, 2007
	dacthal	W	USGS, 2007
Insecticides	diazinon	W	Hoffman et al., 2000
			Struger and Fletcher, 2007
	carbaryl	W	Hoffman et al., 2000
	chlorpyrifos	W	Hoffman et al., 2000
			Struger and Fletcher, 2007
	malathion	W	Hoffman et al., 2000
	pyrethroids	S	Daniels et al., 2000
	dieldrin	S	Black et al., 2000
	trans-nonchlor	S	Black et al., 2000
	cis-, trans-chlordane	S,W	Black et al., 2000
			USGS, 2007
	DDT, DDD, and DDE	S	Black et al., 2000
	Carbofuran	W	Struger and Fletcher, 2007
	Cypermethrin	W	Struger and Fletcher, 2007
Fungicides	Fenpropimorph	S	Daniels et al., 2000
Other	nonlyphenol	S	Daniels et al., 2000
surfactant residue	pentachloroanisole	W	USGS, 2007
Wood preservative	hexahydrohexamethyl-cyclopentabenzopyran	W	USGS, 2007
Synthetic musk	methyldibenzofuran	W	USGS, 2007
Furan	BDE 47	W	USGS, 2007
Personal care products			
Aspirin	painkiller	E	Richardson and Bowron, 1985
Bieomycin	antinoplastic	W,E	Aherne et al., 1990
Caffeine	psychomotor stimulant	E	Richardson and Bowron, 1985
Clofibrate	lipid lowering agent	W	Richardson and Bowron, 1985
Cyclophosphamide	antinoplastic	E	Steger-Hartmann et al., 1996
Dextropropxyphene	painkiller	W	Richardson and Bowron, 1985
Diazepam	anxiolytic	E,W	Waggott, 1981
Erthomycin	analgesic	W	Watts et al., 1983
Dichlofenac	antibiotic	W	Stumpf et al., 1996
Estrogen	hormone	E	Shore et al., 1992
Ethinylestradiol	hormone	W,E,D	Kalbfus, 1995
			Aherne and Briggs, 1989
Ibuprofen	analgesic	W	Stumpf et al., 1996
Ifosfamide	antinoplastic	E	Steger-Hartmann et al., 1996
Indometacin	analgetic	W	Stumpf et al., 1996

Table continued.

Table 15–4. Continued.

Type	Compound	Source	Reference
Methaqualone	hypnotic	E	Richardson and Bowron, 1985
Methotrezate	antinoplastic	W,E,P	Aherne et al., 1985
Morphinan Structure	narcotic	W	Richardson and Bowron, 1985
Norethisterone	hormone	W,E,P	Richardson and Bowron, 1985
Oral contraceptive	hormone	W,E	Aherne et al., 1985
Penicilloyl groups	antibiotic	W,E,P	Richardson and Bowron, 1985
Sulphamethoxazole	antibiotic	W	Watts et al., 1983
Tetracycline	antibiotic	W	Watts et al., 1983
Theophylline	psychomotor stimulant	W	Watts et al., 1983
Testosterone	hormone	WS	Shore et al., 1993
Ciprofloxacin	antibiotic	W	Costanzo et al., 2005
Norfloxacin	antibiotic	W	Costanzo et al., 2005
Cephalexin	antibiotic	W	Costanzo et al., 2005
Cocaine	narcotic	W	Huerta-Fontela et al., 2008
Benzoylecegonine	metabolite (cocaine)	W	Huerta-Fontela et al., 2008
Amphetamine	narcotic	W	Huerta-Fontela et al., 2008
Methamphetamine	narcotic	W	Huerta-Fontela et al., 2008
MDMA (ecstasy)	narcotic	W	Huerta-Fontela et al., 2008
Nicotine	narcotic	W	Huerta-Fontela et al., 2008
Norcocaine	metabolite (cocaine)	W	Zaccato et al., 2007
Morphine	narcotic	W	Zaccato et al., 2007
THC-COOH	narcotic	W	Zaccato et al., 2007
Codeine	narcotic	W	Zaccato et al., 2007
Methadone	narcotic	W	Zaccato et al., 2007

streams with and without WWTF or no decrease in bulk DOC due to the introduction of a WWTF (Daniels et al., 2002; Sickman et al., 2007; Aitkenhead-Peterson et al., 2009). Although the effects of WWTFs on total carbon loads are still being investigated, one of the issues of most concern is the type of compounds being passed through the WWTF into urban streams and rivers and their affects on ecosystem health and drinking water quality. Recent studies have found detectable levels of surfactants, personal-care, pharmaceutical, and narcotic compounds in wastewater effluent and surface waters (Halling-Sorensen et al., 1998; Isobe et al., 2001; Andreozzi et al., 2003; Jones-Lepp et al., 2004; Zaccato et al., 2007). A wide variety of compounds have been identified in surface waters from the pharmaceutical and personal-care categories (Table 15–4). The interest in these compounds is high, despite the low concentrations found, because they are specifically designed to be biologically active, and many have the necessary properties to bioaccumulate and influence biotic and biogeochemical cycles (Halling-Sorensen et al., 1998).

Antibiotics in surfaces waters may pose a significant threat to ecosystem health by altering microbial communities and the biota that depend on them, in addition to fostering antibiotic resistance. An estimated 30 to 90% of an antibiotic is released with urine as an active substance to wastewater treatment facilities (Rang et al., 1999). Findings in the United States and Germany have identified 15 types of antibiotics in streams receiving urban and industrial wastewaters (Ternes et al., 2002). Removal of antibiotics during treatment varies depending on the compound and the treatment methods used. In a study by Costanzo et al. (2005), all three of the tested antibiotics in the receiving sludge were present in the leaving effluent of a wastewater treatment facility in Brisbane, Australia. All three antibiotics were detected 50 m downstream of the plant, but only one was detected 500 m downstream (Costanzo et al., 2005). Whether this was due to dilution and hence limited detection, or biotic uptake, is unknown. Some antibiotics have affected denitrification in laboratory experiments; however, only at levels higher than normally found in surface waters (Costanzo et al., 2005). Costanzo et al. (2005) also examined bacterial resistance to antibiotics and reported that bacteria cultured in a bioreactor displayed resistance to all six antibiotics tested, and bacteria cultured from the receiving waters showed resistance to two of the six antibiotics.

Recently several studies have reported narcotic compounds in both wastewater effluent and surface waters receiving effluent at several locations in both the United States and Europe (Jones-Lepp et al., 2004; Zaccato et al., 2007). A variety of compounds have been detected, with cocaine and methamphetamine the most common (Zaccato et al., 2007). While the concentrations detected are relatively low (nanograms per liter) compared with average bulk DOC loads, the summed loads of illicit drugs can be surprising. Zaccato et al. (2007) measured concentrations and loads of six drugs and metabolites in four Italian rivers and found loads of these compounds ranged from 0.1 to 391 g d^{-1}. These compounds still have pharmacological activities, and the biological and ecological effect of these compounds and their interactions in aquatic ecosystems is still unknown (Zaccato et al., 2007). On the positive side, a study by Huerta-Fontela et al. (2008) found that standard drinking water treatment was effective at removing the majority of illicit compounds tested. A potential benefit of this type of study is to estimate community drug use by comparing illicit drug loading within watersheds. Some illicit drugs are similar to compounds used for medical treatments, which may affect the results of such comparisons (Kaleta et al., 2006; Bones et al., 2007; Zaccato et al., 2007).

Metals

Urbanization often results in increased metal concentrations in the water column and its associated sediment (Bryan, 1974; Wilber and Hunter, 1977; Neal et al., 1997; Horowitz et al., 1999; Neal and Robson, 2000; Fitzpatrick et al., 2007). In Tampa Bay, FL, the concentrations of zinc and copper were correlated (r^2 = 0.85 and 0.75, respectively) with population density (Xian et al., 2007). A large number of metals occur in urban surface waters in a wide range of concentrations. Some of the most commonly occurring metals are cadmium, chromium, copper, lead, manganese, nickel, and zinc (Wilber and Hunter, 1977; Fitzpatrick et al., 2007; Xian et al., 2007). Fitzpatrick et al. (2007) found vanadium, chromium, cobalt, copper, selenium, lead, molybdenum, strontium, and barium in association with urban waters and uranium and arsenic associated with agriculture. Mercury and methyl-mercury concentrations are also elevated in some urban streams during baseflow, and they increase in concentration during storm flows (Mason and Sullivan, 1998; Horowitz et al., 1999). Platinum eroded from catalytic converters is also found in urban surface water and sediments as well as aquatic organisms (Wei and Morrison, 1992; Rauch and Morrison, 1999).

Sources of Metals

Impervious Surfaces

The runoff from impervious surfaces that cover urban landscapes is a notorious contributor to the high concentrations of metals in urban surface waters (Gobel et al., 2007). Gobel et al. (2007) compiled average concentrations of various chemical constituents from different types of impervious surfaces and concluded that the chemistry from the various sources was unique. The metal concentration of runoff from an impervious surface can be influenced by the type and intensity of human activity most commonly occurring on that particular surface, as well as the type of material from which the surface is constructed (Gnecco et al., 2006; Gobel et al., 2007). Nonpoint-source pollution from vehicle traffic on roads, bridges, and parking lots often results in high levels of metals accumulating on these surfaces and subsequently being washed into surface waters (Sartor et al., 1974; Wilber and Hunter, 1977; Helsel et al., 1979; Yousef et al., 1983, 1985; Forman and Alexander, 1998; Mason and Sullivan, 1998). Cars and trucks contain metals that are eroded onto surfaces, including nickel, chromium, lead, and copper from brake linings; zinc, lead, chromium, copper, and nickel from tires; nickel, chromium, copper, and manganese from engine parts; and platinum from catalytic converters (Muschak, 1990; Wei and Morrison, 1992; Mielke et al., 2000). Precipitation washes these impervious surfaces resulting in increased concentrations and loads of heavy metals during storm flow or during seasonal periods with high precipitation (Horowitz et al., 1999; Kang et al., 2009).

Sewage and Wastewater Treatment Facilities

Metals are often removed during the treatment process in WWTF as they settle out with the biosolids, and very small concentrations typically are left in the effluent (Sheikh et al., 1987). However, if the source is industrial or the treatment is inadequate or altogether missing, the untreated sludge and industrial wastewater can potentially impact heavy metal concentrations and loads in the receiving water and their associated sediments (Gonzales et al., 2000; Toze, 2006). Gonzales et al. (2000) found elevated Zn, Pb, Cr, Cu, and Cd concentrations in the sediments of a river receiving untreated urban sewage and olive oil and table olive industrial wastewaters.

In-Stream Chemistry and Biotic Effects

The retention of particulate-bound metals or their release into dissolved forms depends on several factors, including organic matter content and character, sediment characteristics, and speciation (Tada and Suzuki, 1982; Rhoads and Cahill, 1999). Heavy metals lack the transformation or decay mechanisms associated with organic compounds and nutrients; therefore, they are prone to accumulation in sediments or transport downstream (Chapra, 1997). Organic matter has the ability to bind metals. Bed and suspended sediments with high organic matter content often have an estimated 50 to 7500 times higher concentrations of metals than sediments with lower organic matter content (Warren and Zimmerman, 1994; Mason and Sullivan, 1998; Gonzales et al., 2000). Like other ions that are adsorbed and released, the particle size distribution of sediments and their associated charge influence the concentrations of metals adsorbed to sediments (Wilber and Hunter, 1977). Within a stream, metals will accumulate with clays and other fine particles where the velocity of the water slows (Rhoads and Cahill, 1999).

Metals have the potential to bioaccumulate in many aquatic organisms, which can change the abundance and diversity of biota in urban streams (Davis and George, 1987; Lenat and Crawford, 1994; Rauch and Morrison, 1999; Gundacker, 2000). Species shifts are often found from pristine, to agricultural, to urban streams, with urban streams limited to only the most tolerant groups (Whiting and Clifford, 1983; Lenat and Crawford, 1994; Kemp and Spotila, 1997). This effect is so common that the presence of heavy metals in plant and animal tissue has been suggested as an indicator of overall water quality (Davis and George, 1987). Traffic volume on roads and the concentration of metals in runoff have been correlated with mortality of fish and other aquatic organisms, with some measured effects as far downstream from the source as 8 km (Horner and Mar, 1983; Morgan et al., 1983). Researchers suggest that both direct exposure and ingestion of adsorbed metal are possible mechanisms for bioaccumulation and that toxicity may be more strongly associated with sediments rather than water column concentrations (Medeiros et al., 1983; House et al., 1993; Rauch and Morrison, 1999; Paul and Meyer, 2001). Metal speciation also determines the toxicity and bioavailability, with ions and labile complexes being more toxic (Bingham et al., 1984; Rijstenbil and Poortvliet, 1992; Almås et al., 2006).

Nickel

Nickel in surface waters comes from a variety of terrestrial sources, including both anthropogenic and natural rock sources (Beasley and Kneale, 2002). Anthropogenic sources have been credited with doubling the concentration of nickel in fresh waters every decade since 1930 (Sreedevi et al., 1992; Biney et al., 1994). In natural waters with a pH range of 5 to 9 the dominant oxidation state of nickel is Ni^{2+}, and it is highly electronegative (Beasley and Kneale, 2002). Nickel has a high affinity for clays, and its concentration in sediments can be several orders of magnitude higher than those in the water column, but it is also found in soluble salt and organic complexes (Stokes, 1988). Factors promoting the dissolution and migration of nickel in surface waters include low pH, chloride, nitrate, sulfate, and DOC. High pH, phosphate, carbonate, hydroxyl, and hydrogen sulfides limit nickel migration by precipitation and chelation, while organics, iron, manganese, aluminum, and clay limit its migration by absorbance and adsorption (Beasley and Kneale, 2002). Nickel is required by aquatic organisms, but in high concentrations it is harmful to both their survival and productivity (Beasley and Kneale, 2002). Nickel has been shown to affect aquatic populations of commercially important marine and freshwater fish and is a health hazard for humans (Moore and Ramamoorthy, 1984; Chaudhry and Kedarnath, 1985).

Copper

Copper is commonly used in alloys, construction materials, fungicides, the manufacturing of wood preservatives, rayon, paint, textiles, glass, and ceramics and is present in plant and animal waste (Beasley and Kneale, 2002). Brake linings contain copper, and a study by Beasley and Kneale, (2002) suggested that variability in sediment samples may be due to the increased braking by motorists at road construction sites. In natural waters, copper exists in the cupric ion form or complexed with carbonate, chlorides, humic and fulvic acids, or in precipitates of hydroxides, phosphates, or sulfides and can be adsorbed by sediments and particulate matter (Beasley and Kneale, 2002). Speciation of copper is dependent on its concentration, and the hardness, alkalinity, salinity, pH, and concentrations of carbonates, sulfides, phosphates, organic ligands, and other metals in surface waters (Beasley and Kneale, 2002). Copper is toxic to a variety of aquatic organisms, particularly at early stages of development, but it has been shown in laboratory settings that populations of invertebrates can acclimatize to higher concentrations (Leland et al., 1989; Courtney and Clements, 2000).

Zinc

One of the most commonly used metals in the world, zinc is found in a host of products, including galvanized metals, zinc-based alloys, rubber, paints, cosmetics, ceramics, dyes, and wood preservatives. It also enters the environment via activities like manufacturing, purification of fats, mining, purification of metal ores, steel production, coal and waste burning, and waste treatment discharges (Radhakrishnaiah et al., 1993; Beasley and Kneale, 2002). Many zinc salts are highly soluble and have significant toxicity risks to aquatic organisms (Beasely and Kneale, 2002).

Cations and Anions

Many other cations and anions are elevated in urban streams, including chloride, sulfate, calcium, sodium, potassium, and magnesium and overall electrical conductivity (Fig. 15–3 and 15–4) (McConnell, 1980; Smart et al., 1985; Zampella, 1994; Ometo et al., 2000; Paul and Meyer, 2001). A breakdown of urban land uses in the Shimousa Upland, Japan revealed significant positive correlations between base cation concentrations and land use. Concentrations of K^+, Mg^{2+}, and Ca^{2+} were correlated with residential area in a watershed, while Mg^{2+} and Ca^{2+} were correlated with commercial areas, and Ca^{2+} was correlated with urban developing areas (Bahar and Yamamuro, 2008). In contrast, a literature review shows that calcium and magnesium concentrations both tend to decline with the fraction of urban land use in a basin (Fig. 15–3). Similarly, Fitzpatrick et al. (2007) found biogeochemical fingerprints of human impact differed between agricultural and urban land uses for major cations (urban: Na, K, Cl; agriculture: Ca, Mg). Bhatt and McDowell (2007) demonstrated strong relationships between many ion concentrations and the human population density adjacent to the Baghmati River in Nepal, where untreated sewage had a significant impact on the water chemistry. The combined effect of increases in ion concentrations is an increase in electrical conductivity. These increases in ions are so common that some have suggested using chloride concentration or electrical conductivity as an indicator of urban impacts on water chemistry (Wang and Yin, 1997; Herlihy et al., 1998; Paul and Meyer, 2001). Sources of cation and anions in urban stream water include wastewater effluent, irrigation runoff, deposition to impervious surfaces (e.g., atmospheric, dust, vehicle exhaust, and animal waste), infrastructure dissolution, spills, and sediment erosion. Irrigation runoff in urban areas may impact surface water chemistries if the source water for potable use is significantly different than the surface water (Fig. 15–5).

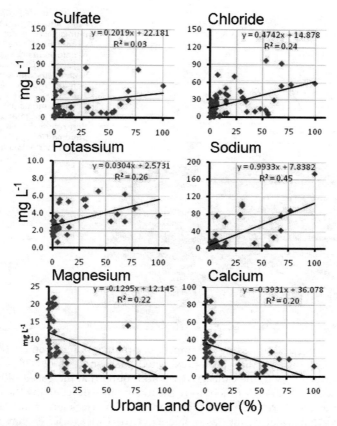

Fig. 15–3. Relationship between surface water concentrations of chloride, sulfate, potassium, sodium, magnesium, and calcium and the percentage of urban land cover in watersheds reported in the literature. Source data are obtained from the following references: Aitkenhead-Peterson et al., 2009; Bahar and Yamamaro, 2008; Bedore et al., 2008; Bhatt and McDowell, 2007; Brett et al., 2005; Chang and Carlson, 2005; Chea et al., 2004; Cunningham et al., 2009; Daniels et al., 2002; Fitzpatrick et al., 2007; Lewis et al., 2007; Liu et al., 2000; Rose and Peters, 2001; Schoonover et al., 2005; Smart et al., 1985; Von Schiller, 2008; Zampella et al., 2007.

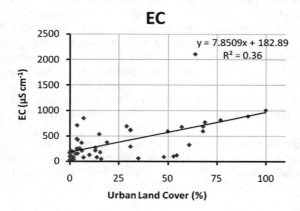

Fig. 15–4. Relationship between surface water electrical conductivity and the percentage of urban land cover in watersheds reported in the literature. Source data are obtained from the following references: Aitkenhead-Peterson et al., 2009; Bahar and Yamamaro, 2008; Bedore et al., 2008; Bhatt and McDowell, 2007; Brett et al., 2005; Chang and Carlson, 2005; Chea et al., 2004; Cunningham et al., 2009; Daniels et al., 2002; Fitzpatrick et al., 2007; Lewis et al., 2007; Liu et al., 2000; Rose and Peters, 2001; Schoonover et al., 2005; Smart et al., 1985; Von Schiller, 2008; Zampella et al., 2007.

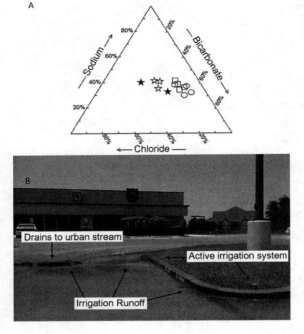

Fig. 15–5. (A) Chemistry of rural and urban streams and municipal tap water. Black stars denote rural sites; white stars denote industrial and urbanizing sites. Open squares identify watersheds with a high proportion of irrigated golf courses and parks, and open circles identify municipal tap water supplies to the area. (B) Irrigation runoff draining into urban streams College Station, TX by way of storm drainage system.

Measuring and assessing the impact of urbanization on ion concentrations may be more difficult than at first expected. Jackson et al. (2008) assessed the density-driven flow in the Chicago River and found that increased salinity from deicing salts and WWTF effluent were the likely cause of reversing the flow direction due to the increased density of the underflow (Jackson et al., 2008). They suggested that water quality assessments of the Chicago River may underestimate (or overestimate) water quality impairment because standard monitoring practices do not account for density-driven underflows (or overflows) and do not adjust their depth of sampling to accommodate this (Jackson et al., 2008).

Sodium and Chloride

Although total salinity can have negative impacts on ecosystem function and aquatic health in streams and rivers, there are certain ions that can be particularly detrimental. Sodium and chloride are naturally occurring constituents of surface waters. Sources in surface waters include saline groundwater, geologic weathering and soil exchange processes, marine aerosols, saltwater intrusion, and atmospheric deposition. Ecological problems with sodium and chloride concentrations generally only occur in freshwater ecosystems, and include toxicity to plants, invertebrates, and fish. Human health concerns relate to the impacts of salt intake on hypertension (Howard and Haynes, 1993; Forman and Alexander, 1998; Wegner and Yaggi, 2001).

Some surface waters are naturally enriched in sodium, but recent findings have indicated a strong anthropogenic enrichment of sodium and chloride in surface waters through the use of deicing salts and wastewater treatment effluent. The Salt Institute (2004) estimated that 18 million Mg (or metric tons) of NaCl are spread on paved surfaces for deicing annually. Deicing salts are used in cooler climates to prevent and treat roads and sidewalks to make vehicle and pedestrian traffic safer. A study in Toronto found that 45% of the salt applied to the catchment is removed through runoff, and the remainder is stored temporarily in shallow subsurface waters (Howard and Haynes, 1993). Salt stored in shallow subsurface waters is released as baseflow throughout the year. They predicted that if salt application continues at the present rate, the average chloride concentrations in groundwater discharging as springs will increase threefold (Howard and Haynes, 1993). Kaushal et al. (2005) found the largest increases in the winter months in chloride concentrations in urban streams when salt was being applied in three northeastern cities; however, chloride concentrations did not return to base level during the summer, demonstrating an increasing trend over several decades. In the colder climate of New Hampshire, chloride concentrations were typically highest during summer low flows, suggesting that pervasive groundwater contamination with road salt has occurred (Daley et al., 2009). Chloride concentrations were found in some Maryland locations to be 25% those of seawater and increasing, rendering many potential drinking water sources non-potable within the next century (Kaushal et al., 2005). Similarly, thirteen lakes in the Twin Cities metropolitan area of Minnesota, USA had sodium and chloride concentrations that were 10 to 25 times higher than non-urban lakes; this increase in concentrations was correlated with road salt applications (Novotny et al., 2008).

Both sodium and chloride are relatively mobile in the soil and deeper geologic environment compared to other major ions, although chloride is somewhat more mobile than sodium due to some ion exchange or adsorption of sodium (Shanley, 1994; Jackson and Jobbagy, 2005; Daley et al., 2009). The change in the Na/Cl ratio may allow researchers to determine the relative importance or contribution of direct inputs of salts, via direct surface drainage, or those that have an indirect underground flow path (Jackson and Jobbagy, 2005). For streams affected by surface runoff or drainage systems, the Na/Cl ratio should be roughly similar to the source material (0.65:1 mass ratio); however, if the dominant flow path involves transport through soils or bedrock, then the ratio of Na/Cl will be lower. Sodium will gradually displace Ca, Mg, K, and protons in the soil, altering soil fertility and uncoupling the flow of Na from Cl (Norrstrom and Bergstedt, 2001; Jackson and Jobbagy, 2005). The Na/Cl ratio in surface waters impacted by extensive road salt application also can be expected to increase with time due to the exhaustion of exchangeable cations in watershed soils.

Solutions and Future Research

Numerous best management practices have been suggested to decrease the loads of metals, nutrients, organic components, and salinity entering urban surface waters. Effective management will have to consider all aspects of the chemical and biochemical cycles governing the storage, transformation, and delivery to surface waters of each type of contaminant. One of the simplest and most effective management strategies is to invest in preserving an undisturbed upstream catchment, particularly if those catchments are a drinking water source (Daily and Ellison, 2002). Another popular management strategy is stream restoration, which may involve a variety of efforts, including bank stabilization, revegetation and reestablishment of riparian zones, removal of contaminated sediments, and rerouting to a more meandering path that decreases flow velocity (Bernhardt and Palmer, 2007). Restoration has become a profitable business; however, restoration of urban streams is more expensive and difficult than their rural counterparts (Lavendel, 2002; Malakoff, 2004; Bernhardt and Palmer, 2007). Restoration projects can be effective at reducing loads. Bukaveckas (2007), for example, found that stream restoration enhanced nutrient uptake by decreasing flow velocity. Macroinvertebrate richness and biomass are correlated with the percentage of stream banks covered with roots or wood on both restored and degraded urban streams (Sudduth and Meyer, 2006). Fish and other mobile taxa are also negatively impacted as impervious surface increases; however, impacts of restoration on these species is still unknown (Wang et al., 2000; Morgan and Cushman, 2005; Bernhardt and Palmer, 2007). Bernhardt and Palmer (2007) suggested that urban stream restoration efforts need to be integrated with watershed management practices to be effective and prevent redegradation of restored streams.

Watershed protection practices designed to improve flow and chemistry range in size and scope. Berke et al. (2003) investigated the impact of the new urbanism development style on water quality in urban areas. The term *new urbanism* involves a development goal of creating urban spaces at the walkable scale to promote equality and quality of life for urban residents (Congress for the New Urbanism, 1996). This style of development may have positive benefits for water and environmental quality. According to Berke et al. (2003), new urbanism communities around the nation also incorporate low-impact development techniques (Chapter 20, Li et al., 2010, this volume), such as pervious pavements, rain gardens, and bioswales, which allow these communities to be effective at improving water chemistry; however, new urban communities redeveloped from an older urban community are generally less successful than those in previously undeveloped spaces. Slightly smaller-scale management practices include bioretention areas, constructed wetlands (Chapter 20, Li et al., 2010, this volume), or the addition of labile carbon sources to reduce nitrogen loading through enhanced denitrification. All three remove pollutants from surface runoff by filtration, settling, and biotic activity. Bioretention systems, or rain gardens, generally include a meter or less of sand, soil, and organic media, a mulch layer, vegetation, slopes that allow for ponding of water, an inlet, and an outlet for water that drains below the media. Water is funneled into the system from an impervious surface. The soil, sand, and organics filter out particulates and debris, and help to adsorb, chelate, and fix organics, metals, and P. Nitrate and ammonium can be taken up by the vegetation or converted to nitrogen gas in anaerobic conditions. Laboratory and field tests indicate that these systems can remove most metals, motor oil and hydrocarbons, pathogens, and suspended solids (Davis et al., 2003; Hsieh et al., 2005; Hong et al., 2006; Rusciano and Obropta, 2007; Chapter 20, Li et al., 2010, this volume). Removal of nutrients, N and P, are more variable despite the presence of vegetation in all these systems. Based on the available literature, Davis et al. (2009) found 30 to 99% (56% average) of the total N load was removed and −240 to 99% (23% average) of the total P load was removed. Constructed wetlands function very similarly to bioretention devices, but are generally more saturated and have vegetation suited to those conditions. Constructed wetlands have also been found to reduce nutrients, metals, and total suspended solids (Mitsch et al., 1989; Strecker et al., 1992; Carlton et al., 2000). Another management practice that does not involve construction is the addition of coarse particulate organic matter, which can be depleted in the benthic material in urban streams not contaminated with sewage (Aldridge et al., 2009; Bernhardt and Palmer, 2007). The organic matter provides a substrate for microbial growth and thereby increases the demand for N and P (Aldridge et al., 2009). For all types of stream restoration and management practices, more studies and long-term monitoring are needed (Bernhardt and Palmer, 2007; Davis et al., 2009). Several key aspects that need to be addressed are the optimization of designs to enhance and maintain the long-term function of these systems and development of designs suitable for local needs (Davis et al., 2009).

Abstaining from using novel chemicals and metals that collect in urban areas would be one of the most fool-proof methods of long-term remediation, but it is unlikely that people will give up their cars and other commodities any time soon. The next best thing may be developing technologies and commodities that do not contain or emit these pollutants. The promise of hydrogen fuel cells that emit only water has potentially enormous benefits for urban water quality. As with any technology, the unforeseen impacts, such as the manufacturing waste and breakdown of components, often become the most detrimental problems. One area that has made great improvements in the developed world is the treatment of wastewater effluent (Bernhardt and Palmer, 2007). A system that has primary, secondary, and tertiary treatment can effectively remove the majority of nutrients. For example, improved wastewater treatment and banning P-based detergents reduced algal blooms in freshwater portions of the Neuse River estuary in North Carolina, USA (Paerl et al., 2004). However, these actions increased eutrophication and hypoxia downstream in the more saline parts of the estuary, where primary production is N-limited (Paerl et al., 2004). The variability in nitrogen and phosphorus removal by retention facilities and the shifting hypoxic zones in the Neuse River illustrate how complicated remediating surface waters in urban areas can be. The biogeochemical cycles of nitrogen and phosphorus are complex and contain intricate interactions among aquatic and terrestrial organisms. Although N and P can be removed from products, such as detergent, they are necessary for human and animal life in urban ecosystems and will always be a part of the food and waste stream. Removal of nitrogen is usually accomplished through uptake by vegetation, conversion to N gas under anaerobic conditions, or chemical precipitation. However, nitrate is prone to leaching and can move through the soil rapidly compared with many other ions. To further complicate matters, anaerobic conditions, which so effectively remove N from the system, can release enormous amounts of dissolved P through the reduction of otherwise very insoluble compounds, such as iron phosphate. This was the cause of the enormous variability (−240 to 99%) in the range of total P removed by bioretention systems (Davis et al., 2009). Similarly, N release in bioretention systems can be strongly influenced by the wetting and drying cycles that control the denitrification or mineralization of N (Hatt et al., 2009). In these systems, use of low-P media and maintaining the most effective moisture conditions can help, but more research is needed (Hatt et al., 2009).

Remediation of excess N and P at the watershed scale and the resulting hypoxic zones at discharge points is difficult. Much effort has been focused on identifying diffuse and point sources, particularly within larger watersheds with multiple land uses (e.g., Hasic and Wu, 2006; Turner et al., 2007). However, remediation efforts have had mixed results. Paerl (2009) argued that only a dual approach will be successful at remediating eutrophication and hypoxic zones. Like the problem in North Carolina, reductions in P without reducing N loads could actually expand the hypoxic zones in the Gulf of Mexico by displacing the dead zone (Scavia et al., 2003). On the other hand, N reductions in freshwater streams and rivers without comparable P reductions may increase N fixation and negate the reduction efforts. Likewise, reducing P without reducing N may displace the effects downstream where P is more available (Paerl et al., 2004; Paerl, 2009). The use of riparian zones to retain nutrients involves issues similar to those associated with bioretention systems. Wet anoxic conditions can release P, and drier aerobic conditions can leach N. Riparian zones of urban streams often have lower water tables that prevent the formation of reducing conditions (Groffman et al., 2002; Chapter 13, Stander and Ehrenfeld, 2010, this volume). While more oxygen may reduce denitrification and the ability of riparian zones to act as sinks, it may actually prevent the chemical reduction of P-bearing minerals and thus prevent large releases of P into surface waters.

The collective knowledge on urban stream chemistry has improved in the last 40 yr, yet questions still seem to outnumber answers. We need comprehensive and controlled catchment- level studies that evaluate the broad ecological impacts of stormwater runoff (Booth and Jackson, 1997). We need a better understanding of the multiple interactions of biogeochemical cycles for both natural and novel compounds, and the effects that these natural and novel compounds have on aquatic life. Developing management practices that are both effective and economically viable is an important research avenue that has yet to be fully developed (Bernhardt and Palmer, 2007). Although recent progress has been made in developing nitrogen budgets, similar budgets for other elements are lacking and thus so is an understanding of the system as a whole. The ability to answer the important questions—What are the inputs to urban ecosystems? Where are these inputs going or accumulating? How and why are they leaving the systems?—will help us to understand how the urban system is functioning.

Conclusions

- Surface water in urbanized settings is influenced by numerous human activities that generally increase sediment, overall salinity, sodium, chloride, nutrients, trace metals, and natural and anthropogenic carbon compounds present in the water.
- Urbanization creates a "loading effect" where elements from many watersheds are concentrated in the urban region. The ability of the urban watershed to retain or lose these elements to surface water depends on the natural biogeochemical cycling of the element, soil and hydrologic conditions, and human intervention.
- The in-stream processing of elements is often altered by stream hydrology, changes in sediment texture, organic matter in sediments and the water column, the presence of other elements, and changes in biomass and type of stream flora and fauna.
- Wastewater treatment, new urbanism, and other best management practices have the potential to alter loads of pollutants and surface water chemistry; however, implementation and adaptation for local conditions is still in progress.

References

Ahn, H., and R.T. James. 2001. Variability, uncertainty and sensitivity of phosphorus deposition load estimates in South Florida. Water Air Soil Pollut. 126:37–51.

Aherne, G.W., and R. Briggs. 1989. The relevance of the presence of certain synthetic steroids in the aquatic environment. J. Pharm. Pharmacol. 41:735–736.

Aherne, G.W., J. English, and V. Marks. 1985. The role of immunoassay in the analysis of microcontaminants in river samples. Ecotoxicol. Environ. Saf. 9:79–83.

Aherne, G.W., A. Hardcastle, and A.H. Nield. 1990. Cytotoxic drugs and the aquatic environment: Estimation of bleomycin in river and water samples. J. Pharm. Pharmacol. 42:741–742.

Aitkenhead-Peterson, J.A., M.K. Steele, N. Nahar, and K. Santhy. 2009. Dissolved organic carbon and nitrogen in urban and rural watersheds of south-central Texas: Land use and land management influences. Biogeochemistry 96:119–129, doi:10.1007/s10533-009-9348-2.

Aldridge, K.T., J.D. Brookes, and G.G. Ganf. 2009. Rehabilitation of stream ecosystem functions through the reintroduction of coarse particulate organic matter. Restor. Ecol. 17:97–106.

Almås, A.R., P. Lombnæs, T.A. Sogn, and J. Mulder. 2006. Speciation of Cd and Zn in contaminated soils assessed by DGT-DIFS, and WHAM/Model VI in relation to uptake by spinach and ryegrass. Chemosphere 62:1647–1655.

Andreozzi, R., M. Raffele, and P. Nicklas. 2003. Pharmaceuticals in STP effluents and solar photodegration in aquatic environment. Chemosphere 50:1319–1330.

Arnscheidt, J., P. Jordan, S. Li, S. McCormick, R. McFaul, H.J. McGrogan, M. Neal, and J.T. Sims. 2007. Defining the sources of low flow phosphorus transfers in complex catchments. Sci. Total Environ. 382:1–13.

Bahar, M., and M. Yamamuro. 2008. Assessing the influence of watershed land use patterns on the major ion chemistry of river waters in the Shimousa Upland, Japan. Chem. Ecol. 24:341–355.

Baker, L.A., D. Hope, and Y. Xu. 2001. Nitrogen balance for the central Arizona-Phoenix (CAP) ecosystem. Ecosystems 4:582–602.

Ball, R.C., N.R. Kevern, and T.A. Haines. 1973. An ecological evaluation of stream eutrophication. Tech. Rep. 36. Inst. of Water Resources, Michigan State Univ., East Lansing.

Banks, H.H., and J.E. Nighswander. 1999. Relative contribution of hemlock pollen to the phosphorus loading of the Clear Lake ecosystem near Minden, Ontario. p. 168–173. In Symposium on Sustainable Management of Hemlock Ecosystems in Eastern North America, Available at http://www.fs.fed.us/ne/newtown_square/publications/technical_reports/pdfs/scanned/gtr267d.pdf. USDA Forest Service, Durham, NH.

Bannerman, R.T., D.W. Owens, R.B. Dodds, and N.J. Hornewer. 1993. Sources of pollutants in Wisconsin stormwater. Water Sci. Technol. 28:241–259.

Beasley, G., and P. Kneale. 2002. Reviewing the impact of metals and PAHs on Macroinvertebrates in urban watercourse. Prog. Phys. Geogr. 26:236–270.

Bedore, P.D., M.B. David, and J.W. Stucki. 2008. Mechanisms of phosphorus control in urban streams receiving sewage effluent. Water Air Soil Pollut. 191:217–229.

Bennett, E.M., T. Reed-Anderson, J.N. Houser, J.R. Gabriel, and S.R. Carpenter. 1999. A phosphorus budget for the Lake Mendota watershed. Ecosystems 2:69–75.

Berke, P.R., J. MacDonald, N. White, M. Holmes, D. Line, K. Oury, and R. Ryznar. 2003. Greening development to protect watersheds, does new urbanism make a difference. J. Am. Plann. Assoc. 69:397–413.

Bernhardt, E.S., L.E. Band, C.J. Walsh, and P.E. Berke. 2008. Understanding, managing, and minimizing urban impacts on surface water. Ann. N.Y. Acad. Sci. 1134:61–96.

Bernhardt, E.S., and M.A. Palmer. 2007. Restoring streams in an urbanizing world. Freshwater Biol. 52:738–751.

Bhatt, M.P., and W.H. McDowell. 2007. Evolution of chemistry along the Bagmati Drainage Network in Kathmandu Valley. Water Air Soil Pollut. 185:165–176.

Biney, C., A.T. Amuzu, D. Calamari, N. Kaba, I.L. Mbome, H. Naeve, P.B.O. Ochumba, O. Osibanjo, V. Radegonde, and M.A.H. Saah. 1994. Review of heavy metals in the African aquatic environment. Ecotoxicol. Environ. Saf. 28:134–159.

Bingham, F.T., G. Sposito, and J.E. Strong. 1984. The effect of chloride on the availability of cadmium. J. Environ. Qual. 13:71–74.

Black, R., A.L. Haggland, and F.D. Voss. 2000. Predicting the probability of detecting organochlorine pesticides and polychlorinated biphenyls in stream systems on the basis of land use in the Pacific Northwest, USA. Environ. Toxicol. Chem. 19:1044–1054.

Bleken, M.A., and L.R. Bakken. 1998. The nitrogen cost of food production: Norwegian society. Ambio 26:134–142.

Bones, J., K.V. Thomas, and B. Paul. 2007. Using environmental analytical data to estimate levels of community consumption of illicit drugs and abused pharmaceuticals. J. Environ. Monit. 9:701–707.

Booth, D.B., and C.R. Jackson. 1997. Urbanization of aquatic systems: Degradation thresholds, stormwater detection, and the limits of mitigation. J. Am. Water Resour. Assoc. 33:1077–1090.

Bowes, M.J., J. Hilton, G.P. Irons, and D.D. Hornby. 2005. The relative contribution of sewage and diffuse phosphorus sources in River Avon catchment, southern England: Implications for nutrient management. Sci. Total Environ. 334:67–81.

Boyer, E.W., C.L. Goodale, N.A. Jaworski, and R.W. Howarth. 2002. Anthropogenic nitrogen sources and relationships to riverine nitrogen export in northeastern U.S.A. Biogeochemistry 57/58:137–169.

Brandt, G.H., E.S. Conyers, M.B. Ettinger, F.J. Lowes, J.W. Mighton, and J.W. Pollack. 1972. An economic analysis of erosion and sediment control methods for watersheds undergoing urbanization. Dow Chemical, Midland, MI.

Brett, M.T., G.B. Arhonitsis, S.E. Mueller, D.M. Hartley, J.D. Frodge, and D.E. Funke. 2005. Non-point-source impacts on stream nutrient concentrations along a forest to urban gradient. Environ. Manage. 35:330–342.

Bryan, E.H. 1974. Concentrations of lead in urban stormwater. J. Water Pollut. Control Fed. 46:2419–2421.

Bukaveckas, P.A. 2007. Effects of channel restoration on water velocity, transient storage, and nutriet uptake in a channelized stream. Environ. Sci. Technol. 41:1570–1576.</jrn>

Burian, S.J., T.N. McPherson, M.J. Brown, G.E. Streit, and H.J. Turin. 2002. Modeling the effects of air quality policy changes on water quality in urban areas. Environ. Model. Assess. 7:179–190.

Burian, S.J., and C.A. Pomeroy. 2010. Urban impacts on the water cycle and potential green infrastructure implications. p. 277–296. In J. Aitkenhead-Peterson and A. Volder (ed.) Urban ecosystem ecology. Agron. Monogr. 55. ASA, CSSA, and SSSA, Madison, WI.

Carlton, J.N., T.J. Grizzard, A.D. Godrej, H.E. Post, L. Lampe, and P.P. Kenel. 2000. Performance of constructed wetlands in treating urban stormwater runoff. Water Environ. Res. 72:295–304.

Chang, H., and T.N. Carlson. 2005. Water quality during winter storm events in Spring Creek, Pennsylvania USA. Hydrobiologia 544:321–332.

Chapra, S.C. 1997. Surface water-quality modeling. McGraw-Hill, Boston, MA.

Chaudhry, H.S., and P. Kedarnath. 1985. Nickel induced hyperglycemia in the freshwater fish, Colisa fasciatus. Water Air Soil Pollut. 24:173–176.

Chea, G., S. Yun, K. Kim, P. Lee, and B. Choi. 2004. Atmospheric versus lithogenic contribution to the composition of first-and second-order stream waters in Seoul and its vicinity. Environ. Int. 30:73–84.

Christensen, H.E., T.T. Lugin Byhl, and B.S. Carroll. 1975. Registry of toxic effects of chemical substances. United States Department of Health, Education and Welfare, Rockville, MD.

Chrost, R.J. 1989. Characterization and significance of β-glucosidase activity in lake water. Limnol. Oceanogr. 34:660–672.

Clements, W.H., J.T. Oris, and T.E. Wissing. 1994. Accumulation and food-chain transfer of fluoranthene and benzo[a]pyrene in *Chironomus riparius* and *Lepomismacrochirus*. Arch. Environ. Contam. Toxicol. 26:261–266.

Congress for New Urbanism. 1996. Charter of new urbanism. Available at http://www.cnu. org/charter (accessed September 2009, verified 18 Mar. 2010). CNU, Chicago, IL.

Conely, D.J., H.W. Paerl, R.W. Howarth, D.F. Boesch, S.P. Seitzinger, K.E. Havens, C. Lancelot, and G.E. Likens. 2009. Controlling eutrophication: Nitrogen and phosphorus. Science 323:1014–1015.

Conley, D.J., C. Humborg, L. Rahm, O.P. Savchuk, and F. Wulff. 2002. Hypoxia in the Baltic Sea and basin-scale changes in phosphorus biogeochemistry. Environ. Sci. Technol. 36:5315–5320.

Costanzo, S.D., J. Murby, and J. Bates. 2005. Ecosystem response to antibiotics entering the aquatic environment. Mar. Pollut. Bull. 51:218–223.

Courtney, L.A., and W.H. Clements. 2000. Sensitivity to acidic pH in benthic invertebrate assemblages with different histories of exposure to metals. J. North Am. Benthol. Soc. 19:112–127.

Cunningham, M.A., C.M. O'Reilly, K.M. Menking, D.P. Gillikin, K.C. Smith, C.M. Foley, S.L. Belli, A.M. Pregnall, M.A. Schlessman, and P. Batur. 2009. The surburban stream syndrome: Evaluating land use and stream impairments in the suburbs. Physical Geography. 30:269–284.

Daily, G.C., and K. Ellison (ed.) 2002. The new economy of nature: The quest to make conservation profitable. Island Press, Washington, DC.

Daley, M.L., J.D. Potter, and W.H. McDowell. 2009. Salinization of urbanizing New Hampshire streams and groundwater: Effects of road salt and hydrologic variability. J. North Am. Benthol. Soc. 28:929–940.

Daniels, W.M., W.A. House, J.E. Rae, and A. Parker. 2000. The distribution of micoorganic contaminants in river bed-sediment. Sci. Total Environ. 253:81–92.

Daniels, M.B., A.A. Montebelo, M.C. Bernardes, J.P.H.B. Ometto, P.B. De Camargo, A.V. Krusche, M.V. Ballester, R.L. Victoria, and L.A. Martinelli. 2002. Effects of urban sewage on dissolved oxygen, dissolved inorganic carbon and organic carbon, and electrical conductivity of small streams along a gradient of urbanization in the Piracicaba river basin. Water Air Soil Pollut. 136:189–206.

Davis, A.P., W.F. Hunt, R.G. Traver, and M. Clar. 2009. Bioretention technology: Overview of current practices and future needs. J. Environ. Eng. 135:109–117.

Davis, J.B., and J.J. George. 1987. Benthic invertebrates as indicators of urban and motor discharge. Sci. Total Environ. 59:291–302.

Davis, M.B., and L.E. Gentry. 2000. Anthropogenic inputs of nitrogen and phosphorus and riverine export for Illinois, USA. J. Environ. Qual. 29:494–508.

Davis, N.M., V. Weaver, K. Parks, and M.J. Lydy. 2003. An assessment of water quality, physical habitat, and biological integrity of an urban stream in Witchita, Kansas, prior to restoration improvements (Phase I). Arch. Environ. Contam. Toxicol. 44:351–359.

Diaz, R.J., and R. Rosenberg. 2008. Spreading dead zones and consequences for marine ecosystems. Science 321:926–929.

Dorney, J.R. 1985. Leachable and total phosphorus in urban street tree leaves. Water Air Soil Pollut. 28:439–443.

Easton, Z.M., P. Gerard-Marchant, M.T. Walter, A.M. Petrovic, and T.S. Steenhuis. 2007. Identifying dissolved phosphorus sources areas and predicting transport from an urban watershed using distributed hydrologic modeling. Water Resour. Res. 43:1–16.

Eganhouse, R.P., B.R.T. Simoneit, and I.R. Kaplan. 1981. Extractable organic matter in urban stormwater runoff. 2. molecular characterization. Environ. Sci. Technol. 15:315–326.

Eisler, R. 1987. Polycyclic aromatic hydrocarbon hazards to fish, wildlife, and invertebrates: A synoptic review. Rep. 85. U.S. Dep. of the Interior, Fish and Wildlife Service, Laurel, MD.

Faerge, J., J. Magid, and F. de Vries. 2001. Urban nutrient balance for Bangkok. Ecol. Modell. 139:63–74.

Faulkner, H., V. Edmonds-Brown, and A. Green. 2000. Problems of quality designed in diffusely polluted urban streams—The case of Pymme's Brook, north London. Environ. Pollut. 109:91–107.

Fiebig, D.M., M.A. Lock, and C. Neal. 1990. Soil water in the riparian zone as a source of carbon for headwater stream. J. Hydrol. 116:217–237.

Finkenbine, J.K., D.S. Atwater, and D.S. Mavinic. 2000. Stream health after urbanization. J. Am. Water Resour. Assoc. 36:1149–1160.

Findlay, S., and R.L. Sinsabaugh. 1999. Unraveling the sources and bioavailability of dissolved organic matter in lotic aquatic ecosystems. Mar. Freshwater Res. 50:781–790.

Fitzpatrick, M.L., D.T. Long, and B.C. Pijanowski. 2007. Exploring the effects of urban and agricultural land use on surface water chemistry, across a regional watershed using multivariate statistics. Appl. Geochem. 22:1825–1840.

Forman, R.T.T., and L.E. Alexander. 1998. Roads and their major ecological effects. Ann. Rev. Ecol. Syst. 29:207–231.

Foster, G.D., E.C. Roberts, Jr., B. Grussner, and D.J. Velinsky. 2000. Hydrogeochemistry and transport of organic contaminants in an urban watershed of Chesapeake Bay (USA). Appl. Geochem. 15:901–915.

Galloway, J.N., F.J. Dentener, D.G. Capone, E.W. Boyer, R.W. Howarth, S.P. Seitzinger, G.P. Asner, C.C. Cleveland, P.A. Green, E.A. Holland, D.M. Karl, A.M. Michaels, J.H. Porter, A.R. Townsend, and C.J. Vöosmarty. 2004. Nitrogen cycles: Past, present, and future. Biogeochemistry 70:153–226.

Gerritse, R.G., J.A. Adney, G.M. Dimmock, and Y.M. Oliver. 1995. Retention of nitrate and phosphate in soils of Darling Plateau in Western Australia: Implications for domestic septic tank systems. Aust. J. Soil Res. 33:353–367.

Georgia Department of Transportation. 1993. Standard specifications: Construction of roads and bridges. Georgia Dep. of Transp., Atlanta.

Gibson, C.A., and J.L. Meyer. 2007. Nutrient uptake in a large urban river. J. Am. Water Resour. Assoc. 43:576–587.

Gnecco, I., C. Berretta, L.G. Lanza, and P. La Barbera. 2006. Quality of stormwater runoff from paved surfaces of two production sites. Water Sci. Technol. 54:177–184.

Gobel, P., C. Dierkes, and W.G. Coldewey. 2007. Storm water runoff concentration matrix for urban areas. J. Contam. Hydrol. 91:26–42.

Gonzales, A.E., M.T. Rodriquez, J.C.J. Sanchez, A.J.F. Espinosa, and F.J.B. De La Rosa. 2000. Assessment of metals in sediments in a tributary of Guadalquivir River (Spain), heavy metal partitioning and relation between the water and sediment system. Water Air Soil Pollut. 121:11–29.

Grimm, N.B., C.L. Crenshaw, and C.N. Dahm. 2005. Nutrient retention and transformation in urban streams. J. North Am. Benthol. Soc. 24:626–642.

Grimm, N.B., J.M. Grove, S.A. Pickett, and C.L. Redman. 2000. Integrated approaches to long-term studies of urban ecological systems. Bioscience 50:571–584.

Groffman, P.M., N.J. Boulware, and W.C. Zipperer. 2002. Soil nitrogen cycling processes in urban riparian zones. Environ. Sci. Technol. 36:4547–4552.

Groffman, P.M., N.L. Law, K.T. Belt, L.E. Band, and G.T. Fisher. 2004. Nitrogen fluxes and retention in urban watershed ecosystems. Ecosystems 7:393–403.

Gundacker, C. 2000. Comparison of heavy metal bioaccumulation in freshwater mollusks of urban river habitats in Vienna. Environ. Pollut. 110:61–71.

Guy, I.P., and G.E. Ferguson. 1962. Sediment in small reservoirs due to urbanization. ASCE Proc. J. 88:27–37.

Guy, I.P., 1963. Residential construction and sedimentation at Kensington, Maryland. Paper presented at Federal Inter-Agency Sedimentation Conference, Jackson, MI.

Haggard, B.E., E.H. Stanley, and R. Hyler. 1999. Sediment–phosphorus relationships in three northcentral Oklahoma streams. Trans. ASAE 42:1709–1714.

Haggard, B.E., E.H. Stanley, and D.E. Storm. 2005. Nutrient retention in a point-source-enriched stream. J. North Am. Benthol. Soc. 24:29–47.

Haggard, B.E., D.E. Storm, and E.H. Stanley. 2001. Effect of a point source input on stream nutrient retention. J. Am. Water Resour. Assoc. 37:1291–1299.

Hale, R.L., and P.M. Groffman. 2006. Chloride effects on nitrogen dynamics in forested and suburban stream debris dams. J. Environ. Qual. 35:2425–2432.

Hall, R.O., Jr., J.L. Tank, D.J. Sobota, P.J. Mulholland, J.M. O'Brien, W.K. Dodds, J.R. Webster, H.M. Valett, G.C. Poole, B.J. Peterson, J.L. Meyer, W.H. McDowell, S.L. Johnson, S.K. Hamilton, N.B. Grimm, S.V. Gregory, C.N. Dahm, L.W. Cooper, L.R. Ashkenas, S.M. Thomas, R.W. Sheibley, J.D. Potter, B.R. Niederlehner, L. Johnson, A.M. Helton, C. Crenshaw, A.J. Burgin, M.J. Bernot, J.J. Beaulieu, and C. Arango. 2009. Nitrate removal in stream ecosystems measured by 15N addition experiments: Total uptake. Limnol. Oceanogr. 54(3):653–665.

Halling-Sorensen, B., S.N. Nielsen, P.F. Lanzky, F. Ingersiev, H.C.H. Lutzhoft, and S.E. Jorgensen. 1998. Occurrence, fate, and effects of pharmaceutical substances in the environment– a review. Chemosphere 36(2):357–393.

Harbott, E.L., and M.R. Grace. 2005. Extracellular enzyme response to bioavailability of dissolved organic C in streams of varying catchment urbanization. J. North Am. Benthol. Soc. 24(3):588–601.

Hasic, I., and J. Wu. 2006. Land use and watershed health in the United States. Land Econ. 82(2):214–239.

Hatt, B.E., T.D. Fletcher, and A. Deletic. 2009. Hydrologic and pollutant removal performance of stormwater biofiltration systems at the field scale. J. Hydrol. 365:310–321.

Hatt, B.E., T.D. Fletcher, C.J. Walsh, and S.L. Taylor. 2004. The influence of urban density and drainage infrastructure on the concentrations and loads pollutants in small streams. Environ. Manage. 34(1):112–124.

Heckrath, G., P.C. Brookes, P.R. Poulton, and K.W.T. Goulding. 1995. Phosphorus leaching from soils containing different phosphorus concentrations in the broadbalk experiment. J. Environ. Qual. 24:904–910.

Helsel, D.R., J.I. Kim, T.J. Grizzard, C.W. Randall, and R.C. Hoehn. 1979. Land use influences on metals in storm drainage. J. Water Pollut. Control Fed. 51:709–717.

Herlihy, A.T., J.L. Stoddard, and C.B. Johnson. 1998. The relationship between stream chemistry and watershed land cover data in the Mid-Atlantic region, U.S. Water Air Soil Pollut. 105:377–386.

Hilton, J., P. Buckland, and G.P. Irons. 2002. An assessment of a simple method for estimating the relative contributions of point and diffuse source phosphorus to in-rier phosphorus loads. Hydrobiologia 472(1–3):77–83.

Hoare, R.A. 1984. Nitrogen and phosphorus in Rotorua urban streams. N.Z. J. Mar. Freshw. Res. 18:451–454.

Hoffman, E.J., J.S. Latimer, G.L. Mills, and J.G. Quinn. 1982. Petroleum hydrocarbons in urban runoff from a commercial land use area. J. Water Pollut. Control Fed. 54:1517–1525.

Hoffman, R.S., P.D. Capel. And, and S.J. Larson. 2000. Comparison of pesticides in eight U.S. urban streams. Environ. Toxicol. Chem. 19:2249–2258.

Holeman, J.N., and A.F. Geiger. 1959. Sedimentation of Lake Barcroft, Fairfax County, Va., USDA Rep. SCS-TP 136.

Hooda, P.S., A.R. Rendell, A.C. Edwards, P.J.A. Withers, M.N. Aitken, and V.W. Truesdale. 2000. Relating soil phosphorus indices to potential phosphorus release to water. J. Environ. Qual. 29:1166–1171.

Hook, A.M., and J.A. Yeakley. 2005. Stormflow dynamics of dissolved organic carbon and total dissolved nitrogen in a small urban watershed. Biogeochemisty 75:409–431.

Hong, E., E.A. Seagren, and A.P. Davis. 2006. Sustainable oil and grease removal from synthetic storm water runoff using bench-scale bioretention studies. Water Environ. Res. 78(2):141–155.

Horner, R.R. 1992. Water quality analysis for Covington master drainage plan. Covington Master Drainage Plan. Vol. II. Appendix 8. King County Dep. of Public Works, Surface Water Management Division. King County Arch., Seattle, WA.

Horner, R.R., and B.W. Mar. 1983. Guide for assessing water quality impacts of highway operations and maintenance. Transp. Res.. Rec. 948:31–39.

Horowitz, A.J., M. Maybeck, Z. Idlafkih, and E. Biger. 1999. Variations in trace element geochemistry in the Seine River Basin based on floodplain deposits and bed sediments. Hydrol. Processes 13:1329–1340.

House, W.A., and L. Donaldson. 1986. Adsorption and coprecipitation of phosphate on Calcite. J. Colloid Interface Sci. 112(2):309–324.

House, W.A., and F.H. Denison. 2002. Exchange of inorganic phosphate between river waters and bed-sediments. Environ. Sci. Technol. 36(20):4295–4301.

House, M.A., J.B. Ellis, E.E. Herricks, T. Hvitved-Jacobsen, J. Seager, L. Lijklema, H. Aalderink, and I.T. Clifforde. 1993. Urban drainage—Impacts on receiving water quality. Water Sci. Technol. 27(12):117–158.

Howard, K.W.F., and J. Haynes. 1993. Groundwater contamination due to road de-icing chemicals—Salt balance implications. Geosci. Can. 20:1–8.

Hsieh, C., A.P. Davis, and B.A. Needelman. 2005. Bioretention column studies of phosphorus removal from urban stormwater runoff. Water Environ. Res. 79(2):177–184.

Hu, F.S., B.P. Finney, and L.B. Brubaker. 2001. Effects of Holocene *Alnus* expansion of aquatic productivity, nitrogen cycling, and soil development in southwestern Alaska. Ecosystems 4:358–368.

Huerta-Fontela, M., M.T. Galceran, and F. Ventura. 2008. Stimulatory drugs of abuse in surface waters and their removal in a conventional drinking water treatment plant. Environ. Sci. Technol. 42(18):6809–6816.

Huisman, H., C.P. Matthijs, and P.M. Visser. 2005. Harmful cyanobacteria. Springer Aquatic Ecology Ser. 3. Springer, Dordrecht, the Netherlands.

Hunter, J.V., T. Sabatino, R. Gromperts, and M.J. MacKenzie. 1979. Contribution of urban runoff to hydrocarbon pollution. J. Water Pollut. Control Fed. 51:2129–2138.

Illinois EPA. 1995. Illinois urban manual. Illinois EPA, Bureau of Water, Watershed Management, Springfield.

Ingall, E.D., R.M. Bustin, and P. Van Cappellen. 1993. Influence of water column anoxia on the burial and preservation of carbon and phosphorus in marine shales. Geochim. Cosmochim. Acta 57:303–316.

IPCC. 2007. Climate change 2007: The physical science basis. Contribution of Working Group I to the Fourth Assessment Report of the Intergovernmental Panel on Climate Change. S. Solomon et al. (ed.) Cambridge Univ. Press, Cambridge, UK.

Isobe, T., H. Nishiyama, A. Nakashima, and H. Takada. 2001. Distribution and behavior of nonylphenol, octylphenol, and nonylphenol monoethoxylate in Tokyo Metropolitan Area: Their association with aquatic particles and sediment distribution. Environ. Sci. Technol. 35:1041–1049.

Jackson, P.R., C.M. Gracia, K.A. Oberg, K.K. Johnson, and M.H. Garcia. 2008. Density currents in the Chicago River: Characterization, effects on water quality, and potential sources. Sci. Total Environ. 401:130–143.

Jackson, R.B., and E. Jobbagy. 2005. From icy roads to salty streets. Proc. Natl. Acad. Sci. USA 102:14487–14488.

Jarvie, H.P., C. Neal, and P.J.A. Withers. 2006. Sewage-effluent phosphorus: A greater risk to river eutrophication than agriculture. Sci. Total Environ. 360:246–253.

Jaworski, N.A., P.M. Groffman, A.A. Keller, and J.C. Prager. 1992. A watershed nitrogen and phosphorus balance: The upper Potomac River basin. Estuaries 15(1):83–95.

Jones-Lepp, T.L., D.A. Alvarez, J.D. Petty, and J.N. Huckins. 2004. Polar organic chemical integrative sampling and liquid chromatography-electrospray/ion-trap mass spectrometry for assessing selected prescription and illicit drugs in treated sewage effluents. Arch. Environ. Contam. Toxicol. 47:427–439.

Kaleta, A., M. Ferdig, and W. Buchberger. 2006. Semiquantitative determination of residues of amphetamine in sewage sludge samples. J. Sep. Sci. 29:1662–1666.

Kalbfus, W. 1995. Belastung bayrischer Gewasser durch synthetische Ostrogene. Vorlrag bei der 50. fachtung des bay. LA for Wasserwirtscafl: Stoffc mit endokriner Wirkung im Wasser (Abstract) (in German).

Kang, J., Y.S. Lee, S.J. Ki, Y.G. Lee, S.M. Cha, K.H. Cho, and J.H. Kim. 2009. Characteristics of wet and dry weather heavy metal discharges in Yeongsan watershed, Korea. Sci. Total Environ. 407:3482–3493.

Kaushal, S.S., P.M. Groffman, G.E. Likens, K.T. Belt, W.P. Stack, V.R. Kelly, L.E. Band, and G.T. Fisher. 2005. Increased salinization of fresh water in the northeastern United States. Proc. Natl. Acad. Sci. USA 102:13517–13520.

Kaye, J.P., P.M. Groffman, N.B. Grimm, L.A. Baker, and R.V. Pouyat. 2006. A distinct urban biogeochemistry? Trends Ecol. Evol. 21(4):192–199.

Kemp, S.J., and J.R. Spotila. 1997. Effects of urbanization on brown trout *Salmo trutta*, other fishes and macroinvertebrates in Valley Creek, Valley Forge, Pennsylvania. Am. Midl. Nat. 138:55–68.

Kennedy, C., J. Cuddihy, and J. Engel-Yan. 2007. The changing metabolism of cities. J. Ind. Ecol. 11:43–59.

King, K.W., J.C. Balogh, K.L. Hughes, and R.D. Harmel. 2007. Nutrient load generated by storm event runoff from a golf course watershed. J. Environ. Qual. 36:1021–1030.

Latimer, J.S., and J.G. Quinn. 1998. Aliphatic petroleum and biogenic hydrocarbons entering Narragansett Bay from tributaries under dry weather conditions. Estuaries 21:91–107.

La Valle, P.D. 1975. Domestic sources of stream phosphate in urban streams. Water Res. 9:913–915.

Lavendel, B. 2002. The business of ecological restoration. Ecol. Res. 20:173–178.

Law, N.L., L.E. Band, and J.M. Grove. 2004. Nutrient input from residential lawn care practices in suburban watersheds in Baltimore County, MD. J. Environ. Plann. Manage. 47:737–755.

Leland, H.V., S.V. Fend, T.L. Dudley, and J.L. Carter. 1989. Effects of copper on species composition of benthic insects in a Sierra Nevada, California. Stream Freshwater Biol. 21:163–179.

Lenat, D.R., and J.K. Crawford. 1994. Effects of land use on water quality and aquatic biota of three North Carolina Piedmont streams. Hydrobiologia 294:185–199.

Lewis, G.P., J.D. Mitchell, C.B. Andersen, D.C. Haney, M.K. Liao, and K.A. Sargent. 2007. Urban influences on stream chemistry and biology in the Big Bushy Creek watershed, South Carolina. Water Air Soil Pollut. 182:303–323.

Li, M.-H., B. Dvorak, and C.Y. Sung. 2010. Bioretention, low impact development, and stormwater management. p. 413–430. *In* J. Aitkenhead-Peterson and A. Volder (ed.) Urban ecosystem ecology. Agron. Monogr. 55. ASA, CSSA, and SSSA, Madison, WI.

Liu, Z., D.E. Weller, D.L. Correll, and T.E. Jordan. 2000. Effects of land cover and geology on stream chemistry in watershed of the Chesapeake Bay. J. Am. Water Resour. Assoc. 36:1349–1365.

MacKenzie, M.J., and J.V. Hunter. 1979. Sources of fates of aromatic compounds in urban stormwater runoff. Environ. Sci. Technol. 13:179–183.

Maestre, A., and R. Pitt. 2005. The National Stormwater Quality Database. Version 1.1. A Compilation and Analysis of NPDES Stormwater Monitoring Information. USEPA Office of Water, Washington, DC.

Mahler, B.J., P.C. Van Metre, T.J. Bashara, J.T. Wilson, and D.A. Johns. 2005. Parking lot sealcoat: And unrecognized source of urban polycyclic aromatic hydrocarbons. Environ. Sci. Technol. 39:5560–5566.

Malakoff, D. 2004. Profile: Dave Rosgen—The river doctor. Science 305:937–939.

Maltby, L., A.B.A. Boxall, D.M. Forrow, P. Calow, and C.I. Betton. 1995a. The effects of motorway runoff on freshwater ecosystems: 2. Identifying major toxicants. Environ. Toxicol. Chem. 14(6):1093–1101.

Maltby, L., D.M. Forrow, A.B.A. Boxall, P. Calow, and C.I. Betton. 1995b. The effects of motorway runoff on freshwater ecosystems: 1. Field study. Environ. Toxicol. Chem. 14:1079–1092.

Maniquiz, M.C., S. Lee, E. Lee, D. Kong, and L. Kim. 2009. Unit loss rate from various construction sites during a storm. Water Sci. Technol. 59:2187–2196.

Marti, E., J. Aumatell, L. Gode, M. Poch, and F. Sabater. 2004. Nutrient retention efficiency in streams receiving inputs from wastewater treatment plants. J. Environ. Qual. 33:285–293.

Mason, R.P., and K.A. Sullivan. 1998. Mercury and methylmercury transport through and urban watershed. Water Res. 32(2):321–330

Mastran, T.A., A.A. Dietrich, D.L. Gallagher, and T.J. Grizzard. 1994. Distribution of polyaromatic hydrocarbons in the water column and sediments of a drinking water reservoir with respect to boating activity. Water Res. 28:2353–2366.

McConnell, J.B. 1980. Impact of urban storm runoff on stream quality near Atlanta, Georgia. EPA-600/2-80-094. USEPA, Washington, DC.

McDaniel, M.D., and M.B. David. 2009. Relationships between benthic sediments and water column phosphorus. J. Environ. Qual. 38:607–617.

McDowell, R.W., and A.N. Sharpley. 2001. A comparison of fluvial sediment phosphorus (P) chemistry in relation to location and potential to influence stream P concentrations. Aquat. Geochem. 7:255–265.

McDowell, R.W., A.N. Sharpley, and G. Folmar. 2003. Modification of phosphorus export from and eastern USA catchment by fluvial sediment and phosphorus inputs. Agric. Ecosyst. Environ. 99:187–199.

McMahon, G., and T. Cuffney. 2000. Quantifying urban intensity in drainage basins for assessing stream ecological condition. J. Am. Water Resour. Assoc. 36:1247–1262.

Meals, D.W., S.N. Levine, D. Wang, J.P. Hoffman, E.A. Cassell, J.C. Drake, D.K. Pelton, H.M. Galarveau, and A.B. Brown. 1999. Retention of spike additions of soluble phosphorus in a northern eutrophic stream. J. North Am. Benthol. Soc. 18(2):185–198.

Medeiros, C., R. Leblanc, and R.A. Coler. 1983. An in situ assessment of the acute toxicity of urban runoff to benthic macroinvertebrates. Environ. Toxicol. Chem. 2:119–126.

Meross, S. 2000. Salmon restoration in an urban watershed: Johnson creek, Oregon. Available at http://www.portlandonline.com/shared/cfm/image.cfm?id=37172 (verified 18 Mar. 2010). Portland Multnomah Progress Board, Portland, OR.

Meybeck, M. 1998. Man and river interface: Multiple impacts on water and particulates chemistry illustrated in the Seine river basin. Hydrobiologia 373:1–20.

Meyer, J.L., M.J. Paul, and W.K. Taulbee. 2005. Stream ecosystem function in urbanizing landscapes. J. North Am. Benthol. Soc. 24:602–612.

Mielke, H.W., C.R. Gonzales, M.K. Smith, and P.W. Mielke. 2000. Quantities and associations of lead, zinc, cadmium, manganese, chromium, nickel, vanadium, and copper in fresh Mississippi delta alluvium and New Orleans alluvial soils. Sci. Total Environ. 246:249–259.

Mitsch, W.J., J.W. Day, and J.W. Gilliam. 2001. Reducing nitrogen loading to the Gulf of Mexico from the Mississippi River basin: Strategies to counter a persistent ecological problem. Bioscience 51:373–387.

Moore, W.M., and S. Ramamoorthy. 1984. Heavy metals in natural waters: Applied monitoring and impact assessment. Springer-Verlag, New York.

Morgan, E., W. Porak, and J. Arway. 1983. Controlling acidic-toxic metal leachates from southern Appalachian construction slopes: Mitigating stream damage. Transp. Res. Rec. 948:10–16.

Morgan, R.P., and S.E. Cushman. 2005. Urbanization effects on stream fish assemblages in Maryland, USA. J. North Am. Benthol. Soc. 24:643–655.

Mortimer, C.H. 1941. The exchange of dissolved substances between mud and water in lakes. Int. J. Ecol. 30:280–329.

Mulholland, P.J., R.O. Hall, Jr., D.J. Sobota, W.K. Dodds, S. Findlay, N.B. Grimm, S.K. Hamilton, W.H. McDowell, J.M. O'Brien, J.L. Tank, L.R. Ashkenas, L.W. Cooper, C.N. Dahm, S.V. Gregory, S.L. Johnson, J.L. Meyer, B.J. Peterson, G.C. Poole, H.M. Valett, J.R. Webster, C. Arango, J.J. Beaulieu, M.J. Bernot, A.J. Burgin, C. Crenshaw, A.M. Helton, L. Johnson, B.R. Niederlehner, J.D. Potter, R.W. Sheibley, and S.M. Thomas. 2009. Nitrate removal in stream ecosystems measured by 15N addition experiments: Denitrification. Limnol. Oceanogr. 54(3):666–680.

Muschak, W. 1990. Pollution of street run-off by traffic and local conditions. Sci. Total Environ. 93:419–431.

Neal, C., and A.J. Robson. 2000. A summary of river water quality data collected within the land-ocean interaction study: Core data for eastern UK rivers draining to the North Sea. Sci. Total Environ. 251/252:585–665.

Neal, C., A.J. Robson, H.A. Jeffery, M.L. Harrow, M. Neal, C.J. Smith, and H.P. Jarvie. 1997. Trace elements inter-relationships for the Humber rivers: Inferences for hydrological and chemical controls. Sci. Total Environ. 194/195:321–343.

Nelson, E.J., and D.B. Booth. 2002. Sediment sources in an urbanizing, mixed land-use watershed. J. Hydrol. 264:51–68.

Norrstrom, A.C., and E. Bergstedt. 2001. The impact of Road de-icing salts (NaCl) on colloid dispersion and bas cation pools in roadside soils. Water Air Soil Pollut. 127:281–299.

Novotny, E.V., D. Murphy, and H.G. Stefan. 2008. Increase of urban lake salinity by road deicing salt. Sci. Total Environ. 406:232–244.

Novotny, V., and H. Olem. 1994. Water quality: Prevention, identification, and management of diffuse pollution. Van Nostrand Rheinhold, New York.

Ometo, J.P.H.B., L.A. Martinelli, M.V. Ballester, A. Gessner, and A. Krusche. 2000. Effects of land use on water chemistry and macroinvertebrates in two streams of the Piracicaba River Basin, southeast Brazil. Freshwater Biol. 44:327–337.

Osborne, L.L., and M.J. Wiley. 1988. Empirical relationships between land use/cover and stream water quality in an agricultural watershed. J. Environ. Manage. 26:9–27.

Paerl, H.W. 2009. Controlling eutrophication along the freshwater-marine continuum: Dual nutrient (N and P) reductions are essential. Estuaries Coasts 32:593–601.

Paerl, H.W., L.M. Valdes, M.F. Piehler, and M.E. Lebo. 2004. Solving problems resulting from solutions: The evolution of a dual nutrient management strategy for the eutrophying Neuse River Estuary, North Carolina, USA. Environ. Sci. Technol. 38:3068–3073.

Patrick, W.H., Jr., and R.A. Khalid. 1974. Phosphate release and sorption by soils and sediments: Effect to aerobic and anaerobic conditions. Science 186:53–55.

Paul, M.J. 1999. Stream ecosystem function along a land use gradient. Ph.D. diss. Univ. Georgia, Athens.

Paul, M.J., and J.L. Meyer. 2001. Streams in the urban landscape. Annu. Rev. Ecol. Syst. 32:333–365.

Peierls, B.L., N.F. Caraco, M.L. Pace, and J.J. Cole. 1991. Human influence on river nitrogen. Nature 350:386–387.

Peterson, B.J., W.M. Wollheim, P.J. Mulholland, J.R. Webster, J.L. Meyer, J.L. Tank, E. Marti, W.B. Bowden, H.M. Valett, A.E. Hershey, W.H. McDowell, W.K. Dodds, S.K. Hamilton, S. Gregory, and D.D. Morrall. 2001. Control of nitrogen export from watersheds by headwater streams. Science 292:86–90.

Pionke, H.B., J.R. Hoover, R.R. Schnabel, W.J. Gburek, J.B. Urban, and A.S. Rogowski. 1988. Chemical–hydrologic interactions in the near-stream zone. Water Resour. Res. 24:1101–1110.

Pizzuto, J.E., W.C. Hession, and M. McBride. 2000. Comparing gravel-bed rivers in paired urban and rural catchments of southeastern Pennsylvania. Geology 28:79–82.

Pollock, J.B., and J.L. Meyer. 2001. Phosphorus assimilation below a point source in Big Creek. p. 506–509. In K.J. Hatcher (ed.) Proc. 2001 Georgia Water Resources Conf. Univ. of Georgia, Athens.

Pouyat, R.V., I.D. Yesilonis, J. Russell-Anelli, and N.K. Neerchal. 2007. Soil chemical and physical properties that differentiate urban land-use and cover types. Soil Sci. Soc. Am. J. 71(3):1010–1090.

Prather, M., D. Ehalt, and F. Dentener. 2001. Atmospheric chemistry and greenhouse gases. In J.T. Houghton et al. (ed.) Climate change 2001: The scientific basis. Third Assessment Report of the Intergovernmental Panel on Climate Change. Cambridge Univ. Press, Cambridge, UK.

Preston, S.D., V.J. Bierman, Jr., and S.E. Silliman. 1989. An evaluation of methods for the estimation of tributary mass loads. Water Resour. Res. 25:1379–1389.

Radhakrishnaiah, K., A. Suresh, and B. Sivaramakrishna. 1993. Effect of sub lethal concentration of mercury and zinc on the energetics of a freshwater fish *Cyprinus carpio* (Linnaeus). Acta Biol. Hung. 44:375–385.

Rang, H.P., M.M. Dale, and J.M. Ritter. 1999. Pharmacology. Churchill Livingstone, Edinburgh.

Rauch, S., and G.M. Morrison. 1999. Platinum uptake by the freshwater isopod *Asellus aquaticus* in urban rivers. Sci. Total Environ. 235:261–268.

Reinelt, L. 1996. Sediment and phosphorus loading from construction sites and residential land areas in King County: A case study of the Laughing Jacobs creek subcatchment. King County Dep. of Public Works, Surface Water Management Div., Seattle

Rhoads, B.L., and R.A. Cahill. 1999. Geomorphological assessment of sediment contamination in an urban stream system. Appl. Geochem. 14:459–483.

Richards, S.R., and R. Knowles. 1995. Inhibition of nitrous-oxide reduction by a component of Hamilton Harbor sediment. FEMS Microbiol. Ecol. 17:39–46.

Richardson, M.L., and J.M. Bowron. 1985. The fate of pharmaceutical chemicals in the aquatic environment—A review. J. Pharm. Pharmacol. 37:1–12.

Rijstenbil, J.W., and T.C.W. Poortvliet. 1992. Copper and zinc in estuarine water: Chemical speciation in relation to bioavailability to the marine planktonic diatom *Ditylum brightwellii*. Environ. Toxicol. Chem. 11:1615–1625.

Robertson, D.M., G. E. Schwarz, D. A. Saad, and R. B. Alexander. 2009. Incorporating uncertainty into the ranking of SPARROW model nutrient yields from Mississippi/Atchafalaya river basin watersheds. J. Am. Water Res. 42:534–549.

Rose, S., and N.E. Peters. 2001. Effects of urbanization on streamflow in the Atlanta area (Georgia, USA): A comparative hydrological approach. Hydrol. Processes 15:1441–1457.

Rusciano, G.M., and C.C. Obropta. 2007. Bioretention column study: Fecal coliform and total suspended solids reduction. Trans. ASABE 50(4):1261–1269.

Salt Institute. 2004. Salt mining statistics. The Salt Institute, Alexandria, VA.

Sartor, J.D., G.B. Boyd, and F.J. Agardy. 1974. Water pollution aspects of street surface contaminants. J. Water Pollut. Control Fed. 66:458–467.

Scavia, D., N.N. Rabalais, R.E. Turner, J. Dubravko, and W.J. Wiseman, Jr. 2003. Predicting the response of Gulf of Mexico hypoxia to variations in Mississippi River nitrogen load. Limnol. Oceanogr. 48:951–956.

Schoonover, J.E., B.G. Lockaby, and S. Pan. 2005. Changes in chemical and physical properties of stream water across an urban–rural gradient in western Georgia. Urban Ecosyst. 8:107–124.

Shanley, J.B. 1994. Effects of ion exchange on stream solute fluxes in a basin receiving highway deicing salts. J. Environ. Qual. 23:977–986.

Sharpley, A.N. 1981. The contribution of phosphorus leached from crop canopy to losses in surface runoff. J. Environ. Qual. 10:160–165.

Sharpley, A.N., S.C. Chapra, R. Wedepohl, J.T. Sims, T.C. Daniel, and K.R. Reddy. 1994. Managing agricultural phosphorus for protection of surface waters—Issues and options. J. Environ. Qual. 23:437–451.

Sheikh, B., R.S. Jaques, and R.P. Cort. 1987. Reuse of tertiary municipal wastewater effluent for irrigation of raw eaten food crops: A 5 year study. Desalinization 67:245–254.

Sickman, J.O., M.J. Zanoli, and H.L. Mann. 2007. Effects of urbanization on organic carbon loads in the Sacramento River, California. Water Resour. Res. 43:1–15.

Shore, L., Y. Kapolnik, B. Ben-Dov, Y. Fridman, S. Winmger, and M. Shemesh. 1992. Effects of estrone and 17β estradiol on vegetative growth of *Medicago sativa*. Physiol. Plant. 84:217–222.

Shore, L.S., M. Gurevitz, and M. Shemesh.1993. Estrogen as an environmental pollutant. Bull. Environ. Contam. Toxicol. 51:361–366.

Shutes, R.B.E. 1984. The influence of surface runoff on the macro-invertebrate fauna of an urban stream. Sci. Total Environ. 33:271–282.

Sloane-Richey, J.S., M.A. Perkins, and K.W. Malueg. 1981. The effects of urbanization and stormwater runoff on the food quality in two salmonid streams. Verh. Int. Ver. Theor. Angew. Limnol. 21:812–818.

Smart, M.M., J.R. Jones, and J.L. Sebaugh. 1985. Stream–watershed relations in the Missouri Ozark Plateau Province. J. Environ. Qual. 14:77–82.

Sonoda, K., and J.A. Yeakley. 2007. Relative effects of land use and near-stream chemistry on phosphorus in an urban stream. J. Environ. Qual. 36:114–154.

Sreedevi, P., B. Sivaramakrishna, A. Suresh, and K. Radhakrishnaiah. 1992. Effect of nickel on some aspects of protein metabolism in the gill and kidney of the freshwater fish, *Cyprinus carpio* L. Environ. Pollut. 77:59–63.

Stander, E.K., and J.G. Ehrenfeld. 2010. Urban riparian function. p. 253–276. *In* J. Aitken-head-Peterson and A. Volder (ed.) Urban ecosystem ecology. Agron. Monogr. 55. ASA, CSSA, and SSSA, Madison, WI.

Steger-Hartmann, T., K. Kummerer, and J. Schecker. 1996. Trace analysis of the antineo-plastic ifosfamide and cyclophosphamide in sewage water by two-step solidphase extraction and gas chromatography-mass spectrometry. J. Chromatogr. A 720:179–184.

Stigliani, W.M., P. Doelman, W. Salomons, R. Schulin, G.R.B. Smidt, and S.E.A.T.M. Van der Zee. 1991. Chemical time bombs: Predicting the unpredictable. Environment 33:4–30.

Strecker, E.W., J.M. Kersnar, E.D. Driscoll, and R.R. Horner. 1992. The use of wetlands for controlling stormwater pollution. Prepared for USEPA. Woodward-Clyde Consultants, Walnut Creek, CA.

Stokes, P. 1988. Nickel in aquatic systems. p. 31–46. *In* H. Sigel (ed.) Metal ions in biological system. Marcel Dekker, New York.

Stumpf, M., T.A. Terries, K. Haberer, P. Seel, and W. Baumann. 1996. Nachweis von Arz-neimittelruckstanden in Klaranlagen und Flicssgewassern. Vom Wasser 86:291–303.

Struger, J., and T. Fletcher. 2007. Occurrence of lawn care and agricultural pesticides in the Don River and Humber River watersheds (1998–2002). J. Great Lakes Res. 33:887–905.

Sudduth, E.B., and J.L. Meyer. 2006. Effects of bioengineered streambank stabilization on bank habitat and macroinvertebrates in urban streams. Environ. Manage. 38:218–226.

Tada, F., and S. Suzuki. 1982. Adsorption and desorption of heavy metals in bottom mud of urban rivers. Water Resour. 16:1489–1494.

Ternes, T.A., M. Meisenheimer, D. McDowell, F. Sacher, H.J. Brauch, B.H. Gulde, G. Preuss, U. Wilme, and N.Z. Seibert. 2002. Removal of pharmaceuticals during drinking water treatment. Environ. Sci. Technol. 36:3855–3863.

Toze, S. 2006. Reuse of effluent water—Benefits and risks. Agric. Water Manage. 80:147–159.

Trimble, S.J. 1997. Contribution of stream channel erosion to sediment yield from an urbanizing watershed. Science 278:1442–1444.

Triska, F.J., J.H. Duff, and R.J. Avanzino. 1993. The role of water exchange between a stream channel and its hyporheic zone in nitrogen cycling at the terrestrial–aquatic interface. Hydrobiologia 251:167–184.

Turner, R.E., N.N. Rabalais, R.B. Alexander, G. McIsaac, and R.W. Howarth. 2007. Characterization of nutrient, organic carbon, and sediment loads and concentrations from the Mississippi River into the Northern Gulf of Mexico. Estuaries Coasts 30(5):773–790.

United Nations Population Division. 2008. World Population Prospects: The 2007 Revision, Highlights (online database). ESA/P/WP.180, revised 26 Feb. 2003. Available at http://www.un.org/esa/population/publications/wup2007/2007wup.htm (accessed 19 Mar. 2010, verified 22 May 2010).

USEPA. 2002. National water quality inventory: 2000 report. Available at http://www.epa.gov/305b/2000report (updated 4 Dec. 2006, accessed July 2009, verified 28 Mar. 2010). USEPA, Office of Water, Washington, DC.

USEPA. 2004. Primer for municipal wastewater treatment systems. EPA 832-R-04-001. USEPA Office of Water, Office of Wastewater Management, Washington, DC.

USEPA. 2003. Information on the toxic effects of various chemicals and groups of chemicals. Available at http://www.epa.gov/R5Super/ecology/html/toxprofiles.htm# (accessed July 2009).

USGS. 1999. The quality of our nation's waters—Nutrients and pesticides. USGS Circ. 1225. USGS, Reston, VA.

USGS. 2007. Use of chemical analysis and assays on semipermeable membrane devices extracts to assess the response of bioavailable organic pollutants in streams to urbanization in six metropolitain areas of the United States. Sci. Invest. Rep. 2007-5113. USGS, Denver, CO.

Van Breeman, N., E.W. Boyer, C.L. Goodale, N.A. Jaworski, K. Paustian, S.P. Seitzinger, K. Lajtha, B. Mayer, D. Van Dam, R.W. Howarth, K.J. Nadelhoffer, M. Eve, and G. Billen. 2002. Where did all the nitrogen go? Fate of nitrogen inputs to large watersheds in the northeastern USA. Biogeochemistry 57/58:267–293.

Von Schiller, D., E. Marti, J.L. Riera, M. Ribot, J.C. Marks, and F. Sabater. 2008. Influence of land use on stream ecosystem functions in a Mediterranean catchment. Freshwater Biol. 53:2600–2612.

Waggott, A. 1981. Trace organic substances in the River Lee. p. 55–99. In W.J. Cooper (ed.) Chemistry in water reuse. Ann Arbor Sci. Publ., Ann Arbor, MI.

Walsh, C.J., T.D. Fletcher, and A.R. Ladson. 2005. Stream restoration in urban catchments through redesigning stormwater systems: Looking to the catchment to save the stream. J. North Am. Benthol. Soc. 24:690–705.

Wang, L., J. Lyons, P. Kanehl, R. Bannerman, and E. Emmons. 2000. Watershed urbanization and changes in fish communities in southeastern Wisconsin streams. J. Am. Water Resour. Assoc. 36:1173–1189.

Wang, X., and Z. Yin. 1997. Using GIS to assess the relationship between land use and water quality at a watershed level. Environ. Int. 23:103–114.

Wark, J.W., and F.J. Keller. 1963. Preliminary study of sediment sources and transport in the Potomac River basin. Tech. Bull. 11. Interstate Comm., Potomac River Basin, Washington, DC.

Warren, L.A., and A.P. Zimmerman. 1994. Suspended particulate oxides and organic matter interactions in trace metal sorption reactions in a small urban river. Biogeochemistry 23:21–34.

Warren-Rhodes, K., and A. Koenig. 2001. Ecosystem appropriation by Hong Kong and its implications for sustainable development. Ecol. Econ. 39:347–359.

Waschbusch, R.J., W.R. Selbig, and R.T. Bannerman. 1999. Sources of phosphorus in stormwater and street dirt from two urban residential basins in Madison, Wisconsin. USGS Water Resour. Invest. Rep. 99-4021. USGS, Reston, VA.

Watts, C.D., M. Craythorne, M. Fielding, and C.P. Steel. 1983. Identification of non-volatile organics in water using field desorption mass spectrometry and high performance liquid chromatography p. 120–131. In G. Angeletti and A. Bjarseth (ed.) 3rd European Symposium on Organic Micropollutents in Water. D.D. Reidel Publ., Dordrecht, the Netherlands.

Wegner, W., and M. Yaggi. 2001. Environmental impacts of road salt and alternatives in the New York City Watershed. Stormwater 2(5). Available at http://www.stormh2o.com/july-august-2001/salt-road-environmental-impacts.aspx (verified 18 Mar. 2010).

Wei, C., and G. Morrison. 1992. Bacterial enzyme activity and metal speciation in urban river sediments. Hydrobiologia 235/236:597–603.

Wilber, W.G., and J.V. Hunter. 1977. Aquatic transport of heavy-metals in urban-environments. Water Resour. Bull. 13:721–734.

Winter, J., and H. Duthie. 2000. Export coefficient modeling to assess phosphorus loading in an urban watershed. J. Am. Water Resour. Assoc. 36:1053–1061.

Whipple, W., and J.V. Hunter. 1979. Petroleum hydrocarbons in urban runoff. Water Resour. Bull. 15:1096–1104.

Whipple, J.A. 1981. An ecological perspective of the effects of monocyclic aromatic hydrocarbon on fishes. p. 89–105. In F.J. Calabazas et al. (ed.) Biological monitoring of marine pollutants. Academic Press, New York.

Whiting, E.R., and H.F. Clifford. 1983. Invertebrates and urban runoff in a small northern stream, Edmonton, Alberta, Canada. Hydrobiologia 102:73–80.

Wolf, L., J. Klinger, H. Hoetzl, and U. Mohrlok. 2007. Quantifying mass fluxes from urban drainage systems to the urban soil-aquifer system. J. Soils Sediments 7:85–95.

Wollheim, W.M., B.A. Pellerin, C.J. Vorosmarty, and C.S. Hopkinson. 2005. N retention in urbanizing headwater catchments. Ecosystems 8:871–884.

Wolman, M.G., and A.P. Schick. 1967. Effects of construction on fluvial sediment, urban and suburban areas of Maryland. Water Resour. Res. 3:451–464.

Xian, G., M. Crane, and J. Su. 2007. An analysis of urban development and its environment impact on the Tampa Bay watershed. J. Environ. Manage. 85:965–976.

Yamamoto, K., M. Fukushima, N. Kakatani, and K. Kuroda. 1997. Volatile organic compounds in urban rivers and their estuaries in Osaka, Japan. Environ. Pollut. 95:135–143.

Yousef, Y.A., M.P. Wanielista, and H.H. Harper. 1985. Removal of highway contaminants by roadside swales. Transp. Res. Rec. 1017:62–68.

Yousef, Y.A., M.P. Wanielista, H.H. Harper, and E.T. Skene. 1983. Impact of bridging on floodplains. Transp. Res. Rec. 948:26–30.

Zaccato, E., S. Castiglioni, R. Bagnati, C. Chiabrando, P. Grassi, and R. Fanelli. 2007. Illicit drugs, a novel group of environmental contaminants. Water Res. 42:961–968. doi:10.1016/j.watres.2007.09.010.

Zampella, R.A. 1994. Characterization of surface water quality along a watershed disturbance gradient. Water Resour. Bull. 30:605–611.

16

Urban Stream Ecology

Allison H. Roy
Michael J. Paul
Seth J. Wenger

Abstract

Urban watersheds characteristically have high impervious surface cover, resulting in high surface runoff and low infiltration following storms. In response, urban streams experience "flashy" stormflows, reduced baseflows, bank erosion, channel widening, and sedimentation. Urban watersheds also typically exhibit high nutrient, total ion, and contaminant concentrations. This harsh physical and chemical environment tends to produce biotic assemblages of low diversity dominated by tolerant and nonnative species. Ecosystem processes in urban streams also differ from nonurban streams, with fast leaf breakdown rates, reduced nutrient uptake, and high respiration rates, although these responses are not universal. The ecology of urban streams is variable among systems, depending on the natural environment (e.g., climate, geology, vegetation) and anthropogenic stresses (e.g., age, type, and extent of development; riparian deforestation; stormwater piping). Thus, management efforts must be geared toward urban stressors specific to each ecosystem if more diverse assemblages and more natural ecosystem processing are desired.

Streams and rivers are directly influenced by the landscapes in which they flow (Hynes, 1975). Both natural conditions (e.g., topography, geology, soils) and human land uses (e.g., agriculture, urbanization, forestry, and mining) determine the abiotic and biotic characteristics of streams draining these landscapes (Vannote et al., 1980; Allan, 2004). Urbanization is a particularly influential land use, causing major effects on stream ecosystems even though cities cover only 2% of the global landscape. For example, catchments with only 2% effective impervious cover (i.e., impervious surfaces directly connected to storm sewers) have exhibited altered water quality, algal biomass, and diatom assemblages relative to similarly sized nonurban catchments (Taylor et al., 2004; Newall and Walsh, 2005; Walsh et al., 2005b). The disproportionate influence of urbanization on streams,

A.H. Roy, USEPA, National Risk Management Research Lab., 26 West Martin Luther King Dr., Cincinnati, OH 45268, presently at Dep. of Biology, Kutztown University, Kutztown, PA 19530 (roy@kutztown.edu); M.J. Paul, Tetra Tech, Inc., Center for Ecological Sciences, Owings Mills, MD 21117 (Michael.Paul@tetratech.com); S.J. Wenger, University of Georgia, River Basin Center, 100 Riverbend Road, Athens, GA 30602 (swenger@uga.edu).

doi:10.2134/agronmonogr55.c16

Table 16–1. Symptoms associated with the urban stream syndrome.†

Hydrology	Organic matter
↑ Frequency of overland flow	↓ Retention
↑ Frequency of erosive flow	↑↓ Standing stock/inputs
↑ Magnitude of high flow	**Fishes**
↓ Lag time to peak flow	↓ Sensitive fishes
↑ Rise and fall of storm hydrograph	↑↓ Tolerant fishes
↑↓ Baseflow magnitude	↑↓ Fish abundance/biomass
Water chemistry	**Invertebrates**
↑ Nutrients (N, P)	↑ Tolerant invertebrates
↑ Toxicants	↓ Sensitive invertebrates
↑ Temperature	**Algae**
↑↓ Suspended sediments	↑ Eutrophic diatoms
Channel morphology	↓ Oligotrophic diatoms
↑ Channel width	↑↓ Algal biomass
↑ Pool depth	**Ecosystem processes**
↑ Scour	↓ Nutrient uptake
↓ Channel complexity	↑↓ Leaf breakdown
↑↓ Sedimentation	

† Adapted from Walsh et al. (2005b).

coupled with the rapid increase in urban land cover globally (Alig et al., 2004; Salvatore et al., 2005), has prompted a large body of research on urban stream ecology and management strategies. Much of this work has been synthesized in papers and reports that attempt to describe the essential characteristics of urban streams, termed the *urban stream syndrome* (Table 16–1; Paul and Meyer, 2001; Center for Watershed Protection, 2003; Walsh et al., 2004, 2005b; Wenger et al., 2009).

Although many urban streams exhibit common symptoms, it has become increasingly clear that the urban stream syndrome is quite variable (Wenger et al., 2009). Regional climatic and geologic conditions, the amount of impervious surface cover in the catchment, the age of development, management of point source pollutants, and other factors all shape the ecology of urban streams (Table 16–2). In this chapter, we describe typical characteristics of urban streams along with known or potential sources of variability in urban stream ecosystems.

The Abiotic Environment

Impervious surfaces increase surface runoff and lower infiltration, resulting in urban streams with "flashy" behavior such as more frequent, higher magnitude flows of shorter duration following storms. For a typical urban stream in a highly impervious catchment, even small storms increase stream flow (Paul and Meyer, 2001; Shuster et al., 2005; Konrad and Booth, 2005). However, stormflow dynamics in urban streams depend on the degree of connectedness between impervious surfaces and drainage pipes; high connectedness to storm sewers exacerbates the stormflow response (Walsh et al., 2005a). Conversely, if impervious surfaces drain to pervious areas (e.g., forested buffers, infiltration areas) or to combined sewers that discharge farther downstream, local stream hydrology will be less altered (Booth and Jackson, 1997; Booth et al., 2002; Walsh et al., 2005b). Where impervious surfaces drain to detention–retention basins designed for flood control, the peak magnitude of flows may be lower than conventional urban drainage, although the total volume of stormflows will be similar and greatly exceed those

Table 16–2. Sources of variability influencing the ecology of urban streams.

Natural
Catchment and stream size
Climate
Geology and geomorphic stability
Soils
Natural vegetation

Anthropogenic
Age of development and infrastructure
Type and intensity of urban land cover
Total and connected impervious surface cover
Total and riparian forest cover
Historical land use (e.g., agriculture, forest)
Type and age of sewer or septic infrastructure
Nature of sewage effluent and level of treatment
Stormwater runoff controls
Point-source discharges
Population density
Traffic intensity
Application of fertilizers, pesticides
Dams and other impoundments
Road and utility crossings
Socioeconomics
Legal systems controlling water quality and quantity

that would occur in otherwise similar streams in undeveloped catchments (Booth and Jackson, 1997). As a result of decreased groundwater recharge associated with reduced infiltration and groundwater withdrawals in urban areas, reduced baseflows are also common in urban areas (Konrad and Booth, 2005). However, lawn watering, leaky infrastructure, and deforestation can offset reduced infiltration, resulting in normal or high magnitude and duration of baseflow conditions (Chapters 14 [Burian and Pomeroy, 2010] and 17 [Jenerette and Alstad, 2010], this volume; Lerner, 2002; Konrad and Booth, 2005; Roy et al., 2009).

Due to increased stormflows, streams in recently urbanized areas typically experience high rates of channel erosion, resulting in enlarged channels (i.e., increased channel widths and depths) (Leopold, 1968; Chapter 14, Burian and Pomeroy, 2010, this volume). Forested buffers and hard engineered bank structures such as gabions may minimize local erosion, but do not reduce erosion downstream (Booth and Jackson, 1997; Hession et al., 2003). Landscape and bank erosion result in high sedimentation in urban stream channels, particularly following new construction (Wolman, 1967; Trimble, 1997). In older urban areas, stream beds may stabilize with coarser bed sediments, although bed scour may remain high (Finkenbine et al., 2000; Henshaw and Booth, 2000). Many urban streams are piped, filled, or placed in concrete-lined, straight channels (Elmore and Kaushal, 2008), resulting in extreme habitat simplification and separation of stream channels from adjacent subsurface and floodplain interactions. Whether through direct or indirect causes, low channel complexity (e.g., low depth variability, sinuosity, substrate variability, and habitat structure) is a universal symptom of urban streams (Walsh et al., 2005b).

The physical habitat of urban streams is linked not only to hydrology, but also to vegetation in the catchment and riparian areas. Urban streams without forested riparian areas have high insolation, high temperatures, and low inputs of wood (Finkenbine et al., 2000; Elosegi and Johnson, 2003; Roy et al., 2005). High temperatures in urban streams may also result from urban heat island effects and runoff from heated pavement (Van Buren et al., 2000; Nelson and Palmer, 2007; Chapter 2, Heisler and Brazel, 2010, this volume). Because storm drain networks connect upland areas to streams, leaves and grass from street trees and residential clippings may result in high inputs of organic material to urban streams, even where there is minimal riparian forest cover (Miller and Boulton, 2005). These responses are not universal, however, and habitat characteristics vary among urban streams due to both natural factors such as climate, geology, vegetation and anthropogenic influences which can include impervious connectivity, riparian canopy cover, street sweeping.

Nearly all urban streams have elevated concentrations of chemical contaminants, many of which are toxic to aquatic life. Common toxicants found in urban streams include heavy metals (especially cadmium, chromium, copper, lead, manganese, nickel, and zinc), pesticides, petroleum byproducts, and pharmaceuticals (Wilber and Hunter, 1977; Kolpin et al., 2002; Gilliom et al., 2006; Chapter 15, Steele et al., 2010, this volume). Catchments with untreated point-source discharges, high impervious surface connectivity, high pesticide application rates, and high population and traffic densities are likely to have the highest contaminant levels (Chapter 15, Steele et al., 2010, this volume). Downstream of insufficiently treated wastewater discharges and failing septic drain fields, streams and rivers typically exhibit increased bacteria concentrations, increased biological and chemical oxygen demand, and decreased dissolved oxygen concentrations (Martí et al., 2004). High conductivity (i.e., high ionic concentrations) is also a common symptom of urban streams, especially in high latitude regions where salt is used as a deicer, although concentrations are dependent on underlying geology (Kaushal et al., 2005). Finally, urban streams typically have elevated nitrogen and phosphorus concentrations, with the highest concentrations in areas with high fertilizer applications, wastewater effluent, septic systems, impervious surface connectivity (Chapter 15, Steele et al., 2010, this volume; Martí et al., 2004; Hatt et al., 2004; Mueller and Spahr, 2006; Bernhardt et al., 2008). A surprisingly large nutrient source is localized atmospheric deposition of combustion-derived nitrogen from automobiles (Chapter 3, Santosa, 2010, this volume; Bernhardt et al., 2008). Although the suite of physical and chemical stressors differs among urban streams, nearly all urban streams exhibit abiotic conditions that markedly differ from natural streams.

Ecosystem Structure and Function

Biotic Assemblages

Urban streams frequently have high algal biomass as a result of high nutrient and light levels, although high production can be offset by flow disturbance, toxicity, grazing, and/or poor light attenuation (Paul and Meyer, 2001; Taylor et al., 2004; Potapova et al., 2005; Catford et al., 2007). Algal assemblages are typically composed of eutrophic species in urban streams, and may be dominated by either diatoms (Munn et al., 2002) or filamentous algae (Taylor et al., 2004). Tolerant taxa typically dominate diatom assemblages in urban streams as a result of poor water quality (Newall and Walsh, 2005; Duong et al., 2007). The limited studies on aquatic macrophytes show low diversity in urban streams, with compositional shifts favoring nonnative species (Paul and Meyer, 2001).

Macroinvertebrates are the most widely studied biotic group and exhibit the most consistent responses to urbanization (Brown et al., 2009). Macroinvertebrate assemblages in urban streams have low diversity, with limited numbers and abundances of sensitive species (e.g., Ephemeroptera, Plecoptera, and Trichoptera) and numerical dominance by tolerant species (e.g., oligochaetes and chironomids; Paul and Meyer, 2001; Walsh et al., 2005b). As a result, measures of macroinvertebrate biotic integrity in urban streams consistently demonstrate fair or poor aquatic health (Roy et al., 2009; Wang and Kanehl, 2003). These characteristic macroinvertebrate assemblages are likely caused by a variety of stressors, including hydrologic alteration, sedimentation, high bed mobility, poor water quality, and riparian deforestation. Native mussels are often absent or in low abundances in urban streams, which may be due to reduced water quality or a lack of fish hosts (Lyons et al., 2007).

Urban streams typically have homogenized fish assemblages, with high richness and abundance of tolerant, cosmopolitan, and nonnative species and relatively low richness and abundance of sensitive, endemic species (Morgan and Cushman, 2005; Walters et al., 2003; Scott, 2006). More so than other biota, fishes are limited by biogeography and dispersal barriers (both natural and anthropogenic), resulting in urban fish assemblages that commonly vary based on region (Brown et al., 2009). Lastly, fishes in urban streams also often exhibit lesions and parasites, indicating exposure to contaminants and stressful conditions (Helms et al., 2005).

There are several other vertebrate groups that rely on streams for all or part of their life cycle that have been studied in urban settings, although to a lesser degree than algae, macroinvertebrates and fishes. Urban streams have low diversity and abundance of herpetofauna, including frogs (Riley et al., 2005), snakes (Burger et al., 2007), and salamanders (Miller et al., 2007). Furthermore, frogs and snakes show increased levels of metals and other contaminants in urban streams (Burger et al., 2007; Stoylar et al., 2008). Riparian birds are also commonly dominated by tolerant species in urban settings (Chapters 4 [Shochat et al., 2010] and 6 [Ehrenfeld and Stander, 2010], this volume; Lussier et al., 2006), although this depends on the quality of riparian habitat and macroinvertebrate assemblages (Mattson and Cooper, 2006).

Exactly why urban streams have a low diversity of native organisms compared with nonurban streams is often unclear. Urban streams tend to have multiple co-occurring stressors, and there are numerous direct and indirect mechanisms by which individual organisms may be impacted by urbanization. Organisms vary in their tolerance to stressors, and sensitive taxa may be directly extirpated by contaminant toxicity, altered flow, and/or stream habitat loss. A number of indirect ecological factors also play a role. For example, growth rates of chironomids can be lower in urban streams than nonurban streams (Rosi-Marshall, 2004), which may reduce food supply to higher trophic levels. Populations of aquatic invertebrate species may be constrained by limitations to survival and dispersal during their adult stages (Smith et al., 2009). For some species, permanent extirpation may be caused by movement barriers that prevent recolonization after disturbances. For example, stream road crossings, which are found in high densities in urban areas, have been shown to serve as barriers to fish (Warren and Pardew, 1998), aquatic insects (Blakely et al., 2006) and other organisms.

Ecosystem Processes

The principal ecosystem processes that occur in streams are leaf decomposition, material cycling, and metabolism (photosynthesis and respiration), all of which are altered in urban systems. Allochthonous organic matter, in the form of leaves and wood, is integral to the energetics of many stream systems (Vannote et al., 1980). The decomposition of this material involves microbes, invertebrates, and vertebrates and is therefore important to stream food webs. Leaf breakdown tends to be rapid in urban streams because of nutrient-enriched microbial decomposition rates (Imberger et al., 2008) and physical abrasion from high flows and suspended sediment levels (Paul et al., 2006). In streams that are severely impacted by toxicants and thus have reduced abundances of shredding macroinvertebrates and microorganisms, leaf breakdown may be reduced (Duarte et al., 2008; Roussel et al., 2008). Chadwick et al. (2006) showed that leaf decomposition peaked at 30 to 40% impervious cover, consistent with physical abrasion and microbial decomposition causing rapid breakdown at intermediate levels of urbanization, and the lack of shredding macroinvertebrates limits decomposition at the highest levels of urbanization. Since breakdown rates vary based on leaf species, urban streams with nonnative riparian vegetation may experience different breakdown rates than those with native species (Chapter 13, Stander and Ehrenfeld, 2010, this volume).

Nutrient cycling involves the biogeochemical transformation of materials between reduced and oxidized states. Alteration of natural cycling can affect decomposition, primary and secondary production, metabolism, and community structure. Urban streams tend to have reduced nutrient uptake. This has been attributed to different factors, including decreased fine benthic organic matter (Meyer et al., 2005), reduced channel complexity, and reduced primary productivity (Grimm et al., 2005). Denitrification is reduced in riparian areas of urban streams that have high channel incision and a low water table because there is less anoxia (Groffman et al., 2003). In urban stream channels, denitrification is primarily confined to organic-rich debris dams and gravel bars (Groffman et al., 2005). Lower denitrification rates likely contribute to long nitrogen uptake lengths in urban streams (Grimm et al., 2005; Gibson and Meyer, 2007). Of course, the generality of these patterns among urban streams is dependent on nutrient concentrations and the presence of anaerobic conditions that serve as denitrification hot spots within stream networks (Grimm et al., 2005; Groffman et al., 2005).

Ecosystem metabolism is a measure of the productivity of an ecosystem and incorporates gross primary production and community respiration. It is, therefore, an indicator of both total carbon supply and processing efficiency in streams. Streams receiving wastewater effluent have high respiration (R) due to elevated nutrients and labile carbon (Izagirre et al., 2008). Streams with reduced canopy cover, high light, and high nutrients may have increased primary productivity (P), although, as discussed previously, algal production is not necessarily elevated in urban streams. Thus, urban stream P/R ratios can either be high (indicating autotrophy) or low (indicating heterotrophy) depending on bed stability, organic material loads, temperature, or any of the aforementioned characteristics (Paul and Meyer, 2001; Wenger et al., 2009). Further research is necessary to be able to better predict metabolism and other ecosystem processes in urban streams.

Managing Urban Stream Ecosystems

Urban stream ecosystems are highly variable depending on local biogeocli-matic conditions, past land use, degree of urbanization, and, most importantly, the nature of the urban stressors. Management efforts must take these factors into account. In locations where the primary source of disturbance is stormwater runoff, tools that minimize connected impervious cover (e.g., downspout discon-nection, planning and zoning tools) or mimic natural hydrologic pathways by infiltrating and abstracting runoff at the source (e.g., infiltration swales, rain bar-rels, green roofs) are key to managing streams (Booth and Jackson, 1997; Walsh et al., 2005a). In cities with constant or episodic direct discharges of untreated wastewater, interception and treatment is a critical management step. Riparian restoration is important for moderating temperatures, stabilizing banks, and pro-viding organic material, although these and other functions of buffers may be impaired in urban streams. Bank stabilization and channel reconfiguration are common restoration tools employed to address localized erosion and sedimenta-tion, but their utility in overall stream mitigation is often constrained by other catchment stressors (Bernhardt and Palmer, 2007). In their synthesis, Wenger et al. (2009) pointed to several other management approaches for urban streams, including controls on fertilizers and toxicants, septic and sewer management policies, dam management policies, water use regulations, and road and utility crossing regulations. Depending on the sources of impairment in a given stream, management tools may simultaneously restore multiple structural and func-tional components of urban streams.

When managing urban streams, one must also consider the pattern of eco-system shifts in response to urban stressors. For example, where the addition of stressors results in a gradual or linear change in ecosystem responses, it is likely that piecemeal mitigation will cause incremental improvements in stream condition (Fig. 16–1A). Alternatively, ecosystems may shift dramatically from one regime to another based on external forces, with internal feedbacks acting to maintain alternative stable states (Fig. 16–1B). In this scenario, urban ecosys-tems that have already undergone dramatic shifts are likely to require significant improvements (i.e., release of constraints) and/or time before experiencing a shift to higher stream quality, assuming the state is reversible (Suding et al., 2004). Management may also be used to increase the resilience of more desirable states,

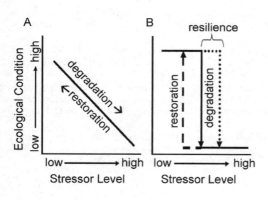

Fig. 16–1. Potential patterns of ecosystem degradation and restora-tion based on (A) gradual or linear response or (B) a stepped or thresh-old response. In (B), forces maintain states of either high or low ecological condition, with ecosystem resilience extending the amount of stress an ecosystem can endure before switch-ing from a high to low ecological state. Restoration may require a reduction in stressors beyond the level that encouraged the degradation shift.

thus creating more self-sustaining ecosystems (Palmer et al., 2005). Studies that assess streams across a gradient of urbanization indicate both linear and threshold relationships between urbanization and ecosystem responses (Walsh et al., 2005b). However, experimental studies that monitor stream responses to catchment development and restoration are necessary to determine whether responses are gradual or stepped and whether there are forces acting to maintain alternate ecosystem states in urban stream ecosystems.

Urban stream ecology is a growing field, and one that involves not just scientists, but also planners, policymakers, engineers, landowners, and others who are directly involved in decisions that affect the quality of these ecosystems. In many countries, society's perspective of streams and rivers has changed drastically in the last century, from a conduit to dilute and remove waste, to an ecosystem that provides important functions. These functions vary widely among urban streams, but may include transportation, recreational opportunities, fisheries, processing of nutrients and carbon, and cultural and aesthetic value. Communities are responsible for making decisions on how to manage their streams and watersheds on the basis of their desired goals, and these decisions will ultimately dictate the structure and function of urban stream ecosystems.

Acknowledgments

We acknowledge and thank the many ecologists who have and continue to conduct research in urban streams. The views expressed in this paper are those of the authors and in no manner represent or reflect current or planned policy by the USEPA.

References

Alig, R.J., J.D. Kline, and M. Lichtenstein. 2004. Urbanization on the US landscape: Looking ahead in the 21st century. Landsc. Urban Plan. 69:219–234.

Allan, J.D. 2004. Landscapes and riverscapes: The influence of land use on stream ecosystems. Annu. Rev. Ecol. Evol. Syst. 35:257–284.

Bernhardt, E.S., L.E. Band, C.J. Walsh, and P.E. Berke. 2008. Understanding, managing, and minimizing urban impacts on surface water nitrogen loading. Ann. N.Y. Acad. Sci. 1134:61–96.

Bernhardt, E.S., and M.A. Palmer. 2007. Restoring streams in an urbanizing world. Freshwater Biol. 52:738–751.

Blakely, T.J., J.S. Harding, A.R. McIntosh, and M.J. Winterbourn. 2006. Barriers to the recovery of aquatic insect communities in urban streams. Freshwater Biol. 51:1634–1645.

Booth, D.B., D. Hartley, and R. Jackson. 2002. Forest cover, impervious-surface area, and the mitigation of stormwater impacts. J. Am. Water Resour. Assoc. 38:835–845.

Booth, D.B., and C.R. Jackson. 1997. Urbanization of aquatic systems: Degradation thresholds, stormwater detection, and the limits of mitigation. J. Am. Water Resour. Assoc. 33:1077–1090.

Brown, L.R., T. Cuffney, J.F. Coles, F. Fitzpatrick, G. McMahon, J. Steuer, A.H. Bell, and J.T. May. 2009. Urban streams across the USA: Lessons learned from studies in nine metropolitan areas. J. North Am. Benthol. Soc. 28:1051–1069.

Burger, J., K.R. Campbell, S. Murray, T.S. Campbell, K.F. Gaines, C. Jeitner, T. Shukla, S. Burke, and M. Gochfeld. 2007. Metal levels in blood, muscle and liver of water snakes (*Nerodia* spp.) from New Jersey, Tennessee and South Carolina. Sci. Total Environ. 373:556–563.

Burian, S.J., and C.A. Pomeroy. 2010. Urban impacts on the water cycle and potential green infrastructure implications. p. 277–296. *In* J. Aitkenhead-Peterson and A. Volder (ed.) Urban ecosystem ecology. Agron. Monogr. 55. ASA, CSSA, and SSSA, Madison, WI.

Catford, J.A., C.J. Walsh, and J. Beardall. 2007. Catchment urbanization increases benthic microalgal biomass in streams under controlled light conditions. Aquat. Sci. 69:511–522.

Center for Watershed Protection. 2003. Impacts of impervious cover on aquatic ecosystems. Watershed Protection Res. Monogr. 1. Center for Watershed Protection, Ellicott City, MD.

Chadwick, M.A., D.R. Dobberfuhl, A.C. Benke, A.D. Huryn, K. Suberkropp, and J.E. Thiele. 2006. Urbanization affects stream ecosystem function by altering hydrology, chemistry, and biotic richness. Ecol. Appl. 16:1796–1807.

Duarte, S., C. Pascoal, A. Alves, A. Correia, and F. Cassio. 2008. Copper and zinc mixtures induce shifts in microbial communities and reduce leaf litter decomposition in streams. Freshwater Biol. 53:91–101.

Duong, T.T., A. Feurtet-Mazel, M. Coste, D.K. Dang, and A. Boudou. 2007. Dynamics of diatom colonization process in some rivers influenced by urban pollution (Hanoi, Vietnam). Ecol. Indicators 7:839–851.

Ehrenfeld, J.G., and E.K. Stander. 2010. Habitat function in urban riparian zones. p. 103–118. In J. Aitkenhead-Peterson and A. Volder (ed.) Urban ecosystem ecology. Agron. Monogr. 55. ASA, CSSA, and SSSA, Madison, WI.

Elmore, A.J., and S.S. Kaushal. 2008. Disappearing headwaters: Patterns of stream burial due to urbanization. Front. Ecol. Environ. 6:308–312.

Elosegi, A., and L.B. Johnson. 2003. Wood in streams and rivers in developed landscapes. p. 337–353. In S.V. Gregory et al. (ed.) Ecology and management of wood in world rivers. Symp. 37. Am. Fisheries Soc., Bethesda, MD.

Finkenbine, J.K., J.W. Atwater, and D.S. Mavinic. 2000. Stream health after urbanization. J. Am. Water Resour. Assoc. 36:1149–1160.

Gibson, C.A., and J.L. Meyer. 2007. Nutrient uptake in a large urban river. J. Am. Water Resour. Assoc. 43:576–587.

Gilliom, R.J., J.E. Barbash, C.G. Crawford, P.A. Hamilton, J.D. Martin, N. Nakagaki, L.H. Nowell, J.C. Scott, P.E. Stackelberg, G.P. Thelin, and D.M. Wolock. 2006. The quality of our nations's waters—Pesticides in the nation's streams and ground water, 1992–2001. USGS Circ. 1291. USGS, Reston, VA.

Grimm, N.B., R.W. Sheibley, C.L. Crenshaw, C.N. Dahm, W.J. Roach, and L.H. Zeglin. 2005. N retention and transformation in urban streams. J. North Am. Benthol. Soc. 24:626–642.

Groffman, P.M., D.J. Bain, L.E. Band, K.T. Belt, G.S. Brush, J.M. Grove, R.V. Pouyat, I.C. Yesilonis, and W.C. Zipperer. 2003. Down by the riverside: Urban riparian ecology. Front. Ecol. Environ. 1:315–321.

Groffman, P.M., A.M. Dorsey, and P.M. Mayer. 2005. N processing within geomorphic structures in urban streams. J. North Am. Benthol. Soc. 24:613–625.

Hatt, B.E., T.D. Fletcher, C.J. Walsh, and S.L. Taylor. 2004. The influence of urban density and drainage infrastructure on the concentrations and loads of pollutants in small streams. Environ. Manage. 34:112–124.

Heisler, G.M., and A.J. Brazel. 2010. The urban physical environment: Temperature and urban heat islands. p. 29–56. In J. Aitkenhead-Peterson and A. Volder (ed.) Urban ecosystem ecology. Agron. Monogr. 55. ASA, CSSA, and SSSA, Madison, WI.

Helms, B.S., J.W. Feminella, and S. Pan. 2005. Detection of biotic responses to urbanization using fish assemblages from small streams in western Georgia, USA. Urban Ecosyst. 8:39–57.

Henshaw, P.C., and D.B. Booth. 2000. Natural restabilization of stream channels in urban watersheds. J. Am. Water Resour. Assoc. 36:1219–1236.

Hession, W.C., J.E. Pizzuto, T.E. Johnson, and R.J. Horwitz. 2003. Influence of bank vegetation on channel morphology in rural and urban watersheds. Geology 31:147–150.

Hynes, H.B.N. 1975. The stream and its valley. Verh. Int. Ver. Theor. Angew. Limnol. 19:1–15.

Imberger, S.J., C.J. Walsh, and M.R. Grace. 2008. More microbial activity, not abrasive flow or shredder abundance, accelerates breakdown of labile leaf litter in urban streams. J. North Am. Benthol. Soc. 27:549–561.

Izagirre, O., U. Agirre, M. Bermejo, J. Pozo, and A. Elosegi. 2008. Environmental controls of whole-stream metabolism identified from continuous monitoring of Basque streams. J. North Am. Benthol. Soc. 27:252–268.

Jenerette, G.D., and K.P. Alstad. 2010. Water use in urban ecosystems: Complexity, costs, and services of urban ecohydrology. p. 353–372. In J. Aitkenhead-Peterson and A. Volder (ed.) Urban ecosystem ecology. Agron. Monogr. 55. ASA, CSSA, and SSSA, Madison, WI.

Kaushal, S.S., P.M. Groffman, G.E. Likens, K.T. Belt, W.P. Stack, V.R. Kelly, L.E. Band, and G.T. Fisher. 2005. Increased salinization of fresh water in the northeastern United States. Proc. Natl. Acad. Sci. USA 102:13517–13520.

Kolpin, D.W., E.T. Furlong, M.T. Meyer, E.M. Thurman, S.D. Zaugg, L.B. Barber, and H.T. Buxton. 2002. Pharmaceuticals, hormones, and other organic wastewater contaminants in US streams, 1999–2000: A national reconnaissance. Environ. Sci. Technol. 36:1202–1211.

Konrad, C.P., and D.B. Booth. 2005. Hydrologic changes in urban streams and their ecological significance. p. 157–177. In L.R. Brown et al. (ed.) Effects of urbanization on stream ecosystems. Symp. 47. Am. Fisheries Soc., Bethesda, MD.

Leopold, L.B. 1968. Hydrology for urban planning. USGS, Washington, DC.

Lerner, D.N. 2002. Identifying and quantifying urban recharge: A review. Hydrogeol. J. 10:143–152.

Lussier, S.M., R.W. Enser, S.N. Dasilva, and M. Charpentier. 2006. Effects of habitat disturbance from residential development on breeding bird communities in riparian corridors. Environ. Manage. 38:504–521.

Lyons, M.S., R.A. Krebs, J.P. Holt, L.J. Rundo, and W. Zawiski. 2007. Assessing causes of change in the freshwater mussels (Bivalvia: Unionidae) in the Black River, Ohio. Am. Midl. Nat. 158:1–15.

Martí, E., J. Aumatell, L. Godé, M. Poche, and F. Sabater. 2004. Nutrient retention efficiency in streams receiving inputs from wastewater treatment plants. J. Environ. Qual. 33:285–293.

Mattson, B.J., and R.J. Cooper. 2006. Louisiana waterthrushes (Seiurus motacilla) and habitat assessment as cost-effective indicators of instream biotic integrity. Freshwater Biol. 51(10):1941–1958.

Meyer, J.L., M.J. Paul, and W.K. Taulbee. 2005. Stream ecosystem function in urbanizing landscapes. J. North Am. Benthol. Soc. 24:602–612.

Miller, W., and A.J. Boulton. 2005. Managing and rehabilitating ecosystem processes in regional urban streams in Australia. Hydrobiologia 552:121–133.

Miller, J.E., G.R. Hess, and C.E. Moorman. 2007. Southern two-lined salamanders in urbanizing watersheds. Urban Ecosyst. 10:73–85.

Morgan, R.P., and S.F. Cushman. 2005. Urbanization effects on stream fish assemblages in Maryland, USA. J. North Am. Benthol. Soc. 24:643–655.

Mueller, D.K., and N.E. Spahr. 2006. Nutrients in rivers and streams across the nation 1992–2001. USGS Invest. Rep. 2006-5107. USGS, Reston, VA.

Munn, M.D., R.W. Black, and S.J. Gruber. 2002. Response of benthic algae to environmental gradients in an agriculturally dominated landscape. J. North Am. Benthol. Soc. 21:221–237.

Nelson, K.C., and M.A. Palmer. 2007. Stream temperature surges under urbanization and climate change: Data, models, and responses. J. Am. Water Resour. Assoc. 43:440–452.

Newall, P., and C.J. Walsh. 2005. Response of epilithic diatom assemblages to urbanization influences. Hydrobiologia 532:53–67.

Palmer, M.A., E.S. Bernhardt, J.D. Allan, P.S. Lake, G. Alexander, S. Brooks, J. Carr, S. Clayton, C.N. Dahm, J.F. Shah, D.L. Galat, S.G. Loss, P. Goodwin, D.D. Hart, B. Hassett, R. Jenkinson, G.M. Kondolf, R. Lave, J.L. Meyer, T.K. O'Donnell, L. Pagano, and E. Sudduth. 2005. Standards for ecologically successful river restoration. J. Appl. Ecol. 42:208–217.

Paul, M.J., and J.L. Meyer. 2001. Streams in the urban landscape. Annu. Rev. Ecol. Evol. Syst. 32:333–365.

Paul, M.J., J.L. Meyer, and C.A. Couch. 2006. Leaf breakdown in streams differing in catchment land use. Freshwater Biol. 51:1684–1695.

Potapova, M., J.F. Coles, E.M. Giddings, and H. Zappia. 2005. A comparison of the influences of urbanization on stream benthic algal assemblages in contrasting environmental settings. p. 333–359. In L.R. Brown et al. (ed.) Effects of urbanization on stream ecosystems. Symp. 47. Am. Fisheries Soc., Bethesda, MD.

Riley, S.P.D., G.T. Busteed, L.B. Kats, T.L. Vandergon, L.F.S. Lee, R.G. Dagit, J.L. Kerby, R.N. Fisher, and R.M. Sauvajot. 2005. Effects of urbanization on the distribution and abundance of amphibians and invasive species in southern California streams. Conserv. Biol. 19:1894–1907.

Rosi-Marshall, E.J. 2004. Decline in the quality of suspended particulate matter as a food resource for chironomids downstream of an urban area. Freshwater Biol. 49:515–525.

Roussel, H., E. Chauvet, and J.M. Bonzom. 2008. Alteration of leaf decomposition in copper-contaminated freshwater mesocosms. Environ. Toxicol. Chem. 27:637–644.

Roy, A.H., A.L. Dybas, K.M. Fritz, and H.R. Lubbers. 2009. Urbanization affects the extent and hydrologic permanence of headwater streams in a midwestern U.S. metropolitan area. J. North Am. Benthol. Soc. 28:911–928.

Roy, A.H., C.L. Faust, M.C. Freeman, and J.L. Meyer. 2005. Reach-scale effects of riparian forest cover on urban stream ecosystems. Can. J. Fish. Aquat. Sci. 62:2312–2329.

Salvatore, M., F. Pozzi, E. Ataman, B. Huddleston, and M. Bloise. 2005. Mapping global urban and rural population distributions. Environmental and Natural Resources Working Paper 24. FAO, Rome.

Santosa, S.J. 2010. Urban air quality. p. 57–74. In J. Aitkenhead-Peterson and A. Volder (ed.) Urban ecosystem ecology. Agron. Monogr. 55. ASA, CSSA, and SSSA, Madison, WI.

Shochat, E., S. Lerman, and E. Fernández-Juricic. 2010. Birds in urban ecosystems: Population dynamics, community structure, biodiversity, and conservation. p. 75–86. In J. Aitkenhead-Peterson and A. Volder (ed.) Urban ecosystem ecology. Agron. Monogr. 55. ASA, CSSA, and SSSA, Madison, WI.

Scott, M.C. 2006. Winners and losers among stream fishes in relation to land use legacies and urban development in the southeastern US. Biol. Conserv. 127:301–309.

Shuster, W.D., J. Bonta, H. Thurston, E. Warnemuende, and D.R. Smith. 2005. Impacts of impervious surface on watershed hydrology: A review. Urban Water J. 2:263–275.

Smith, R. F., L. C. Alexander and W. O. Lamp. 2009. Dispersal by terrestrial stages of stream insects in urban watersheds: A synthesis of current knowledge. J. North Am. Benthol. Soc. 28: 1022–1037.

Stander, E.K., and J.G. Ehrenfeld. 2010. Urban riparian function. p. 253–276. In J. Aitkenhead-Peterson and A. Volder (ed.) Urban ecosystem ecology. Agron. Monogr. 55. ASA, CSSA, and SSSA, Madison, WI.

Steele, M.K., W.H. McDowell, and J.A. Aitkenhead-Peterson. 2010. Chemistry of urban, suburban, and rural surface waters. p. 297–340. In J. Aitkenhead-Peterson and A. Volder (ed.) Urban ecosystem ecology. Agron. Monogr. 55. ASA, CSSA, and SSSA, Madison, WI.

Stoylar, O.B., N.S. Loumbourdis, H.I. Falfushinska, and L.D. Romanchuk. 2008. Comparison of metal bioavailability in frogs from urban and rural sites of western Ukraine. Arch. Environ. Contam. Toxicol. 54:107–113.

Suding, K.N., K.L. Gross, and G.H. Houseman. 2004. Alternative states and positive feedbacks in restoration ecology. Trends Ecol. Evol. 19:46–53.

Taylor, S.L., S.C. Roberts, C.J. Walsh, and B.E. Hatt. 2004. Catchment urbanisation and increased benthic algal biomass in streams: Linking mechanisms to management. Freshwater Biol. 49:835–851.

Trimble, S.W. 1997. Contribution of stream channel erosion to sediment yield from an urbanizing watershed. Science 278:1442–1444.

Van Buren, M.A., W.E. Watt, J. Marsalek, and B.C. Anderson. 2000. Thermal enhancement of stormwater runoff by paved surfaces. Water Res. 34:1359–1371.

Vannote, R.L., W.G. Minshall, K.W. Cummins, J.R. Sedell, and C.E. Cushing. 1980. The river continuum concept. Can. J. Fish. Aquat. Sci. 37:130–137.

Walsh, C.J., T.D. Fletcher, and A.R. Ladson. 2005a. Stream restoration in urban catchments through redesigning stormwater systems: Looking to the catchment to save the stream. J. North Am. Benthol. Soc. 24:690–705.

Walsh, C.J., A.W. Leonard, A.R. Ladson, and T.D. Fletcher. 2004. Urban stormwater and the ecology of streams. Coop. Res. Ctr. for Freshwater Ecol. and Coop. Res. Ctr. for Catchment Hydrol., Canberra, Australia.

Walsh, C.J., A.H. Roy, J.W. Feminella, P.D. Cottingham, P.M. Groffman, and R.P. Morgan, II. 2005b. The urban stream syndrome: Current knowledge and the search for a cure. J. North Am. Benthol. Soc. 24:706–723.

Walters, D.M., D.S. Leigh, and A.B. Bearden. 2003. Urbanization, sedimentation, and the homogenization of fish assemblages in the Etowah River Basin, USA. Hydrobiologia 494:5–10.

Wang, L., and P. Kanehl. 2003. Influences of watershed urbanization and instream habitat on macroinvertebrates in cold water systems. J. Am. Water Resour. Assoc. 39:1181–1196.

Warren, M.L., and M.G. Pardew. 1998. Road crossings as barriers to small-stream fish movement. Trans. Am. Fish. Soc. 127:637–644.

Wenger, S.J., A.H. Roy, C.R. Jackson, E.S. Bernhardt, T.L. Carter, S. Filoso, C.A. Gibson, W.C. Hession, S.S. Kaushal, E. Martí, J.L. Meyer, M.A. Palmer, M.J. Paul, A.H. Purcell, A. Ramirez, A.D. Rosemond, K.A. Schofield, E.B. Sudduth, and C.J. Walsh. 2009. Twenty-six key research questions in urban stream ecology: An assessment of the state of the science. J. North Am. Benthol. Soc. 28:1080–1098.

Wilber, W.G., and J.V. Hunter. 1977. Aquatic transport of heavy metals in bottom sediments of the Saddle River. Water Resour. Bull. 15:790–800.

Wolman, M.G. 1967. A cycle of sedimentation and erosion in urban river channels. Geogr. Ann. 49A:385–395.

17

Water Use in Urban Ecosystems: Complexity, Costs, and Services of Urban Ecohydrology

G. Darrel Jenerette
Karrin P. Alstad

Abstract

The urban water component of the global freshwater cycle does not use amounts of water comparable with agriculture; however, cities are hotspots of water use and have water footprints often extending orders of magnitude larger than their physical footprints. To better understand this water use we propose an *urban ecohydrology* perspective as a new framework for considering urban water dynamics that emphasizes the combination of hydrological and ecosystem science tools to characterize the fluxes, constraints, services, and feedbacks between urban ecosystem dynamics and urban ecohydrological processes occurring across a broad range of scales. We present a conceptual model of urban ecohydrology and use as an example the complexities of water supply and demand fluctuations within the Los Angeles metropolitan region. On the basis of the general model and description of the Los Angeles system we then generate several hypotheses that might differentiate the importance of individual water fluxes and their contribution to ecosystem services based on environmental, socioeconomic, and cultural gradients among cities.

Total water withdrawn for human uses has almost tripled in the last fifty years from 1382 km³ yr⁻¹ in 1950 to 3973 km³ yr⁻¹ and is projected to further increase to 5235 km³ yr⁻¹ by 2025 (Clarke and King, 2004), an amount more than nine times the annual flow of the Mississippi River (Goudie, 2000). Currently, this withdrawal rate represents more than one-half of the available fresh water supplies feasibly accessible by societies, and this available portion is decreasing (Postel et al., 1996; Vorosmarty and Sahagian, 2000). The amounts of water used in the urban water component of the global freshwater cycle are not comparable with agricultural water use (Gleick, 2003), but cities have water footprints often extending orders of magnitude larger than the physical footprint of a city (Luck et al., 2001; Jenerette et al., 2006b). Population growth trends indicate that by 2030, more than 60% of the world's population is expected to live in cites. Migration from rural to urban

G.D. Jenerette (Darrel.Jenerette@ucr.edu) and K.P. Alstad (karrin.alstad@ucr.edu), Dep. of Botany and Plant Sciences, University of California, Riverside, CA 92571.

doi:10.2134/agronmonogr55.c17

areas is especially high among countries having low and mid levels of development (United Nations Population Fund, 2007). Thus, cities are essential locations for maintaining the health and comfort of the majority of humans (Cohen, 2006). Because urbanization is increasing rapidly, the growing urban water uses are an increasingly larger proportion of total water consumption relative to agricultural consumption (Falkenmark, 1998; Gleick, 2003). The global expansion of urbanization has furthermore increased the competition for water sources and driven deep rivalries in agriculture and ecological water allocations (United Nations, 2008). The availability of water provides for many ecosystem services, those biophysical processes directly related to the maintenance of health, comfort, aesthetic, economic, and other requirements of human and societal functioning (Costanza et al., 1997; Daily et al., 1997, 2009; Carpenter et al., 2009; Chapter 18, Aitkenhead-Peterson et al., 2010, this volume). To ensure the ongoing availability of ecosystem services, cities have a critical need to (i) distribute large but regulated volumes of water with constrained ranges of chemical constituencies, (ii) efficiently remove water for sewage treatment, (iii) prevent floods, and (iv) provide for instream and downstream uses. The challenges associated with water distribution lead to complex hierarchical organization structures that simultaneously compete and cooperate in the allocation of water resources depending on the scale of organization. Urban water use varies substantially across global climate, topographic, biologic, economic, and social and cultural gradients (Gleick, 2003; Vörösmarty et al., 2005; Jenerette and Larsen, 2006; Hoekstra and Chapagain, 2007; United Nations, 2008). Advancing understanding about how water is partitioned among diverse urban users, the services provided by the water, and the costs associated with water delivery will be essential for ensuring sustainable human health and development throughout the world. Increasing urban populations, global warming, and more erratic precipitation patterns are recognized as critical challenges for many cities in ensuring sustainable urban water deliveries (Vorosmarty et al., 2000; Jenerette and Larsen, 2006). Adding to these complications, changes to the hydrologic cycle through human redistribution of water can lead to large changes in the global carbon cycle, with potential implications for atmospheric greenhouse gas concentration and future climate dynamics (Jenerette and Lal, 2005).

We propose a new framework for considering urban water dynamics—the *urban ecohydrology* perspective—to emphasize the combination of hydrological and ecosystem science tools to characterize the fluxes, constraints, services, and feedbacks between urban ecosystem dynamics and urban ecohydrological processes occurring across a broad range of scales. *Ecosystem science* is a subdiscipline of ecology examining ecological units as integrated systems of biotic and abiotic components. *Urban ecosystem science* is an expanded ecological subdiscipline that considers the ecological, human, and political–economic–regulatory influences on material and energy fluxes within a city (Pickett et al., 1997, 2001; Grimm et al., 2000, 2008a). *Ecohydrology* is a rapidly growing integration of ecosystem and hydrologic sciences that emphasizes the coupling between hydrologic cycles and ecological functioning to understand how life depends on and affects the partitioning and chemical constituency of water on the continental surfaces (Rodriguez-Iturbe, 2000; Bond, 2003; Scanlon et al., 2005). Within an urban context the ecohydrologic perspective can be a useful bridge between engineering approaches to managing water flows and ecological approaches to understand the biological, physical, and societal system encompassing the dynamics of water flows. Unlike urban hydrology, with its predominantly engineering focus for meeting water delivery and removal needs (Chapter 14, Burian and Pomeroy, 2010, this volume), urban ecohydrology examines the coupling between hydrologic and ecosystem functioning as specifically related to cities. In identifying sustainable development pathways, ecohydrology is well suited to link the ecosystem services associated with water fluxes to the potential costs arising from the use of these services.

We use an urban ecohydrologic approach to develop a general systems model explaining the drivers of variation in the water supply, uses, and fluxes across the broad ranges of decision and biophysical process scales affecting a city. In this model we explicitly recognize how the controls on water fluxes respond to processes and decisionmaking on scales ranging from individuals to international treaties. The urban ecohydrologic approach examines the hierarchies associated across these wide-ranging scales of processes. We present our conceptual model and use as an example the complexities of water supply and demand fluctuations within southern California. The Los Angeles, CA metropolitan region (LA) is a canonical example of urban water complexity and provides rich examples of multiple-scaled ecohydrologic dynamics. Based on the general model and description of the LA system we then generate several hypotheses that might differentiate among cities the importance of individual water fluxes and their contribution to ecosystem services based on environmental, socioeconomic, and cultural gradients. We conclude by describing how urban water use decisions affect urban sustainability and discussing a range of options available to cities to attain more sustainable water use dynamics.

Urban Ecohydrological Systems: A General Model with Examples from Los Angeles

An urban ecohydrological system includes an accounting of all sources of water influxes that supply a city, how the water is used to supply ecosystem services, how the water is removed, and how multiple stakeholders interact in deciding water allocation (Fig. 17–1). The supply of water varies extensively, including precipitation occurring within the physical footprint of the city to virtual water associated with international trade of goods. The use of urban water throughout a city is associated with a broad range of ecosystem services, including human health–related uses like drinking water and cooling, industrial and commercial economic activity, and aesthetic and recreational benefits resulting from irrigated gardening and landscaping. An insidious component of water uses is often extensive leakage fluxes; these leakages can also directly impact ecosystem service production. Outflows of urban water include evaporation within the city, downstream or ocean releases, and virtual fluxes associated with the export of goods and services. The ability to efficiently drain water is critical for cities in the removal of wastes and flood prevention. Finally, hierarchical linkages between multiple users has often led to complex decisionmaking, with the same agents at times cooperating and competing for water deliveries in both water scarce and traditionally water replete regions.

The LA metropolitan region provides an example of the complexity in an urban ecohydrological system. Several comprehensive descriptions of the history

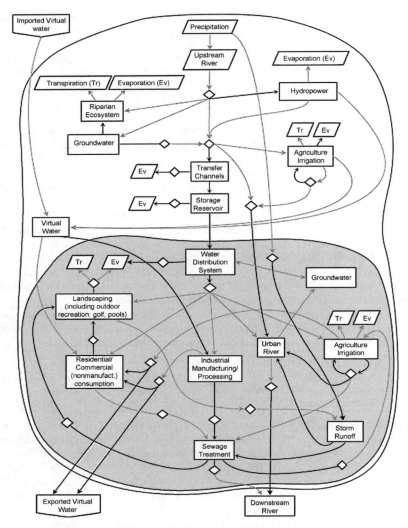

Fig. 17–1. The urban ecohydrological system. Each box describes a possible pool, source, or loss of water associated with urban fluxes. Each diamond is a unique decision point where humans influence the partitioning of water among pools. Dynamics occurring within the entire water footprint of the city are included in the urban ecohydrological system. This water boundary of the water footprint is shown. The white background describes the upstream water fluxes, and the gray area describes the fluxes of water occurring directly within the physical footprint of the city.

and complexity of LA water are available (Kahrl, 1982; Reisner, 1993; Hundley, 2001); here we will only highlight key aspects of the dynamics associated with LA water resources to elucidate the general model. The LA region extends from the Pacific Ocean on the west to the Transverse Ranges through San Gabriel and the San Jacinto and Peninsular ranges on the north and east, and to Mexico's border to the south. The LA metropolitan region is the highest density region of California with nearly 1000 people km^{-2} (U.S. Census Bureau, 2000). More than 50% of the California population lives in the LA region. Major urban centers within the LA region include Pasadena, Los Angeles, San Diego, Irvine, and Riverside. Within the metropolitan region, often demarcated as the combined statistical area including the five counties of Los Angeles, Orange, San Bernardino, Riverside, and Ventura, there is a strong coastal to inland gradient creating broad differences in topography, local climates, and ecological communities. Local government institutions within the metropolitan region include more than 100 distinct municipalities (more than 40 with a population larger than 100,000), five counties, hundreds of neighborhoods and districts, and two separate water distribution councils. All of this variability affects how end users, including individuals, use water.

Water Sources

Cities can directly acquire water from the catchments including the physical footprint of the city, the groundwater aquifers directly below the city, and interbasin transfers through aqueducts across large distances. For most cities the extent of area required to supply water is orders of magnitude larger than the physical footprint of the city (Luck et al., 2001; Jenerette and Larsen, 2006). Recently, virtual waters have been identified as a water source associated with the exchange of goods and services. The relatively recent inclusion of virtual water fluxes into regional water budgeting was originally proposed to account for quantities of water used to produce an agricultural crop that is subsequently exported to a different region (Allan, 1998) and has been expanded to include other ecosystem services, including fiber, wood, and energy (Ioris et al., 2004; Hoekstra and Hung, 2005; Hoekstra and Chapagain, 2007; Chapagain and Hoekstra, 2008). In arid environments virtual water fluxes can be a sizeable proportion of the overall water budget and contribute to trading and water use decisions (Allan, 1998; Chapagain et al., 2006; Ma et al., 2006; Varis, 2006). In response to severe water limitations in some regions, there is recent interest in other potentially large water sources: desalinization and wastewater recycling (Asano and Cotruvo, 2004; Hinkebein and Price, 2005; Toze, 2006; Dolnicar and Schafer, 2009). Currently, only a small proportion of urban fresh water is derived from desalinization projects because of the large costs of energy associated with desalinization. Recycling water is a relatively recent approach, with only a few cities in the United States having gray water systems or reintroduction of treated sewage into municipal storage facilities. Treated sewage, when recycled, is commonly applied for landscape and agriculture production within the urban boundary or used to recharge groundwater aquifers. The application of secondary treated municipal wastewater to irrigated agricultural crops has been widely accepted as a viable method of waste disposal (Pettygrove and Asano, 1985). However, treated wastewater applications to agricultural crops pose a risk of groundwater pollution, and regulations within the United States require careful monitoring of these applications (Jetten et al., 1997; Bouwer, 2000).

Along with securing an adequate supply of water, cities also need to ensure the quality of the water available for public consumption or commercial, industrial, or agricultural processes. Common water quality problems include elevated levels of nitrate, heavy metals, and organic toxins. Recently, many urban water supplies have been shown to contain chronic low levels of pharmaceutical drugs

(Stackelberg et al., 2004). Another potential problem with water quality can be increased levels of dissolved organic matter, which can lead to the formation of trihalomethanes when chlorine is used as a disinfection agent (Singer, 1999; Chow et al., 2007). Trihalomethanes in drinking water have been associated with increased cancer risks and pregnancy complications (Morris et al., 1992; Dodds et al., 1999).

The water supply of the LA region exemplifies the complexity of water sources. The natural and engineered watershed of the LA region includes the Sacramento River draining northern California and the Sierra mountains, the Owens River draining the southern Sierra mountains, the Colorado River draining a large proportion of the southwestern United States, the watershed of the several rivers flowing through the physical footprint of the metropolitan region including the Los Angeles and Santa Anna Rivers, and the multiple groundwater basins within the metropolitan region. For the city of Los Angeles, approximately 89% of the water supply originates from outside the hydrologic region via aqueduct transfers. Approximately 11% is extracted from local groundwater sources, primarily the San Fernando basin, and less than 1% comes from recycled water (Los Angeles Department of Water and Power, 2005). Groundwater provides nearly 15% of total water supplies in normal years and a larger percentage in drought years. During extreme periods of water stress extraction levels can exceed recharge rates by up to 30%. Currently no accounting for virtual water fluxes has been performed for the LA region, although this region imports many goods and services from far reaching sources and it is likely virtual water represents a substantial component of the LA water budget. Each of the LA system water sources has been subject to recent flow reductions from within California as a result of litigation for environmental remediation and/or increased competition between users dependent on the same source basin. Water transfers to LA via the Colorado River Aqueduct also face eminent reductions due to increasing demand from the other lower basin states and increasing water quality issues. Further exacerbating water delivery constraints are the recent invasions of species that can foul inlets and delivery pipes—Quagga mussels (*Dreissena rostriformis bugensis*), first identified in Lake Mead in 2007, and zebra mussels (*Dreissena polymorpha*), first identified in San Justo Reservoir in 2008 (California Department of Fish and Game). Many comprehensive treatises on the development of water sources for LA have been written and provide more extensive details of water supply history in this region (Kahrl, 1982; Reisner, 1993; Hundley, 2001).

Water Uses

Urban water uses are extensive and diverse. To better understand how and why water is used in a city we describe uses in terms of the ecosystem services provided by the water. An ecosystem services framework is quickly becoming useful for describing the collection of ecological processes that direct relate to societal needs (Carpenter et al., 2009; Luck et al., 2009; Nelson et al., 2009). We have identified 19 ecosystem services directly associated with urban ecohydrology: aesthetics, biodiversity, climate regulation, disease control, drinking, energy, erosion regulation, fiber production, fire protection, flood control, food production, pest regulation, pollution removal, recreation, sanitation, water capture, water purification, water storage, and wood production (Table 17–1). Beyond the ecosystem services provided, water is also used in industrial applications associated

Table 17–1. Urban ecohydrologic pools, fluxes, and associated ecosystem services

Pool or flux of urban water	Mechanism of flux	Associated ecosystem services
Precipitation	Native	Source for all hydrologic services
Upstream water sources	Native, engineered	Water capture, water purification, recreation, food production, industry, wood production, fiber production
Groundwater	Native, engineered	Water storage, riparian biodiversity, food production, industry
Hydropower	Virtual, engineered	Energy, aesthetics, recreation, food production, water capture
Irrigation	Engineered	Food, aesthetics, recreation, climate regulation
Storage basins	Engineered	Storage, flood control, recreation
Sewer systems	Engineered	Pollution removal, flood control, erosion regulation, pest regulation, disease control, aesthetics
Landscaping	Engineered, native	Climate regulation, aesthetics
Imported/exported goods	Virtual	Wood production, fiber production, food production
Agricultural irrigation	Engineered	Wood production, fiber production, food production, fire protection
Residential/commercial	Engineered	Drinking, sanitation
Native ecosystems	Native, engineered	Flood protection, pollution removal, erosion regulation, pollination removal, food, aesthetics, recreation, spiritual,
Leakage	Unintended	Food production, aesthetics, recreation, pest regulation, disease control

with sanitation, material processing, mining, manufacturing, and cooling. An additional use of water derives from the dependence on transpiration fluxes associated with vegetation and therefore most ecosystem services—transpiration provides direct services through localized cooling associated with energy required for vaporization (latent heat), which can lead to mitigations of the urban heat island (Chapters 1 [Shepherd et al., 2010] and 2 [Heisler and Brazel, 2010], this volume). In many heat-prone regions transpiration fluxes may have a societal benefit by potentially reducing health risks and peak energy demands associated with extreme heat events. Several studies have suggested that urban uses of water and their sensitivity to climate forcing can vary substantially within a city (Jenerette et al., 2007; Balling et al., 2008). Additional uses of water may not be directly associated with any ecosystem service, such as endangered species habitat protection, but these are also critical uses that should be included in allocation decisions (Jackson et al., 2001; Baron et al., 2002).

Irrigation is a dominant component of water use for many cities. Agricultural production accounts for a large portion of regional water use, and beyond traditional food crops, irrigation is used substantially for landscaping within cities. The amount of irrigation water needed to promote the growth of planted vegetation is partly related to the water-use efficiency of the crop or landscaping species, the amount of water that is lost in transpiration from the leaf stomata per amount of carbon gained across the stomata opening of the leaves. Irrigation efficiency is also associated with the amount of evaporated water lost during plant growth, and the degree of evaporative loss is a function of various factors, including the environmental conditions (vapor pressure deficit, temperature), soil type, mulch or cover type, and irrigation methods. Thus, for a given crop or landscape species, the irrigation efficiency is assessed by the amount of water transpired versus amount of water evaporated. More broadly, water-use efficiency for the production of different ecosystem services could be estimated. Potentially, water-use efficiency–based approaches could help in allocating among different uses or in deciding how best to use water for a given ecosystem service to maximize the production of societal needs with minimal water needs.

The LA system uses extensive amounts of water, with large fluxes associated with irrigation of agricultural lands and recreational landscaping activities. The 2000 USGS compilation of freshwater usage for the five-county LA region documents 497 ML d^{-1} for domestic, 280 ML d^{-1} for industrial, 6651 ML d^{-1} for irrigation, 125 ML d^{-1} for aquaculture, 223 ML d^{-1} for livestock, 18 ML d^{-1} for mining, and 12 ML d^{-1} for thermoelectric (Hutson et al., 2004). Within these highly urbanized counties, irrigation uses are the dominant allocation for water, 85% of the total water used. Since LA is generally a very hot, dry environment, it has a large evaporative demand, and any vegetated surface has the potential to transpire extensive amounts of water. Average estimated potential evapotranspiration derived from California Irrigation Management Information System (CIMIS) network for LA exceeds 1200 mm yr^{-1}. Within the city of Los Angeles, the Department of Water and Power estimated that in 2003–2004, 68% of water used was for domestic household uses, 19% of water was used by industrial and commercial uses, and irrigation accounted for less than 1% of water used (Department of Water and Power, 2005). Likely much of the domestic household use was associated with individual homeowner landscaping, again highlighting unknowns in ecohydrologic fluxes for systems that are well monitored and documented.

Water Leakage

Associated with most planned urban water fluxes are leakages from pipes in transit, metering errors, unintended water flows, and theft—of these causes leakage is often the primary contributor (Brothers, 2001; Colombo and Karney, 2002; Tabesh et al., 2009). The amount of lost or "unaccounted for" water is typically in the range of 20 to 30% of deliveries (Colombo and Karney, 2002). In addition to environmental and economic losses caused by leakage, leaky pipes pose a public health risk because leaks are potential entry points for contaminants if a pressure drop occurs in the system and can cause unintended surface water availability. The degree of leakage within a water distribution system is dependent on several factors, including mains pressure, local climate and topography, local value of water, age of the system, main type, and soil types. Commonly, leakage results from poor condition of water mains. Leaks can be managed, and delivery reductions are generally used to reduce pressure in the water supply, thereby reducing the leak. Replacement of the mains is a costly investment that may not be affordable without external subsidies, even though potential cost savings and certain water savings would be realized over the lifetime of the replacement,

California the Department of Water Resources suggests leakage rates vary from between 5 and 30% of all deliveries; the Department of Water and Power, serving only the city of Los Angeles estimated leaks to be less than 11% of total deliveries for the year 2003–2004 (Department of Water and Power, 2005). While several leaks are a chronic, leaks are also manifest as acute events with the failure of individual water mains. The urban ecohydrological system view and potential impacts to leaks is highlighted by the repeated water main failures in the summer of 2009 in Los Angeles. While the cause is still under investigation, much evidence suggests that landscape irrigation restrictions led to spikes in water demand during the few days designated as watering days, which overloaded the capacity of many older water mains.

Evaporation is commonly excluded from leakage estimates, although this can be a substantial flux of water out of the urban system, with potentially large losses occurring in surface storage basins. Evaporative losses, those not associated with transpiration but occurring from the soil or water surface, can be large. Ongoing research challenges are improving our understanding of the partitioning of surface evaporation and transpiration fluxes from the total evapotranspiration term (Williams et al., 2004) and developing strategies to minimize the nontranspiration losses (Fig. 17–1). Within Los Angeles little information is available on the evaporative losses as compared with the transpiration fluxes. The CIMIS network, which provides the most detailed information on potential evapotranspiration, and other similar meteorological services do not provide a partitioning between evaporation and transpiration.

Water Removal

In conjunction with water delivery throughout a city, an essential component of urban ecohydrology includes the efficient removal of water, both precipitation runoff within the city and water contaminated with residential, commercial, or industrial sewage. Altered surface materials (concrete/asphalt) within a city alter flowpaths of surface water, and runoff is often increased in urban ecosystems. Pollutants are carried with storm water runoff over these impermeable surfaces and through concrete conduits, which can be directly transferred back into urban streams and to downstream areas (or to the ocean) (Davis et al., 2001). Water removal and sewer systems serve dual purposes of ensuring rapid drainage of urban landscapes to prevent flooding and removing polluted water from the city. Flooding is a major problem for many cities, and as the recent events in New Orleans highlight, may become a larger problem with increased meteorological anomalies. Waste waters from residential and industrial manufacturing are commonly transferred to centralized sewage treatment facilities in most cities in developed countries. Many cities discharge their treated waste water into stream systems, which eventually serve as intake sources for downstream cities.

The city of Los Angeles Department of Public Works includes more than 10,000 km of sewage pipes and 46 sewage pumping stations. The LA system collects contaminated water and discharges it into the Pacific Ocean; only a small fraction of the water currently treated is recycled for reuse within LA. For flood protection, the city of Los Angeles Department of Public Works storm drainage system includes more than 3500 catchment basins and a 2400-km network of drainage routes that deliver water directly into the Pacific Ocean independent of the sewer system. The reported drainage averages about 378 million L d^{-1} of water during dry periods and reaches peak flows of approximately 3.78 billion L d^{-1} following storms. During storm events this water can be highly polluted with agricultural, industrial, transportation, and sewage overflow waste.

Hierarchical Linkages

Urban ecohydrology spans a broad range of spatial and temporal scales. Many complex hierarchical relationships can be identified in water sources and uses. An individual municipality is only one organizational level that must simultaneously act cooperatively and competitively with others in determining the dynamics of urban ecohydrology. The infrastructure for many urban water delivery projects in the United States often involves cooperation between multiple states and federal support and regulation. The far-reaching capacity of cities to acquire water has frequently led to intense competition for water among competing interests—within stream and riparian needs vs. human needs, urban uses vs. agricultural uses, between municipalities, and competition within a city. Much of the complexity in urban ecohydrology derives from resolving the capacity to deliver large quantities of water and the many competing desires for water uses. Agents responsible for water use range from individuals to federal agencies. An understanding of water use decision making needs to incorporate this diversity of scales, their nestedness, and organization. In securing access to water, cities compete with agricultural users distributed throughout entire regions, other cities, nonurban industrial users, and the needs for instream functioning. Within a city there also arises competition among stakeholders as to the allocation of urban water resources. While the cost of water to individual users has been considered negligible, water use patterns can vary strongly within a city (Balling et al., 2008), suggesting that this cost is associated with user decision making.

The LA region has historically competed and cooperated for water throughout the state of California and with other states in the western United States. The Colorado River canal system delivering water to the LA region is a massive interbasin transfer system that brings together the water needs of seven western states and requires extensive federally supported infrastructure projects and regulations. Within California, the LA region brings in water from throughout the entire state. Now that these inter- and intrastate systems are working, the users of the system are frequently competing for appropriate shares of the available water. The allocation schemes result in complex market and centralized activities to portion the water. Agricultural users pay much lower rates for water than private business and residential users. Los Angeles rate-blocks are reduced in regions with hotter temperatures and larger plots. Marketization offers a decentralized approach for deciding water use patterns and dealing with the large number of competing interests, although it is likely this approach alone will lead to many disparities in use and availability.

Global Variation in Urban Ecohydrological Systems

Cities differ substantially in their demand for water and the local availability of water resources (Luck et al., 2001; Jenerette and Larsen, 2006). The urban ecohydrological systems model we have presented provides a general framework to describe the structure of urban water fluxes in any city. Using this initial framework the variation among cities in the rates of fluxes, storage capacity of pools, costs associated with water fluxes, and organization required for water redistribution can be systematically examined. In addition to the differences in total inputs and outputs among cities, the relative allocation among different ecosystem services and resources also varies among cities and can be similarly examined. The variability in urban ecohydrologic systems results in differences in the production of ecosystem services and their distribution throughout and between cities. For example, commercial agriculture is a large component of Phoenix, AZ metropolitan region water uses, while in the Seattle, WA metropolitan region agriculture is a much smaller sector of water use (Hutson et al., 2004).

Within the U.S. Southwest, irrigation water uses are a major proportion of residential water uses in the Los Angeles, CA and Phoenix, AZ metropolitan regions but much less so for the cities of Tucson, AZ and Albuquerque, NM. All of these cities are within an arid climate but have contrasting amounts of total water flowing through them and contrasting allocations between uses.

Understanding variability in urban ecohydrological systems is a needed focus for urban ecological studies and more broadly for improving policies to promote sustainability of societal–ecosystem linkages. The majority of urban ecological studies have focused on developing knowledge at scales within individual cities, and the field has only recently begun asking questions at scales describing multiple cities. Initial comparative urban studies have begun developing hypotheses describing the variation among cities arrayed along multiple environmental and social gradients (Decker et al., 2000; Luck et al., 2001; Jenerette and Larsen, 2006; Pataki et al., 2006; Grimm et al., 2008b; Schneider and Woodcock, 2008). We extend this hypothesis generation to further the understanding of urban ecohydrologic fluxes. We do not intend for this list of hypotheses to be complete, but rather an initial review of potentially important drivers of variation in urban ecohydrological systems.

- *Urban ecohydrological systems increase in size and complexity with greater aridity.* Aridity is an overriding driver of urban ecohydrology. Aridity in a city is directly associated with decreased local water production and increased evaporative demand. The per-capita area required to supply water for a city increases substantially with aridity, as has been shown repeatedly through calculations of urban water footprints. The estimation of the degree of increase in footprint size with increased aridity is also a function of the regional aridity beyond the physical footprint of the city (Luck et al., 2001; Jenerette et al., 2006b). Geographic patterns can have the effect of reducing the size of the water footprint (Luck et al., 2001). For example, a relatively arid urban system might be located adjacent to a relatively mesic rural environment, such as a large mountain range, thus allowing arid cities access to extensive nearby waters. A larger urban ecohydrological system is associated with increases in complexity. More stakeholders are directly affected, from the people within a city to those at the extent of the water footprint. More interactions occur between cities. Greater governmental subsidies and regulations are required at municipality, state, and national levels. Thus, aridity will lead to increases in both the effective size and complexity of urban ecohydrological systems.

- *The socioeconomic influence on the availability of fresh water and the associated ecosystem services will increase with aridity.* The results of economic-based water redistributions bridges across scales, from international to individual decisions. Disparities in economic capacity and ability to redistribute water to arid cities have the potential to create conflicts at several scales—between neighbors within communities to between countries. The concept of a socioeconomic influence on the ability to address water shortages among water-stressed countries has been a working premise of the UN-Habitat Collaborative, and they have recognized clear patterns between high-, middle-, and low-income countries in both the status of water infrastructure and also in the capacity of the institutional and management systems.

- *Increased economic capacity will be associated with increased water deliveries related to recreational and aesthetic ecosystem services over services directly related to human health and economic activity.* As a corollary to the luxury hypothesis, which describes increased vegetation cover and biodiversity in neighbor-hoods having high incomes (Hope et al., 2003), we expect high-income cities to require a larger proportion of water for landscaping activities.

- *Leakage fluxes will decrease with economic capacity of a city.* Engineering low-leak urban water delivery systems and maintaining these systems as they age requires substantial economic resources. Cities with fewer resources will not have the capacity to reduce leaks in the system. This trend of increased urban water infrastructure leakage with decreased economic capacity can generally be observed in the results of a water-use efficiency analysis con-ducted by the European Environment Agency. Germany, with a relatively high gross national product, has less than 5% leakage, while Bulgaria, a country with a comparatively low gross national product, losses 50% of its water supply to leakage.

- *Sprawling urban development will have larger total water flows than more compact cities.* Sprawling developments are associated with a greater dispersion of inhabitants and concomitant increases in residential lot size. The increase in housing lot size, decrease in landscape heterogeneity, and high require-ment for vegetation irrigation for all contribute to increased water needs in more sprawling cities. The increased dispersion will require larger and more complex water distribution networks, and total leakage fluxes will increase, even if the proportional rate of leakage per length of piping is independent between cities.

- *Resilience of ecohydrologic dynamics will increase with diversity of water sources.* The availability of multiple water sources will provide a city with increased opportunities for dealing with periods of water scarcity. In contrast, cities that have only a single water source will be much more sensitive to disruptions of supply by this source. This hypothesis suggests, perhaps counter-intuitively, that cities in the western United States such as LA may be more resilient than large cities in the more mesic eastern United States. The diversity of water sources and flexibility of water routing decisions within California has led to suggestions that this state may be especially resilient to water stresses (Smith and Mendelsohn, 2006).

Urban Ecohydrology and Sustainable Development

Sustainable development can be broadly defined as maintaining the current capacity of coupled societal–ecological systems to provide critical ecosystem and economic services without degrading the future capacity for such services (Chapin et al., 1996; Kates et al., 2001; Clark and Dickson 2003; Turner et al., 2003). Sustainability is not achieved by a single solution, but requires ongoing rede-velopment and adaptation in response to changing ecological and economic conditions (Holling, 2001; Chapin et al., 2006). In describing urban sustainable development, an ecosystem service framework is an essential tool for evaluat-ing the importance of a given ecological process for society and making rational choices about the role of alternative ecological activities. Urban sustainability depends on a functioning ecohydrologic system providing diverse ecosystem services to the city. The continued availability of water is directly and indirectly associated with most ecosystem services urban residents depend on. Ensuring the sustainable supply of fresh water sufficient to meet individual city demands has been and continues to be a concern for many cities (Niemczynowicz, 1996; Kupel, 2003; Fitzhugh and Richter, 2004). Specific solutions needed to meet these growing urban water demands will differ depending on the geographic factors, cultural desires, and economic capacity of the city. Sustainable solutions must include approaches for managing not only with processes within the physical footprint of the city but should further connections extending to rural and wild-land regions within the urban footprint. The inclusion of these broader levels of organization will describe multiple cities interacting within a network of needs and national and international scales of demand. The urban ecohydrological systems approach we have developed here explicitly includes functioning and water uses in these "hinterland" regions as essential to dynamics within a city.

Challenges associated with achieving sustainable development of urban ecohydrological systems are extensive. A key challenge is the large potential shifts in supply and demand associated with multiple interactive global and regional changes. Changes in climate, land use, atmospheric chemistry, species communities, and disturbances are all expected to occur through global- and regional-scale processes (Vitousek, 1994; Easterling et al., 2000; Tudhope et al., 2001). Overall, water balance model projections suggest that while the global precipitation supply may not change, most regions are expected to have substantial alterations in the distribution of precipitation (Douville et al., 2002). In conjunction with the changes in the distribution of precipitation, increased warming will increase evaporation rates (Gregory et al., 1997) and thereby reduce overall water availability. Additionally, many regions may also have extensive reductions in water supply. In the southwestern United States there is increasing evidence of extensive future drought conditions (Cook et al., 2004; Seager et al., 2007). Different land-use distributions between high and low water-demand activities will directly affect the use of water and associated ecosystem services. Species changes and atmospheric chemistry changes can alter vegetation water-use efficiencies (Drake et al., 1997). Associated with these human intensive uses, there is increasing awareness that a significant percentage of available water must be conserved for aquatic systems for instream and ecosystem service and nonservice related needs (Ricciardi and Rasmussen, 1999; Naiman and Turner, 2000; Baron et al., 2002). Each of these changes will affect the availability of water resources and the allocation between ecosystem services by cities.

Effectively achieving water management and use that promotes sustainable development will depend on the local and regional conditions of each city. A growing body of international policymakers (UN and others) has proposed an Integrated Urban Water Resource Management approach (IUWRM; United Nations Environment Programme, 2003), intended to guide multifaceted urban water source decisions (United Nations, 2008). The IUWRM tenets are similar to, or were derived from, other environmental management concepts, such as "sustainable resource management," "pillar-based ecosystem approach," and "total urban water cycle management." All of these constructs address the balance involved in sustaining environmental resources while meeting urban water demands. For most cities sustainable development solutions will include reductions in total water fluxes and reallocations of water into uses that produce high water use–efficient ecosystem services. To reduce flows of water several tools are available that can have immediate economic advantages; the tools available for whole city responses to variability in use and supply include individual, institutional, grass-roots, and government actions. Expanding the use of virtual water supplies in arid regions for the generation of food and fiber needs is already beginning in several arid cities (Hoekstra and Hung, 2005; Chapagain et al., 2006; Ma et al., 2006). Installing functional metering systems and pricing structures that favor reduced water use can provide direct incentives to users to reduce water demands. In the United States some pricing rates reward higher water use with discounted rates rather than increasing rates (Jenerette et al., 2006a). Planting vegetation with locally native species that require minimal irrigation can reduce overall landscaping management costs and substantially reduce water uses (Hilaire et al., 2008). Demand management incentive schemes in conjunction with water recycling and pressure and leakage management initiatives are a few examples of least-cost planning strategies being adopted by water authorities to achieve water balance without expanding the water infrastructure asset requirements. Developing water recycling and gray-water reuse options are more complex solutions that have larger initial costs, but could substantially reduce the overall water needs of a city. At the furthest extreme, desalinization can provide new freshwater supplies but have high initial and continual costs in infrastructure and energy that make its use prohibitive. Another strategy has been to increase the runoff capture capacity by adding new reservoirs, since increased storage capacity provides some adaption to the increased variability and higher peaks in source water flows resulting from climate changes (Tanaka et al., 2006; Vicuna et al., 2007; Medellin-Azuara et al., 2008).

Ongoing marketizations of water rights are beginning to provide some flexibility in exchanging water rights between municipalities and agricultural water users (Draper et al., 2003; Hanak, 2007; Hansen et al., 2008). However, much controversy currently exists in the equitability of these contracts (Hundley, 2001; Hanak, 2007). Addressing the challenges of ensuring sustainable water resources for cities is at the forefront for many municipal decisionmakers. The urban ecohydrological systems approach should help provide tools for better understanding how an individual city uses water; how this use is related to similar cities along cultural, socioeconomic, and environmental gradients, and how to identify solutions to water resource issues.

References

Aitkenhead-Peterson, J.A., M.K. Steele, and A. Volder. 2010. Services in natural and human dominated ecosystems. p. 373–390. In J. Aitkenhead-Peterson and A. Volder (ed.) Urban ecosystem ecology. Agron. Monogr. 55. ASA, CSSA, and SSSA, Madison, WI.

Allan, J.A. 1998. Virtual water: A strategic resource global solutions to regional deficits. Ground Water 36:545–546.

Asano, T., and J.A. Cotruvo. 2004. Groundwater recharge with reclaimed municipal wastewater: Health and regulatory considerations. Water Res. 38:1941–1951.

Balling, R.C., P. Gober, and N. Jones. 2008. Sensitivity of residential water consumption to variations in climate: An intraurban analysis of Phoenix, Arizona. Water Resour. Res. 44: W10401, doi:10.1029/2007WR006722.

Baron, J.S., N.L. Poff, P.L. Angermeier, C.N. Dahm, P.H. Gleick, N.G. Hairston, R.B. Jackson, C.A. Johnston, B.D. Richter, and A.D. Steinman. 2002. Meeting ecological and societal needs for freshwater. Ecol. App. 12:1247–1260.

Bond, B. 2003. Hydrology and ecology meet—And the meeting is good. Hydrol. Processes 17:2087–2089.

Bouwer, H. 2000. Integrated water management: Emerging issues and challenges. Agric. Water Manage. 45:217–228.

Brothers, K.J. 2001. Water leakage and sustainable supply—Truth or consequences? J. Am. Water Works Assoc. 93:150–152.

Burian, S.J., and C.A. Pomeroy. 2010. Urban impacts on the water cycle and potential green infrastructure implications. p. 277–296. *In* J. Aitkenhead-Peterson and A. Volder (ed.) Urban ecosystem ecology. Agron. Monogr. 55. ASA, CSSA, and SSSA, Madison, WI.

Carpenter, S.R., H.A. Mooney, J. Agard, D. Capistrano, R.S. DeFries, S. Diaz, T. Dietz, A.K. Duraiappah, A. Oteng-Yeboah, H.M. Pereira, C. Perrings, W.V. Reid, J. Sarukhan, R.J. Scholes, and A. Whyte. 2009. Science for managing ecosystem services: Beyond the Millennium Ecosystem Assessment. Proc. Natl. Acad. Sci. USA 106:1305–1312.

Chapagain, A.K., and A.Y. Hoekstra. 2008. The global component of freshwater demand and supply: An assessment of virtual water flows between nations as a result of trade in agricultural and industrial products. Water Int. 33:19–32.

Chapagain, A.K., A.Y. Hoekstra, and H.H.G. Savenije. 2006. Water saving through international trade of agricultural products. Hydrol. Earth Syst. Sci. 10:455–468.

Chapin, F.S., A.L. Lovecraft, E.S. Zavaleta, J. Nelson, M.D. Robards, G.P. Kofinas, S.F. Trainor, G.D. Peterson, H.P. Huntington, and R.L. Naylor. 2006. Policy strategies to address sustainability of Alaskan boreal forests in response to a directionally changing climate. Proc. Natl. Acad. Sci. USA 103:16637–16643.

Chapin, F.S., M.S. Torn, and M. Tateno. 1996. Principles of ecosystem sustainability. Am. Nat. 148:1016–1037.

Chow, A.T., R.A. Dahlgren, and J.A. Harrison. 2007. Watershed sources of disinfection byproduct precursors in the Sacramento and San Joaquin rivers. Cal. Environ. Sci. Technol. 41:7645–7652.

Clark, W.C., and N.M. Dickson. 2003. Sustainability science: The emerging research program. Proc. Natl. Acad. Sci. USA 100:8059–8061.

Clarke, R., and J. King. 2004. The water atlas. New Press, New York.

Cohen, B. 2006. Urbanization in developing countries: Current trends, future projections, and key challenges for sustainability. Technol. Soc. 28:63–80.

Colombo, A.F., and B.W. Karney. 2002. Energy and costs of leaky pipes: Toward comprehensive picture. J. Water Resour. Plann. Manage. 128:441–450.

Cook, E.R., C.A. Woodhouse, C.M. Eakin, D.M. Meko, and D.W. Stahle. 2004. Long-term aridity changes in the western United States. Science 306:1015–1018.

Costanza, R., R. d'Arge, R. de Groot, S. Farber, M. Grasso, B. Hannon, K. Limburg, S. Naeem, R.V. O'Neill, J. Paruelo, R. G. Raskin, P. Sutton, and M. van den Belt. 1997. The value of the world's ecosystem services and natural capital. Nature 387:253–260.

Daily, G.C., S. Alexander, P.R. Ehrlich, L. Goulder, J. Lubchenco, P.A. Matson, H.A. Mooney, S.L. Postel, S.H. Schneider, D. Tilman, and G.M. Woodwell. 1997. Ecosystem services: Benefits supplied to human societies by natural ecosystems. Ecol. Soc. Am., Washington, DC.

Daily, G.C., S. Polasky, J. Goldstein, P.M. Kareiva, H.A. Mooney, L. Pejchar, T.H. Ricketts, J. Salzman, and R. Shallenberger. 2009. Ecosystem services in decision making: Time to deliver. Front. Ecol. Environ. 7:21–28.

Davis, A.P., M. Shokouhian, and S.B. Ni. 2001. Loading estimates of lead, copper, cadmium, and zinc in urban runoff from specific sources. Chemosphere 44:997–1009.

Decker, E.H., S. Elliott, F.A. Smith, D.R. Blake, and F.S. Rowland. 2000. Energy and material flow through the urban ecosystem. Ann. Rev. Energy Environ. 25:685–740.

Department of Water and Power. 2005. Urban water management plan: Fiscal year 2003–2004 annual update. Department of Water and Power, Los Angeles, CA.

Dodds, L., W. King, C. Woolcott, and J. Pole. 1999. Trihalomethanes in public water supplies and adverse birth outcomes. Epidemiology 10:233–237.

Dolnicar, S., and A.I. Schafer. 2009. Desalinated versus recycled water: Public perceptions and profiles of the accepters. J. Environ. Manage. 90:888–900.

Douville, H., F. Chauvin, S. Planton, J.F. Royer, D. Salas-Melia, and S. Tyteca. 2002. Sensitivity of the hydrological cycle to increasing amounts of greenhouse gases and aerosols. Clim. Dyn. 20:45–68.

Drake, B. G., M. A. Gonzàlez-Meler, and S. P. Long. 1997. More efficient plants: A consequence of rising atmospheric CO_2? Annu. Rev. Plant Physiol. Plant Mol. Biol. 48:609–639.

Draper, A.J., M.W. Jenkins, K.W. Kirby, J.R. Lund, and R.E. Howitt. 2003. Economic-engineering optimization for California water management. J. Water Resour. Plan. Manage. 129:155–164.

Easterling, D.R., G.A. Meehl, C. Parmesan, S.A. Changnon, T.R. Karl, and L.O. Mearns. 2000. Climate extremes: Observations, modeling, and impacts. Science 289:2068–2074.

Falkenmark, M. 1998. Preparing for the future: Water for a growing population. J. Water Serv. Res. Technol.-Aqua 47:161–166.

Fitzhugh, T.W., and B.D. Richter. 2004. Quenching urban thirst: Growing cities and their impacts on freshwater ecosystems. Bioscience 54:741–754.

Gleick, P.H. 2003. Water use. Annu. Rev. Environ. Resour. 28:275–314.

Goudie, A. 2000. The human impact on the natural environment. 5th ed. MIT Press, Cambridge, MA.

Gregory, J.M., J.F.B. Mitchell, and A.J. Brady. 1997. Summer drought in northern midlatitudes in a time-dependent CO_2 climate experiment. J. Clim. 10:662–686.

Grimm, N.B., S.H. Faeth, N.E. Golubiewski, C.L. Redman, J.G. Wu, X.M. Bai, and J.M. Briggs. 2008a. Global change and the ecology of cities. Science 319:756–760.

Grimm, N.B., D. Foster, P. Groffman, J.M. Grove, C.S. Hopkinson, K.J. Nadelhoffer, D.E. Pataki, and D.P.C. Peters. 2008b. The changing landscape: Ecosystem responses to urbanization and pollution across climatic and societal gradients. Front. Ecol. Environ. 6:264–272.

Grimm, N.B., J.M. Grove, S.T.A. Pickett, and C.L. Redman. 2000. Integrated approaches to long-term studies of urban ecological systems. Bioscience 50:571–584.

Hanak, E. 2007. Finding water for growth: New sources, new tools, new challenges. J. Am. Water Resour. Assoc. 43:1024–1035.

Hansen, K., R. Howitt, and J. Williams. 2008. Valuing risk: Options in California water markets. Am. J. Agric. Econ. 90:1336–1342.

Heisler, G.M., and A.J. Brazel. 2010. The urban physical environment: Temperature and urban heat islands. p. 29–56. In J. Aitkenhead-Peterson and A. Volder (ed.) Urban ecosystem ecology. Agron. Monogr. 55. ASA, CSSA, and SSSA, Madison, WI.

Hilaire, R.S., M.A. Arnold, D.C. Wilkerson, D.A. Devitt, B.H. Hurd, B.J. Lesikar, V.I. Lohr, C.A. Martin, G.V. McDonald, R.L. Morris, D.R. Pittenger, D.A. Shaw, and D.F. Zoldoske. 2008. Efficient water use in residential urban landscapes. HortScience 43:2081–2092.

Hinkebein, T.E., and M.K. Price. 2005. Progress with the desalination and water purification technologies U.S. roadmap. Desalination 182:19–28.

Hoekstra, A.Y., and A.K. Chapagain. 2007. Water footprint of nations: Water use by people as a function of their consumption pattern. Water Resour. Manage. 21:35–48.

Hoekstra, A.Y., and P.Q. Hung. 2005. Globalisation of water resources: International virtual water flows in relation to crop trade. Global Environ. Change-Human. Policy Dimen. 15:45–56.

Holling, C.S. 2001. Understanding the complexity of economic, ecological, and social systems. Ecosystems 4:390–405.

Hope, D., C. Gries, W.X. Zhu, W.F. Fagan, C.L. Redman, N.B. Grimm, A.L. Nelson, C. Martin, and A. Kinzig. 2003. Socioeconomics drive urban plant diversity. Proc. Natl. Acad. Sci. USA 100:8788–8792.

Hundley, N. 2001. The great thirst. Univ. of California Press, Los Angeles.

Hutson, S.S., N.L. Barber, J.F. Kenny, K.S. Linsey, D.S. Lumia, and M.A. Maupin. 2004. Estimated use of water in the United States in 2000. USGS, Reston, VA.

Ioris, A.A.R., S. Merrett, and T. Allan. 2004. Virtual water in an empty glass: The geographical complexities behind water scarcity. Water Int. 29:119–121.

Jackson, R.B., S.R. Carpenter, C.N. Dahm, D.M. McKnight, R.J. Naiman, S.L. Postel, and S.W. Running. 2001. Water in a changing world. Ecol. Appl. 11:1027–1045.

Jenerette, G.D., S.L. Harlan, A. Brazel, N. Jones, L. Larsen, and W.L. Stefanov. 2007. Regional relationships between surface temperature, vegetation, and human settlement in a rapidly urbanizing ecosystem. Landscape Ecol. 22:353–365.

Jenerette, G.D., and R. Lal. 2005. Hydrologic sources of carbon cycling uncertainty throughout the terrestrial–aquatic continuum. Glob. Change Biol. 11:1873–1882.

Jenerette, G.D., and L. Larsen. 2006. A global perspective on changing sustainable urban water supplies. Global Planet. Change 50:202–211.

Jenerette, G.D., W.A. Marussich, and J. Newell. 2006a. Linking ecological footprints with ecosystem service valuation in the urban consumption of freshwater. Ecol. Econ. 59:38–47.

Jenerette, G.D., W. Wu, S. Goldsmith, W. Marussich, and W.J. Roach. 2006b. Contrasting water footprints of cities in China and the United States. Ecol. Econ. 57:346–358.

Jetten, M.S.M., S.J. Horn, and M.C.M. van Loosdrecht. 1997. Towards a more sustainable municipal wastewater treatment system. Water Sci. Technol. 35:171–180.

Kahrl, W. 1982. Water and power. Univ. of California Press, Berkeley.

Kates, R.W., W.C. Clark, R. Corell, J.M. Hall, C.C. Jaeger, I. Lowe, J.J. McCarthy, H.J. Schellnhuber, B. Bolin, N.M. Dickson, S. Faucheux, G.C. Gallopin, A. Grubler, B. Huntley, J. Jager, N.S. Jodha, R.E. Kasperson, A. Mabogunje, P. Matson, H. Mooney, B. Moore, T. O'Riordan, and U. Svedin. 2001. Environment and development—Sustainability science. Science 292:641–642.

Kupel, D.E. 2003. Fuel for growth: Water and Arizona's urban environment. The Univ. Arizona Press, Tucson.

Los Angeles Department of Water and Power. 2005. Urban water management plan. Los Angeles Dep. of Water and Power, Los Angeles, CA.

Luck, G.W., R. Harrington, P.A. Harrison, C. Kremen, P.M. Berry, R. Bugter, T.P. Dawson, F. de Bello, S. Diaz, C.K. Feld, J.R. Haslett, D. Hering, A. Kontogianni, S. Lavorel, M. Rounsevell, M.J. Samways, L. Sandin, J. Settele, M.T. Sykes, S. van den Hove, M. Vandewalle, and M. Zobel. 2009. Quantifying the contribution of organisms to the provision of ecosystem services. Bioscience 59:223–235.

Luck, M.A., G.D. Jenerette, J.G. Wu, and N.B. Grimm. 2001. The urban funnel model and the spatially heterogeneous ecological footprint. Ecosystems 4:782–796.

Ma, J., A.Y. Hoekstra, H. Wang, A.K. Chapagain, and D. Wang. 2006. Virtual versus real water transfers within China. Philos. Trans. R. Soc. B Biol. Sci. 361:835–842.

Medellin-Azuara, J., J.J. Harou, M.A. Olivares, K. Madani, J.R. Lund, R.E. Howitt, S.K. Tanaka, M.W. Jenkins, and T. Zhu. 2008. Adaptability and adaptations of California's water supply system to dry climate warming. Clim. Change 87:S75–S90.

Morris, R.D., A.M. Audet, I.F. Angelillo, T.C. Chalmers, and F. Mosteller. 1992. Chlorination, chlorination by-products, and cance: A metaanalysis. Am. J. Public Health 82:955–963.

Naiman, R.J., and M.G. Turner. 2000. A future perspective on North America's freshwater ecosystems. Ecol. Appl. 10:958–970.

Nelson, E., G. Mendoza, J. Regetz, S. Polasky, H. Tallis, D.R. Cameron, K.M.A. Chan, G.C. Daily, J. Goldstein, P.M. Kareiva, E. Lonsdorf, R. Naidoo, T.H. Ricketts, and M.R. Shaw. 2009. Modeling multiple ecosystem services, biodiversity conservation, commodity production, and tradeoffs at landscape scales. Front. Ecol. Environ. 7:4–11.

Niemczynowicz, J. 1996. Megacities from a water perspective. Water Int. 21:198–205.

Pataki, D.E., R.J. Alig, A.S. Fung, N.E. Golubiewski, C.A. Kennedy, E.G. McPherson, D.J. Nowak, R.V. Pouyat, and P.R. Lankao. 2006. Urban ecosystems and the North American carbon cycle. Glob. Change Biol. 12:2092–2102.

Pettygrove, G., and T. Asano (ed.) 1985. Irrigation with reclaimed municipal wastewater: A guidance manual. Lewis Publ., Chelsea, MI.

Pickett, S.T.A., W.R. Burch, S.E. Dalton, T.W. Foresman, J.M. Grove, and R. Rowntree. 1997. A conceptual framework for the study of human ecosystems in urban areas. Urban Ecosyst. 1:185–199.

Pickett, S.T.A., M.L. Cadenasso, J.M. Grove, C.H. Nilon, R.V. Pouyat, W.C. Zipperer, and R. Costanza. 2001. Urban ecological systems: Linking terrestrial ecological, physical, and socioeconomic components of metropolitan areas. Annu. Rev. Ecol. Syst. 32:127–157.

Postel, S.L., G.C. Daily, and P.R. Ehrlich. 1996. Human appropriation of renewable fresh water. Science 271:785–788.

Reisner, M. 1993. Cadillac desert: The American West and its disappearing water. Rev. ed. Penguin, New York.

Ricciardi, A., and J.B. Rasmussen. 1999. Extinction rates of North American freshwater fauna. Conserv. Biol. 13:1220–1222.

Rodriguez-Iturbe, I. 2000. Ecohydrology: A hydrologic perspective of climate–soil–vegetation dynamics. Water Resour. Res. 36:3–9.

Scanlon, B.R., D.G. Levitt, R.C. Reedy, K.E. Keese, and M.J. Sully. 2005. Ecological controls on water-cycle response to climate variability in deserts. Proc. Natl. Acad. Sci. USA 102:6033–6038.

Schneider, A., and C.E. Woodcock. 2008. Compact, dispersed, fragmented, extensive? A comparison of urban growth in twenty-five global cities using remotely sensed data, pattern metrics and census information. Urban Stud. 45:659–692.

Seager, R., M.F. Ting, I. Held, Y. Kushnir, J. Lu, G. Vecchi, H.P. Huang, N. Harnik, A. Leetmaa, N.C. Lau, C.H. Li, J. Velez, and N. Naik. 2007. Model projections of an imminent transition to a more arid climate in southwestern North America. Science 316:1181–1184.

Shepherd, J.M., J.A. Stallins, M.L. Jin, and T.L. Mote. 2010. Urbanization: Impacts on clouds, precipitation, and lightning. p. 1–28. *In* J. Aitkenhead-Peterson and A. Volder (ed.) Urban ecosystem ecology. Agron. Monogr. 55. ASA, CSSA, and SSSA, Madison, WI.

Singer, P.C. 1999. Humic substances as precursors for potentially harmful disinfection by-products. Water Sci. Technol. 40:25–30.

Smith, J., and R. Mendelsohn (ed.) 2006. The impact of climate change on regional systems: A comprehensive analysis of California. Edward Elgar Publ., Cheltenham, UK.

Stackelberg, P.E., E.T. Furlong, M.T. Meyer, S.D. Zaugg, A.K. Henderson, and D.B. Reissman. 2004. Persistence of pharmaceutical compounds and other organic wastewater contaminants in a conventional drinking-watertreatment plant. Sci. Total Environ. 329:99–113.

Tabesh, M., A.H.A. Yekta, and R. Burrows. 2009. An integrated model to evaluate losses in water distribution systems. Water Resour. Manage. 23:477–492.

Tanaka, S.K., T.J. Zhu, J.R. Lund, R.E. Howitt, M.W. Jenkins, M.A. Pulido, M. Tauber, R.S. Ritzema, and I.C. Ferreira. 2006. Climate warming and water management adaptation for California. Clim. Change 76:361–387.

Toze, S. 2006. Reuse of effluent water—Benefits and risks. Agric. Water Manage. 80:147–159.

Tudhope, A.W., C.P. Chilcott, M.T. McCulloch, E.R. Cook, J. Chappell, R.M. Ellam, D.W. Lea, J.M. Lough, and G.B. Shimmield. 2001. Variability in the El Nino—Southern oscillation through a glacial-interglacial cycle. Science 291:1511–1517.

Turner, B.L., R.E. Kasperson, P.A. Matson, J.J. McCarthy, R.W. Corell, L. Christensen, N. Eckley, J.X. Kasperson, A. Luers, M.L. Martello, C. Polsky, A. Pulsipher, and A. Schiller. 2003. A framework for vulnerability analysis in sustainability science. Proc. Natl. Acad. Sci. USA 100:8074–8079.

United Nations. 2008. Status report on integrated water resource management (IWRM) and water efficiency plans for CSD16. Task Force on IWRM Monitoring and Reporting.

United Nations Environment Programme. 2003. Integrated urban resources management strategy—Water. Available at http://www.unep.or.jp/ietc/focus/pdf/iuwrm.pdf (verified 26 Mar. 2010).

United Nations Population Fund. 2007. State of world populations 2007. Unleashing the potential of urban growth. United Nations Population Fund.

U.S. Census Bureau. 2000. United States Census 2000. Available at http://www.census.gov/main/www/cen2000.html (verified 18 Mar. 2010).

Varis, O. 2006. Megacities, development and water. Int. J. Water Resour. Dev. 22:199–225.

Vicuna, S., E.P. Maurer, B. Joyce, J.A. Dracup, and D. Purkey. 2007. The sensitivity of California water resources to climate change scenarios. J. Am. Water Resour. Assoc. 43:482–498.

Vitousek, P.M. 1994. Beyond global warming: Ecology and global change. Ecology 75:1861–1876.

Vörösmarty, C., E. Douglas, P. Green, and C. Revenga. 2005. Geospatial indicators of emerging water stress: An application to Africa. Ambio 34:230–236.

Vorosmarty, C.J., P. Green, J. Salisbury, and R.B. Lammers. 2000. Global water resources: Vulnerability from climate change and population growth. Science 289:284–288.

Vorosmarty, C.J., and D. Sahagian. 2000. Anthropogenic disturbance of the terrestrial water cycle. Bioscience 50:753–765.

Williams, D.G., W. Cable, K. Hultine, J.C.B. Hoedjes, E.A. Yepez, V. Simonneaux, S. Er-Raki, G. Boulet, H.A.R. de Bruin, A. Chehbouni, O.K. Hartogensis, and F. Timouk. 2004. Evapotranspiration components determined by stable isotope, sap flow and eddy covariance techniques. Ag. Forest Meteorol. 125:241–258.

18

Services in Natural and Human Dominated Ecosystems

Jacqueline A. Aitkenhead-Peterson
Meredith K. Steele
Astrid Volder

Abstract

Ecosystem services have been defined as the benefits that human populations derive directly or indirectly from ecosystem functions. These services range from the obvious, such as provision of food and fiber and timber for homes, to the less obvious, such as microbial nutrient cycling in soil and environmental cooling by vegetation. This chapter discusses the services, resources, and benefits provided in natural and human-dominated ecosystems.

The term *ecosystem* was first used by Sir Arthur George Tansley in his landmark paper about the use and abuse of vegetational concepts and terms (Tansley, 1935). In a later presidential address to the British Ecological Society he defined ecosystem as a concept "which includes the inorganic as well as the living components in the whole to be considered" (Tansley, 1939). This definition encompasses all processes within the ecosystem, and many of these processes are necessary to provide essential services to keep the ecosystem functioning. Ecosystem functions are processes within ecosystems that may or may not have an effect on the human population; they are properties inherent to the ecosystem.

Ecosystem services have been defined as "the benefits human populations derive, directly or indirectly from ecosystems functions" (Costanza et al., 1997). Thus, the defining criterion for an ecosystem "service" is that the ecosystem function must provide some direct or indirect benefit to humans. These benefits could range from very quantifiable benefits, in monetary terms, such as providing timber or purification of water, to benefits that are more difficult to quantify, such as providing habitat to wildlife. However, many of the other components, both

J.A. Aitkenhead-Peterson (jpeterson@ag.tamu.edu) and M.K. Steele (mkbilek@gmail.com), Dep. of Soil and Crop Sciences, Texas A&M University, 2474 TAMU, College Station, TX 77843; A. Volder, Dep. of Horticultural Sciences, Texas A&M University, 2133 TAMU, College Station, TX 778432133 (a-volder@tamu.edu).

doi:10.2134/agronmonogr55.c18

living (nonhuman) and nonliving within an ecosystem also receive benefits at some cost, which in the natural world is energy.

The environmentalist ethic was championed by Aldo Leopold (Leopold, 1949). The works of Leopold and his ideas of land ethics and good environmental stewardship are key elements to this philosophy. A more human focused, utilitarian view appears to have been adopted from biblical scriptures (e.g., Genesis 1:25–27; 9:2–3; Deuteronomy 14:3–26). One major difference between the two points of view is that in the non–human dominated or natural ecosystem the services are typically obtained from within the local ecosystem, whereas in the human-dominated system some of the benefits are obtained within the local ecosystem, and other benefits are imported from other ecosystems. For example, paper products are derived from timber, which originates in natural or forested ecosystems. Silk, cotton, or wool products for clothing originate from agricultural ecosystems, but these ecosystems may be many thousands of miles away from the human-dominated ecosystem that will use them. The additional energy costs associated with importing services from outside the local ecosystem leads to an energy usage many times greater than the energy supplied by the local ecosystem itself. Frolke et al. (2007) suggested in a study of the 29 largest Baltic cities that the ecosystem support areas were between 500 and 1000 times larger than the areas of the cities themselves—they defined this as the *ecological footprint*. Urban areas, with their concentrated population and high rate of imports generally have much greater ecological footprints than rural areas.

Intrinsic Value of Ecosystems, Functions, and Services

Under the environmentalist approach intrinsic value is placed on all organisms (including humans), the abiotic components (climate, water, and geology), and the processes that result from interactions between components. According to Hettinger and Throop (1999), *balance* and *equilibrium* are related concepts, where "a system is in equilibrium if the various forces acting on it are sufficiently balanced that the system is constant and orderly with respect to those features under consideration." Equilibrium can be both static and dynamic. For example, an old growth stand of trees in a temperate forest would be static compared with underbrush that burns and then regrows with cyclic fires. One of the key arguments against this view is that there may be no such thing as equilibrium in either the short or long term or that ecosystems are in a constant state of transition in relation global and local forces. The question then becomes, how does one place a value on services that promote an ideal condition if no ideal condition exists? Campbell (1983) described two philosophies: the first is an emotional or spiritual view in which a person has a love for the diversity of life and nature, and the second is a practical view that recognizes human dependence on nature and our ignorance of its complex systems and leads to the conclusion that nature should be disturbed as little as possible until we have a better understanding.

From the ecological perspective, if all ecosystem members are deemed to have value, then all abiotic processes and actions and all biotic organisms that promote the survival of a specific ecosystem type would be considered a service. In an ecosystem these processes may be cooperative or competitive interactions. Science has identified many major processes within an ecosystem that promote the maintenance of systems as we understand them. However, given the extreme diversity and complexity of ecosystems that exist, it would be arrogant to think that our understanding is anything more than a limited framework of these services. There is often a certain amount of personification or anthropogenic value placed on abiotic elements. The value of these elements and their services only exist in their relation to an entity which has purpose. For example, water is often graded or classified on its quality, but that classification is based on how well it supports various aquatic biotic community members or human activities. The water itself does not have a goal to be clean—it is not self aware, and its condition is a result of human labeling. The situation is similar for soil and air quality. The following is an overview of general ecosystem functions that have value in maintaining ecosystems but may not necessarily have a direct benefit to human society.

Cycling of nutrients through life and death is fundamental to ecosystem processes and thus the services provided. Vegetation is an integral part of most natural ecosystems, and it affects and is affected by the other components within its ecosystem. For example, climatic elements, including precipitation, sunlight or solar radiation, and atmospheric carbon dioxide all affect the success or failure of vegetative species and maintain characteristic species composition. The atmosphere provides CO_2, which it received from the decomposition of dead plants and animals in the biotic soil and from the respiration of living plants and animals (e.g., Bowden et al., 1993; Sulzman et al., 2005). Photosynthesis occurs when this CO_2 is combined with water and energy from solar radiation, necessary for all vegetative growth. The plant returns O_2 to the atmosphere and scavenges particulates from the air around it. Rainfall events clean dry deposits from the plants, and water is necessary for growth of all vegetation.

Soil represents the medium in which vegetation grows. It provides a strong infrastructure to hold roots and the plants in place and a source of nutrients for the plants. The cycling of nutrients necessary for vegetative growth is performed by the biotic component of soil, which includes invertebrates such as arthropods, nematodes, mollusks, annelids plus fungi, protozoa and bacteria. The biotic component of soil needs the water provided by precipitation to maintain soil moisture in its habitat. Soil moisture can be considered the master variable that may explain different patterns of soil microbial community composition either by altering the connectivity of the heterogeneity of habitats in the soil matrix or by impacting microbial stress physiology (Lennon et al., 2008). When there is too much water, soil pore spaces will be entirely filled with water, which will cause anaerobic conditions, favoring a group of bacteria that use terminal electron acceptors such as nitrate (denitrifyers), sulfate (sulfate reducing bacteria), or organic molecules (fermenters) instead of oxygen. Too little water and the plant will not be able to extract water from the soil and will wilt (wilting point), while bacterial processes are inhibited. Under ideal conditions, that point between field capacity (i.e., soil pore spaces are filled with water) and wilting point (i.e., there is not enough water in the pore spaces to maintain vegetation growth), the pore space will be filled with a mixture of air and water. The chemistry of the incoming water is important because of the supply of further nutrients to the soil heterotrophs as precipitation gains carbon and nutrients from its interaction with vegetation as throughfall and flushing of upper organic layers of the soil (Aitkenhead-Peterson et al., 2003; Chapter 7, Pouyat et al., 2010, this volume). Warmth, supplied by radiation, is beneficial to soil microorganisms and enables them to cycle carbon and nutrients faster (e.g., Elsgaard and Jorgensen, 2002).

Soil benefits through its interaction with vegetation. While the plant is living, it provides the biotic soil with organic acids and carbon compounds through rhizodeposits, and its roots create channels that facilitate water and oxygen diffusion down the soil profile (Kuzyakov, 2002). Rhizodeposits are also important for the formation of aggregates in the mineral soil (Rillig et al., 2002), which increase pore space and habitat for the biotic component of soil. When the plant dies it provides the biotic soil with a nutrient-rich substrate. Water benefits from its interaction with vegetation and soil through cleansing. An example of this includes wetland species that take up excess nutrients dissolved in the water (e.g., Johnson, 1991). Infiltration of water through soil horizons will also cleanse water as ions are adsorped onto soil mineral surfaces, resulting in a relatively low ionic strength solution that reaches the groundwater table. Vegetation provides food for the herbivores and soil provides a means of a permanent habitat for some invertebrates and some mammals such as moles (e.g., *Talpa europaea* L.) and temporary habitat or shelter for animals such as armadillos (e.g., *Dasypus novemcinctus* L.), aardvarks (*Orycteropus afer* Pallus), European badgers (*Meles meles* L.), and fox (*Vulpes vulpes* L.), but more importantly soil is a repository for recycling their waste material. Animals benefit vegetation by their roles in pollination (Koltowski et al., 1999) and seed dispersal (e.g., Herrera, 1989; Vanderwall, 1992; Schupp, 1993).

Monetary Value of Ecosystem Services

This section examines the benefits or services obtained by humans and in some cases, where calculated, the monetary cost of that service.

Services, Resources, or Benefits Obtained by Humans

Excellent reviews of ecosystem services provided to humans can be found in Daily (1997), Bolund and Hunhammar (1999), and Jim and Chen (2009). Services obtained from natural ecosystems are undervalued by the human society; take for example the "Tragedy of the Commons" (Hardin, 1968). For the most part, these services are not traded in the marketplace, so there is no price change to signal change in the supply or condition of these services (Daily, 1997b). Many ecosystem services have been ignored or not fully realized until their disruption or loss highlighted their importance. For example, deforestation revealed the role that forests serve in water cycle regulation (Myers, 1997). Furthermore, while a 50% decline of honey bees [*Apis mellifera* (L.)] in the USA between 1947 and 1994 resulted in a loss of honey, the loss of the pollination service the bees provide may be more important (Nabhan and Buchmann, 1997).

Vegetation and Soil

Vegetation provides a cleansing service for both air and water. In many cities constructed wetlands are being used to treat sewage water. Wetland plants can assimilate large amounts of nutrients, trap sediments, and slow down the flow of water, thereby allowing small particulates to settle. Constructed wetlands and wetland restoration have been successful in increasing biodiversity and lowering costs of sewage treatments (Ewel, 1997). Urban greening in China was set with Mao's policy of "afforesting the motherland," which laid the foundation of urban greening in cities (Jim and Chen, 2009), but despite this long history of urban greening, the study of urban forests was not initiated until after the 1990s (Wang,

1995; Li et al., 2005). Trees are excellent in cleaning the air in the lower troposphere, which is important in Chinese cities—because of the recent and rapid increase of cars, urbanization, and industrialization, air quality is declining rapidly (Zhuang, 2003). The cleaning that trees provide is caused by scavenging particulates and gases from the air (Chapter 3, Santosa, 2010, this volume). Givoni (1991) suggested that the scavenging or filtering capacity of vegetation increases with a larger leaf area and thus is usually higher for trees than bushes or grasses. Conifer species have a better scavenging capacity than deciduous trees, mainly because their needles do not shed in the winter, when air quality can be worse, particularly in urban areas that still use coal for heating. In Chicago, trees have been estimated to remove 5575 tonnes of particulates from the air (Nowak, 1994), which equates to providing more than 9 million U.S. dollars of air cleaning service. In Bejing the removal capacity of atmospheric sulfur dioxide by urban forests is estimated at 2192 t yr^{-1} (Leng et al., 2004). The value of removing air pollutants using 19,944 urban trees in Shenyang in northeast China was $16,318 yr^{-1}, based on western urban forest studies (He et al., 2003). Urban forests make significant contributions in the removal of sulfur dioxide, nitrogen oxides, ozone, carbon monoxides, and particulates in the air within a city (Jim and Chen, 2009).

Riparian forests and vegetated floodplains provide the service of reducing floods. Floodplain means just that, an area which experiences occasional or periodic flooding that provides unconsolidated sediments rich in nutrients to the area. The extent of inundation of floodplains and riparian forests depends in part on the flood magnitude, which is defined by the return period or recurrence interval. Species that are endemic to floodplains tend to be tolerant of root disturbance and are faster growing than nonriparian trees. Human alterations of these important vegetated areas reduce the ecosystem service that they provide. Loss of riparian vegetation in urban ecosystems may result in narrowing of the stream channel, compromising in-stream processing (Sweeney et al., 2004). Historically towns and settlements were built on floodplains because they were close to a water source and the ground was fertile for farming. Furthermore, river transportation was a key factor in the founding of many communities. Thus riparian ecosystems generally have been more influenced by human presence than most other types of ecosystems.

Vegetation provides the service of noise reduction. Noise from traffic and construction can create health problems for humans in the urban ecosystem. Dense shrubbery of 5-m width can reduce noise by 2 dB(A) and a 50-m plantation of trees will reduce noise by 3 to 6 dB(A) (Naturvardsverket, 1996). In a study in Varanasi, Pathak et al. (2008) reported that *Hibiscus rosa-sinensis* L. reduced traffic noise at both low and high frequencies. Hedges, which typically have been used as wind breaks, have also been shown to reduce noise (Herrington, 1976). Plant leaves absorb acoustic energy by the energy transfer of the kinetic energy of the vibrating air molecule to vibration of the plant leaf (Pathak et al., 2008).

Cooling, heating, and shade are other important services that vegetation provides within an urban ecosystem. Urban forests can modify solar radiation, wind speed, air temperature, relative humidity, and terrestrial reradiation (Grimmond et al., 1994; McPherson et al., 1997; Miller, 1997). The cooling effect of evapotranspiration can significantly reduce ambient air temperatures within a city (Chen, 1998; Cai et al., 2002; Wang et al., 2007). A single tree can transpire 450 L of water each day, consuming 1000 MJ of heat energy to drive the evaporative process. Trees can therefore lower city temperatures considerably (Hough, 1989). Different vegetation contributes different evapotranspirative cooling effects. For example, *Platanus* × *hispanica* Munchh. absorbed an average of 7009 kj m^{-2} d^{-1} during June to August, lowering air temperatures by 0.47°C, *Prunus triloba* Lindl. absorbed 6559 kj m^{-2} d^{-1} and reduced temperature by 0.44°C and Kentucky Bluegrass (*Poa pratensis* L.) absorbed 8010 kj m^{-2} d^{-1} and reduced temperature by 0.53°C (Ding, 2007, as cited in Jim and Chen, 2009). Shading by trees is important in urban subdivisions that contain single-family homes but is less important in multistory apartment complexes in the urban core. Vegetation shading can decrease fossil fuel energy used for heating and cooling houses, representing between $50 to $90 per dwelling each year (McPherson et al., 1997). Fuel wood is also important for heating, particularly in northern cities, but this is usually sourced from outside the urban ecosystem.

Recreation is important for humans residing in urban ecosystems. Botkin and Beveridge (1997) argued that "vegetation is essential to achieving the quality of life that creates a great city and makes it possible for people to live a reasonable life within an urban environment." More than 90% of the citizens of Stockholm visit parks at least once during the year and 45% do so every week (Bolund and Hunhammar, 1999). From the results of a study performed in Stockholm, Transek (1993) reported that citizens were willing to pay extra to live near parks, forests, or open water. With a working week of 40 h or less in the 21st century, humans have time on their hands to play sports such as golf or enjoy one of the many athletic fields springing up in urban ecosystems. While athletic fields, games fields, or courts tend to be turfgrass the golf course has evolved to contain plantings of trees, shrubs, plants, and typically a water feature. Yet another form of recreation for dwellers of the urban ecosystem has been allotment or community gardens. Here the gardens provide food, exercise, community spirit, and, if tended well, an aesthetically pleasing green space (Chapter 10, Iaquinta and Drescher, 2010, this volume). Thus, the recreational aspects of all urban greenspaces, with possibilities to play and rest, are perhaps the highest valued ecosystem service in cities. Vegetated greenspaces are important for the psychology of urban dwellers. Ulrich (1991) reported on a study of the response of humans to different environments. They found that when subjects were exposed to undisturbed ecosystems their stress levels decreased rapidly, but when exposed to urban ecosystems their stress levels remained high or increased. An earlier study examined the recovery of hospital patients in rooms facing a park compared with those in rooms facing a building (Ulrich, 1984). Those patients with rooms facing a park needed 50% less pain-relieving medication than patients with rooms facing a building and had a 10% faster recovery (Ulrich, 1984).

One service that vegetation provides, which is often forgotten, is the removal of carbon dioxide and the production of oxygen during photosynthesis. Vegetation with a larger leaf area index and high photosynthetic rate contribute to more oxygen production and carbon dioxide fixation in an urban ecosystem. For example, Yang (1996, as cited in Jim and Chen, 2009) reported that *Aleurites moluccana* (L.) Willd. released 66 g m^{-2} d^{-1} O$_2$ and fixed 91 g m^{-2} d^{-1} CO$_2$ compared with *Nerium indicum* Mill., which released 7.6 g m^{-2} d^{-1} O$_2$ and fixed 10.3 g m^{-2} d^{-1} CO$_2$. In a further study conducted in Bejing the photosynthetic rates of 65 common tree species were measured (Chen et al., 1998, as cited in Jim and Chen, 2009). The authors concluded that urban forests in Bejing could release

23,000 t O_2 d^{-1} and fix 33,000 t CO_2 d^{-1}. Season also has an effect on oxygen production and carbon dioxide sequestration (Cai et al., 2002; Lu et al., 2006, as cited in Jim and Chen, 2009), with most benefit derived in the summer followed by the spring and fall seasons.

There are some negative aspects to trees in cities. Some species commonly planted in urban ecosystems, such as pine (*Pinus* spp.), oak (*Quercus* spp.), and willow (*Salix* spp.), emit volatile organic compounds that may contribute to urban smog or ozone formation (Winer et al., 1983; Chameides et al., 1988; Chapters 3 [Santosa, 2010] and 9 [Volder, 2010], this volume). These volatile organic compounds include isoprene and monoterpene compounds, which can vary considerably between tree species. For example, *Quercus* spp. emit 11.9 μg C g^{-1} h^{-1} at 30°C, whereas *Platanus* × *acerifolia* (Aiton) Willd. emits 35.0 μg C g^{-1} h^{-1} at the same temperature, and isoprene emissions from *Koelreuteria paniculata* Laxm. can reach 49.0 μg C g^{-1} h^{-1} at 30°C (McPherson et al., 1998). Using existing models, McPherson et al. (1998) made the argument that in California, there is no convincing evidence that planting 500,000 trees in Sacramento would produce cost savings when only air quality benefits are considered.

Other major services that vegetation provides to the urban dweller are pharmaceuticals (Ma et al., 2005), food, clothing such linen and cotton, and timber for construction, fuel, and paper products. These are examples of services from vegetation imported into the urban ecosystem. While these goods have monetary value and are paid for in cash, the cost to the ecosystem from which they were obtained is not always apparent. Approximately 57% of the top name-brand drugs contain at least one major active compound extracted from plant material. For example, analgesics such as aspirin and codeine are derived from salicylate-rich willow bark and sap of the poppy (*Papaver bracteatum* Lindl.). Food, linen, and cotton can be grown relatively locally in agricultural ecosystems or obtained from similar agroecosystems in different countries but are nevertheless imported into the urban ecosystem. The cost to the consumer will not typically include the cost of remediating high nutrient, pesticide, and herbicide concentrations in surface and groundwater or the reduction in soil microbial community composition.

By converting large areas into agricultural land humans have changed the kinds and amounts of ecosystem services that would typically be provided (Daily, 1997). Poor agricultural practices involving overfertilization, applications of pesticides and herbicides, and irrigation with water sometimes two orders of magnitude higher in dissolved solids than would normally be present in rain water have destroyed the ecological service of soil fertility, biotic regulation, nutrient recycling, assimilation of wastes, sequestration or fixing of carbon dioxide, and the maintenance of genetic information that soil would normally provide (Vitousek et al., 1986; Pimentel, 1988: Ehrlich and Erlich, 1992; Paoletti et al., 1992; Odum, 1993; Vitousek, 1994). Soils in agricultural systems are considered in much the same way as soils in the urban ecosystem, a structure to hold vegetation. As soil fertility has declined in agricultural fields so have crop yields. Additions of manures and inorganic fertilizers increased, as did intensive cropping practices and genetic engineering of crop plants to produce higher yields. In 1952, 1904 MJ energy, 15 kg N, 36 kg fodder, and 226 h of labor produced 101 kg protein, 81 kg fat, and 292 kg carbohydrate. In 1992 to produce similar quantities of protein, fat, and carbohydrate 4014 MJ energy, 44 kg N, 200 kg fodder, and 36 h of labor were needed (Bjorklund et al., 1999). Thus, roughly three times more N and six times more fodder were used and one-sixth of the labor was needed. The difference in input and labor is met by fossil fuels. One of the major losses to soils in the agricultural system is organic matter. Lilliesköld and Nilsson (1997) estimated that 1 million tonnes of organic matter are lost in Swedish agricultural soils each year, resulting in an overall decline during the last 40 years of approximately 9000 kg ha^{-1}.

The surface or skin of an urban ecosystem is typically concrete or asphalt roads and pavements and roofs of wood, asphalt, or asbestos shingles, clay tile, slate, metal or concrete—all of these are impermeable to water. A move toward green roofs (Chapter 19, Rowe and Getter, 2010, this volume), urban greenspace (Chapter 21, Beer, 2010, this volume), allotments and community gardens (Chapter 10, Iaquinta and Drescher, 2010, this volume), and urban landscaping and forestry (Chapters 9 [Volder, 2010] and 11 [Volder and Watson], this volume) will increase the proportion of soil relative to impervious surfaces in urban ecosystems. Soil in an urban ecosystem is typically a constructed soil with a recipe to benefit water infiltration and retain moisture. The benefit of soil to the human in the urban ecosystem is its use as a medium for landscaping, such as turfgrass for parks and recreation, and as a medium for vegetable growth in town allotments or city community vegetable gardens.

Water

Water is a necessity for life for all humans. Sources of water, typically treated to USEPA drinking water standards (USEPA, 2009), come from surface water or groundwater. Surface water supplies include reservoirs, lakes, and rivers. Groundwater supplies come from deep confined or shallow unconfined aquifers. Humans pay money to receive clean drinkable water in urban ecosystems of developed countries. The water is used for drinking; cooking; bathing; washing clothes, dishes, and vehicles; and irrigating landscapes and allotments or community gardens used for vegetables (Chapter 17, Jenerette and Alstad, 2010, this volume). Water is also used for processing food and beverages; there are few drinks that do not contain water. Water is also important to the human for recreation, such as boating, swimming, and fishing. Other goods that fresh water supplies are freshwater commercial fisheries, sports fishing, and waterfowl hunting, which are billion dollar industries in the United States. All of these services are threatened by point-source discharge of sewage and industrial effluents to streams and rivers. All municipalities rely on the diluting capacity of natural waters, but dilution becomes less effective as the concentrations of nutrients and metals in freshwaters have increased between discharge points through nonpoint-source urban and agricultural runoff and more water is abstracted for municipal supply and agricultural irrigation. Water transportation on inland waterways for the movement of goods from one place to another is important in many parts of the world (Biswas, 1987).

Incoming water to an urban ecosystem in the form of precipitation is run off directly into surface waters or retained in bioretention ponds (Chapters 14 [Burian and Pomeroy, 2010] and 20 [Li et al., 2010], this volume), quickly evaporated from the impervious surfaces, or, in vegetated areas, infiltrates the soil thus contributing to groundwater supplies or is evapotranspired. In cities with little vegetation 60% of the precipitation is directed to storm water drains to surface water (Chapter 14, Burian and Pomeroy, 2010). Modern humans recognized the need to keep water clean and produced the Clean Water Act 1972 to "restore and maintain the chemical, physical and biological integrity of the nation's waters so that they can support the protection and propagation of fish, shellfish, and wildlife and recreation in and on the water." However, the difference in surface water quality in human-dominated ecosystems, and thus the service it provides, compared with the water quality in an upland forested ecosystem cannot be compared in terms of quality for drinking water (Chapter 15, Steele et al., 2010, this volume). Human demand for water has increased in the past few decades because of population growth and dietary changes (Postel and Carpenter, 1997). There is no substitute for fresh water. Although the technology exists for the desalinization of sea water, this technique is energy intensive and would increase the cost of freshwater supplies by four to eight times the cost of current urban supplies (Postel and Carpenter, 1997).

Animals

Animals are sometimes considered beneficial and sometimes considered pests. The domesticated animal such as the cat, dog, bird, or rat for the most part enjoys a better life than its peers, or so we like to think. While domesticated animals benefit humans by reducing stress and encouraging exercise, they also have other important roles. For example, dogs have been used by humans for centuries for hunting and herding. The modern dog performs many duties, including assistance for blind and deaf humans. Social or therapy dogs provide unconditional love to humans who cannot have a dog because of disability, illness, or age. Military dogs provide a service by acting as a sentry to warn of the presence or approach of strangers, scouts in detecting snipers or enemy forces, or mine-dogs trained to detect trip wires, booby traps, metallic, and non-metallic mines. Dogs are also employed in search and rescue and explosives detection. Police dogs are trained to sniff out drugs, explosives, and firearms. Other domesticated animals are used for food, including cattle, pigs, goats, chickens, and lambs.

Wildlife, or undomesticated, animals benefit humans in an urban ecosystem too in many ways. For example, insects pollinate urban landscapes and human-influenced ecosystems such as agricultural fields and orchards. Bees are responsible of the pollination of 15 to 30% of the food humans eat in the USA. The recent decline of bee species (e.g., Otterstatter and Thomson, 2008) may be the result of habitat loss, but invasive species, emerging disease, pesticide use, and climate change may also have the potential to reduce bee populations (Brown and Paxton, 2009).

McCleery (Chapter 5, 2010, this volume) suggested that a new urban wildlife might be evolving given the abundant food in garbage, den places in urban structures, and the lack of predation. Most urban wildlife mammals are nocturnal and so are rarely seen by humans in urban ecosystems. Rarely do mammals such as bears interact with humans in the urban core, but in suburban and rural areas, lack of food in the wild has encouraged them to visit trash cans and bird feeders. The most common encounters between humans and wildlife in the urban ecosystem is that between human and insects, and it is usually not a beneficial interaction for either. Examples in the urban core include mosquitoes, which breed in stagnant pools of water such as containers that can collect rain or irrigation water in homeowner gardens. These containers include clogged gutters, leaf-filled drains, drain outlets from air-conditioners, plastic wading pools, dog dishes, soft drink cans, plastic bags, old tires, bird baths, potted plant saucers, and standing water in tire ruts, stumps, and tree holes. Encephalitis in various forms such as the West Nile virus, which is carried by mosquitoes and can be transferred to humans, is endemic to the United States and increasing in incidence.

Climate

Climate in terms of precipitation and sunlight provides a much needed service to humans for water supply and sunlight for warmth and photosynthetic reactions for food. Other climatic factors that provide a service include wind which provides the service of seed dispersal and crop pollination of basic cereal crops (Sagoff, 2008).

As a potential renewable energy source the climate provides precipitation, which in turns enables the use of hydropower systems, and more recently the technology to use solar radiation and wind power. Selection of regions where solar, wind, or hydropower would be most successful will be key to the success of renewable energy. Hydropower as an energy source has been used for many decades (Baun, 1919). It is estimated that one-fifth of the world's energy is generated by hydropower. Human-induced climate change may result in spatial variation in the distribution and volume of precipitation, rendering some regions better equipped to use hydropower than others. Solar power may eventually be another form of bulk energy input into human ecosystems (Wang and Qiu, 2009). However, the quantity, intensity, and temporal distribution of solar irradiation may affect electricity production costs considerably (Tapiador, 2009). Solar power is useful in many situations; for example, solar-driven reverse osmosis for water supply is proving very cost effective in many remote areas (Ghermandi and Messalem, 2009). The energy of the wind may well be the answer to energy supplies in many areas globally (Changliang and Zhanfeng, 2009). Based on global average annual wind speeds, there is the potential to generate 106 million gigawatt-hours of electricity a year from wind, which is five times the total amount of electricity generated at the global scale (Edwards, 2008). The Global Wind Energy Council based in Belgium predicts that the global wind market will grow more than 15% by 2012 (Edwards, 2008). Currently Great Britain, Germany, and Denmark have large offshore wind programs (Tapiador, 2009). The wake effect of offshore wind farms is such that future planning needs to account for decreasing wind power from nearby wind farms (Tapiador, 2009). Benefits of wind energy also include environmental cooling (Tapiador, 2009).

Humans have modified the climate by increasing CO_2 in the atmosphere through burning fossil fuels and clearing forests for development. Climate change refers to a change in climate with time, whether due to natural variability or as a result of human activity (IPCC, 2007). There is growing evidence that climate change is affecting hydrological systems, including increased runoff and earlier spring peak discharge in many glacier- and snow-fed rivers, and that spring events are occurring earlier, such as leaf-out, bird migration and egg laying, and poleward and upward shifts in the ranges of vegetation and animal species (IPCC, 2007).

Geology

Geological formations are important to the human as a resource for building materials, fuels, and decoration of either the human habitat or the human self. Granite, gneiss, marble, limestone, sandstone, and slate are the most often used rocks for building homes, churches, and government buildings in city centers. For example, Hot Springs in South Dakota, USA is almost entirely built of red sandstone; the Tennessee State Supreme court building is constructed of an Ordovician limestone. The U.S. post office in Athens, GA, USA is built of marble.

Coal and lignite have been used for centuries for home heating in Europe and in coal-fired power stations for the generation of electricity globally. There are approximately 650 coal-fired power stations in the United States, with the state of Ohio boasting 119 stations generating 65% of the total electrical generating capacity in the state. China is the world's largest consumer of coal, with 69% of its electricity from coal-fired operations. The Middle East is the world's smallest consumer of coal. Reliance on coal appears to be waning in Europe and Eurasia, increasing significantly in Asia, and increasing steadily in the USA (Fig. 18–1). Energy, in the form of geothermal energy is also trapped in geological deposits and is currently used in several countries, such as Japan, Iceland, and New Zealand, as a form of renewable energy (Smith, 2008). It is the only renewable energy source that is unaffected by climate change (Tapiador, 2009). Although the major hotspots are known, satellite identification of lower grade geothermal sources is an emergent research field (Tapiador, 2009).

Metals are also mined for a multitude of human uses. For example, gold is used as jewelry or coinage and in electronics, dentistry, medicine, aerospace industry, glassmaking, gilding, and goldleaf. Metals are essential components in human-dominated ecosystems. Important systems rely on one or multiple metals—transportation systems, including highways, bridges, railroads, airports and vehicles; homes and buildings, including construction materials, electrical utilities, and plumbing; and food production and distribution.

While the majority of mining operations are not contained within the traditional urban ecosystem setting, the impact of humans on other ecosystems in the search for

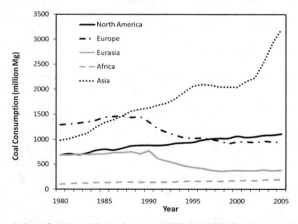

Fig. 18–1. Consumption of coal worldwide between 1980 and 2006. Coal includes anthracite, subanthracite, bituminous, subbituminous, lignite, brown coal, and for Estonia, oil shale (adapted from U.S. Energy Information Administration, 2008).

geological materials is well known. For example, acid mine drainage formed through the weathering of iron pyrite (Evangelou and Zhang, 1995). Oil and gas abstraction contaminates ground and surface waters with saline water (Pettyjohn, 1973; Murphy et al., 1988). Drilling of wells for ground water, which if too intensely pumped, will result in a cone of depression so that surface waters have no baseflow or surface water will flow into the aquifer (Sophocleous, 2002). Mining of coral reef limestone for lime and cement manufacture is a threat to coral reef habitats (Clark and Edwards, 1995) resulting in reduced fish abundance and diversity and loss of shoreline and mangrove forests through increased erosion rates (Dulvy et al., 1995).

More recently geology has provided an important service to humans though injection of surplus CO_2 into geological formations (Kharaka et al., 2006; Bachu, 2008). Sedimentary basins have the characteristics necessary for accepting injected CO_2 (Bachu, 2008). While this technology will help ameliorate atmospheric CO_2 concentrations, it does not come without risk. Injection pressure may result in geomechanical effects such as opening of preexisting bedrock fracture, rock fracturing or fault activation, and induced microseismicity or localized earthquakes (Bachu, 2008). Furthermore, leakage of injected CO_2 may pose risks because of its potential consequences. Carbon dioxide injection into a 24-m-deep sandstone of a regional brine and oil reservoir in the U.S. Gulf coast resulted in a sharp drop in pH, which led to a pronounced increase in alkalinity and iron. Potential problems include dissolution of carbonate minerals leading to leaks of brine and toxic compounds (Kharaka et al., 2006).

Changing Ecosystem Services

Probably the biggest change in the service an ecosystem provides to the human is seen in soil. Biotic soil components have been so affected by humans that the "normal" service provided has changed drastically. Chapin et al. (2000) stated

> Human alteration of the global environment has triggered the sixth major extinction event in the history of life and caused widespread changes in the global distribution of organisms. These changes in biodiversity alter ecosystem processes and change the resilience of ecosystems to environmental change. This has profound consequences for services that humans derive from ecosystems. The large ecological and societal consequences of changing biodiversity should be minimized to preserve options for future solutions to global environmental problems.

Chapin et al. (2000) suggested that changing diversity ranged from genetic diversity within organisms to the diversity of the landscape.

Deposition or dumping of heavy metals onto urban soils may decrease microbial diversity and function. Kandeler et al. (1995) introduced different metals onto three different soil types in a pot study to quantify microbial enzyme function. Enzyme functions differed—enzymes involved in C cycling were least affected, but those involved in cycling of nitrogen, phosphate, and sulfate were considerably decreased (Kandeler et al., 1995). Kaye et al. (2005) examined C and N cycling adjacent to urban, agricultural, and unmanaged shortgrass steppe soils in Colorado. They reported that belowground C allocation was between 2.5 and 5 times higher under urban soils than the other soils tested. Also, measuring microbial community structure, they suggested that land use type had a large impact on microbial biomass but a small impact on the relative abundance of the taxonomic groups of microorganisms (Kaye et al., 2005). Deposition of hydrocarbons in urban ecosystems may also limit activity of soil microbes and invertebrates (White and McDonnell, 1988).

Cost of a Service

In our discussion of ecosystems services it is clear that ecosystem functions require an exchange of energy. One good example is the soil microbe–plant relationship, where both benefit from the relationship. Soil processes such as mineralization provide nutrients to the plants, while the presence of the plant stimulates soil processes via the exudation of energy-rich carbon compounds (Kuzyakov, 2002). Animals, including humans, spend energy to acquire enough energy (in the form of food) to survive. For example, a carnivorous mammal may stalk prey all day to eat, or a herbivore may forage all day to consume enough energy- and protein-rich food to survive. The cost to acquire this food is the energy expended in hunting or foraging for it. Before the industrial revolution a greater proportion of the human population lived in rural areas where they farmed and hunted to provide food for their family. Energy was expended through work and gained through eating the product of their labor.

As humans have evolved through time, money, not personal energy, is the currency more-often used in developed countries. The world wildlife fund (WWF) is leading the development and approach to secure financing for conservation termed "Payment for Environmental Services" or PES. For this system, costs and benefits are converted into economic value. The system requires three steps—(i) an assessment of the ecosystem services that flow from a particular area, (ii) identification of who benefits, and (iii) an estimate of the economic value of this benefit—whereby the landowner is rewarded for conserving the source of the ecosystem service. The actual cost of ecosystem services and how these costs might be decided is described very well by Sagoff (2008). Although rather "tongue-in-cheek" in places, Sagoff's paper is well worth reading and really puts the service an ecosystem provides into context.

A change in thinking is needed. Ecosystem services should be thought of as multiple resources that interact, and the individual resource and interactions among components should be conserved. Ecosystem services should not be considered a service that the planet will provide indefinitely. If this change in thinking is not made, there will be a decline in that resource and hence service, and no amount of monetary or technological influence will be able to repair it. Human wants rather than needs have led to a situation of more waste than the urban and surrounding rural ecosystems can handle. Ecosystem services provided by climate, geological formations, soil, water, vegetation, and animals have been fully strained and can no longer provide the full range of benefits that humanity has enjoyed in the past. Several questions remain. Should planners build urban centers with a smaller footprint and therefore less time to travel from the urban ecosystem to a relatively undisturbed natural ecosystem? Should we build urban ecosystems in sprawling forms, such as the Fort Worth/Dallas or Houston cities, and within them incorporate ample greenspace, wildlife corridors, and water features for the human and wildlife residents? Should ecosystems be conserved so that the services they provide continue? Finally, if the services within an ecosystem fail, can we, as humans, use money and technology to replace those services?

References

Aitkenhead-Peterson, J.A., W.H. McDowell, and J.C. Neff. 2003. Sources, production, and regulation of allochthonous dissolved organic matter to surface waters. p. 26–70. In S. Findlay and R.L. Sinsabaugh (ed.) Aquatic ecosystems: Interactivity of dissolved organic matter. Academic Press, San Diego, CA.

Bachu, S. 2008. CO_2 storage in geological media: Role, means, status and barriers to deployment. Prog. Energ. Combust. Sci. 34:254–273.

Baun, E.. 1919 Reservoirs for low pressure-hydropower plants, their economic viability and their coal saving value. Z. Vereines Deutsch. Ing. 63:856–860.

Beer, A.R. 2010. Greenspaces, green structure, and green infrastructure planning. p. 431–448. In J. Aitkenhead-Peterson and A. Volder (ed.) Urban ecosystem ecology. Agron. Monogr. 55. ASA, CSSA, and SSSA, Madison, WI.

Biswas, A.K. 1987. Inland waterways for transportation of agricultural, industrial and energy products. Int. J. Water Res. Dev. 3:9–22.

Bjorklund, J., K. Limburg, and T. Rydberg. 1999. Impact of production intensity on the ability of agricultural landscape to generate ecosystem services: An example from Sweden. Ecol. Econ. 29:269–291.

Bolund, P., and S. Hunhammar. 1999. Ecosystem services in urban areas. Ecol. Econ. 29:293–301.

Botkin, D.B., and C.E. Beveridge. 1997. Cities as environments. Urban Ecosyst. 1:3–19.

Bowden, R.D., K.J. Nadelhoffer, R.D. Boone, J.M. Melillo, and J.B. Garrison. 1993. Contributions of aboveground litter, belowground litter and root respiration to total soil respiration in a temperate mixed hardwood forest. Can. J. For. Res. 23:1402–1407.

Brown, M.J.F., and R.J. Paxton. 2009. The conservation of bees: A global perspective. Apidologie (Celle) 40:410–416.

Burian, S.J., and C.A. Pomeroy. 2010. Urban impacts on the water cycle and potential green infrastructure implications. p. 277–296. In J. Aitkenhead-Peterson and A. Volder (ed.) Urban ecosystem ecology. Agron. Monogr. 55. ASA, CSSA, and SSSA, Madison, WI.

Cai, Y., N. Zhu, and H. Han. 2002. Photosynthesis and transpiration of urban tree species in Harbin. p. 186–194. In X. He and Z. Ning (ed.) Advances in urban forest ecology. (In Chinese.) China Forestry Publishing House, Beijing.

Campbell, E.K. 1983. Beyond anthropocentrism. J. Hist. Behav. Sci. 19:54–67.

Chen, Z. 1998. The analysis of current composition of urban forests in Beijing. (In Chinese.) J. Chinese Landscape Arch. 14:58–60.

Chen, Z., X. Su, S. Liu, R. Gu, and Y. Li. 1998. A study of ecological benefits of urban forests in Beijing (2). (In Chinese.) J. Chinese Landscape Arch. 14:51–53.

Changliang, X., and S. Zhanfeng. 2000. Wind energy in China: Current scenario and future perspectives. Renew. Sustain. Energ. Rev. 13:1966–1974.

Chameides, W.L., R.W. Lindsay, J. Richardson, and C.S. Kiang. 1988. The role of biogenic hydrocarbons in urban photochemical smog: Atlanta as a case study. Science 241:1473–1475.

Chapin, F.S., III, E.S. Zavaleta, V.T. Eviners, R.L. Naylor, P.M. Vitousek, H.L. Reynolds, D.U. Hooper, S. Lavorel, O.E. Sala, S.E. Hobbie, M.C. Mack, and S. Diaz. 2000. Consequences of changing biodiversity. Nature 405:234–242.

Clark, S., and A.J. Edwards. 1995. Use of artificial reef structures to rehabilitate reef flats degraded by coral mining in the Maldives. Bull. Mar. Sci. 55:724–744.

Costanza R, R. d'Arge, R. de Groot, S. Farber, M. Grasso, B. Hannon, K. Limburg, S. Naeem, R.V. O'Neill, J. Paruelo, R.G. Raskin, P. Sutton, and M. van den Belt. 1997. The value of the world's ecosystem services and natural capital. Nature 387:253–260.

Daily, G.C. (1997) Nature's services. Social dependence on natural ecosystem services. Island Press. Washington, DC.

Daily, G.C. 1997b. Valuing and safeguarding earth's life-support system. p. 365–374. In G.C. Daily (ed.) Nature's services. Social dependence on natural ecosystem services. Island Press, Washington, DC.

Ding, X. 2007. Ecological effects of the main plant species in urban forest. (In Chinese.) J. Central South Univ. For. Technol. 27:142–146.

Dulvy, N.K., D. Stanwell-Smith, W.R.T. Darwall, and C.J. Horrill. 1995. Coral mining at Mafia Island, Tanzania: A management dilemma. Ambio 24:358–365.

Edwards, R. 2008. Renewable energy: Anywhere the wind blows. New Scientist 2677: 32–35.

Ehrlich, P.R., and A.H. Erlich. 1992. The value of biodiversity. Ambio 21:219–226.

Elsgaard, L., and L.W. Jorgensen. 2002. A sandwich-designed-temperature-gradient incubator for studies of microbial temperature responses. J. Microbiol. Methods 49:19–29.

Evangelou, V.P., and Y.L. Zhang. 1995. A review: Pyrite oxidation mechanisms and acid mine drainage prevention. Crit. Rev. Environ. Sci. Technol. 25:141–199.

Ewel, K.C. 1997. Water quality improvement by wetlands. p. 329–344. In G.C. Daily (ed.) Natures services. Societal dependence on natural ecosystems. Island Press. Washington, DC.

Frolke, C., A. Jansson, J. Larsson, and R. Costanza. 2007. Ecosystem appropriation of cities. Ambio 26:167–172.

Ghermandi, A., and R. Messalem. 2009. Solar-driven desalination with reverse osmosis: The state of the art. Desal. Water Treat. 7:285–296.

Givoni, B. 1991. Impact of planted areas on urban environmental quality: A review. Atmos. Environ. 25B(3):289–299.

Grimmond, S., C. Souch, R. Grant, and G. Heisler. 1994. Local scale energy and water exchange in a Chicago neighborhood. p. 41–61. In E.G. McPherson et al. (ed.) Chicago's urban forest ecosystem: Results of the Chicago urban forest climate project. General Tech. Rep. NE-186. USDA Forest Service Northeastern Forest Exp. Stn., Radnor, PA.

Hardin, G.J. 1968. Tragedy of the commons. Science 162:1243–1248.

He, X., Z. Hu, Y. Jin, and W. Chen. 2003. The structure of urban forest in Shenyang city and their ecological benefits. (In Chinese.) J. Chinese Urban Forest. 1:25–32.

Hettinger, N., and W. Throop. 1999. Refocusing ecocentrism. Environ. Ethics 21:3–21.

Herrera, C.M. 1989. Seed dispersal by animals—A role in angiosperm diversification. Am. Nat. 133:309–322.

Herrington, L.P. 1976. Effect of vegetation on the propagation of noise in the outdoors. p. 229–233. In USDA Forest Service General Tech. Rep. 25. U.S. Rocky Mountain Forest Range Exp. Stn., Fort Collins, CO.

Hough, M. 1989. City form and natural process. Routledge, London.

Iaquinta, D.L., and A.W. Drescher. 2010. Urban agriculture: A comparative review of allotment and community gardens. p. 199–226. In J. Aitkenhead-Peterson and A. Volder (ed.) Urban ecosystem ecology. Agron. Monogr. 55. ASA, CSSA, and SSSA, Madison, WI.

IPCC. 2007. Climate Change 2007: Synthesis report. Contribution of Working Groups I, II, and III to the Fourth Assessment Report of the Intergovernmental Panel on Climate Change. IPCC, Geneva, Switzerland.

Jenerette, G.D., and K.P. Alstad. 2010. Water use in urban ecosystems: Complexity, costs, and services of urban ecohydrology. p. 353–372. In J. Aitkenhead-Peterson and A. Volder (ed.) Urban ecosystem ecology. Agron. Monogr. 55. ASA, CSSA, and SSSA, Madison, WI.

Jim, C.Y., and W.Y. Chen. 2009. Ecosystem services and valuation of urban forests in China. Cities 26:187–194.

Johnson, C.A. 1991. Sediment and nutrient retention by freshwater wetlands: Effects on surface water quality. Crit. Rev. Environ. Sci. Technol. 21:491–565.

Kandeler, F., C. Kampichler, and O. Horak. 1995. Influence of heavy metals on the functional diversity of soil microbial communities. Biol. Fertil. Soils 23:299–306.

Kaye, J.P., R.L. McCulley, and I.C. Burke. 2005. Carbon fluxes, nitrogen cycling, and soil microbial communities in adjacent urban, native and agricultural ecosystems. Glob. Change Biol. 11:575–587.

Kharaka, Y.K., D.R. Cole, S.D. Hovorka, W.D. Gunter, K.G. Knauss, and B.M. Freifeld. 2006. Gas–water–rock interactions in Frio Formation following CO_2 injection: Implications for the storage of greenhouse gases in sedimentary basins. Geology 34:577–580.

Koltowski, Z., S. Pluta, B. Jablonski, and K. Szlanowska. 1999. Pollination requirements of eight cultivars of black current (*Ribes nigrum* L.). J. Hort. Sci. Technol. 74:472–474.

Kuzyakov, Y. 2002. Review: Factors affecting rhizosphere priming effects. Z. Pflanzenernahr. Bodenkd. 165:382–396.

Leng, P., X. Yang, F. Su, and B. Wu. 2004. Economic valuation of urban greenspace ecological benefits in Beijing city. (In Chinese.) J. Beijing Agric. Col. 19(4):25–28.

Lennon, J.T., Z.T. Aanderud, and C.A. Klausmeier. 2008. Maintenance of microbial diversity in soils: Assessing the importance of habitat heterogeneity and physiological stress with theory and experiments. Abstract OOS 5-5. 93rd E.S.A. Annual Meeting, Milwaukee, WI. 3–8 Aug. 2008.

Leopold, A. 1949. A sand county almanac. Oxford Univ. Press, New York.

Li, M.-H., B. Dvorak, and C.Y. Sung. 2010. Bioretention, low impact development, and stormwater management. p. 413–430. *In* J. Aitkenhead-Peterson and A. Volder (ed.) Urban ecosystem ecology. Agron. Monogr. 55. ASA, CSSA, and SSSA, Madison, WI.

Li, F., R. Wang, X. Liu, and X. Zhang. 2005. Urban forest in China: Development patterns, influencing factors and research prospects. Int. J. Sustain. Dev. World Eco. 12:197–204.

Lilliesköld, M., and J. Nilsson. 1997. Kol i marken: Konsekvenser av markanvändning i skogs- och jordbruk. (In Swedish with English summary.) Naturvårdsverket, Rap. 4789.

Lu, G., Z. Yin, J. Gu, D. Meng, H. Wu, and Y. Li. 2006. A research on the function of fixing carbon and releasing oxygen of afforestation trees along the main road in Dalian city. (In Chinese.) J. Agric. Univ. Hebei 29:49–51.

Ma, J.K.-C., R. Chikwamba, P. Sparrow, R. Fische, R. Mahoney, and R.M. Twyman. 2005. Plant derived pharmaceuticals– the road forward. Trends Plant Sci. 10:580–585.

McCleery. 2010. Urban mammals. p. 87–102. *In* J. Aitkenhead-Peterson and A. Volder (ed.) Urban ecosystem ecology. Agron. Monogr. 55. ASA, CSSA, and SSSA, Madison, WI.

McPherson, E.G., D. Nowak, G. Heisler, S. Grimmond, C. Souch, R. Grant, and R. Rowntree. 1997. Quantifying urban forest structure, function, and value: The Chicago Urban Forest Climate Project. Urban Ecosyst. 1:49–61.

McPherson, E.G., K.I. Scott, and J.R. Simpson. 1998. Estimating cost effectiveness of residential yard trees for improving air quality in Sacrament, California, using existing models. Atmos. Environ. 32:75–84.

Miller, R.W. 1997. Urban forestry: Planning and managing urban green spaces. 2nd ed. Prentice Hall, Upper Saddle River, NJ.

Murphy, E.C., A.E. Kehew, and W.A. Groenewold. 1988. Leachate generated by an oil-and-gas brine pond site in North Dakota. Ground Water 26:31–38.

Myers, N. 1997. The world's forests and their ecosystem services. p. 215–236. *In* G.C. Daily (ed.) Nature's services. Social dependence on natural ecosystem services. Island Press, Washington, DC.

Nabhan, G.P., and S.L. Buchmann. 1997. Services provided by pollinators. p. 133–150. *In* G.C. Daily (ed.) Nature's services: Social dependence on natural ecosystem services. Island Press, Washington, DC.

Naturvardsverket. 1996. Vagtrafikbuller. Nordiska berakningsmodeller (Roadnoise. Nordic calculation models). (In Swedish.) Rep. 4653. Stockholm.

Nowak, D.J. 1994. Air pollution removal by Chicago's urban forest. p. 63–82. *In* E.G. McPherson et al. (ed.) Chicago's urban forest ecosystem: Results of the Chicago urban forest climate project. General Tech. Rep. NE-186. USDA Forest Service Northeastern Forest Exp. Stn., Radnor, PA.

Odum, E.P. 1993. Ecology and our endangered life-support system. Sinauer Associates, Sunderland, MA.

Otterstatter, M.C., and J.D. Thomson. 2008. Does pathogen spillover from commercially reared bumble bees threaten wild pollinators? PLoS One 3:E2771.

Pathak, V., B.D. Tripathi, and V.K. Mishra. 2008. Dynamics of traffic noise in a tropical city Varanasi and its abatement through vegetation. Environ. Monit. Assess. 146:67–75.

Paoletti, M.G., D. Pimentel, B.R. Stinner, and D. Stinner. 1992. Agroecosystem biodiversity: Matching production and conservation biology. Agric. Ecosyst. Environ. 40:3–24.

Pettyjohn, W.A. 1973. Hydrogeologic aspects of contamination by high chloride wastes in Ohio. Water Air Soil Pollut. 2:35–48.

Pimentel, D. 1988. Industrialized agriculture and natural resources. In P.R. Ehrlich and J.P. Holdren (ed.) The Cassandra Conference. Resources and the human predicament. Texas A&M Univ. Press, College Station, TX.

Postel, S., and S. Carpenter. 1997. Freshwater ecosystem services. p. 195–214. In G.C. Daily (ed.) Nature's services. Social dependence on natural ecosystem services. Island Press, Washington, DC.

Pouyat, R.V., K. Szlavecz, I.D. Yesilonis, P.M. Groffman, and K. Schwarz. 2010. Chemical, physical, and biological characteristics of urban soils. p. 119–152. In J. Aitkenhead-Peterson and A. Volder (ed.) Urban ecosystem ecology. Agron. Monogr. 55. ASA, CSSA, and SSSA, Madison, WI.

Rillig, M.C., S.F. Wright, and V.T. Eviner. 2002. The role of arbuscular mycorrhizal fungi and glomalin in soil aggregation: Comparing effects of five plant species. Plant Soil 238:325–333.

Rowe, D.B., and K.L. Getter. 2010. Green roofs and garden roofs. p. 391–412. In J. Aitken-head-Peterson and A. Volder (ed.) Urban ecosystem ecology. Agron. Monogr. 55. ASA, CSSA, and SSSA, Madison, WI.

Sagoff, M. 2008. On the economic value of ecosystem services. Environ. Values 17:239–257.

Santosa, S.J. 2010. Urban air quality. p. 57–74. In J. Aitkenhead-Peterson and A. Volder (ed.) Urban ecosystem ecology. Agron. Monogr. 55. ASA, CSSA, and SSSA, Madison, WI.

Schupp, E.W. 1993. Quantity, quality and the effectiveness of seed dispersal by animals. Vegetatio 108:15–29.

Smith, J. 2008. Renewable energy: Power beneath our feet. New Scientist 2677.

Sophocleous, M. 2002. Interactions between groundwater and surface water: The state of the science. Hydrogeol. J. 10:52–67.

Steele, M.K., W.H. McDowell, and J.A. Aitkenhead-Peterson. 2010. Chemistry of urban, suburban, and rural surface waters. p. 297–340. In J. Aitkenhead-Peterson and A. Volder (ed.) Urban ecosystem ecology. Agron. Monogr. 55. ASA, CSSA, and SSSA, Madison, WI.

Sulzman, E.W., J.B. Brant, R.D. Bowden, and K. Lajtha. 2005. Contribution of aboveground litter, belowground litter and rhizophere respiration to total soil CO_2 efflux in an old growth coniferous forest. Biogeochemistry 73:231–256.

Sweeney, B.W., T.L. Bott, J.K. Jackson, L.A. Kaplan, J.D. Newbold, L.J. Standley, W.C. Hession, and R.J. Horwitz. 2004. Riparian deforestation, stream narrowing, and loss of stream ecosystem services. Proc. Natl. Acad. Sci. USA 101:14132–14137.

Tansley, A.G. 1935. The use and abuse of vegetational concepts and terms. Ecology 16:284–307.

Tansley, A.G. 1939. British ecology during the past quarter-century: The plant community and the ecosystem. J. Ecol. 27:513–530.

Tapiador, F.J. 2009. Assessment of renewable energy potential through satellite data and numerical models. Energ. Environ. Sci. 2:1121–1220.

Transek. 1993. Vardering av miljo faktorer (Valuation of environmental aspects). (In Swedish.) Transek AB, Solna.

Ulrich, R.S. 1984. View through a window may influence recovery from surgery. Science 224:420–421.

Ulrich, R.S. 1991. Effects of health facility interior design on wellness: Theory and scientific research. J. Health Care Des. 3:97–109.

U.S. Energy Information Administration. 2008. World energy consumption. Available at http://www.eia.doe.gov/emeu/international/coalconsumption.html (verified 22 May 2010).

USEPA. 2009. National primary drinking water regulations. Available at http://www.epa.gov/safewater/consumer/pdf/mcl.pdf (verified 26 Mar. 2010).

Vanderwall, S.B. 1992. The role of animals in dispersing a wind-dispersed pine. Ecology 73:614–621.

Vitousek, P.M. 1994. Beyond global warming: Ecology and global change. Mac Arthur Award Lecture. Ecology 74:1861–1876.

Vitousek, P.M., P.R. Ehrlich, A.H. Ehrlich, and P.A. Matson. 1986. Human appropriation of the products of photosynthesis. Bioscience 36(6):368–373.

Volder, A. 2010. Urban plant ecology. p. 179–198. *In* J. Aitkenhead-Peterson and A. Volder (ed.) Urban ecosystem ecology. Agron. Monogr. 55. ASA, CSSA, and SSSA, Madison, WI.

Volder, A., and W.T. Watson. 2010. Urban forestry. p. 227–240. *In* J. Aitkenhead-Peterson and A. Volder (ed.) Urban ecosystem ecology. Agron. Monogr. 55. ASA, CSSA, and SSSA, Madison, WI.

Wang, M. 1995. The research and development of urban forestry. (In Chinese.) Sci. Silvae Sinicae 31:460–466.

Wang, Q., and H.N. Qiu. 2009. Situation and outlook of solar energy utilization in Tibet, China. Ren. Sustain. Energ. Rev. 13:2181–2186.

Wang, X., Y. Hu, X. Liu, F. Gao, and Q. Zhu. 2007. Microclimate modification of urban grassland in Beijing. (In Chinese.) J. Guangxi Normal Univ. Nat. Sci. 25:23–27.

White, C.S., and M.J. McDonnell. 1988. Nitrogen cycling processes and soil characteristics in an urban versus rural forest. Biogeochemistry 5:243–262.

Winer, A.M., D.R. Fitz, and P.R. Miller. 1983. Investigation of the role of natural hydrocarbons in photochemical smog formation in California. Final Rep. Calif. Air Resources Bd., Sacramento, CA

Yang, S. 1996. Urban ecology. (In Chinese.) Science Press, Beijing.

Zhuang, G. 2003. Air quality, urbanization, and automobilization in China. (In Chinese.) Scientific Chinese 2003(8): 28–29.

19

Green Roofs and Garden Roofs

D. Bradley Rowe
Kristin L. Getter

Abstract

Implementing green roofs is one way to mitigate the many problems caused by population growth and urbanization. They impact the environmental, economic, and social issues of sustainable urban sites and serve as urban ecosystems that can partially offset some of the negative impacts of urban areas. Establishing plant material on rooftops provides numerous ecological and economic benefits, including stormwater management, energy conservation, mitigation of the urban heat island effect, carbon sequestration, increased longevity of roofing membranes, habitat for wildlife, mitigation of noise and air pollution, and a more aesthetically pleasing environment in which to work and live. This chapter will introduce how green roofs can improve urban ecology and how plant selection influences the magnitude of these benefits.

Green roofs, or vegetated roofs, involve growing plants on rooftops, thus partially replacing the vegetated footprint that was destroyed when the building was constructed. They are categorized as "intensive" or "extensive" systems depending on the design intent of the roof. *Intensive green roofs*, sometimes referred to as *garden roofs*, are designed to be similar to landscaping found at ground level, and as such require substrate depths greater than 10 cm (4 in). They are characterized by a variety of plant materials and are generally designed to be accessible (Fig. 19–1 and 19–2). In contrast, *extensive green roofs* have shallower substrate depths and require minimal maintenance. Because of their shallower media depth (<10 cm [4 in]), plant species are limited to herbs, grasses, mosses, and drought tolerant succulents such as *Sedum* species. In addition, extensive green roofs can be built on a sloped surface (Fig. 19–3 and 19–4). Due to building weight restrictions and costs, shallow substrate extensive green roofs are much more common than deeper intensive roofs (Dunnett and Kingsbury, 2004; Getter and Rowe, 2006; Oberndorfer et al., 2007; Snodgrass and Snodgrass, 2006).

D.B. Rowe (rowed@msu.edu) and K.L. Getter (smithkri@msu.edu), Dep. of Horticulture, Michigan State University, East Lansing, MI 48824.

doi:10.2134/agronmonogr55.c19

Fig. 19–1. An intensive green roof on the Schlossle Galerie in Pforzheim, Germany (photograph: Brad Rowe).

Fig. 19–2. An aerial view of the intensive and extensive green roof on the convention center of the Church of Jesus Christ of Latter-day Saints in Salt Lake City, UT (photograph: Brad Rowe).

Fig. 19–3. Portion of a 4.2-ha (10.4-acre) extensive green roof on an assembly plant at Ford Motor Company in Dearborn, MI. Plant material consists of 13 species and cultivars of *Sedum* (photograph: Brad Rowe).

Fig. 19–4. An extensive green roof on a T-Mobile office building in Stuttgart, Germany (photograph: Brad Rowe).

Stormwater Runoff Quantity and Quality

One of the most important ecological services that green roofs provide is the role they play in managing both the quantity and quality of stormwater runoff. As urban areas expand, constructed surfaces increase in the form of buildings, roads, and parking lots. These built structures are impervious to water since they cannot absorb precipitation. In the United States, it is estimated that 10% of residential developments and 71 to 95% of industrial areas and shopping centers are covered with impervious surfaces (Ferguson, 1998). Excessive runoff from these surfaces increases the likelihood that stormwater will exceed channel capacities, leading to flooding and ultimately property damage and human harm. High volumes of stormwater runoff can also overwhelm municipal sewer systems. The problems associated with impervious surfaces and stormwater are discussed elsewhere (Chapters 14 [Burian and Pomeroy, 2010], 15 [Steele et al., 2010], and 20 [Li et al., 2010], this volume.

In many areas of the world green roofs are typically implemented because of their ability to reduce stormwater runoff (Carter and Rasmussen, 2006; Liesecke, 1998; Mentens et al., 2006; Schade, 2000; Villarreal and Bengtsson, 2005). Modern green roofs became popular in Stuttgart, Germany in the early 1980s as a means to mitigate stormwater rather than investing tax dollars into expanding the municipal stormwater system. In a green roof system, much of the precipitation is captured in the media or vegetation and will eventually evaporate from the soil surface or be released back into the atmosphere by transpiration. Green roofs can reduce runoff by 50 to 100% depending on the type of green roof system (Carter and Jackson, 2007; DeNardo et al., 2005; Getter et al., 2007; Hilten et al., 2008; VanWoert et al., 2005a). Water retention depends on design factors such as substrate depth, composition, roof slope, and plant species, as well as preexisting substrate moisture and the intensity and duration of rainfall (Fig. 19–5). For example, for a 14-mo period, VanWoert et al. (2005a) reported that an extensive green roof retained 96, 83, and 52% of all rainfall for rain events measuring less than 2, 2 to 6, and >6 mm, respectively. In addition, runoff is delayed because it takes time for the media to become saturated and for the water to drain through the media. By releasing runoff over a

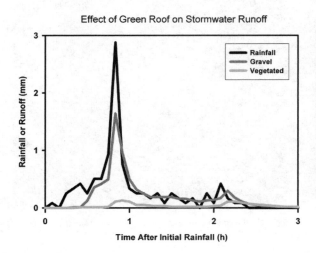

Fig. 19–5. Typical hydrograph from a 1.2-cm (0.5-in) October rain event in East Lansing, MI. Note the overall reduction in runoff, the reduced peak, and the time delay of runoff. Substrate composition and depth, plant selection, pre-existing substrate moisture, and the intensity and duration of the rain event all influence the stormwater retention capacity of a green roof. (Source: Michigan State University Green Roof Research Program, www.hrt.msu.edu/greenroof [verified 18 May 2010].)

longer period of time, green roofs can help keep stormwater sewer systems from overflowing and reduce potential erosion downstream.

In the United States there are over 700 communities using combined sewer systems (USEPA, 2008b), which consist of a single pipe carrying wastewater and stormwater to treatment plants. When stormwater exceeds capacity, the untreated sewage can flow out of relief points into rivers, an event termed a *combined sewage overflow* (CSO). In New York City, about half of all rainfall events result in a CSO event. These CSO events dump 40 billion gallons of untreated wastewater into New York's surface waters annually (Cheney, 2005). Even in communities with separate stormwater managements systems, impervious surfaces still contaminate waterways by collecting pollutants such as oil, heavy metals, salts, pesticides, and animal wastes. During runoff events, these contaminants may wash into waterways. Research supports the link between runoff from impervious surfaces and the reduction of water quality in streams.

Because of their filtering capacity, green roofs are generally attributed with improving the quality of the water that runs off them, especially when compared with a conventional roof (Berndtsson et al., 2006; Hathaway et al., 2008; USEPA, 2009). However, the effect of green roofs on water quality depends on management practices. Potential contaminants in the media components (primarily organic matter) and added fertilizers may result in excess nutrient runoff (Hathaway et al., 2008; Retzlaff et al., 2008; USEPA, 2009). Some cities have modeled green roof stormwater storage against building new stormwater sewer systems. In Toronto, implementing only 6% of roof surfaces as green roofs would impact stormwater retention equivalent to building a $60 million (CAD) storage tunnel (Peck, 2005). In Washington, DC, if 20% of buildings had green roofs, it would add more than 71 million L (19 million gallons) to the city's stormwater storage capacity and store approximately 958 million L (253 million gallons) of rainwater in an average year (Deutsch et al., 2005).

One might assume that an intensive green roof would retain more stormwater than an extensive green roof because of the deeper growing substrate and larger canopy to intercept rain, but this is not always the case. Since extensive green roofs are not designed for public visitation they tend to cover close to 100% of the area. In contrast, intensive green roofs often have a large portion of the roof consisting of pavers or other impervious surfaces that do not provide any benefits. An area consisting of pavers does not retain stormwater and actually hastens runoff compared to a gravel ballasted roof. They also are likely to absorb heat.

Energy Conservation and the Urban Heat Island

Because urbanization increases runoff, the quantity of water available for evaporative cooling is reduced. A great deal of incoming solar energy that would have been used to evaporate water is instead transformed into sensible heat (Barnes et al., 2001). The loss of vegetation in urban areas also results in less shading and evaporative cooling. Coupled with the heat-absorbing properties of impervious surfaces like conventional rooftops and concrete, the increased sensible heat load results in higher internal building temperatures and ambient air temperatures. Since the albedo of urban surfaces is generally 10% lower than the albedo of rural surfaces (Oliver, 1973), urban areas can have higher ambient air temperatures, a phenomenon known as the urban heat island effect (UHIE). (For more information on the UHIE see Chapters 1 [Shepherd et al., 2010] and 2 [Heisler and Brazel, 2010], this volume.)

According to the U.S. Environmental Protection Agency (USEPA, 2003), urban air temperatures can be up to 5.6°C (10°F) warmer than the surrounding country-side. Even night air temperatures are warmer because built surfaces absorb heat and radiate it back during the evening hours. In Berlin, temperatures on a clear windless night were reported to be 9°C (16°F) higher than in the countryside (Von Stulpnagel et al., 1990). In the summer, this translates into higher air condition-ing use. For every 0.6°C (1°F) increase in air temperature, peak utility load may increase by 2% (USEPA, 2003). But, the UHIE does more than just increase power usage. Excess heat can result in harm to human health, including physiological disruptions, organ damage, and death (USEPA, 2003). The U.S. Department of Health and Human Services Centers for Disease Control and Prevention esti-mated that for the 24-yr period between 1979 and 2003, 8015 deaths resulted from excessive heat exposure in the United States, which is more deaths than those from hurricanes, lightning, tornadoes, floods, and earthquakes combined during that time (Department of Health and Human Services, 2006). Green roofs can play a role in reducing the urban heat island. Conventional roof surfaces have albedos ranging from 0.05 to 0.25 (USEPA, 2005a), whereas albedos of green roofs gener-ally range from 0.7 to 0.85, depending on water availability (Gaffin et al., 2005). Covering 50% of roofs in Toronto with vegetation would cool the city by up to 3.6°F (Bass et al., 2003). This would not only reduce energy required to power air conditioners, but also would influence human health. In many areas the primary reason for choosing a green roof is due to the energy conservation they provide to the individual building. Green roofs shade and insulate buildings, which when combined with the evapotranspiration of the substrate and plant material reduce summer air temperatures just above a green roof as well as indoor temperatures underneath a green roof (Connelly and Liu, 2005; Santamouris et al., 2007; Sailor, 2008; Takebayashi and Moriyama, 2007; Fig. 19–6). However, the amount of shad-ing is highly dependent on the types of plants chosen, because leaf area index has a significant impact on the shading effect (Wong et al., 2003).

Green roofs reduced indoor temperatures 3 to 4°C (6–8°F) when outdoor tem-peratures were between 25°C and 30°C (77–86°F) (Peck et al., 1999). Every decrease in internal building air temperature of 0.5°C (1°F) may reduce electricity use for air conditioning by up to 8% (Dunnett and Kingsbury, 2004). Likewise, during a

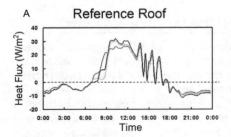
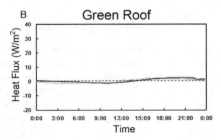

Fig. 19–6. Comparison of heat gain and loss through a standard conventional reference roof and a green roof. Green roofs can reduce energy requirements in buildings during the summer and to a lesser extent in the winter. (Source: Karen Liu, National Research Council, Institute for Research in Construction.)

30-d warm fall period in British Columbia, total heat flux through a reference roof and green roof was 2.634 kWh m^{-2} (0.245 kWh ft^{-2}) and 0.7 kWh m^{-2} (0.065 kWh ft^{-2}) respectively, a 70% reduction (Connelly and Liu, 2005). In Portland, OR, heat flux through a green roof was reduced 13% in winter and 72% in summer relative to a conventional gravel ballast roof (Spolek, 2008). Of course, heat flux depends any many factors such as media composition, depth, and moisture content; plant species; supplementary irrigation; local climate; and building type and construction details. Although green roofs reduce heat transfer through the roof, thus reducing the consumption of energy for heating and cooling the building, the roof/envelope ratio impacts green roof performance. In a study conducted in Toronto, a 250 by 250 m green roof with 50,000 W internal loading was found to save 73, 29, and 18% energy for a one-, two-, and three-story building, respectively (Martens et al., 2008). Most of the insulation benefits result from cooling during the summer months. Saiz-Alcázar and Bass (2005) reported reductions in energy for summer cooling of 6.2 to 6.4%, while the reduction in winter heating was only 0.12 to 0.2%. This is because the insulation properties of the substrate are greater when air space exists in the pores as opposed to when they are saturated, which is normally the case during winter. Also, large air conditioning systems require less power when intake air is cooler. When intake air exceeds 35°C (95°F), power requirements for the air conditioner increase, and cooling capacity drops (Leonard and Leonard, 2005). This decrease in roof temperature is one reason why photovoltaic solar panels were found to be 6% more efficient when placed over a green roof (Köhler et al., 2007; Fig. 19–7).

Fig. 19–7. Solar panels are more efficient when located over a green roof because of cooler roof surface temperatures (photograph: Brad Rowe).

Thus, the benefits obtained from green roofs may be greater in climatic regions with warmer temperatures and intense rain events relative to northern climates. Simmons et al. (2008) compared six different extensive green roof designs with conventional (black) roofs and reflective (white) roofs in Texas and found differences among all of them. On a warm day when ambient air temperature reached 33°C, the membranes on the black and white reflective roofs reached 68°C and 42°C, respectively, while membrane temperatures on the green roof systems ranged from 31 to 38°C (Simmons et al., 2008).

Since buildings consume 39% of total energy use and 71% total electricity consumption (U.S. Green Building Council Research Committee, 2007), green roof implementation on a wide scale could significantly impact energy consumption. Laberge (2003) estimated that for Chicago City Hall, energy savings alone could result in $4,000 annually for heating and cooling combined. If all of Chicago had green roofs, the savings could be $100,000,000 annually (Laberge, 2003).

Carbon Sequestration

Although green roofs are often adopted for energy savings and heat island mitigation, rarely has this technology been promoted for its ability to mitigate climate change, even though they do reduce carbon emissions. By lowering demand for electricity for air conditioning use, less carbon dioxide is released from power plants. Sailor (2008) integrated green roof energy balance into Energy Plus, a building energy simulation model supported by the U.S. Department of Energy. His simulations found 2% reductions in electricity consumption and 9 to 11% reductions in natural gas consumption. Based on his model of a generic building with a 2000-m^2 green roof, these annual savings ranged from 27.2 to 30.7 GJ of electricity saved and 9.5 to 38.6 GJ of natural gas saved, depending on climate and green roof design. When considering the national averages of CO_2 produced for generating electricity and burning natural gas (USEPA, 2007; USEPA, 2008a), these figures translate to 2.3 to 2.6 kg CO_2 m^{-2} of green roof in electricity and 0.24 to 0.97 kg CO_2 m^{-2} of green roof in natural gas each year.

The above estimates pertain only to the avoided energy use by reducing the flux of energy through the building shell. There are also indirect reductions of energy use due to the implementation of heat island reduction (HIR) strategies that achieve ambient cooling. Akbari and Konopacki (2005) simulated many different HIR strategies for a variety of climates and building types in the United States. While not specifically focusing on green roofs, they found that electricity savings due to indirect HIR consisted on average of 25% of energy consumption savings.

As an example of these potential emission savings using Sailor's (2008) model, the campus of Michigan State University in East Lansing, MI has 1.1 km^2 of flat roof surface. If all of these roofs were greened, they could avoid 3640,263 kg CO_2 emitted per year in electricity and natural gas consumption combined. This is the equivalent of taking 661 vehicles off the road each year (USEPA, 2005b). These figures depend on climate, green roof design, and source of fuel for electricity and gas. However, the components of a green roof have a CO_2 "cost" in terms of the manufacturing process. Embodied energy is a term used to describe total energy consumed, or carbon released, of a product over its life cycle. Many life cycle analysis studies ignore the unique components of a green roof by making the assumption that the waterproofing membranes, drainage layers, and substrate all have a carbon cost similar to the traditional roofs' gravel ballast (Alcázar and Bass, 2006; Kosareo and Ries, 2007). One cradle-to-gate study analyzed many building materials from the beginning (cradle) through the entire production process up until the product leaves the factory (gate) (Hammond and Jones, 2008). Assuming a generic industry root barrier, drainage layer, and 6.0 cm of substrate consisting of half sand and half heat expanded slate by volume, total embodied energy for the green roof components are 23.6 kg CO_2 m^{-2} of green roof. This carbon cost can be "made up" in approximately 10 yr via emission savings.

In addition, green roofs may also be a tool for further reducing carbon footprints due to the presence of plants and soils. The process of photosynthesis removes carbon dioxide from the atmosphere and stores carbon in plant biomass. Carbon is transferred to the substrate via plant litter and exudates. The length of time that this carbon remains in the soil before decomposition has yet to be quantified for green roofs, but immediately after initial green roof installation, net primary production should exceed decomposition making this human-made ecosystem a carbon sink, at least in the short term. Once equilibrium has been reached whereby carbon gain is offset by respiratory losses and decomposition, this ecosystem will likely no longer be sequestering new carbon, but will still be storing more carbon than the original barren roof. One study on a 6.0-cm substrate depth extensive green roof in Michigan with four *Sedum* species found that at the end of two growing seasons aboveground plant material and root biomass stored an average of 168 g C m^{-2} and 106.7 g C m^{-2}, respectively, with species influencing this value (Getter et al., 2009b). Substrate carbon content averaged 912.8 g C m^{-2}. In total, this entire extensive green roof system held 1187.4 g C m^{-2} in combined plant material and substrate.

Extended Lifespan of Roofing Membranes

A plant canopy and the growing substrate help shield traditional bituminous roofing membranes from ultraviolet radiation and wide day–night temperature fluctuations. For example, a study conducted in Toronto, Canada reported the membrane temperature on a convention roof reached 70°C (158°F) in the afternoon, whereas the roof membrane under the green roof was only 25°C (77°F) (Liu and Baskaran, 2003). The daily expansion and contraction of the roofing membrane due to swings in day–night temperatures stress the membrane, resulting in fatigue and eventual failure. Green roofs are estimated to last 45 yr or longer in terms of mechanical lifespan (Kosareo and Ries, 2007) compared to a conventional roof with a typical lifespan of 20 yr. An excellent example is the roof of the water treatment facility in Zurich, Switzerland, that was installed in 1914 and repaired for the first time in 2005, a period a 91 yr. The extended lifespan results in less waste that must go to a landfill and translates into a better return on investment than traditional roofs (Clark et al., 2008).

Noise Reduction

The vegetation and growing substrate on a green roof will absorb sound waves to a greater degree than a hard surface. Van Renterghem and Botteldooren (2008) reported that green roofs reduced noise pollution by 10 dB. Likewise, when comparing an extensive green roof to a conventional roof, Connelly and Hodgson (2008) reported a reduction in sound transmission of 5 to 13 dB over the low- and mid-frequency range (50–2000 Hz) and 2 to 8 dB at higher frequencies through the green roof. This can be favorable in buildings near airports, industrial areas, and in urban settings. Some of the health problems associated with excessive noise include hearing impairment, hypertension and ischemic heart disease, annoyance, sleep disturbance, and decreased school performance (Passchier-Vermeer and Passchier, 2000).

Mitigation of Air Pollution

It is common knowledge that vegetation can mitigate levels of air pollution by taking up gaseous pollutants through their stomates and intercepting particulate matter with their leaves. Green roofs have been found to reduce air pollution by 85 kg ha^{-1} yr^{-1} (Yang et al., 2008), and it is estimated that 2000 m^2 of unmown grass on a green roof can remove up to 4000 kg of particulate matter (Johnson and Newton, 1996). In practical terms, for approximately every 1.6 km (1 mile) driven, gasoline powered automobiles produce approximately 0.01 g of particulate matter. Assuming a vehicle is driven 16 093 km (10,000 miles per year, then 0.1 kg (0.22 lb) is spewed into the atmosphere. One square meter of green roof could offset the annual particulate matter emissions of one car (City of Los Angeles Environmental Affairs Department, 2006).

Regarding specific pollutants, sulfur dioxide and nitrous acid were reduced 37 and 21%, respectively, directly above a newly installed green roof (Yok Tan and Sia, 2005). At the University of Michigan, Clark et al. (2005) estimated that if 20% of all industrial and commercial roof surfaces in Detroit, MI were traditional extensive sedum green roofs, more than 800,000 kg (889 tons) per year of NO$_2$ (or 0.5% of that areas emissions) would be removed. In addition, an air quality model for greening all rooftops in Chicago predicted a reduction of 417,309.26 kg (460 tons) of nitrogen oxide and 517,100.61 kg (570 tons) of sulfur oxide emissions per year (Laberge, 2003).

Although trees found at ground level or on intensive garden roofs play a much larger role in improving air quality than grasses or succulents that are often found on extensive green roofs, the added loading requirements and cost of intensive roofs make it unlikely that they will be implemented on a large scale (Currie and Bass, 2008). Shallow green roofs can augment the urban forest, but cannot replace it. Even so, if only 20% of all existing "green roof ready" buildings in Washington, DC installed green roofs, the resulting plantings would remove the same amount of air pollution as 17,000 street trees (Deutsch et al., 2005).

Benefits to Human Health

Green roofs improve urban air quality and by extension public health and quality of life as air pollution can cost billions of dollars per year from hospital stays, employee absenteeism, and lost productivity. Nitrogen oxides (NO$_x$) result from combustion of fossil fuels and can form ground level ozone, causing serious human respiratory problems, including premature death, as well as a reduction in crop yields, all of which have economic impacts (USEPA, 1998). Clark et al. (2008) found that green roofs yield an annual benefit of $0.45 to $1.70 m^{-2}

($0.04–$0.16 ft^{-2}) in terms of NO$_x$ uptake per year. In addition, when humans view green plants and nature, it has beneficial health effects as well as improved health and worker productivity. Employees that had a view of nature were found to be less stressed, have lower blood pressure, report fewer illnesses, and experience greater job satisfaction (Kaplan et al., 1988; Ulrich, 1984).

Provide Habitat for Wildlife

Since rooftops are generally undisturbed areas they offer places where habitat can be created for microorganisms, insects, birds, and other animals (Baumann, 2006; Brenneisen, 2006). Constructing a green roof can provide a new habitat or a refuge in urban areas that are lacking diversity. They can also act as wildlife corridors linking existing habitats (Kadas et al., 2008). In a study of 17 green roofs in Basel, Switzerland, 78 spider and 254 beetle species were identified where 18% of those spiders and 11% of the beetles were listed as endangered or rare (Brenneisen, 2003). In addition, many birds have been recorded utilizing green roofs in Germany, Switzerland, and England (Brenneisen, 2003; Gedge, 2003). Even relatively new green roofs can provide habitat. The 42,900-m^2 (10.4-acre) green roof on the Ford Motor Company assembly plant in Dearborn, MI consists of a mix of 13 *Sedum* species planted in <7.6 cm (3 in) of media. Within 2 yr of initial plant establishment, 29 insect species, 7 spider species, and 2 bird species were identified on the Ford roof (Coffman and Davis, 2005; Fig. 19–8).

However, not all green roofs are created the same. Kadas et al. (2008) collected and quantified invertebrates (primarily spiders and beetles) on various types of green roofs during 3 yr and reported that substrate type and depth are the two major factors in terms of species abundance and diversity. The diversity of spiders was highest on the deeper biodiverse roofs and much lower on the shallow sedum-based roofs. Initially, sedum roofs generally supported a more abundant, but less diverse invertebrate population than biodiverse roofs. However, with time the invertebrate diversity builds up on a deeper substrate roof along with a diverse plant community. This is partly due to the lack of structural plant diversity of the sedum roofs. Structural diversity provides the microhabitats for various flora and fauna to live, which increases the overall

Fig. 19–8. A killdeer nesting on the Ford green roof in Dearborn, MI (photograph courtesy of Ford Motor Company).

Fig. 19–9. A log in combination with varying substrate depths and plant species provides wildlife habitat on a green roof in Sins, Switzerland (photograph: Brad Rowe).

diversity (Fig. 19–9). Invertebrates rely on the structural diversity of various species of plants, and the plant species present depends on substrate depth.

Because of weight and cost restraints, many green roofs contain a shallow substrate layer which limits the biodiversity of both plants and wildlife. With ecological services in mind, green roofs can be designed to provide habitat for rare and endangered species affected by land-use change (Brenneisen, 2008). In some locations, such as Basel, Switzerland, green roofs are mandatory on all new flat-roofed buildings, and habitat must be considered (Brenneisen, 2008). Design criteria include varying substrate depths and using local natural soils.

Ecological Importance of Plant Selection

The plant taxa selected have a major impact on the ecological services provided by a green roof (Getter and Rowe, 2006; Oberndorfer et al., 2007). They can influence stormwater retention, the water quality of stormwater runoff, energy savings, carbon sequestration, biodiversity, and aesthetic appeal, but selecting plants for green roofs is often difficult due to the harsh urban environment and because plants are also more susceptible to extremes in temperature and drought due to their shallow substrate and elevation above ground. Because of their deeper substrate depths, intensive roofs allow more numerous species choices. In contrast, drought tolerance is one of the most limiting factors in extensive green roof systems, given their shallow substrate depths (<10 cm) and usual reliance on natural precipitation events to sustain plant life. If the benefits of green roofs are to be realized, then green roof plant performance is extremely important.

Substrate Composition and Depth

Substrate composition influences plant selection. High levels of compost are not recommended because it will decompose and result in substrate shrinkage (Beattie and Berghage, 2004) and can result in increased N and P in the runoff (Hathaway et al., 2008). Also, it is not feasible or practical to continually replace the substrate on a rooftop. Substrate composition will depend on what materials are available locally and can be formulated for the intended plant selection, climatic zone, and anticipated level of maintenance.

Likewise, substrate depth also has a major impact on the plants that can be grown. Deeper substrates are necessary for woody species, grasses, and many annual or perennial flowering plants. Shallower substrate depths will dry out faster (Dunnett and Nolan, 2004; VanWoert et al., 2005b), but some taxa are naturally found growing under these conditions, such as the genus *Sedum*. But even for such a well-suited genus, substrate depth can influence the rate of substrate coverage and subsequent plant growth (Durhman et al., 2007; Getter and Rowe, 2008; Getter and Rowe, 2009; Getter et al., 2009a). Deeper substrates are beneficial for both increased water holding capacity (VanWoert et al., 2005a,b) and as a buffer for overwintering survival because shallow substrates are more subject to fluctuations in temperature (Boivin et al., 2001).

In a 6-yr study, Dunnett et al. (2008a) reported the greatest survival, diversity, size, and flowering performance of plant species occurred at a media depth of 20 cm relative to 10 cm. Even when drought tolerant plant species are selected, the limiting factor for plant survival on most green roofs is soil moisture. Species richness (i.e., mean number of taxa per subplot) decreased with time at both substrate depths, but the rate of decline was greater at 10 cm (Dunnett et al., 2008a). Some species performed better at the 10- or 20-cm depth. The low-growing species such as sedum that are typical of shallow extensive roofs were not as competitive at 20 cm. Likewise, the perennial plants that normally possess greater biomass could not survive as well at 10 cm. Despite the cultural limitations of shallow substrate depths, they are often desirable because buildings must be structurally strong enough to support the added weight of the green roof.

Biodiversity

The genus *Sedum* is a popular choice among extensive green roofing projects due to its tolerance for drought (Durhman et al., 2006; Wolf and Lundholm, 2008), shallow substrate adaptability (Durhman et al., 2007; Emilsson, 2008), persistence (Köhler, 2006; Monterusso et al., 2005; Rowe et al., 2006), solar radiation intensity (Getter et al., 2009a), and ability to limit transpiration (Kluge, 1977; Lee and Kim, 1994) and store water (Gravatt, 2003; Teeri et al., 1986). Studies have shown that many species of sedum can survive up to 2 yr without water (Kirschstein, 1997; Teeri et al., 1986; VanWoert et al., 2005b). Besides *Sedum* spp., other succulents such as *Delosperma* spp., *Euphorbia* spp., and *Sempervivum* spp. can be appropriate choices for extensive roofs because they possess many of the same characteristics as *Sedum* (Snodgrass and Snodgrass, 2006). Many succulents such as *Sedum* spp. have been identified as exhibiting some form of Crassulacean acid metabolism (CAM) (Sayed et al., 1994; Kluge, 1977; Gravatt and Martin, 1992; Gravatt, 2003; Teeri et al., 1986; Lee and Kim, 1994). CAM plants limit transpirational water loss by keeping their stomata closed during the day. They take up CO_2 during the night, store it as an organic acid, and then use it the following day as the source for the normal photosynthetic carbon reduction cycle (Cushman, 2001).

By definition, green roofs are not generally considered native sites, but artificial systems where the growing substrates and rooftop environmental conditions are quite different than what is found at ground level. However, there is often a desire to use native plant species because of their real and perceived ecological benefits, such as their longevity without the use of pesticides, fertilizers, or irrigation (USEPA, 2008c). These plants and the wildlife that coevolved with them through time rely on each other, and the introduction of exotic species may be detrimental to the ecosystem. Researchers in Michigan evaluated 18 native taxa on nonirrigated extensive green roof platforms (Monterusso et al., 2005). After 3 yr, only four of the species still existed. The majority of the plants tested were considered to be drought tolerant, but their survival in a native environment relies on deep tap roots to obtain moisture. In a shallow extensive roof, these roots can still grow sideways, but periods of drought resulted in death without supplemental irrigation. Deeper growing substrates, supplemental irrigation, or alternate plant choices could alleviate this problem. Another strategy to improve native plant survival may be to amend the growing substrate with mycorrhizae fungi to more closely represent their native soil (Sutton, 2008). Another option would be to incorporate the species that did survive into a standard *Sedum* roof that is not irrigated to add diversity and aesthetic appeal.

An excellent example of an extensive green roof native plant community is the convention center of the Church of Jesus Christ of Latter-day Saints in Salt Lake City, UT (Fig. 19–10). However, to maintain the roof's high visual appeal,

Fig. 19–10. An extensive green roof native plant community on the convention center of the Church of Jesus Christ of Latter-day Saints in Salt Lake City, UT (photograph: Brad Rowe).

the roof requires irrigation and intense maintenance. Another example is located in northeastern Switzerland, where nine orchid species and other rare and endangered plant species exist on a 90-yr-old green roof (Brenneisen, 2004). It is generally assumed that native plants will provide greater biodiversity benefits and are more ecological than nonnative plants (Kendle and Rose, 2000). Nonnative plants are often characterized as invasive and too successful, but at the same time are said to require higher levels of maintenance to survive local environmental conditions. In reality, both native and nonnative plant species can be invasive, a trait that has more to do with their ability to propagate effectively either by seed or by vegetative means than with geographic origin.

Regarding wildlife, it is generally true that native plants will support a greater number of species (invertebrates) because they have coexisted through time (Dunnett, 2006). However, nonnatives may provide a food source at times when natives are not flowering. Geophytes (bulbs, corms, and tubers) can provide visual interest and nectar sources at a time of year when little else may be flowering, regardless of whether they are native (Nagase and Dunnett, 2008). A study in England found that the presence of nonnative plants in landscapes made no significant difference in biodiversity of invertebrates when they compared 70 gardens of all types, sizes, and locations for a period of 3 yr. The major factor that determined biodiversity was the structure of the vegetation, in other words, many different plant species providing layers of vegetation (Dunnett, 2006; Smith et al., 2006). Plant selection should include not only numerous species, but also variations in season of flowering, plant height, and spread.

Stormwater and Energy

Vegetation can influence stormwater retention, runoff water quality, and building energy efficiency. Plant canopies intercept rainfall, shade the roof and reduce evaporation from the soil, and transpire. Researcher in England compared grasses, forbs, and sedum and reported that the grasses were most effective in reducing stormwater runoff followed by the forbs (Dunnett et al., 2008b). However, *Sedum rupestre* L. and *Sedum spurium* M. Bieb. 'Coccineum' with their upright structure reduced runoff more than creeping species such as *Sedum acre* L. and *Sedum album* L. (Dunnett et al., 2008b). Species possessing greater height and diameter were associated with less runoff.

Maintenance practices can also influence runoff quality. Fertilizers and pesticides should be used with caution. Emilsson (2004) studied three fertilization levels—low and medium rates of controlled release fertilizer and controlled release fertilizer in combination with water soluble applications—on two types of green roof vegetation in Sweden (old prevegetated mats and fertilization of newly established green roofs). All had the potential for nutrients leaching into stormwater runoff. In addition, fungicides, herbicides, and insecticides should be used sparingly or not at all because of the potential for premature degradation of roof membranes and runoff concerns. Organic matter in the substrate can also result in increased N and P in the runoff (Hathaway et al., 2008).

In terms of building heat flux, green roofs save energy as the substrate acts as an insulation layer, the plant canopy shades the roof, and the plants provide transpirational cooling. VanWoert et al. (2005b) showed that even though several *Sedum* spp. survived for 88 d without water, their evapotranspiration rates dropped to nearly zero by Day 4. Such low rates of evapotranspiration would likely diminish the potential of a green roof system to provide cooling to the building underneath. If sufficient substrate moisture was available, either naturally or through supplemental irrigation, then taxa exhibiting higher transpiration rates may result in greater evaporative cooling potential.

Aesthetics

Since we are human, expectations of aesthetics must be addressed. Many species have dormant periods where the green roof may not appear green. For example, many native prairie grasses and perennials will normally dry and brown in the summer. Although, a natural occurrence, some may find this to be unacceptable. Providing deeper substrates and higher moisture levels provides the possibility for a greater number of species to survive, but also allows weeds to invade. On a shallow sedum roof, most grasses and herbs cannot survive due to the lack of moisture, thus making them relatively low maintenance roofs because very little weeding is necessary. Visual appearance may not be a concern on extensive green roofs if the roof is not normally visible and was installed primarily for its functional attributes such as stormwater retention.

Many people consider naturalistic landscapes on roofs to be weedy and not aesthetically pleasing. However, people will generally accept naturalistic landscapes if it is obvious that the landscape was designed and is being maintained (Nassauer, 1995)—a good example at ground level would be to mow the outer edges of a native prairie. Just as with any landscape at ground level, plants that survive initially on a green roof may not continue to exist in the long term because of variability in climate and other factors. Although, ecological services and aesthetic appeal are important criteria on many roofs, the chosen plants must first be capable of surviving.

Conclusions

Green roofs offer an ecosystem service that can help mitigate the problems of our built environment. They can counteract the destruction of natural habitats and provide environmental benefits such as stormwater retention, a reduction in energy usage, and increased biodiversity. Since roofs represent 21 to 26% of urban areas (Wong, 2005), they provide a unique opportunity to use these typically unused spaces to reclaim habitat that was lost due to construction while also aiding in the protecting of our environment through more sustainable practices.

References

Akbari, H., and S. Konopacki. 2005. Calculating energy-saving potentials of heat island reduction strategies. Energy Policy 33:721–756.

Alcázar, S.S., and B. Bass. 2006. Life cycle assessment of green roofs—Case study of an eight-story residential building in Madrid and implications for green roof benefits. *In* Proc. 4th North American Green Roof Conference: Greening Rooftops for Sustainable Communities. Boston, MA. 11–12 May 2006. The Cardinal Group, Toronto, Canada.

Barnes, K., J. Morgan, and M. Roberge. 2001. Impervious surfaces and the quality of natural built environments. Available at http://chesapeake.towson.edu/landscape/impervious/download/Impervious.Pdf (verified 10 May 2010). Report prepared for Towson University's NASA/Raytheon/Synergy project. Towson Univ., Baltimore, MD.

Bass, B., E.S. Krayenhoff, A. Martilli, R.B. Stull, and H. Auld. 2003. The impact of green roofs on Toronto's urban heat island. *In* Proc. 1st North American Green Roof Conference: Greening Rooftops for Sustainable Communities, Chicago, IL. 30 May 2003. The Cardinal Group, Toronto, Canada.

Baumann, N. 2006. Ground-nesting birds on green roofs in Switzerland. Preliminary observations. Urban Habitats 4:37–50.

Beattie, D.J., and R. Berghage. 2004. Green roof media characteristics: The basics. *In* Proc. 2nd North American Green Roof Conference: Greening Rooftops for Sustainable Communities, Portland, OR. 2–4 June 2004. The Cardinal Group, Toronto, Canada.

Berndtsson, J.C., T. Emilsson, and L. Bengtsson. 2006. The influence of extensive vegetated roofs on runoff water quality. Sci. Total Environ. 355:48–63.

Boivin, M., M. Lamy, A. Gosselin, and B. Dansereau. 2001. Effect of artificial substrate depth on freezing injury of six herbaceous perennials grown in a green roof system. Horttechnology 11:409–412.

Brenneisen, S. 2003. The benefits of biodiversity from green roofs—Key design consequences. *In* Proc. of 1st North American Green Roof Conference: Greening Rooftops for Sustainable Communities, Chicago, IL. 29–30 May 2003. The Cardinal Group, Toronto, Canada.

Brenneisen, S. 2004. Green roofs—How nature returns to the city. Acta Hortic. 643:289–293.

Brenneisen, S. 2006. Space for urban wildlife: Designing green roofs as habitats in Switzerland. Urban Habitats 4:27–36.

Brenneisen, S. 2008. Benefits for biodiversity. *In* Proc. World Green Roof Congress, London. 17–18 Sept. 2008.

Burian, S.J., and C.A. Pomeroy. 2010. Urban impacts on the water cycle and potential green infrastructure implications. p. 277–296. *In* J. Aitkenhead-Peterson and A. Volder (ed.) Urban ecosystem ecology. Agron. Monogr. 55. ASA, CSSA, and SSSA, Madison, WI.

Carter, T., and C.R. Jackson. 2007. Vegetated roofs for stormwater management at multiple spatial scales. Landsc. Urban Plan. 80:84–94.

Carter, T.L., and T.C. Rasmussen. 2006. Hydrologic behavior of vegetated roofs. J. Am. Water Res. Assoc. 42:1261–1274.

Cheney, C. 2005. New York City: Greening Gotham's rooftops. p. 130–133. *In* EarthPledge (ed.) Green Roofs: Ecological design and construction. Schiffer Books, Atglen, PA.

City of Los Angeles Environmental Affairs Department. 2006. Report: Green roofs—Cooling Los Angeles.

Clark, C., B. Talbot, J. Bulkley, and P. Adriaens. 2005. Optimization of green roofs for air pollution mitigation. *In* Proc. 3rd North American Green Roof Conference: Greening Rooftops for Sustainable Communities, Washington, DC. 4–6 May 2005. The Cardinal Group, Toronto, Canada.

Clark, C., P. Adriaens, and F.B. Talbot. 2008. Green roof valuation: A probabilistic economic analysis of environmental benefits. Environ. Sci. Technol. 42:2155–2161.

Coffman, R.R., and G. Davis. 2005. Insect and avian fauna presence on the Ford assembly plant ecoroof. *In* Proc. 3rd North American Green Roof Conference: Greening Rooftops for Sustainable Communities, Washington, DC. 4–6 May 2005. The Cardinal Group, Toronto, Canada.

Connelly, M., and K. Liu. 2005. Green roof research in British Columbia—An overview. *In* Proc. 3rd North American Green Roof Conference: Greening Rooftops for Sustainable Communities, Washington, DC. 4–6 May 2005. The Cardinal Group, Toronto, Canada.

Connelly, M., and M. Hodgson. 2008. Sound transmission loss of green roofs. *In* Proc. 6th North American Green Roof Conference: Greening Rooftops for Sustainable Communities, Baltimore, MD. 30 April–2 May 2008. The Cardinal Group, Toronto, Canada.

Currie, B.A., and B. Bass. 2008. Estimates of air pollution mitigation with green plants and green roofs using the UFORE model. Urban Ecosyst. 11:409–422.

Cushman, J.C. 2001. Crassulacean acid metabolism. A plastic photosynthetic adaptation to arid environments. Plant Physiol. 127:1439–1448.

DeNardo, J.C., A.R. Jarrett, H.B. Manbeck, D.J. Beattie, and R.D. Berghage. 2005. Stormwater mitigation and surface temperature reduction by green roofs. Trans. ASAE 48:1491–1496.

Department of Health and Human Services. 2006. Extreme heat: A prevention guide to promote your personal health and safety. Available at http://emergency.cdc.gov/disasters/extremeheat/heat_guide.asp (verified 18 May 2010). Centers for Disease Control and Prevention, Atlanta, GA.

Deutsch, B., H. Whitlow, M. Sullivan, and A. Savineau. 2005. Re-greening Washington, DC: A green roof vision based on environmental benefits for air quality and storm water management. *In* Proc. 3rd North American Green Roof Conference: Greening Rooftops for Sustainable Communities, Washington, DC. 4–6 May 2005. The Cardinal Group, Toronto, Canada.

Dunnett, N., and N. Kingsbury. 2004. Planting green roofs and living walls. Timber Press, Portland, OR.

Dunnett, N., and A. Nolan. 2004. The effect of substrate depth and supplementary watering on the growth of nine herbaceous perennials in a semi-extensive green roof. Acta Hortic. 643:305–309.

Dunnett, N. 2006. Green roofs for biodiversity: Reconciling aesthetics with ecology. *In* Proc. of 4th North American Green Roof Conference: Greening Rooftops for Sustainable Communities. Boston, MA. 11–12 May 2006. The Cardinal Group, Toronto, Canada.

Dunnett, N., A. Nagase, and A. Hallam. 2008a. The dynamics of planted and colonizing species on a green roof over six growing seasons 2001–2006: Influence of substrate depth. Urban Ecosyst. 11:373–384.

Dunnett, N., A. Nagase, and R. Booth. 2008b. Influence of vegetation composition on runoff in two simulated green roof experiments. Urban Ecosyst. 11:385–398.

Durhman, A.K., D.B. Rowe, and C.L. Rugh. 2006. Effect of watering regimen on chlorophyll fluorescence and growth of selected green roof plant taxa. HortScience 41:1623–1628.

Durhman, A.K., D.B. Rowe, and C.L. Rugh. 2007. Effect of substrate depth on initial coverage, and survival of 25 succulent green roof plant taxa. HortScience 42:588–595.

Emilsson, T. 2004. Impact of fertilisation on vegetation development and water quality. *In* Proc. 2nd North American Green Roof Conference: Greening Rooftops for Sustainable Communities, Portland, OR. 2–4 June 2004. The Cardinal Group, Toronto, Canada.

Emilsson, T. 2008. Vegetation development on extensive vegetated green roofs: Influence of substrate composition, establishment method and species mix. Ecol. Eng. 33:265–277.

Ferguson, B.K. 1998. Introduction to stormwater: Concept, purpose, design. John Wiley and Sons, New York.

Gaffin, S., C. Rosenzweig, L. Parshall, D. Beattie, R. Berghage, G. O'Keefe, and D. Braman. 2005. Energy balance modeling applied to a comparison of white and green roof cooling efficiency. *In* Proc. 3rd North American Green Roof Conference: Greening Rooftops for Sustainable Communities, Washington, DC. 4–6 May 2005. The Cardinal Group, Toronto, Canada.

Gedge, D. 2003. From rubble to redstarts. *In* Proc. 1st North American Green Roof Conference: Greening Rooftops for Sustainable Communities, Chicago, IL. 29–30 May 2003. The Cardinal Group, Toronto, Canada.

Getter, K.L., and D.B. Rowe. 2006. The role of green roofs in sustainable development. HortScience 41:1276–1285.

Getter, K.L., and D.B. Rowe. 2008. Media depth influences sedum green roof establishment. Urban Ecosyst. 11:361–372.

Getter, K.L., and D.B. Rowe. 2009. Substrate depth influences sedum plant community on a green roof. HortScience 44:401–407.

Getter, K.L., D.B. Rowe, and J.A. Andresen. 2007. Quantifying the effect of slope on extensive green roof stormwater retention. Ecol. Eng. 31:225–231.

Getter, K.L., D.B. Rowe, and B.M. Cregg. 2009a. Solar radiation intensity influences extensive green roof plant communities. Urban For. Urban Green. 8(4):269–281.

Getter, K.L., D.B. Rowe, G.P. Robertson, B.M. Cregg, and J.A. Andresen. 2009b. Carbon sequestration potential of extensive green roofs. Environ. Sci. Technol. 43:7564–7570.

Gravatt, D.A. 2003. Crassulacean acid metabolism and survival of asexual propagules of *Sedum wrightii*. Photosynthetica 41:449–452.

Gravatt, D.A., and C.E. Martin. 1992. Comparative ecophysiology of five species of *Sedum* (Crassulaceae) under well-watered and drought-stressed conditions. Oecologia 92:532–541.

Hammond, G.P., and C.I. Jones. 2008. Embodied energy and carbon in construction materials. Energy 161:87–98.

Hathaway, A.M., W.F. Hunt, and G.D. Jennings. 2008. A field study of green roof hydrologic and water quality performance. Trans. ASABE 51:37–44.

Heisler, G.M., and A.J. Brazel. 2010. The urban physical environment: Temperature and urban heat islands. p. 29–56. *In* J. Aitkenhead-Peterson and A. Volder (ed.) Urban ecosystem ecology. Agron. Monogr. 55. ASA, CSSA, and SSSA, Madison, WI.

Hilten, R.N., T.M. Lawrence, and E.W. Tollner. 2008. Modeling stormwater runoff from green roofs with HYDRUS-1D. J. Hydrol. (Amsterdam) 358:288–293.

Johnson, J., and J. Newton. 1996. Building green, a guide for using plants on roofs and pavement. The London Ecology Unit, London.

Kadas, G.J., D. Gedge, and A.C. Gange. 2008. Can green roofs provide invertebrate habitat in the urban environment. *In* Proc. World Green Roof Congress, London. 17–18 Sept. 2008. The Cardinal Group, Toronto, Canada.

Kaplan, S., J.F. Talbot, and R. Kaplan. 1988. Coping with daily hassles: The impact of the nearby natural environment. Project Report. USDA Forest Service, North Central Forest Experiment Station, Urban Forestry Unit Cooperative. Agreement 23-85-08.

Kendle, A.D., and J.E. Rose. 2000. The aliens have landed! What are the justifications for 'native only' policies in landscape plantings. Landsc. Urban Plan. 47:19–31.

Kirschstein, C. 1997. Die dürreresistenz einiger Sedum-arten. Abgeleitet aus der Bedeutung der Wurzelsaugspannung-Teil 1. Stadt und Grün 46:252–256.

Kluge, M. 1977. Is *Sedum acre* L. a CAM plant? Oceologia 29:77–83.

Köhler, M. 2006. Long-term vegetation research on two extensive green roofs in Berlin. Urban Habitats 4:3–26.

Köhler, M., W. Wiartalla, and R. Feige. 2007. Interaction between PV-systems and extensive green roofs. *In* Proc. of 5th North American Green Roof Conference: Greening Rooftops for Sustainable Communities, Boston, MA. 29 April–1 May 2007. The Cardinal Group, Toronto, Canada.

Kosareo, L., and R. Ries. 2007. Comparative environmental life cycle assessment of green roofs. Build. Environ. 42:2606–2613.

Laberge, K.M. 2003. Urban oasis: Chicago's City Hall green roof. *In* Proc. 1st North American Green Roof Conference: Greening Rooftops for Sustainable Communities, Chicago, IL. 29–30 May 2003. The Cardinal Group, Toronto, Canada.

Lee, K.S., and J. Kim. 1994. Changes in crassulacean acid metabolism (CAM) of *Sedum* plants with special reference to soil moisture conditions. J. Plant Biol. 37:9–15.

Leonard, T., and J. Leonard. 2005. The green roof and energy performance—Rooftop data analyzed. *In* Proc. of 3rd North American Green Roof Conference: Greening Rooftops for Sustainable Communities, Washington, DC. 4–6 May 2005. The Cardinal Group, Toronto, Canada.

Li, M.-H., B. Dvorak, and C.Y. Sung. 2010. Bioretention, low impact development, and stormwater management. p. 413–430. *In* J. Aitkenhead-Peterson and A. Volder (ed.) Urban ecosystem ecology. Agron. Monogr. 55. ASA, CSSA, and SSSA, Madison, WI.

Liesecke, H.J. 1998. Das retentionsvermögen von dachbegrünungen. (Water retention capacity of vegetated roofs.) Stadt und Grün 47:46–53.

Liu, K., and B. Baskaran. 2003. Thermal performance of green roofs through field evaluation. *In* Proc. 1st North American Green Roof Conference: Greening Rooftops for Sustainable Communities, Chicago, IL. 29–30 May 2003. The Cardinal Group, Toronto, Canada.

Martens, R., B. Bass, and S.S. Alcazar. 2008. Roof-envelope ratio on green roof energy performance. Urban Ecosyst. 11:399–408.

Mentens, J., D. Raes, and M. Hermy. 2006. Green roofs as a tool for solving the rainwater runoff problem in the urbanized 21st century? Landsc. Urban Plan. 77:217–226.

Monterusso, M.A., D.B. Rowe, and C.L. Rugh. 2005. Establishment and persistence of *Sedum* spp. and native taxa for green roof applications. HortScience 40:391–396.

Nagase, A., and N. Dunnett. 2008. Extensive green roofs using geophytes in the UK: Effect of substrate depth and covering plants. *In* Proc. World Green Roof Congress, London. 17–18 Sept. 2008. The Cardinal Group, Toronto, Canada.

Nassauer, J. 1995. Messy ecosystems, orderly frames. Landscape J. 14:161–170.

Oberndorfer, E., J. Lundholm, B. Bass, M. Connelly, R. Coffman, H. Doshi, N. Dunnett, S. Gaffin, M. Köhler, K. Lui, and B. Rowe. 2007. Green roofs as urban ecosystems: Ecological structures, functions, and services. Bioscience 57:823–833.

Oliver, J.E. 1973. Climate and man's environment: An introduction to applied climatology. John Wiley and Sons, New York.

Passchier-Vermeer, W., and W.F. Passchier. 2000. Noise exposure and public health. Environ. Health Perspect. 108(Suppl. 1):123–131.

Peck, S.W. 2005. Toronto: A model for North American infrastructure development. p. 127–129. *In* EarthPledge. Green roofs: Ecological design and construction. Schiffer, Atglen, PA.

Peck, S.W., C. Callaghan, M.E. Kuhn, and B. Bass. 1999. Greenbacks from green roofs: Forging a new industry in Canada. Report. Canada Mortgage and Housing Corp., Ottawa, ON.

Retzlaff, W., S. Ebbs, S. Alsup, S. Morgan, E. Woods, V. Jost, and K. Luckett. 2008. What is that running off my green roof? *In* Proc. 6th North American Green Roof Conf. Greening Rooftops for Sustainable Communities, Baltimore, MD. 30 April–2 May 2008. The Cardinal Group, Toronto, Canada.

Rowe, D.B., M.A. Monterusso, and C.L. Rugh. 2006. Assessment of heat-expanded slate and fertility requirements in green roof substrates. Horttechnology 16:471–477.

Sailor, D.J. 2008. A green roof model for building energy simulation programs. Energy Build. 40:1466–1478.

Saiz-Alcázar, S., and B. Bass. 2005. Energy performance of green roofs in a multi storey residential building in Madrid. *In* Proc. of 3rd North American Green Roof Conference: Greening Rooftops for Sustainable Communities, Washington, DC. 4–6 May 2005. The Cardinal Group, Toronto, Canada.

Santamouris, M., C. Pavlou, P. Doukas, G. Mihalakakou, A. Synnefa, A. Hatzibiros, and P. Patargias. 2007. Investigating and analysing the energy and environmental performance of an experimental green roof system installed in a nursery school building in Athens, Greece. Energy 32:1781–1788.

Sayed, O.H., M.J. Earnshaw, and M. Cooper. 1994. Growth, water relations, and CAM induction in *Sedum album* in response to water stress. Biol. Plant. 36:383–388.

Schade, C. 2000. Wasserrückhaltung und Abflußbeiwerte bei dünnschichtigen extensivbegrünungen. (Water retention and runoff in extensive green roofs.) Stadt und Grün 49:95–100.

Shepherd, J.M., J.A. Stallins, M.L. Jin, and T.L. Mote. 2010. Urbanization: Impacts on clouds, precipitation, and lightning. p. 1–28. *In* J. Aitkenhead-Peterson and A. Volder (ed.) Urban ecosystem ecology. Agron. Monogr. 55. ASA, CSSA, and SSSA, Madison, WI.

Simmons, M.T., B. Gardiner, S. Windhager, and J. Tinsley. 2008. Green roofs are not created equal: Hydrologic and thermal performance of six different extensive green roofs and reflective and non-reflective roofs in a sub-tropical climate. Urban Ecosyst. 11:339–348.

Smith, R., P.H. Warren, K. Thompson, and K. Gaston. 2006. Urban domestic gardens (VI): Environmental correlates of invertebrate species richness. Biodivers. Conserv. 15:2415–2438.

Snodgrass, E.C., and L.L. Snodgrass. 2006. Green roof plants: A resource and planting guide. Timber Press, Portland, OR.

Spolek, C. 2008. Performance monitoring of three ecoroofs in Portland, Oregon. Urban Ecosyst. 11:349–359.

Steele, M.K., W.H. McDowell, and J.A. Aitkenhead-Peterson. 2010. Chemistry of urban, sub-urban, and rural surface waters. p. 297–340. *In* J. Aitkenhead-Peterson and A. Volder (ed.) Urban ecosystem ecology. Agron. Monogr. 55. ASA, CSSA, and SSSA, Madison, WI.

Sutton, R.K. 2008. Media modifications for native plant assemblages on green roofs. *In* Proc. 6th North American Green Roof Conference: Greening Rooftops for Sustainable Communities, Baltimore, MD. 30 April–2 May 2008. The Cardinal Group, Toronto, Canada.

Takebayashi, H., and M. Moriyama. 2007. Surface heat budget on green roof and high reflection roof for mitigation of urban heat island. Build. Environ. 42:2971–2979.

Teeri, J.A., M. Turner, and J. Gurevitch. 1986. The response of leaf water potential and crassulacean acid metabolism to prolonged drought in *Sedum rubrotinctum*. Plant Physiol. 81:678–680.

Ulrich, R.S. 1984. View through a window may influence recovery from surgery. Science 224:420–421.

USEPA. 1998. NO$_x$: How nitrogen oxides affect the way we live and breathe. EPA-456/F-98-005. USEPA Office of Air Quality Planning and Standards Research, Washington, DC.

USEPA. 2003. Cooling summertime temperatures: Strategies to reduce urban heat islands. EPA 430-F-03-014. USEPA, Washington, DC.

USEPA. 2005a. Cool roofs. Available at http://www.epa.gov/heatisland/index.htm (verified 27 Jan. 2010).

USEPA. 2005b. Emission facts: Greenhouse gas emissions from a typical passenger vehicle. EPA420-F-05-004. USEPA Office of Transportation and Air Quality, Washington, DC.

USEPA. 2007. Inventory of U.S. greenhouse gas emissions and sinks: Fast facts 1990–2005. Conversion factors to energy units (Heat Equivalents) heat contents and carbon content coefficients of various fuel types. EPA-430-R-07-002. USEPA, Washington, DC.

USEPA. 2008a. Climate leaders greenhouse gas inventory protocol core module guidance: Indirect emissions from purchases/sales of electricity and steam. EPA-430-K-03-006. USEPA, Washington, DC.

USEPA. 2008b. Combined sewer overflows: Demographics. Available at http://cfpub.epa.gov/npdes/cso/demo.cfm (verified 18 May 2010).

USEPA. 2008c. Green landscaping. Available at http://www.epa.gov/greenacres/index.html#Benefits. (verified 18 May 2010).

USEPA. 2009. Green roofs for stormwater runoff control. EPA-600-R-09-026. USEPA, Washington, DC.

U.S. Green Building Council Research Committee. 2007. A national green building research agenda. Available at http://www.usgbc.org/ShowFile.aspx?DocumentID=3402 (verified 10 May 2010).

Van Renterghem, T., and D. Botteldooren. 2008. Numerical evaluation of sound propagating over green roofs. J. Sound Vibrat. 317:781–799.

VanWoert, N.D., D.B. Rowe, J.A. Andresen, C.L. Rugh, R.T. Fernandez, and L. Xiao. 2005a. Green roof stormwater retention: Effects of roof surface, slope, and media depth. J. Environ. Qual. 34:1036–1044.

VanWoert, N.D., D.B. Rowe, J.A. Andresen, C.L. Rugh, and L. Xiao. 2005b. Watering regime and green roof substrate design affect Sedum plant growth. HortScience 40:659–664.

Villarreal, E.L., and L. Bengtsson. 2005. Response of a *Sedum* green-roof to individual rain events. Ecol. Eng. 25:1–7.

Von Stulpnagel, A., M. Horbert, and H. Sukopp. 1990. The importance of vegetation for the urban climate. p. 175–193. *In* H. Sukopp et al. (ed.) Urban ecology. SPB Academic Publishing, The Hague, The Netherlands.

Wolf, D., and J.T. Lundholm. 2008. Water uptake in green roof microcosms: Effects of plant species and water availability. Ecol. Eng. 33:179–186.

Wong, E. 2005. Green roofs and the Environmental Protection Agency's heat island reduction initiative. *In* Proc. 3rd North American Green Roof Conference: Greening

Rooftops for Sustainable Communities, Washington, DC. 4–6 May 2005. The Cardinal Group, Toronto, Canada.

Wong, N.H., Y. Chen, C.L. Ong, and A. Sia. 2003. Investigation of thermal benefits of rooftop garden in the tropical environment. Build. Environ. 38:261–270.

Yang, J., Q. Yu, and P. Gong. 2008. Quantifying air pollution removal by green roofs in Chicago. Atmos. Environ. 42:7266–7273.

Yok Tan, P., and A. Sia. 2005. A pilot green roof research project in Singapore. *In* Proc. 3rd North American Green Roof Conference: Greening Rooftops for Sustainable Communities, Washington, DC. 4–6 May 2005. The Cardinal Group, Toronto, Canada.

Bioretention, Low Impact Development, and Stormwater Management

Ming-Han Li
Bruce Dvorak
Chan Yong Sung

Abstract

This chapter provides a historical overview of stormwater management in the United States, the development and adoption of the low impact development (LID) approach, and a detailed review of the bioretention best management practice (BMP). The stormwater management historical overview introduces readers to the evolution of water-related regulations, from flood control (water quantity management), to water quality improvement, to the present holistic ecosystem protection. The LID section highlights a variety of techniques that achieve a "low impact to the environment" where urban development takes place. The last section of the chapter covers the bioretention BMP on aspects of current research methods, underlying principles, design considerations, performances, and challenges in implementing bioretention.

Overview of the Stormwater Management Movement in the United States

Stormwater management, as a major component of urban ecosystems, is becoming increasingly important in the 21st century as development continues to sprawl and alter the quantity and quality of surface water in urban watersheds. Because there is ever more demand for clean fresh water, the protection of water resources by preserving the integrity of ecosystems and creating built environments that mimic nature's filtering function has emerged as a new paradigm for managing urban stormwater. For many decades stormwater has been regarded as a flood hazard. Beginning in the early 20th century, new and coordinated efforts were initiated as a way to regulate stormwater and take positive steps toward regaining its deserving role as a valuable asset to urban ecosystems.

M.-H. Li (minghan@tamu.edu) and B. Dvorak (bdvorak@tamu.edu), Dep. of Landscape Architecture and Urban Planning, Texas A&M University, College Station, TX 77843-3137; C.Y. Sung, Texas Transportation Institute, Texas A&M University, College Station, TX 77843-3135 (cysung@tamu.edu).

doi:10.2134/agronmonogr55.c20

Fig. 20–1. Highway between Mounds, IL and Cairo, IL during the The Great Mississippi River Flood of 1927. Photographed on March 25, 1927. River stage at Cairo, IL was 16.1 m (52.8 feet). Original source: "The Floods of 1927 in the Mississippi Basin", Frankenfeld, H.C., 1927 Monthly Weather Review Supplement No. 29. Credit: National Oceanic and Atmospheric Administration (NOAA) Central Library (archival photography by Steve Nicklas).

The development of stormwater management strategies in the United States over the past century evolved as two major phases: (i) a quantity control phase and (ii) a quantity and quality control phase. In the first three quarters of the 20th century, stormwater quantity (i.e., flood control) was of major concern to governments, industry, farmers, and residents in rural, urban, and suburban areas. During this time, there were several devastating floods that caused serious damage to public works, industrial plants, farm lands, and hundreds of thousands of homes. As devastating as these floods were, they raised the public awareness of stormwater management issues, which in turn expedited the development of policy and other flood control strategies. Through great efforts by the U.S. Congress and local authorities during the last century, stormwater management has gradually reached widespread regulation in an attempt to control flooding. Water quality issues, however, were in need of attention and soon became a top priority.

Early concerns about stormwater quality focused merely on drinking water. Urban and industrial discharges were treated at treatment plants that received both sewage and urban runoff. These combined sewers were often overloaded with a mixture of raw sewage and urban stormwater runoff. As a result, untreated overflow was allowed to discharge into waterways. In addition, with the invention and use of chemical fertilizers in the mid 1900s, pollution of U.S. waters became intolerable. During the last quarter of the 20th century, water pollution became a vital issue, which soon led to the enactment of a series of acts and practices (Table 20–1).

Key issues that influenced the development of stormwater management strategies are summarized as follows:

1. Congressional acceptance of limited federal responsibility for flood control began in 1927 following major floods on the Mississippi River (Fig. 20–1). Beginning with the Flood Control Act of 1936, after more widespread flooding, the U.S. Congress accepted national responsibility for the problem of reducing flood damages, and the U.S. Army Corps of Engineers was given the responsibility for developing flood-control engineering works. The Army Corps of Engineers introduced three principal methods of controlling floods: (i) increasing the carrying capacity of channels with "channel

Table 20–1. U.S. policies and practices in the evolution in stormwater management.

Time	Emphasis	Key legislation
1930s to present	• structural flood control measures • drainage facilities • erosion control	• Flood Control Act of 1936 • Watershed Protection and Flood Prevention Act of 1954

1970s to present	• floodplain management and other
	  nonstructural flood damage
	  reduction measures
	• watershed management
		• Water Pollution Control Act of 1972
		• Flood Disaster Protection Act of 1973
		• Water Quality Act of 1987

1990s to present	• water quality	

improvement" projects, (ii) reducing flood flows with reservoirs and detention basins, and (iii) building levees, flood walls, and conduits.

2. In 1935, the Soil Conservation Service (SCS) (now Natural Resources Conservation Service of the U.S. Department of Agriculture) was established to conduct soil surveys, carry out erosion control measures, furnish financial aid and technical assistance to individuals and agencies, and acquire lands. However, SCS did not directly get involved in flood control until the passage of the Public Law 566, which initiated the Small Watershed Program. The SCS began construction of numerous small earth-filled dams along headwater streams. During the 1950s, while SCS focused on upstream reservoirs, the U.S. Army Corps dealt mostly with downstream reservoirs. An important act at that time that aided the planning and construction of smaller scaled projects was the Watershed Protection and Flood Prevention Act of 1954.

3. In 1968, a revolutionary solution in response to the rising cost of taxpayer funded disaster relief for flood victims and the increasing amount of damage caused by floods was the National Flood Insurance Program (NFIP) administered by the Federal Emergency Management Agency (FEMA). The NFIP is a Federal program enabling property owners in participating communities to purchase insurance protection against losses from flooding. Participation in the NFIP is based on an agreement between local communities and the federal government that states if a community will adopt and enforce a floodplain management ordinance to reduce future flood risks to new construction, the federal government will make flood insurance available within the community as a financial protection against flood losses. The NFIP was broadened and modified with the passage of the Flood Disaster Protection Act of 1973, and further modified by the National Flood Insurance Reform Act of 1994. Because of the NFIP, local governments began to play a significant role in flood control. During the 1970s, while federal agencies were responsible for the flood control and floodplain management, local communities were in charge of watershed management to solve drainage problems. Community-scale flood control facilities such as detention basins began to appear as a result of local government participation in the NFIP. Most developments before the 1970s did not have any stormwater controls. The term *detention basin* (or *detention pond*) did not appear in earlier literature. For example, Luna B. Leopold in his *Hydrology for Urban Land Planning—A Guidebook on the Hydrologic Effects of Urban Land Use* (Leopold, 1968) used "storage" to describe the effect of flood detention by reducing peak discharge. Several of Leopold's stormwater management strategies, such as storage ponds and open drainage swales, evolved to become what is considered standard practice today.

4.   The Federal Water Pollution Control Act of 1972, known as the Clean Water Act (CWA), is the first legislation governing the protection and improvement of water quality in the United States. In 1987, the CWA was again amended by the Congress (Water Quality Act) to require implementation of a comprehensive national program for addressing nonagricultural sources of stormwater discharges. Since then, U.S. stormwater management efforts began to emphasize both water quantity and quality issues. Under the amended CWA, the National Pollutant Discharge Elimination System (NPDES) is a permitting program that has a significant impact on state governments and municipalities. In addition, NPDES stated that the existing flood control facilities must be examined to determine if retrofitting the device to provide additional pollutant removal from stormwater is feasible. During this time, developments were beginning to preserve or protect floodplains and wetlands.

5.   In the 1990s, the concept of low impact development (LID) began to spread and become popular. Although large detention–retention basins are still the standard practice in many areas of the United States, the paradigm for managing stormwater is shifting from large-scale offsite control toward on-site control and treatment. More detail concerning LID is introduced in the subsequent sections.

Into the 21st century, stormwater quantity and quality control began to be treated together. No longer could either one be a single focus or be ignored. Federally enacted regulations would soon be implemented by authorized state agencies and local municipalities to ensure safer living environments and cleaner waters.

Low Impact Development

Low impact development is an approach to land development (or redevelopment) that mimics natural processes as a way to manage stormwater on site, that is, close to its source. Initially developed and implemented by Prince George's County, Maryland, in the early 1990s, LID techniques are innovative ways to manage stormwater runoff. "The overall goal of LID stormwater treatment is to mimic pre-development hydrologic conditions through the use of a variety of structural and nonstructural practices that detain, retain, percolate, and evaporate stormwater (National Association of Home Builders, 2003, p. 29)." Low impact development employs site design technologies such as "preserving and recreating natural landscape features, minimizing effective imperviousness to create functional and appealing site drainage that treat stormwater as a resource rather than a waste product" (USEPA, 2009). Applied on a broad scale, LID is also a land planning practice that can maintain or restore a watershed's hydrologic and ecological functions. Common LID techniques include bioretention, rain gardens, green roofs (Chapter 19, Rowe and Getter, 2010, this volume), rainwater harvesting, and porous pavements (Fig. 20–2).

Bioretention

Bioretention is a BMP developed in the early 1990s in Prince George's County, Maryland. This BMP is an on-site stormwater storage and infiltration facility that uses permeable engineered soils like sand, organic matter, and vegetation for treating runoff from paved surfaces such as parking lots, streets, and highways

Fig. 20–2. Examples of low impact development techniques. (A) Rainwater cistern at the Lady Bird Johnson Wildflower Center, Austin, TX (photograph: Ming-Han Li); (B) bioretention swales at the High Point neighborhood, Seattle, WA (photograph courtesy of Abby Hall, USEPA); (C) high albedo concrete and permeable concrete trenches (middle) in a back alley, Chicago, IL (photograph courtesy of Abby Hall, USEPA); and (D) green roof on city hall building, Chicago, IL, July 2004 (photograph: Bruce Dvorak).

(Fig. 20–3). The term *bioretention* can be confusing to engineers who understand the terms *detention pond* and *retention pond* as "dry pond" and "wet pond," respectively. Different from commonly known retention ponds, bioretention does not hold a permanent water body in its facility. Pooled water above bioretention typically drains within 12 h after a rain event. Bioretention BMPs may be termed *rain gardens* when they are used for residential applications. Other variations related to the bioretention BMP include biofiltration, bioinfiltration, and bioswale. Since its emergence in the 1990s, bioretention has gained popularity among stormwater professionals. Many bioretention projects have been explored and implemented nationwide. The results of these experimental projects report effective stormwater pollutant removal, particularly in heavy metals. Because of the significant water quality benefit, the USEPA is advocating that the bioretention technique be included in the USEPA's LID program (USEPA, 2009). The U.S. Green Building Council (USGBC) also identifies bioretention BMP as a preferred site practice in Leadership in Energy and Environmental Design (LEED) certification. In addition, the Sustainable Sites Initiative (SSI), a site sustainability rating tool being jointly developed by the American Society of Landscape Architects, the Lady Bird Johnson Wildflower Center, and the U.S. Botanic Garden, recommends the use of bioretention in its guidelines and performance benchmarks report (American Society of Landscape Architects, 2009).

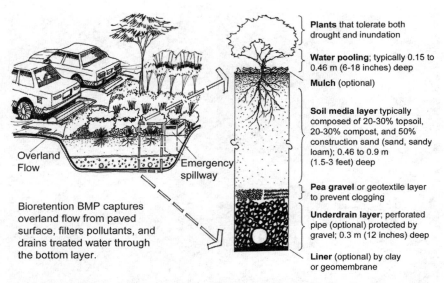

Plants that tolerate both drought and inundation

Water pooling; typically 0.15 to 0.46 m (6-18 inches) deep

Mulch (optional)

Soil media layer typically composed of 20-30% topsoil, 20-30% compost, and 50% construction sand (sand, sandy loam); 0.46 to 0.9 m (1.5-3 feet) deep

Pea gravel or geotextile layer to prevent clogging

Underdrain layer; perforated pipe (optional) protected by gravel; 0.3 m (12 inches) deep

Liner (optional) by clay or geomembrane

Overland Flow

Emergency spillway

Bioretention BMP captures overland flow from paved surface, filters pollutants, and drains treated water through the bottom layer.

Fig. 20–3. Detail of bioretention best management practice (illustration: Ming-Han Li).

Overview of Research Method

The Prince George's County *Bioretention Manual* is by far the most cited reference in bioretention-related literature. The manual was first published in 1993 and later revised in 2002 (Prince George's County, 2002). Since the 1980s, after the first bioretention project was built in Prince George's County, Maryland, several studies have been conducted on evaluating the pollutant removal performance. There are generally three types of testing methods: (i) batch and column testing, (ii) pilot testing, and (iii) field testing.

Batch and column testing is bench-scale testing conducted in a controlled laboratory setting, typically for physical-chemical characteristics such as adsorption by and desorption from soil (e.g., Harter, 1983; Davis et al., 2001; Hunt, 2003; Hong et al., 2006; Blecken et al., 2007; Rusciano and Obropta, 2007). The knowledge about soil chemistry is well established.

Pilot testing is a reduced scale experiment conducted in a laboratory setting (Fig. 20–4). Pilot testing is an important step to study different conditions and variables. For example, Kim et al. (2003) used pilot testing to search for the optimum design for mulch materials. Davis et al. (2001) used two pilot-scale bioretention cells to study flow characteristics and pollutant removal. In sum, pilot testing is used frequently and produces consistent, promising results in bioretention research.

Field testing includes designing, constructing, and monitoring full-scale bioretention BMPs. This type of testing requires a multidisciplinary approach, higher costs, and longer periods of time for monitoring, and therefore, has not been widely conducted (Hunt et al., 2006). Some studies have conducted field installation and monitoring of bioretention projects, including Chavez et al. (2007), Davis (2007, 2008), Dietz and Clausen (2005), and Hunt et al. (2008). In summary, field testing is a final step needed to begin developing design guidelines and

Fig. 20–4. Bioretention pilot testing boxes. (Research sponsored by Texas Department of Transportation and led by Ming-Han Li at Texas A&M University/Texas Transportation Institute; source: Ming-Han Li.)

learn about regional issues such as plant performance and maintenance require-ments across different ecological regions.

Underlying Principles

Bioretention employs a simple, site-integrated, terrestrial-based design that pro-vides the opportunity to infiltrate, filter, and store runoff and allow for its uptake and evapotranspiration by vegetation. As shown in Fig. 20–3, bioretention facili-ties capture stormwater runoff to be filtered through an engineered soil media. Once inflow flow rate exceeds the infiltration capacity of soil media, stormwa-ter begins to pool at the surface. When using the engineered soil mix with an underdrain, ponding will last for less than 12 h.

Another conceptual element of bioretention is the control of stormwater run-off close to the source. Unlike conventional detention and retention ponds that are deep and large, and placed at the lower part of a drainage basin, bioretention BMPs are typically shallow depressions located in upland areas. Thus, bioreten-tion mitigates stormwater runoff close to the site where it is generated.

Bioretention uses its biomass to retain nutrients and other pollutants and depends on the natural cleansing processes that occur in the soil–mulch–plant matrix. Mechanisms of pollutant removal include physical (e.g., settling and fil-tration), chemical (e.g., adsorption), and biological (e.g., plant uptake) methods. Proper bioretention designs allow these processes and cycles to occur and main-tain a functioning system.

Clogging is the most cited problem for early experimental bioretention designs. This problem has improved through the development of new soil media preparation and design. For highly erosive areas and sites with active construc-tion, a forebay should be integrated to capture excessive sediments before they enter the bioretention areas (Hunt, 2003). Poor nitrate-nitrogen removal was also reported in early laboratory testing (Davis et al., 2001, 2003; Hsieh and Davis, 2005). The suggested change for improving nitrate-nitrogen reduction is to provide anaerobic condition to increase denitrification by adding an internal sat-uration area with a carbon source such as newspaper.

Design Considerations

Water quantity and quality control are two major considerations in designing bioretention. Other factors could include long-term maintenance and landscape aesthetics that are appropriate to the surrounding context. In water quantity control design, factors that affect peak discharge or hydrology should be considered. This is similar to the process of designing conventional detention–retention ponds and may include selecting a design storm, delineating drainage basins, calculating discharge rates, sizing storage volumes, and sizing pipes. In water quality control design, factors may include characteristics of target stormwater (i.e., pollutant composition and concentration), concern of groundwater contamination, and plant adaptability and survival. More consideration of design detail is described below.

Bioretention BMPs are typically designed to treat the "first flush," that is, the initial 13 to 25 mm (1/2–1 inch) of the runoff, which carries highly concentrated pollutants (Barbosa and Hvitved-Jacobsen, 1999). Runoff exceeding this capacity is bypassed via an emergency spillway. As such, about 5% of the drainage area is needed for bioretention BMPs. This is particularly implementable in urban environments where available land is limited and high in cost. Also, because of the small volume of the first flush, water pooling in bioretention BMPs does not need to be deeper than 0.3 to 0.46 m (12–18 inches). Again, this shallow pool of water is usually completely drained within 12 h, reducing concerns of safety and health to the public.

Bioretention typically includes several layers of different materials. Figure 20–3 illustrates the vertical profile of a bioretention BMP. The top layer is vegetation. Unlike other BMPs, survival of vegetation under specific climate conditions is an important factor to be considered in the design of bioretention. Since bioretention BMPs collect water from the surrounding areas, plants must tolerate inundation that creates an anoxic condition in the root zone. The Prince George's County (2002) *Bioretention Manual* recommends water pooling on the surface to be discharged within 12 h (preferably 6 h). To ensure this discharge rate, Prince George's County (2002) further recommends that soil used for bioretention should have an infiltration rate of 25 mm h^{-1} (1 inches h^{-1}) or greater, in which sandy soils are appropriate for such application. Another concern as important as inundation is drought tolerance. While sandy soils are appropriate to use in bioretention to drain facilities, at the same time they will not hold moisture for a long time. Thus, plants selected for bioretention must also tolerate long dry conditions. Several bioretention manuals have been developed, including Prince George's County (2002), Wisconsin Department of Natural Resources (2003), and Puget Sound Action Team (2005); each manual provides lists of vegetation suitable for the area. Designers should consult regional sources for their area to ensure compatibility with climate and the availability of plants in the local nursery market.

On the surface layer mulch can be applied to physically adsorb pollutants (particularly metals) and provide nutrients for vegetation. However, thick mulch may interrupt the exchange of air between atmosphere and soil and can result in suffocation of plant roots (Prince George's County, 2002). In addition, mulch is a source of nutrients and might cause high nitrogen and phosphorus concentrations in the effluent. Prince George's County (2002) recommends that the mulch layer be a maximum of 51 to 76 mm (2–3 inches).

A key component of bioretention is the soil media. The *Bioretention Manual* (Prince George's County, 2002) suggests a soil texture of 50 to 60% sand, 20 to 30% compost, and 20 to 30% topsoil. If groundwater contamination is a concern, compost should be reduced or excluded. Hsieh and Davis (2005) also found that 0.55 to 0.76 m (1.8–2.5 feet) of sand and sandy loam (infiltration rate of 38 mm h^{-1} [1.5

inches h[-1]) shows the best pollutant removal performance. They recommended that clay content not exceed 5% because not only does excessive clay decrease the infiltration rate, it also creates preferential paths for runoff (Hsieh and Davis, 2005). Pollutants quickly saturate soil particles along the preferential paths. As a result, the pollutant removal effectiveness decreases shortly after the installation.

Underdrain pipes may or may not be used, depending on the site drainage condition. If in situ soil contains too much clay content and/or groundwater is too shallow, bioretention requires an underdrain pipe to ensure proper discharge of the filtered water. The *Bioretention Manual* (Prince George's County, 2002) advises the underdrain pipe if in situ soil has infiltration rates greater than 25 mm h[-1] (1 inches h[-1]) and if the water table is within 0.6 m (2 feet) below the bottom of the bioretention. If bioretention includes the underdrain pipe, it should be covered either by pea gravel or geotextile fabric to prevent pipe clogging from soil particles. Plastic or clay liner can be added at the bottom of bioretention when case groundwater contamination is a concern.

Hydrologic Performance

Similar to conventional detention ponds, bioretention is designed to attenuate peak discharge by temporarily holding runoff during a storm event. Figure 20–5 shows typical hydrographs at the inlet and the outlet of a bioretention BMP. In a field demonstration study, Hunt et al. (2007) found that bioretention reduced the magnitude of peak discharges by 96% Davis (2008) also found that median peak flow reductions are 44% in a typical bioretention BMP and 63% in a bioretention BMP with a saturation zone at the bottom. By analyzing the water balance of bioretention, Sharkey and Hunt (2005) estimated that an unlined bioretention BMP removed surface runoff by 93% during summer and 44% during winter.

Water Quality Performance

Bioretention BMPs are generally effective for removing heavy metals, oil and grease, and fecal coliforms, but they are only moderately effective for removal of nutrient pollutants. Table 20–2 summarizes pollutant removal efficiencies reported by researchers using different types of experiments.

Soluble forms of nutrients, such as nitrate and nitrite (NO_{3+2}–N), are difficult to separate from water by physical filtering processes. Nutrients can leach out from various sources of organic matter in bioretention, including vegetation, mulch, and soil. As a result, negative nutrient removal rate, that is, an increase in nutrient concentrations in effluents, can occur. Organic matter, however, is essential for vegetation growth and microbial activities, which permanently remove nutri-

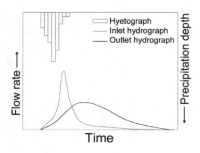

Fig. 20–5. Hydrologic performance of bioretention.

Table 20–2. Bioretention's water quality performance.

Source	Experiment type	Total suspended solids	Fecal coliform	E. Coli	Cu	Pb	Zn	Fe	Oil and grease
					\multicolumn{5}{c}{%}				
Culbertson and Hutchinson (2004)	Lab (bare ground)	–	–	–	–	–	–	–	–
	Lab (daylily planted)	–	–	–	–	–	–	–	–
	Lab (switchgrass planted)	–	–	–	–	–	–	–	–
Davis et al. (2001)	Lab	–	–	–	>99	>99	>99	–	–
Davis et al. (2003)	Field	–	–	–	43–97	54–70	64–>95	–	–
Davis et al. (2006)	Lab	–	–	–	–	–	–	–	–
	Field	–	–	–	–	–	–	–	–
Davis (2007)	Field (without IWS)†	54	–	–	77	84	69	–	–
	Field (with IWS)	59	–	–	83	88	27	–	–
Dietz and Clausen (2005)	Field	–	–	–	–	–	–	–	–
Dietz and Clausen (2006)	Field (without IWS)	–	–	–	–	–	–	–	–
	Field (with IWS)	–	–	–	–	–	–	–	–
Dougherty et al. (2007)	Field (without IWS)	–	–	–	–	–	–	–	–
	Field (with IWS)	–	–	–	–	–	–	–	–
Hong et al. (2006)	Lab (mulch only)	–	–	–	–	–	–	80	80
Hsieh and Davis (2005)	Lab	2–>96	–	–	–	66–>98	–	>96	>96
	Field (without IWS, with liner)	−103	–	–	–	>94	–	>99	>99
	Field (with IWS, with liner)	10	–	–	–	>95	–	>99	>99
Hsieh et al. (2007)	Lab	>94	–	–	–	–	–	–	–
Hunt et al. (2006)	Field (without IWS, low P-index soil)	–	–	–	–	–	–	–	–
	Field (without IWS, high P-index soil)	−170	–	–	99	81	98	–	–
Hunt et al. (2007)‡	Field (with IWS, medium P-index soil)	60	69	71	54	31	77	−330	–
Kim et al. (2003)	Lab (IWS)	–	–	–	–	–	–	–	–
Rusciano and Obropta (2007) ‡	Lab	91.6	95.9	–	–	–	–	–	–
Sharkey and Hunt (2005)	Field (without IWS, without liner)	–	–	–	–	–	–	–	–
	Field (without IWS, with liner)	–	–	–	–	–	–	–	–
	Field (with IWS)	–	–	–	–	–	–	–	–
Zhang et al. (2006)	Lab (sand only)	–	–	–	–	–	–	–	–
	Lab (5% fly ash)	–	–	–	–	–	–	–	–

Table 20–2. Continued.

Source	Experiment type	NO_{3+2}–N	NH_3–N	TKN	Organic N	Total N	Total P	BOD_5
					%			
Culbertson and Hutchinson (2004)	Lab (bare ground)	−200	–	–	–	–	−700	–
	Lab (daylily planted)	−155	–	–	–	–	−867	–
	Lab (switchgrass planted)	−37	–	–	–	–	−400	–
Davis et al. (2001)	Lab	24	79	68	–	68	81	–
Davis et al. (2003)	Field	–	–	–	–	–	–	–
Davis et al. (2006)	Lab	96	–	94	–	96	92	–
	Field	15–16	–	52–67	–	49–59	65–87	–
Davis (2007)	Field (without IWS)	95	–	–	–	–	77	–
	Field (with IWS)	90	–	–	–	–	79	–
Dietz and Clausen (2005)	Field	35.4	84.6	31.2	21.3	32	−110.6	–
Dietz and Clausen (2006)	Field (without IWS)	81	86	22	6	68	−104	–
	Field (with IWS)	87	69	5	−9	69	−98	–
Dougherty et al. (2007)	Field (without IWS)	–	–	–	–	32–71	32–71	–
	Field (with IWS)	–	–	–	–	14–30	15–30	–
Hong et al. (2006)	Lab (mulch only)	–	–	–	–	–	–	–
Hsieh and Davis (2005)	Lab	1–43	2–26	–	–	–	4–85	–
	Field (without IWS, with liner)	31	37	–	–	–	0	–
	Field (with IWS, with liner)	0.1	44	–	–	–	–	–
Hsieh et al. (2007)	Lab	–	–	–	–	–	47–68	–
Hunt et al. (2006)	Field (without IWS, low P-index soil)	13	86	45	–	40	65	–
	Field (without IWS, high P-index soil)	75	−0.99	−4.9	–	40	−240	–
Hunt et al. (2007) ‡	Field (with IWS, medium P-index soil)	−5	73	44	–	32	31	63
Kim et al. (2003)	Lab (IWS)	70–90	–	–	–	–	–	–
Rusciano and Obropta (2007) ‡	Lab	–	–	–	–	–	–	–
Sharkey and Hunt (2005)	Field (without IWS, without liner)	26	77	27	–	27–52	38	–
	Field (without IWS, with liner)	0.52	84	57	–	60	53	–
	Field (with IWS)	–	–	–	–	52	25	–
Zhang et al. (2006)	Lab (sand only)	–	–	–	–	–	2	–
	Lab (5% fly ash)	–	–	–	–	–	85	–

† IWS, internal water storage.

‡ Only event mean concentrations were reported.

ents filtered in soil media. Currently there is no standard for the optimum level of organic matter in the soil media. It should be determined based on site conditions.

The performance of N removal is not always guaranteed. The NO_{3+2}–N and NH_3–N removal rates range from 0 to 95% and –1 to 85% , respectively, as summarized from many field monitoring studies. Even in cases where bioretention showed a higher removal rate of either NO_{3+2}–N or NH_3–N, the mass of total N in effluent did not significantly change, suggesting that bioretention simply changes the chemical species from one to another. For instance, Hunt et al. (2006) found that one bioretention BMP showed a high removal rate for NO_{3+2}–N (75%), but a low rate for NH_3–N (–1%), while the other was high in NH_3–N (86%), but low in NO_{3+2}–N (13%). Total N removal rate, however, was low in both bioretention BMPs (around 40%). Other studies also reported relatively low total N removal rates (50% on average).

The denitrification process occurring in saturated soils is known to be a pathway to remove soil N (Chapter 13, Stander and Ehrenfeld, 2010, this volume). In the absence of oxygen, denitrifying bacteria use nitrate as an electron donor for respiration. The process converts nitrate into nitrous oxide and finally to nitrogen gas. Since Kim et al. (2003) first introduced an internal water storage (IWS) zone at the bottom of bioretention, many studies have applied an IWS zone to create an anaerobic environment to promote denitrification. However, only a few studies showed significant effects of the IWS. The insignificant effect of IWS might be due to the short retention time of stormwater in the IWS zone, which does not allow denitrification to take place. For instance, Kim et al. (2003) found that if water is retained for more than 1 wk in the IWS zone, the nitrates are almost completely removed from the water. No significant removal by IWS zone was observed if the residence time was less than a day. These findings imply that inclusion of an IWS zone may not be effective in practice because stormwater needs to be discharged from bioretention within a day.

Because soil media are easily saturated by P, the removal rate of P depends on initial concentration of P in soil media (Hunt et al., 2006). About half of the previous studies reported elevated P concentrations in effluents of bioretention. The removal rate for total P varies from –240 to 87%. One of the engineering solutions is to mix fly ash with the soil media. Zhang et al. (2006) found that cations in fly ash precipitate phosphate into solids, such as calcium phosphate. However, fly ash rapidly decreased the soil infiltration rate, so that they recommended limiting fly ash up to 5% of the soil media. Media mixed with 5% fly ash removed 85% of total P compared to 2% without fly ash amendment.

Bioretention can effectively remove heavy metals. Removal rates of copper, lead, and zinc range from 43 to 99%, 54 to 99%, and 31 to 99% by mass, respectively. The mulch layer adsorbs most metals. Dietz and Clausen (2006) found that 98% of Cu, 36% of Pb, and 16% of Zn of total mass removed by bioretention are adsorbed on mulch.

Bioretention also effectively removes oil and grease. In a laboratory study, Hong et al. (2006) reported that 30 mm (1.2 inches) of mulch layer removes 80% of oil and grease. A field demonstration study also showed a similar result. Hsieh and Davis (2005) found that bioretention removed 99% of oil and grease from parking lot runoff. Soil microbial activity permanently decomposes hydrocarbons filtered on soil particles.

Bioretention shows a relatively good removal of total suspended solids. Removal rate is between 60 and 90%, but some studies reported that an increase in suspended solids may occur with newly constructed bioretention. Fine particles in media are washed off with effluent (Hunt et al., 2006). Leaching of suspended solids gradually decreases as soils stabilize.

Removal of polycyclic aromatic hydrocarbon (PAH) has also been studied. Diblasi et al. (2009) found that bioretention BMP removed 87% of PAH leached out of sealant on a campus parking lot. The removal occurs in the top crust of soil layer.

Bioretention effectively removes fecal coliform and *Escherichia coli*. Laboratory experiments showed a 96% removal rate for fecal coliform (Rusciano and Obropta, 2007). Hunt et al. (2007) reported in a field demonstration study that bioretention removes 69% of fecal coliforms and 71% of *E. coli*., respectively.

Role of Vegetation

Compared to a simple sand media basin with no plants, bioretention has a longer lifetime because vegetation uptakes nutrients and heavy metals captured in soil media. Davis et al. (2001) discovered that the concentrations of Cu, Pb, and Zn in plant tissues increased by a factor of 6.3, 7.7, and 8.1, respectively, after 31 synthetic stormwater runoff experiments. From the results they estimated that, in the long run, vegetation uptakes 90% of N filtered by soil media (Davis et al., 2006).

Another role of vegetation is to enhance the water and pollutant removal performances of root systems. Culbertson and Hutchinson (2004) compared the performances of bioretention with and without vegetation and found that the vegetation improves both water and NO_3^- removal rates by up to a factor of 2 and 3, respectively. Vegetation also provides shade and dissipates heat by increasing evapotranspiration and therefore decreases effluent temperature (Jones et al., 2007).

Plants used in bioretention also have the opportunity to provide a visual context that default groundcovers such as turf grass do not typically provide. Many plants that function in bioretention BMPs have desirable visual characteristics in their form, color, bloom, and or texture. Many herbaceous species provide colorful blooms to attract the human eye, as well as some desirable wildlife species such as butterflies and song birds (Dunnett and Clayden, 2007). Bioretention BMPs can then be approached in visual terms as either a garden, with its appropriate maintenance requirements, or more like a meadow or wildflower garden, with a less demanding but different set of maintenance expectations.

Challenges Ahead

Bioretention represents a new concept and technique in managing stormwater runoff. As with any new technology, bioretention does not emerge without problems. Davis et al. (2009) provided their insights into questions needing further investigation, such as maximum pooling depth, minimum fill media depth, fill media composition and configuration, underdrain configuration, pretreatment option, and vegetation selection. Because they mostly focus on engineering respects, we want to highlight, from landscape planning and ecosystem management perspectives, some important challenges and opportunities in an attempt to evoke more attention to further advance our understanding and expand the potential of bioretention.

Detention to Bioretention

Current stormwater management practices such as detention–retention ponds are still considered conventional solutions. Ironically, detention ponds have been long recognized as ineffective in managing stormwater runoff and negative in terms of public perception. According to Debo and Ruby (1982), who explored the negative impacts of detention ponds within Georgia metropolitan regions, all interviewed county engineers stated that most detention ponds were installed solely to be compliant with ordinance requirements, and they acknowledged that many detention ponds are not operating as intended. In addition, detention ponds have unfavorable influences on public perception and quality of life in communities as a result of maintenance problems, health and safety issues, and aesthetic and visual disamenity. This raises the question, what is the potential for converting a detention pond to a bioretention BMP or rain garden?

Large-Scale Application

The notion of on-site control for bioretention somehow implies that bioretention BMPs and their treated drainage basins may be small. Many urban built environments are associated with highway infrastructure and tend to be large. A single highway interchange (e.g., see Fig. 20–6) can easily occupy hundreds of hectares. These interchanges are usually not exploited for stormwater benefits and represent a potential for bioretention. The question is, will bioretention work in a large-scale watershed?

Roadside Application

Increasing evidence indicates that nutrients are major constituents in highway runoff, particularly in roadsides adjacent to farm lands (Fig. 20–7). This is problematic because highway departments are typically responsible for maintaining roadsides and treating runoffs collected in roadside swales even if high nutrient contents may originate from agricultural practice. The current preference for managing nutrients is to use retention ponds or constructed wetlands. These are both expensive to construct and maintain. This is particularly true in areas where the natural hydrology makes maintaining a permanent water pool difficult without supplemental water. Despite recent promising work on bioretention to remove nutrients from stormwater, most of it has focused on relatively small

Fig. 20–6. Large highway interchange has a potential for bioretention (US 183 and Mopac in Austin, TX; source: Ming-Han Li).

Fig. 20–7. Typical rural road-side adjacent to farm lands (source: Ming-Han Li).

drainage basins such as parking lots. A Portugal study by Barbosa and Hvitved-Jacobsen (2001), a Virginia Department of Transportation study (Yu and Langan, 1999), and a Caltrans project (Alderete and Scharff, 2005, 2007) are among the few studies dealing with highway stormwater runoff. Bioretention research needs to be extended to the larger scale of the highway environment to determine the practical limits of contributing drainage basins, the relative life of the bioretention vegetative system, and the costs of construction and maintenance.

Fire Ants

Many people living in the south and southeast United States are familiar with red imported fire ants (*Solenopsis invicta* Buren). The fire ant was accidentally introduced in about 1930 at the port of Mobile, AL from South America (Vinson, 1997; Forys et al., 2002). In the absence of natural controls, they have spread throughout the entire southeastern United States, from the East Coast to central Texas, and north into Tennessee (Vinson, 1997) and are listed as invasive species on The National Invasive Species Information Center website. Because fire ants prefer habitats of coarse soils like sand and need to have moisture for survival (Milks et al., 2007), a typical bioretention design will be favored by fire ants. Fire ants build large mounds at high densities, and the channel networks below mounds are extended as deep as 60 cm (Green et al., 1999). Thus, the potential impact of fire ants residing in bioretention BMPs is that the soil structure and hydrology will be greatly altered, in which runoff may not have sufficient residence time due to quick draining through fire ants' channel networks. Also according to Lafleur et al. (2005), the presence of fire ants increases soil fertility. Nitrogen, phosphorus, and other cation concentrations are higher in soils adjacent to fire ant mounds. Although increased soil fertility can improve plant growth, nutrients accumulated in fire ant nest can have adverse effects on effluent water quality. More research is needed to investigate the impact of fire ants on the effectiveness of bioretention in areas where fire ants are common.

Maintenance

As with any technology, intended design function will not last forever without maintenance. Bioretention is no exception. Much bioretention maintenance is related to aesthetics, such as pruning, removing trash, adding mulch, and mowing (Hunt and Lord, 2006). More important maintenance is to ensure that the infiltration rate does not decrease with age; this involves replacing the mulch layer and top 2.5 to 5 cm of soil media once every 2 to 3 yr. In addition, removing and replacing some dead plants will be needed. The challenge of such a maintenance requirement and intensity is how bioretention can be applied beyond the rain garden that tends to be small in size. Regardless of the application type (e.g., parking lot, urban street, or roadside), maintenance must be included as part of the life cycle of bioretention applications to guarantee long-term success.

Conclusions

Bioretention is a new technology developed to manage stormwater by working with natural systems rather than against them. Starting with flood control in the early 20th century, then large detention–retention ponds for stormwater control, and lately LID techniques that mimic natural hydrology, the United States has come a long way in addressing stormwater management. With an increased demand for sustainable living environments, development of better and more sustainable stormwater management technology will only continue to evolve. More and more disciplines will get involved in developing new approaches, both at the planning and site scales, for holistic stormwater management.

References

Alderete, D., and M. Scharff. 2005. Case study: The design of a bioretention area to treat highway runoff and control sediment. p. 137–143. In Proceedings of IECA Conference 36. The International Erosion Control Association, Denver, CO.

Alderete, D., and M. Scharff. 2007. Storm water treatment bioretention basins. In Proceedings of IECA Conference 38. International Erosion Control Association, Denver, CO.

American Society of Landscape Architects. 2009. Lady Bird Johnson Wildflower Center, and United States Botanic Garden. The Sustainable Sites Initiative: Guideline and Performance Benchmarks. ASLA, Washington, DC.

Barbosa, A.E., and T. Hvitved-Jacobsen. 1999. Highway runoff and potential for removal of heavy metals in an infiltration pond in Portugal. Sci. Total Environ. 235:151–159.

Barbosa, A.E., and T. Hvitved-Jacobsen. 2001. Infiltration pond design for highway runoff treatment in semiarid climates. J. Environ. Eng. 27:1014–1022.

Blecken, G.-T., Y. Zinger, T.M. Muthanna, A. Deletic, T.D. Fletcher, and M. Viklander. 2007. The influence of temperature on nutrient treatment efficiency in stormwater biofilter systems. Water Sci. Technol. 56:83–91.

Chavez, R.A., G.O. Brown, and D.E. Storm. 2007. Bioretention cell design and construction specifications. ASABE Paper 072268. ASABE, St. Joseph, MI.

Culbertson, T.L., and S.L. Hutchinson. 2004. Assessing bioretention cell function in a Midwest continental climate. ASABE Paper 047051. ASABE, St. Joseph, MI.

Davis, A.P. 2007. Field performance of bioretention: Water quality. Environ. Eng. Sci. 24:1048–1064.

Davis, A.P. 2008. Field performance of bioretention: Hydrology impact. J. Hydrol. Eng. 13:90–95.

Davis, A.P., W.F. Hunt, R.G. Traver, and M. Clar. 2009. Bioretention technology: Overview of current practice and future needs. J. Environ. Eng. 135:109–117.

Davis, A.P., M. Shokouhian, H. Sharma, and C. Minami. 2001. Laboratory study of biological retention for urban stormwater management. Water Environ. Res. 73:5–14.

Davis, A.P., M. Shokouhian, H. Sharma, and C. Minami. 2006. Water quality improvement through bioretention media: Nitrogen and phosphorus removal. Water Environ. Res. 78:284–293.

Davis, A.P., M. Shokouhian, H. Sharma, C. Minami, and D. Winogradoff. 2003. Water quality improvement through bioretention: Lead, copper, and zinc removal. Water Environ. Res. 75:73–82.

Debo, T.N., and H. Ruby. 1982. Detention basins—An urban experience. Public Works 113:42–43.

Diblasi, C.J., H. Li, A.P. Davis, and U. Ghosh. 2009. Removal and fate of polycyclic aromatic hydrocarbon pollutants in an urban stormwater bioretention facility. Environ. Sci. Technol. 43:83–91.

Dietz, M.E., and J.C. Clausen. 2005. A field evaluation of rain garden flow and pollutant treatment. Water Air Soil Pollut. 167:123–138.

Dietz, M.E., and J.C. Clausen. 2006. Saturation to improve pollutant retention in a rain garden. Environ. Sci. Technol. 41:1335–1340.

Dougherty, M., C. LeBleu, E. Brantley, and C. Francis. 2007. Evaluation of bioretention nutrient removal in a rain garden with an internal water storage (IWS) layer. ASABE 077085. ASABE, St. Joseph, MI.

Dunnett, N., and A. Clayden. 2007. Rain gardens: Managing water sustainably in the garden and designed landscape. Timber Press, Portland, OR.

Forys, E.A., C.R. Allen, and D.P. Wojcik. 2002. Influence of the proximity and amount of human development and roads on the occurrence of the red imported fire ant in the lower Florida Keys. Biol. Conserv. 108:27–33.

Green, W.P., D.E. Pettry, and R.E. Switzer. 1999. Structure and hydrology of mounds of the imported fire ants in the southeastern United States. Geoderma 93:1–17.

Harter, R.D. 1983. Effect of soil pH on adsorption of lead, copper, zinc, and nickel. Soil Sci. Soc. Am. J. 47:47–51.

Hong, E., E.A. Seagren, and A.P. Davis. 2006. Sustainable oil and grease removal from synthetic stormwater runoff using bench-scale bioretention studies. Water Environ. Res. 78:141–155.

Hsieh, C.-H., and A.P. Davis. 2005. Evaluation and optimization of bioretention media for treatment of urban storm water runoff. J. Environ. Eng. 131:1521–1531.

Hsieh, C.-H., A.P. Davis, and B.A. Needleman. 2007. Bioretention column studies of phosphorus removal from urban stormwater runoff. Water Environ. Res. 79:177–184.

Hunt, W.F. 2003. Pollutant removal evaluation and hydraulic characterization for bioretention stormwater treatment devices. PhD. diss. The Pennsylvania State Univ., University Park.

Hunt, W.F., A.R. Jarrett, J.T. Smith, and L.J. Sharkey. 2006. Evaluating bioretention hydrology and nutrient removal at three field sites in North Carolina. J. Irrig. Drain. Eng. 132:600–608.

Hunt, W.F., and W.G. Lord. 2006. Bioretention performance, design, construction, and maintenance. AGW-588-05. North Carolina Coop. Ext., Raleigh.

Hunt, W.F., J.T. Smith, and J. Hathaway. 2007. Hall Marshall bioretention: Final report. Storm Water Services, Charlotte, NC.

Hunt, W.F., J.T. Smith, S.J. Jadlocki, J.M. Hathaway, and P.R. Eubanks. 2008. Pollutant removal and peak flow mitigation by a bioretention cell in urban Charlotte, NC. J. Environ. Eng. 134:403–408.

Jones, M.P., W.F. Hunt, and J.T. Smith. 2007. The effect of urban stormwater BMPs on runoff temperature in trout sensitive waters. ASABE Paper 077096. ASABE, St. Joseph, MI.

Kim, H., E.A. Seagren, and A.P. Davis. 2003. Engineered bioretention for removal of nitrate from stormwater runoff. Water Environ. Res. 75:355–367.

Lafleur, B., L.M. Hooper-Bùi, E.P. Mumma, and J.P. Geaghan. 2005. Soil fertility and plant growth in soils from pine forests and plantations: Effect of invasive red imported fire ants *Solenopsis invicta* (Buren). Pedobiologia 49:415–423.

Leopold, L.B. 1968. Hydrology for Urban Land Planning-A Guidebook on the Hydrologic Effects of Urban Land Use. Geological Survey Circular 554 U.S. Dep. of Interior, Washington, DC.

Milks, M.L., J.R. Fuxa, A.R. Richter, and E.B. Moser. 2007. Multivariate analyses of the factors affecting the distribution, abundance and soil form of Louisiana fire ants, *Solenopsis invicta*. Insectes Sociaux 54:283–292.

National Association of Home Builders. 2003. The practice of low impact development. Publ. H-21314CA. U.S. Dep. of Housing and Urban Dev, NAHB, Washington, DC.

Prince George's County. 2002. Bioretention manual. Dep. Environ. Res. Watershed Protection Branch, Prince George's County, MD.

Puget Sound Action Team. 2005. Low impact development: Technical guidance manual for puget sound. Publ. PSAT 05-03. Washington State Univ., Olympia.

Rowe, D.B., and K.L. Getter. 2010. Green roofs and garden roofs. p. 391–412. *In* J. Aitkenhead-Peterson and A. Volder (ed.) Urban ecosystem ecology. Agron. Monogr. 55. ASA, CSSA, and SSSA, Madison, WI.

Rusciano, G.M., and C.C. Obropta. 2007. Bioretention column study: Fecal coliform and total suspended solids reductions. Trans. ASABE 50:1261–1269.

Sharkey, L.J., and W.F. Hunt. 2005. Design implications on bioretention performance as a stormwater BMP: Water quality and quantity. ASAE Paper 052201. ASABE, St. Joseph, MI.

Stander, E.K., and J.G. Ehrenfeld. 2010. Urban riparian function. p. 253–276. *In* J. Aitkenhead-Peterson and A. Volder (ed.) Urban ecosystem ecology. Agron. Monogr. 55. ASA, CSSA, and SSSA, Madison, WI.

USEPA. 2009. Low impact development (LID). Available at http://www.epa.gov/owow/nps/lid (verified 18 May 2010).

Vinson, S.B. 1997. Invasion of the red imported fire ant (Hymenoptera: Formicidae): Spread, biology, and impact. Am. Entomol. 43:23–39.

Wisconsin Department of Natural Resources. 2003. Rain gardens. A how-to manual for homeowners. WDNR, Madison, WI.

Yu, S.L., and T.E. Langan. 1999. Controlling highway runoff pollution in drinking water supply reservoir watersheds: A final report. VTRC 00-R7. Virginia Transportation Research Council, Charlottesville.

Zhang, W., G.O. Brown, and D.E. Storm. 2006. Enhancement of phosphorus removal in bioretention cells by soil amendment. ASABE Paper 062303. ASABE, St. Joseph, MI.

Greenspaces, Green Structure, and Green Infrastructure Planning

Anne R. Beer

Abstract

This chapter aims to inform scientists working on urban environmental issues about how the planning process views greenspace. There is now real optimism that town planners and other professionals involved in city planning, and indeed some politicians, have returned to the realization that environmental issues are as important as the social and economic issues that have concerned them most in recent decades. It is well over 20 years since people began to ask questions about how cities could become more sustainable in the way they function (Brundtland, 1987). However, it is only recently, with all the concerns about climate change, that environmental aspects have assumed an overriding importance. Plans based on understanding how the urban environment functions have been lacking, so in many cities we have not made the real changes needed to enhance the environment, for instance, by using the attributes of greenspaces to reduce CO_2 emissions and enhance biodiversity. The question of how the form and functioning of cities can be changed to create the least possible adverse impact on their local environment through the way they are planned, designed, laid out, and managed is only just being asked. With more than one-half of the world's population now living in towns (UN-HABITAT, 2008), such changes are crucial. Land use plans must parallel efforts to change social behavior patterns and the way we economic decisions are taken, so that less damage is caused to the environment by our actions and so we can build more sustainable cities in the future. Cities can be understood as complex dynamic ecosystems (Tjallingii, 1995; Piracha and Marcotullio, 2003); like all ecosystems they are in a state of perpetual flux (Walker et al., 1999; Scheffer et al., 2001; Steiner 2002). These changes are as much the result of the preexisting landscape and its natural processes attempting to reassert themselves as of the way in which buildings and their associated infrastructure are laid out on the surface and used by the people who inhabit them.

The planning of a city's green structure in a holistic manner needs to include a plan for the unbuilt and unsealed land which recognizes the stresses that will be put on this land by the needs of the inhabitants and the buildings that support life in the city. Plans should aim to enhance a city's environmental sustainability as part of an overall spatial planning strategy. Planning in this way creates the potential for effective change on the ground in relation to the location, extent, use, design, management, and financing of a city's open land. Greenspaces then become central to city planning, rather than the "leftover bits of land" of the

A.R. Beer, Professor Emeritus, Dep. of Landscape, University of Sheffield, Sheffield, UK (anne.beer@btinternet.com).

doi:10.2134/agronmonogr55.c21

past. Until recently greenspaces (other than formal parks, recreation areas, and gardens) in most city planning schemes have frequently denoted open spaces with no particular function and, therefore, with no financial resources allocated to them and often with no specific agency having full responsibility. The recent development of a more holistic approach inclusive of all the open land in a city now makes it possible to explain to decisionmakers that a properly planned green structure is an essential component of the land use planning process. As such it can help an urban area become more environmentally sustainable.

Greenspaces influence the way a city functions—how the city interacts with its local physical and natural environment, and how it develops as a place that local inhabitants accept as a satisfactory setting for their daily lives. It is often not realized that greenspace can account for a large proportion of the land surface within any city. For instance, studies in Europe have shown:

- Two-thirds of Oslo's 454 km^2 is natural, comprising forest, waterways, or agricultural land (City of Oslo, 2003).
- Forty-nine percent of Vienna is classed as greenspace, with only 5% being formal city parks and open spaces (Erhart, 2005).
- Thirty-six percent of Warsaw is recognized as green structure (Kaliszuk, 2005).
- Two-thirds of London's land area is occupied by greenspaces and water. Of this, about a third is private gardens; a third parks or in sports use; and a third other wildlife habitats, such as grasslands, woodlands, and rivers (London Wildlife Trust, 2008).

This chapter first sets out some definitions of the terminology used in relation to greenspace and its planning. An overview of the ways in which a city's greenspace can interact with environmental factors follows, indicating how greenspace might be used in the planning process to help in resolving some of the environmental problems relating to sustainability of urban areas. Finally, readers are directed to where more information on the recent adoption of Green Infrastructure Planning in England can be found.

Some Definitions

Those involved in the study of green and open spaces within a city come from a range of science, social science, and economics specialties, as well as from the professions concerned with planning, building, and managing the different components of a city. Multidisciplinary work is essential if such spaces are to be used to their best effect in making cities more environmentally, socially, and economically sustainable in the future, so there is an overwhelming need for the specialists and the planners to find a means of communicating. Multidisciplinary work is, of course, the answer; without it, the potential inherent in the proper planning and management of a city's greenspaces will never be fully realized. However, it is only part of the answer, as a fundamental problem persists: how to synthesize the information and knowledge of the vast range of disciplines involved, so that a green structure plan can be drawn up which is detailed enough to be implementable over time, as a city changes and expands. Only in this way can we achieve a more sustainable city form for the future. This is a vast task, but without it, no effective planning briefs can be developed for specific areas of land, and there can be no assurance that the financial means will be found to make it happen and to maintain it once implemented.

Before proceeding further it is essential to define what is meant here by the use of certain words and phrases, as usage varies from discipline to discipline and country to country. Those working in multidisciplinary teams are encouraged to define meanings before progressing too far with any green structure plan.

Greenspace

The term *greenspace* as it is used here encompasses those areas of a city that are not built over and where the surface is not sealed. The following classification is based on that developed by members of an international project part funded by the European Union which ran from the mid 1990s to 2005 (Lundgren Alm, 1996). Greenspace includes:

- Formally Designated Greenspaces found on any town plan—the parks, public gardens, recreation areas, woodlands, and other spaces open to the public, such as burial places, as well as local pocket parks and community gardens.

- Other Actual Greenspaces—those spaces with controlled and limited access, as well as those left unallocated on many town plans, but all of which function for the city in environmental terms in the same way as the formal greenspaces. This includes the farmland, woodland, and presently unallocated unused land found on any urban–rural interface; the privately owned greenspaces such as domestic (including roof gardens) and other historical gardens and estates, as well as areas identified as of special landscape value for cultural reasons; the grounds of educational and health institutions, and those industrial and commercial premises where part of their site is kept unbuilt or used for recreation by workers; the allotment sites (in Europe these are often termed *people's gardens*); the land left open alongside transport corridors; the areas identified and protected because of their special landscape, ecological, or other scientific interest; and water bodies such as marshes, ponds, lakes and rivers, with their associated flood plains and special wetland habitats.

Each greenspace has distinctive natural characteristics deriving from its preexisting landscape and subsequent modifications through human agricultural and then urban use of the land. Each area has both biotic (plants and wildlife) and abiotic (soil, relief, water, and climate) characteristics. Whether it is linked to other open spaces determines whether it can function as part of a corridor in support of biodiversity and water flow, or whether it is an island of green surrounded by built forms and can act only as a stepping stone for the movement of plants and wildlife.

Galen Cranz's (1989) book *The Politics of Park Design: A History of the Urban Park in America* contains a wealth of information that does much to explain the patterns of green structure in American cities to those attempting to re-plan our cities in a more sustainable way.

Green Structure

The green and open spaces of a city and its surroundings can be mapped into a Geographic Information System (GIS) using data from city plans and air photographs and typologies such as that outlined above. The map shows the spatial distribution of greenspace that forms a city's green structure. This structure can then be linked to a vast range of other environmental data, in particular ecological data (Tjallingii, 1995), to begin the process of assessing the environmental potential and the problems inherent in the present distribution of greenspace.

Such studies are a necessary precursor to planning for the future use and management of the existing green structure and identifying where modification is needed as the city grows. A green structure needs to function as a whole, both within and around a city, to achieve the maximum benefit for environmental sustainability (Beer, 2005).

Green Infrastructure Planning

Urban greenspaces in and around a city create a "blue–green" (i.e., water and plants) structure, in contrast to the "gray" structure (i.e., buildings and their associated sealed surfaces and service runs) (Tjallingii, 1995). This structure is as important in enabling a city to function over time as the buildings themselves; it constitutes a green infrastructure that can be planned, designed, and managed for the benefit of nature and people. If well planned it can aid the reduction of CO_2 emissions and enhance biodiversity, as well as add considerably to the quality of life in a city. As Roe and Mell (2007) put it... "Like the term 'built infrastructure'—commonly understood to be a combination of grey (roads, railways, utility lines, sewers etc.) and social infrastructures (hospitals, schools, prisons, etc.)—the term green infrastructure is now used to indicate something we must have rather than something that is nice to have....". In her book *Metrogreen: Connecting Open Space in North American Cities*, Ericson (2006) discussed recent thinking with regard to green infrastructure planning in the United States and Canada. Her book contains several greenspace case studies paired for comparative analysis—Toronto and Chicago, Calgary and Denver, and Vancouver and Portland—and examines the motivations and objectives for connecting open spaces across metropolitan areas.

Urban Greenspaces Are Multifunctional

Since the late 1990s it has become common to accept that all land classed as greenspace in and around a city is multifunctional (COST C11, 2005). This concept of multifunctionality has facilitated the development of different methods with regard to the planning, management, and financing of urban greenspace.

If a city's green structure is to work holistically in support of the functioning of a city, its many multifunctional spaces need to be managed so that they "work together" to create a structure capable of providing the benefits discussed in the following sections.

Providing for Water Management

The natural rivers, streams, ponds, lakes, and the humanmade water systems that keep a city alive. Using the natural preexisting water system that underlies every city to best effect so as to conserve and restore water and to prevent floods requires a major rethink of how water has been and could be handled in the future to the benefit of inhabitants. This was pointed out by Spirn (1984) as early as the 1980s, but in many cities little has been done. Water shortages are likely to become increasingly common as climate change accelerates, as are flash floods from downpours if the predicted increase in instability in the weather behavior happens. How water is dealt with is perhaps the top priority of city green structure planning and will be a major contributor of the financial resources required for change to happen in the way greenspaces are laid out and managed.

Enhancing Water Infiltration into Aquifers and Holding Flood Water Locally

Water and its use are becoming increasingly important to greenspace planning. Over 25 yr ago work started in Germany on local area water management and the use of urban greenspaces in the cleansing process (Hahn and Zeisel, 1988). How well such schemes still work gives a useful insight into what is possible and the costs involved. In the Netherlands, Tjallingii et al. (1995) made an early attempt to link green structure and water planning, but their ideas made little impact on city planners and engineers at the time. However, these ideas were eventually implemented in the Leidsche Rijn development near Utrecht. Surface water was circulated from the builtup areas into a lake in the green structure; the water then flowed through a wetland that filtered out the sediments and nutrients before the purified water was returned to the builtup area for reuse. The issues relating to flooding have become of more interest to urban planners recently, and the importance of keeping the natural flood land "un-built" is now backed up by government advice. Local authorities in the United Kingdom are advised to incorporate Sustainable Urban Drainages Systems (Woods Ballard and Kellagher, 2007) into new developments by the UK Environment Agency (2008). Although politicians are still surprisingly reluctant to enforce a blanket ban on building on flood plains, the Environmental Agency's advice relates to new residential areas, as well as leisure, commercial, and industrial operations, and also draws attention to how such schemes can be retrofitted (see Planning Policy Statement 25, Department of Communities and Local Government, 2006). Citywide green structure plans will need to deal with this issue and recognize that building flood protection to make building possible in one area only means that the problem is pushed somewhere else. The problem of excess water does not just disappear—it becomes a problem that another local government area then has to solve. Green roofs are now found on housing developments and commercial sites in Britain but they are used to best effect in Germany (Dunnet, 2007; Dunnet and Kingsbury, 2008). Multiple benefits are offered in addition to reducing rainwater runoff; such roofs reduce energy use, extend the life of the roof, and provide valuable habitat or amenity benefits (see Chapter 19, Rowe and Getter, 2010, this volume).

Supporting Biodiversity—Habitats, Plants, and Wildlife

Sandström (2008) in his booklet *Biodiversity and Green Infrastructure in Urban Landscapes* examined the extent to which the planning process takes biodiversity into account. He also looked at the level of understanding of ecological issues relating to greenspaces among the planning professionals involved in green infrastructure planning. His findings go some way to explain the problems ecologists can meet in trying to get their ideas incorporated into the planning process. It is not generally taken into account, for instance, that biodiversity is not just about the plants that grow on the surface and the wildlife supported by them, but also the range of bacteria and fungi present in a healthy soil, which to a large extent determine the potential of any soil to grow food crops (see also Chapter 7, Pouyat et al., 2010, this volume). As cities grow, vegetation structure changes, and there is increased fragmentation of preexisting natural habitats. Of particular significance for urban biodiversity is the size of the open spaces, the quality of their management, and their fragmentation (Nyhuus, 2005), as well as the extent of medium and large domestic gardens. A domestic garden in the UK is recognized as the whole area surrounding the house and includes trees (native and ornamental), lawn, and flower and vegetable gardens. Although not natural habitats, mature well-established gardens are now recognized as having the highest levels of biodiversity in some cities (Gaston et al., 2005). They are particularly important supports for birdlife, although as Sandström et al. (2006) showed in newly built towns, it was the planted woodlands that created the best support. Very often biodiversity is higher in residential areas than in the surrounding agricultural landscape.

Growing Trees and Shrubs for Improved Air

Trees and shrubs absorb CO_2 and clean the air of dust and pollution. Some cities such as Oslo have developed with a "forest identity." These forests are important cultural landscapes (Nyhuus, 2005). Such examples show that cities and forests can be mutually beneficial (Randrup, 2006). Their presence also ensures a healthy recreational environment, supporting a wide range of human activities. Urban forests and woodland also help to clean air, as in Stuttgart and Oslo, where the cities have been replanned during the last 50 years to take advantage of the fact that cooled air sinks downhill through well-treed valleys and takes the dirty air from the builtup areas and the city center out beyond the city on the other side (Spirn, 1980; Nyhuus, 2005). In Örebro, Sweden, an area of new residential development was planned in the 1980s to have adjacent woodland for biomass to supply smallscale combined heat and power plants, and as Sandström et al. (2006) showed, these woodland areas now show the added advantage of an enhanced level of biodiversity. The concept of city forests is playing an increasingly important role in sustainability programs at the local and city level. City forests have to be managed, and this generates income that can cover costs, if done well.

Growing Food—Allotments and Community Gardens

In past centuries before bulk transport was possible, urban and periurban farming was the norm. However, greenspaces in Western cities are now rarely used effectively for food growing, despite the increase in demand for allotments in northern and western Europe, such as in Berlin, which has more than 80,000 community gardeners on city land and a long waiting list to obtain a plot (United Nations Development Programme, 1996). Although the urban fringe of metropolitan areas in the United States is productively farmed (United Nations Development Programme, 1996), in some parts of Europe farming in such locations is considered problematic (Gallent et al., 2004). The growing of food in areas such as allotments or community gardens using gardening methods is increasingly accepted by the public as a means of reducing an individual's carbon footprint, and demand for such land could well continue to increase. Modern farming practices have so lowered the fertility of many soils that they cannot support food crops without using vast quantities of chemical fertilizer, which can only be produced from oil and then has to be transported many miles from the chemical plants to the fields where it is spread by fuel-hungry machines. Such an unsustainable level of use of oil, which is now recognized as a diminishing resource, requires counter action to be initiated without delay. City greenspaces, with their readily accessible supply of organic wastes that can be processed into compost locally, are one of the best locations for sustainable food production, in particular for production of vegetables and fruit. However, for optimal yield the soil has to be of sufficient quality, and it takes time for the required soil structure to develop on disturbed soils even when compost is freely available. Such compost is produced in Oslo's greenspaces and recycled back into city domestic gardens by selling it to the public (Nyhuus, 2005).

Managing Local Climate Modification—Temperature and Air Flow

The role of trees and greenspace in general in modifying local climate is clear (Spirn, 1984; Arnfield, 2003), but it tends to be considered a planning issue only where climate is relatively extreme. However, as described in the section on growing trees and shrubs above, Stuttgart (Spirn, 1980) and Oslo (Nyhuus, 2005) are examples of European cities using trees to clean the air. With increasing awareness of the likely impact of global warming, the possibility of manipulating the microclimate within cities will inevitably become of more interest. Greenspaces and the plants that grow within them help to reduce the urban heat island effect and so reduce energy consumption. In very hot weather sealed surfaces can add to the discomfort by radiating heat back, but woodlands, in particular those with multiple layers under the canopy, can help reduce heat buildup (Arnfield, 2003). The presence of trees can even generate a slight cooling breeze in such conditions, due to the temperature variations from top to bottom. Trees and plants can also shelter buildings from cold winds and thus slightly modify energy use. These attributes can be used to enhance sustainability at the detailed design stage for building schemes (Beer, 2000). As Ann Spirn (1984) pointed out in her book *The Granite Garden*, across every city there are many microclimates. At the local level the variations are caused by "small heat islands, micro-inversions, pockets of air pollution and local differences in wind behaviour." She identified three main open space zones in the centers of American cities: street canyons, paved plazas, and parks. The first two she diagnosed as creating conditions that people could not enjoy for much of the time as winds sweep along the streets or down from the walls of adjoining buildings in plazas. It was only the small parks of the city center where conditions could be found that allowed people to enjoy being outside—tree-filled parks were like woodlands with shade, giving protection from the sun's rays, and the whole area absorbing less heat during the day time and losing it rapidly once the sun had gone. The air in such parks also tended to be cleaner. Spirn (1984) noted that the landscape design styles used for new parks were changing and many at the time of her research contained large areas of paving, perhaps to reduce maintenance costs, but the microclimatic benefits of greenspace were lost by doing this. Unfortunately, many landscape designers even now persist in designing spaces that are mainly hard paved. Writing as long ago as 1984 Spirn declared, "To manipulate the climate for health, comfort and (to reduce) energy consumption is imperative. Yet designers and planners of modern cities seldom do so. The builders of ancient cities addressed these issues with more concern and more skill." Unfortunately the planning process has had no mechanism to take account of these matters in recent decades. Rather, efforts have focused on stimulating development opportunities in the rush for ever more economic growth, even where this has meant inappropriate and unsustainable developments on the ground that will inevitably result in substantial costs to society to correct.

Promoting Health via Greenspace

Increasing the proximity of urban dwellers to local greenspace and the linking of these greenspaces into networks can promote health by encouraging exercise.

By the 2000s the link between people's health (both psychological and physical) and the presence or absence of nearby greenspace was well established, including the ways in which providing greenways might stimulate physical exercise, thereby reducing the health costs for society. Grahn (2005) outlined work undertaken in Sweden that indicated that if greenspaces are to be sufficiently well used to encourage people to exercise, then they must be near the home and the design should allow for varied experiences as the user walks around. Mell (2007) stated, "Innovative, accessible and connective spaces are now seen as a vital factor in combating health problems in the UK, and in conjunction with initiatives such as 'Walking the way to health', the UK government is attempting to redress the growing problems of ill health." Mell's literature review identified the positive relationship between psychological and physical well-being and the physical environment in which people reside. Recent work at the Netherlands Institute for Health Services Research, Utrecht (Verheij et al., 2008) has supported this. There is now large-scale epidemiological work supporting the association between urban–rural health differences and the availability of greenspace. It is argued that this would fit with the theories of environmental psychologists such as Kaplan and Kaplan (1989) and that availability of this new evidence would justify renewed attention for greenspace in urban planning (Verheij et al., 2008). Work by Mitchell and Popham (2009) adds further support to the contention that there would be major savings in long-term costs to society if urban dwellers could have easy access to safe, attractive and useable greenspace near their homes.

Supporting Human Well-Being—Quality of Life

Greenspaces can support human well-being through enabling people to have different experiences from those in builtup areas, including, for example, peacefulness; fascination with natural and seminatural habitats and wildlife, a view with "no buildings in sight"; a long, distant view; adventurous play using natural objects, not human-made equipment; seeing landscapes of varying character and ages and historical interest; making art and looking at art, whether as a highly designed landscape or a sculptural object in the greenspace; sitting on rocks beside streams or skimming pebbles over still water. What is possible within any green structure depends on the local landscape and its features, and also the imagination of the planners and designers and those who finance the project. ECOTEC (2006), in a report for the Northern Way consortium (composed of regional bodies in the north of England), identified several quality of life factors related to greenspace. They saw woodland regeneration as involved in developing a sense of pride in local distinctiveness and quality, and parks and greenspaces as potentially being highly valued by residents in creating quality neighborhoods. The possibility that involving local people in greenspace management could enhance social cohesion and be a part of capacity building, with environmental volunteering providing a focus for community action and engagement was also addressed by ECOTEC. Experts from many fields— sociologists, ethnologists, geographers, planners, and landscape architects—have contributed to our understanding of the importance of access to greenspace as part of the quality of life in the city. It is a body of research that has grown quickly over the last two decades, including work by psychologists such as Kaplan and Kaplan (1989) on how people experience the urban environment, as well as by geographers such as Burgess et al. (1988a,b) and Cooper Marcus and Francis (1997), who conducted research on the interface between human needs and how open space in the urban environment is planned and designed. Some of these early studies are summarized in *Environmental Planning for Site Development* (Beer et al., 2000). People's preferences and needs in relation to the provision of greenspace in general and for specific types of greenspace near their homes will always need further local study to inform local greenspace planning and design (Beer et al., 2003). What will work well in one country or city will not necessarily be applicable in other countries or cities; cultural and climatic differences are particularly relevant. For instance, outdoor spaces are used very differently in northern Europe, where people seek the sun, than in the south, where shade is sought. Well-being also involves educational aspects in terms of environmental education to increase people's potential pleasure in the use of greenspace and level of concern for its care and cultural awareness by developing a local link to an area's environmental history. In Britain there has been increased interest at the national level in how to make urban greenspace more useable, as can be seen in the report of the National Audit Office, Office of the Deputy Prime Minister (2006), *Enhancing Urban Green Space.*

Recreational Activity, Both Organized and Informal

This is the most commonly recognized use of greenspace—it is the only one that planning authorities have provided for through the locally prevailing standards of open space provision (Turner, 1992). It consists of parks, formal and private gardens, playing fields and play areas, and even graveyards. The most up to date information for planners and designers about people and their use of open spaces can be found on the Project for Public Spaces, New York website (http://www.pps.org [verified 18 May 2010]). It provides the most comprehensive information on making places that people will enjoy using, including many examples of greenspace design and management, as well as urban open space projects throughout the United States and elsewhere. In Britain, government funded agencies have recently been issuing guidance to local authorities in relation to greenspaces. For instance, the Commission for Architecture and the Built Environment produced *Start with the Park: Creating Sustainable Urban Greenspaces in Areas of Housing Growth and Renewal* (CABE Space, 2005b), which gives guidance on planning and designing recreation spaces. Another national level report was *Guidance on Providing Accessible Natural Greenspace in Towns and Cities: A Practical Guide to Accessing the Resource and Implementing Local Standards for Provision*, which was prepared for English Nature by Harrison et al. (1995) and assessed by Pauleit et al. (2003) for its relevance to present greenspace planning issues.

Supporting the Local Economy

Both economic development and tourism are seen as having strong links with the quality of local greenspace. *Does Money Grow on Trees?* by CABE Space (2005a) deals with the economics of some greenspace issues. Land and property values have been shown to be influenced by the proximity to greenspace (Halleux, 2005). At a time of economic difficulty in the 1970s and 1980s in the UK and elsewhere, job creation schemes were particularly successful in using greenspaces as a focus, and if properly organized, such schemes could again be used to good effect to enhance the green structure during the present economic downturn. There are also many long-term jobs directly related to making and managing greenspaces in any city. Greenspace needs to be planned to act as a buffer for agricultural land outside the city, so that farming can be as efficient and economically viable as possible (Gallent et al., 2004).

Supporting the Other Infrastructures that Service City Life

The green structure also has to carry parts of the transportation infrastructure—railways, highways, local roads, cycle-ways, and waterways; the energy infrastructure—electricity, gas, and oil supply systems; the drinking water infrastructure—the pipes and holding facilities; and the sewerage infrastructure—sewage processing in big city facilities or in the small units now favored in parts of Europe as a more sustainable water management solution (Sieker et al., 2006), and also holding and absorbing the solid dried wastes from sewage works.

All of the above functions relating to greenspaces in cities and to green structure indicate how important it is that it is properly planned for in the future. Its neglect has left many cities with greater environmental problems than they might otherwise have had to face. In the past each greenspace was often designated, planned, designed, financed, and maintained as if it were single use. This made it difficult for an understanding to develop of how greenspaces could be involved in making a city more sustainable. City form and greenspace are totally intertwined, and to plan one without the other can only create more unsustainable development and environmental problems. National, regional, and local planning systems need to be reorganized if this is to happen rather than just be talked about.

Funding Green Structure

No country seems to have worked out how best to fund urban green structure. It is clear that implementation of a green structure through the process of green infrastructure planning will necessitate crossing the long-established boundaries between different professional responsibilities and different local government departments; it is potentially a minefield unless there are clear guidelines and an agreed end product. Well-planned multifunctional green structure projects are not cheap, but the gains of sustainability from their implementation could well outweigh the costs of continuing in the present way. Many funding agencies will be involved, and strong financial controls will need to be in place, but creating multifunctional spaces can result in substantial savings for individual agencies, thus reducing the cost to society. Multifunctional spaces should lead to substantial benefits for local people in terms of health and welfare in particular and in people's general perception of an improved quality of life in urban areas.

Green Structure and Urban Planning—Examples from Europe

Whether the present greenspace structure of any particular city can effectively fulfill all the potential environmental functions required to enhance sustainability depends on making the best use of the multifunctionality of greenspace. This is a land use planning and design problem. Of particular interest are the links between green structure, its planning, and ecology (Jongman and Pungetti, 2004) and how the landscape of each part of the greenstructure should be designed or conserved and how it should be managed. The COST C11 Action (2005), a European Union funded research project that undertook investigations into green structure and urban planning in 15 European countries from 2000 to 2005, took this problem, which is at the heart of how biodiversity and the planning processes can relate, as its main focus. The team found many similarities in these countries relating to concerns for the future of cities and the part that might be played by greenspace in enhancing sustainability, and by extension, biodiversity. In studying cities across Europe they identified cultural and historical differences that had influenced urban form and saw how this influenced the present green structure. The role played by the planning processes in each country was also crucial, as were the mechanisms for paying for anything related to greenspace provision and management. Other important factors the team saw influencing the pattern of green structure were variations in the local climate, and other natural and physical environmental factors (Werquin et al., 2005).

The COST C11 team considered that adopting an ecological emphasis to green structure planning and management could contribute to making the urban environment an attractive and healthy place to live (Pauleit, 2005). The team recognized that the specific green structure of a city results from an interaction between natural and human processes through time (Pauleit and Kaliszuk, 2005). They proposed a data gathering process to enable an understanding to develop the distinctive green structure pattern of each city and saw this as a precursor to making plans for the future green structure of any city. A first requirement for any team responsible for developing a locally applicable approach to ecologically sound green infrastructure planning is to understand the present pattern, of both the built environment and the greenspaces, and the reasons for it and the pressures for change (Hendrix et al., 1988) and then to develop as good an understanding of the present situation as the data allow with regard to biodiversity across a city, before moving to an analysis stage to identify the potential inherent in the present situation to create a more sustainable green structure.

Before moving to an analysis and planning stage it is necessary to gather sufficient information to understand:

The Pre-Urban Landscape

The pre-urban landscape includes the landform and the water courses and bodies, the forests and woodlands, meadows, pastures, and arable land that occupied the area before urbanization. As an example, the research project showed that the topography, woodlands, wetlands, and shoreline strongly influenced the pattern of greenspace within Oslo and Helsinki (Nyhuus, 2005; Vähä-Piikkiö and Maijala, 2005); both these cities have grown out into forest and along the seashore. As the urban area grew, the green fingers of linear open space that had developed within the city boundaries followed the water courses and also encompassed areas of rocky outcrops and other unbuildable land. In both cities any further expansion will involve careful management of the forest–builtup area interface. In contrast, cities such as Munich, Utrecht, and Vienna expanded into preexisting open farmland, and only a few greenspace corridors were formed along rivers and their associated flood plains. In these cities almost all buildable land was urbanized, with the exception of that preserved for it historic and cultural importance, and only a few scattered remnants of the preexisting landscape can now be seen (Pauleit, 2005; Erhart, 2005).

The Urban Layer

Understanding the urban layer includes mapping the existing greenspaces and classifying them by use and their biodiversity. Nature does not discriminate in its

use of greenspace, so all greenspaces (taken to be unsealed land capable of supporting plants) are mapped and categorized, whether private domestic gardens
or greenspaces, commercial or institutional, or abandoned (derelict). The pattern
of greenspace within the city is in most cases related to the phases in the historical development of the particular city and the extent to which densification of
the city core had taken place. The differences in biodiversity found across a city's
greenspaces are very variable, since through time so many factors, both natural
and those influenced by people, can have affected the land and the habitats that
have developed on it. As an example, much is known about the biodiversity of
Munich (Pauleit, 2005) from biotype mapping and subsequent studies on individual habitat types undertaken in the 1980s (Duhme and Pauleit, 2000). This has
improved our understanding of the benefits of the city's green structure for biodiversity and the environment in general.

The Infrastructure Servicing the City

Roads, railway lines, canals, and even the defensive walls of some old cities have
been important influences on their patterns of growth. In some cities these structures are now associated with linear public greenspace (Utrecht and Vienna),
whereas in others the land beside them has been left and nature has reclaimed it,
so that wildlife corridors have formed in what would otherwise be hostile environments. There are major problems for wildlife when their natural movements
are blocked by traffic, and this is another factor for green structure planners to
take into account by working with ecologists.

The Ownership of Green Structure

Knowing who owns and undertakes the landscape management of each area of
greenspace is vital to developing and testing future greenspace planning scenarios. There are great variations between cities. For instance, More than 70% of
Helsinki is owned by the city, including most greenspace, whereas in Munich
only 10% is owned by the city.

Pressures

The COST C11 project (Werquin et al., 2005) identified a wide range of pressures
on urban green structure (Pauleit and Kaliszuk, 2005):

- The low quantity of greenspace and the fragmentation of the existing greenspaces. In any future replanning of a city's green structure the challenges are
  how greenspace can be introduced where there is little or none and how can
  this effort be funded, if it is to be used to help solve urban problems such as
  air pollution, noise, flash flooding, and low biodiversity.

- The loss of greenspace and associated biodiversity. This is due partly to
  pressure to make cities more compact, a policy which was introduced in
  an attempt to reduce energy consumption by lessening the transportation
  problems. As an example, in Britain unused spaces within a city have been
  designated for building because the government policy has encouraged
  densification in existing residential areas (Office of the Deputy Prime Minister, 2005a). This policy also enables developers to build on private garden
  land, and given that such land is particularly rich in biodiversity (Gaston
  et al., 2005), this policy is problematic. As densification of gardens is associated with a greater area of land being sealed, flash flooding problems
  increase, which has added to the environmental problems of cities. The

question then is how can a planning process be developed which acknowledges that environmental factors are as important as economic and social factors in decision making?

- The loss of protected cultural and historic landscape and protected habitats within the green structure continues. In relation to ecology, the loss is often due to inappropriate management regimes or changes in the landscape around the protected site, such as allowing building up to the boundary of a protected area. Here the question is how to ensure the survival of existing biodiversity through proper land and landscape management of the green structure. In relation to cultural and historic landscapes, the problem is often inappropriate management regimes.

- The changes in landscape quality can be perceived in the surrounding rural landscapes as cities grow. Much rural fringe agricultural landscape is already degraded and its areas of ecological value fragmented by modern agricultural practice. The question of how far out from a city the urban green structure should reach if it is going to reverse these trends is going to be important—whether green infrastructure should become a green belt limiting growth or be structured to allow continuing growth.

Green Infrastructure Planning—Examples from England

In many cases the present green structure will be inadequate for all the tasks described here; this is why green infrastructure planning is now being proposed for adoption to guide the holistic planning of the green structure of urban areas. The approaches being taken in the UK are well described in a broad ranging paper by Roe and Mell (2007), "Green Infrastructure and Landscape Planning: Collaborative Projects in the North East of England."

In the last 4 yr, several studies have been undertaken in the UK at regional and city region level to investigate how the use of GIS can aid the development of green infrastructure plans. Two are highlighted here, one studying the Manchester city region and the other the northeast region of England. John Handley's team at the University of Manchester has worked to develop a methodology capable of dealing with the range of problems found in relation to the landscape and greenspace planning of old industrial cities and their city regions. Their methodology for dealing with a very complex builtup urban landscape and the findings of that study are well described in an article, "Characterising the Urban Environment of UK Cities and Towns" (Gill et al., 2008). Maggie Roe of the University of Newcastle worked with a team set up by the North East Community Forest organization (Davies et al., 2006) to develop a holistic approach to landscape and green structure planning over a large area, including many varying sized settlements (Roe and Mell, 2007; Davies et al., 2006). These approaches are very different, but both are based on developing a GIS to store the required data and, after developing appropriate models, to examine the impact of various scenarios.

Referencing "the latest" work in any professional subject area is difficult, as few papers are formally published (even those commissioned at national government level). Often they are issued instead as digital reports and contained on government department websites. They are, however, so influential they cannot be ignored. Several reports have been commissioned relating to green infrastructure and London's particular growth problems. The first was *Creating Sustainable Communities: Greening the Thames Gateway Implementation Plan* (Office of the Deputy Prime Minister, 2005b) and more recently that on "ecotowns" by the Department of Communities and Local Government (2009). Another useful source of information in this very fast expanding field is Natural England, a governmental agency set up by central government with responsibility "to conserve and enhance the natural environment, for its intrinsic value, the well-being and enjoyment of people and the economic prosperity that it brings." This agency has the most direct responsibility for developing ideas on green infrastructure. It regularly publishes its work online, and the latest green infrastructure publications relating to developments in England can be found at http://www.naturalengland. org.uk (verified 20 May 2010). CABE is another useful source of information on sustainability in urban areas (http://www.cabe.org.uk, verified 20 May 2010).

Conclusions

The development of ideas and concepts relating to environmental conservation and management, greenways (Ahern 1995, 2007; Walmsley, 2006), landscape planning (McHarg, 1969; Hough, 1984; Fabos, 1995; Benson and Roe, 2007), and landscape ecology (Steiner, 2006), plus the more recent realization of the importance of understanding greenspace as multifunctional, have all provided the basis for the development of the concepts involved in using green infrastructure as an urban planning tool.

By the early 2000s the concept of green infrastructure planning had spread to the UK from the United States and Europe. It is now increasingly understood as a mechanism for coping with the application of government policies developed for regional and city levels with the intention of making urban areas better places to live, work, and play. It is seen as a holistic mechanism for a more effective greenspace planning (CABE Space, 2003; Natural England, 2009). In particular, England's Community Forest organizations, the Countryside Agency, and Natural England have helped to promote the idea of green infrastructure planning, and there are now a number of practitioners and partner organizations who are willing to champion the concept.

As Mell (2007) suggested, "as an appropriate mechanism for sustainable places, green infrastructure is not a quick fix solution but should be viewed as part of a long-term process of developing livable spaces. Designed appropriately and developed with ecological, economic and social factors in mind green infrastructure can be a valuable component of the urban form for successful renewal. In the same manner that communications, housing or transport infrastructures cannot develop better places to live individually, together this is possible."

Acknowledgments

The ideas expressed here owe much to the discussions over two decades between members of a European research group. The group initially came together because of an interest in the green structure planning undertaken in Breda in the late 1980s under the guidance of landscape planner Sjeff Langevelt. It then met in various countries to discuss the issues of how ecologically sound planning might influence city form. Between 2000 and 2005 the research group was funded by the EU-COST Action C11 to investigate the link between green structure and urban planning. The members of the COST team were: B. Duhem (Chairman), and A.C. Werquin (France); E. Erhart and K. Wagner (Austria); A. Van Herzele and P. Hanocq and J.M. Halleux (Belgium); I. Hanouskova (the Czech Republic), K. Attwell and U. Reeh and S. Guldager (Denmark); K. Lapinte and O. Maijala and I. Vähä-Piikkiö and M. Eronen (Finland); B. Oppermann, S. Pauleit (Germany); L. Martincigh, M.

Meriggi and G. Scudo (Italy); K. Zaleckis (Lithuania); S. Tjallingii and M. Buizer and C. Aalbers and P. Schildwacht and J. Tatenhove (the Netherlands); S. Nyhuus and U. Elfsen and K. Jorgensen and E. Plathe and G. Grundt (Norway); E. Kaliszuk and B. Szulczewska (Poland); J.M. Chapa (Spain); B. Malbert and K. Bjornberg and G. Lindholm and A. Stahle and P. Grahn (Sweden); A. Beer and C. Harrison and P. Draper (UK).

References

Ahern, J. 1995. Greenways and urban planning strategy. Landsc. Urban Plan. 33:131–155.

Ahern, J. 2007. Green infrastructure for cities: The spatial dimension. Proc. of the 3rd Fábos Landscape Planning and Greenways Symposium. Univ. of Massachusetts, Amherst.

Arnfield, A.J. 2003. Two decades of urban climate research: A review of turbulence, exchanges of energy and water, and the urban heat island. Int. J. Climatol. 23:1–26.

Beer, A. 2000. Spaces for people. p. 143–211. In A.R. Beer and C. Higgins (ed.) Environmental planning for site development. 2nd ed. Taylor and Francis, London.

Beer, A. 2005. The green structure of Sheffield. p. 40–51. In Green structure and urban planning. Office for Official Publications of the European Communities, Luxembourg.

Beer, A., T. Delshammer, and P. Schildwacht. 2003. A changing understanding of the role of greenspace in high-density housing: A European perspective. Built Environ. 29(2):132–143.

Beer, A.R., and C. Higgins (ed.) Environmental planning for site development. 2nd ed. Taylor and Francis, London.

Benson, J., and M. Roe (ed.) 2007. Landscape and sustainability. 2nd ed. Spon Press, London.

Brundtland, G. (ed.) 1987. Our common future: World Commission on Environment and Development. Oxford Univ. Press, Oxford, UK.

Burgess, J., C. Harrison, and M. Limb. 1988a. People, parks and the urban green: A study of popular meanings and values for open spaces. Urban Stud. 25:455–473.

Burgess, J., M. Limb, and C. Harrison. 1988b. Exploring environmental values through the medium of small groups: Theory and practice. Environ. Plan. 20:309–326.

CABE Space. 2003. Planning green infrastructure. CABE Space, London.

CABE Space. 2005a. Does money grow on trees? Available at http://www.cabe.org.uk/files/does-money-grow-on-trees.pdf (verified 20 May 2010). CABE Space, London.

CABE Space. 2005b. Start with the park: Creating sustainable urban green spaces in areas of housing growth and renewal. CABE Space, London.

City of Oslo. 2003. Surveying natural habitats and biological diversity and classifying their value. Oslo.

COST C11 Action. 2005. Green structure and urban planning, EU-COST Action C11 final report. Office for Official Publications of the European Communities, Luxembourg.

Cooper Marcus, C., and C. Francis (ed.) 1997. People places: Design guidelines for urban open space. 2nd ed. Van Nostrand Reinhold, New York.

Cranz, G. 1989. The politics of park design—A history of urban parks in America. MIT Press, Cambridge, MA.

Davies, C., C. McGloin, R. MacFarlane, and M. Roe. 2006 Green Infrastructure Planning Guide Project. Final Report to Countryside Agency, English Nature, Forestry Commission and Groundwork, London.

Department of Communities and Local Government. 2006. Planning Policy Statement 25: Development and flood risk. Available at http://www.communities.gov.uk/publications/planningandbuilding/pps25floodrisk (verified 6 Jue 2010).

Department of Communities and Local Government. 2009. Planning Policy Statement: Ecotowns a supplement to Planning Policy Statement 1. Available at http://www.communities.gov.uk/publications/planningandbuilding/pps-ecotowns (verified 6 Jue 2010).

Duhme, F., and S. Pauleit. 2000. Naturschutzprogramm für München, Landschaftsökologisches Rahmenkonzept. Geogr. Rundsch. 44:554–561.

Dunnet, N. 2007. Rain gardens: Managing water sustainably in garden and designed landscape. Timber Press, London.

Dunnet, N., and N. Kingsbury. 2008. Planting green roofs and living walls. Timber Press, London.

ECOTEC. 2006. City regions green infrastructure strategic planning: Raising the quality of the north's city regions. Available at http://www.thenorthernway.co.uk/page.asp?id=409. (verified 6 June 2010).

Erhart, E. 2005. Vienna. p. 200–205. In A.C. Werquin et al. (ed.) COST C11 Action final report, green structure and urban planning. Office for Official Publications of the European Communities, Luxembourg.

Ericson, D. 2006. Metrogreen: Connecting open space in North American cities. Island Press, Washington.

Fabos, J. 1995. Introduction and overview: The greenway movement, uses and potentials of greenways. Landsc. Urban Plan. 33:1–13.

Gallent, N., J. Anderson, M. Bianconi, and F. Osment. 2004. Vision for a sustainable, multifunctional rural–urban fringe. Final Rep. to the Countryside Agency (now Natural England). Bartlett School of Planning and LDA Design, London.

Gaston, K., R. Smith, K. Thompson, and P. Warren. 2005. Urban domestic gardens (II): Experimental tests of methods for increasing biodiversity. Biodiversity Conserv. 14:395–413.

Gill, S., J. Handley, R. Ennosb, S. Pauleit, N. Theuraya, and S. Lindley. 2008. Characterising the urban environment of UK cities and towns. Landsc. Urban Plan. 87:210–222.

Grahn, P. 2005. Eight experienced qualities in urban open spaces. p. 240–248. In A.C. Werquin et al. (ed.) COST C11 Action final report, green structure and urban planning. Office for Official Publications of the European Communities, Luxembourg.

Hahn, T., and E. Zeisel. 1988. Dezentrale Wasseraufbereitung, eine Pflanzenkläranlage in der Diskussion. p. 135–153. In Ökologe und Stadterneuerung, Deutscher Gemeundeverlag, Cologne.

Halleux, J. 2005. Valuing green structure, the use of hedonic models. p. 385–392. In A.C. Werquin et al. (ed.) COST C11 Action final report, green structure and urban planning. Office for Official Publications of the European Communities, Luxembourg.

Harrison, C., J.A. Millward, and G. Dawe. 1995. Accessible natural greenspace in towns and cities: A review of appropriate size and distance criteria. Guidance for the preparation of strategies for local sustainability, English Nature, London. Available at http://www.naturalengland.org.uk (verified 20 May 2010).

Hendrix, W., J. Fabos, and J. Price. 1988. An ecological approach to landscape planning using geographic information system technology. Landsc. Urban Plan. 15:211–225.

Hough, M. 1984. City form and natural process. Routledge, London.

Jongman, R., and G. Pungetti (ed.). 2004. Ecological networks and greenways; concept, design, implementation. Cambridge Univ. Press, New York.

Kaliszuk, E. 2005. Warsaw. p. 206–215. In A.C. Werquin et al. (ed.) COST C11 Action final report, green structure and urban planning. Office for Official Publications of the European Communities, Luxembourg.

Kaplan, R., and S. Kaplan. 1989. The experience of nature: A psychological perspective. Cambridge Univ. Press, New York.

London Wildlife Trust. 2008. Living London—London's Wildlife Trust's annual review 2007–2008. London.

Lundgren Alm, E. 1996. Stadsgrönskan: Integrerat eller separerat stadsbyggnadselement? Chalmers tekniska högskola. Gothenbourg.

McHarg, I. 1969. Design with nature. Natural History Press, Garden City, NY.

Mell, I. 2007. Green infrastructure planning: What are the costs for health and well-being? Environmental, cultural, economic and social sustainability. Int. J. Environ. Cultural Econ. Social Sustain. 3(5):117–124.

Mitchell, R., and F. Popham. 2009. Effect of exposure to natural environment on health inequalities: An observational population study. Lancet 9650:1655–1660.

National Audit Office, Office of the Deputy Prime Minister. 2006. Enhancing urban green space. Report by the Comptroller and Auditor General, The Stationery Office, London.

Natural England. 2009 Green infrastructure strategies: An introduction for local authorities and their partners. Available at http://www.naturalengland.org.uk (verified 18 May 2010).

Nyhuus, S. 2005. Oslo. p. 184–191. *In* A.C. Werquin et al. (ed.) COST C11 Action final report, green structure and urban planning. Office for Official Publications of the European Communities, Luxembourg.

Office of the Deputy Prime Minister. 2005a. Planning policy statement 1: Delivering sustainable development. HMSO, London.

Office of the Deputy Prime Minister. 2005b. Creating sustainable communities: Greening the Thames gateway implementation plan. HMSO, London Available at: http://www.communities.gov.uk/documents/thamesgateway/pdf/146685.pdf (verified 18 May 2010).

Pauleit, S. 2005. Munich. p. 177–183. *In* A.C. Werquin et al. (ed.) COST C11 Action final report, green structure and urban planning. Office for Official Publications of the European Communities, Luxembourg.

Pauleit, S., and E. Kaliszuk. 2005. Green structure patterns. p. 137–140. *In* A.C. Werquin et al. (ed.) COST C11 Action final report, green structure and urban planning. Office for Official Publications of the European Communities, Luxembourg.

Pauleit, S., P. Slinn, J. Handley, and S. Lindley. 2003. Promoting the natural greenstructure of towns and cities: English nature's accessible natural greenspace standards model. Built Environ. 29:157–170.

Piracha, A.L., and P.J. Marcotullio. 2003. Urban ecosystems analysis: Identifying tools and methods. United Nations Univ. Inst. of Ad. Studies (UNU/IAS), Tokyo.

Pouyat, R.V., K. Szlavecz, I.D. Yesilonis, P.M. Groffman, and K. Schwarz. 2010. Chemical, physical, and biological characteristics of urban soils. p. 119–152. *In* J. Aitkenhead-Peterson and A. Volder (ed.) Urban ecosystem ecology. Agron. Monogr. 55. ASA, CSSA, and SSSA, Madison, WI.

Randrup, T.B. 2006. Integrated green space planning and management. Urban For. Urban Green. 4:91.

Roe, M., and I. Mell. 2007. Green infrastructure and landscape planning: Collaborative projects in the north east of England. *In* Proceedings of the 18th Int. annual ECLAS Conf., Belgrade.

Rowe, D.B., and K.L. Getter. 2010. Green roofs and garden roofs. p. 391–412. *In* J. Aitkenhead-Peterson and A. Volder (ed.) Urban ecosystem ecology. Agron. Monogr. 55. ASA, CSSA, and SSSA, Madison, WI.

Sandström, U. 2008. Biodiversity and green infrastructure in urban landscapes. Swedish EIA Centre, Dep. of Urban and Rural Development, Swedish University of Agricultural Sciences, Sweden.

Sandström, U., P. Angelstam, and G. Mikuinski. 2006. Ecological diversity of birds in relation to the structure of urban green space. Landsc. Urban Plan. 77:39–53.

Scheffer, M., S. Carpenter, J. Foley, C. Folke, and B. Walker. 2001. Catastrophic shifts in ecosystems. Nature 413:591–596.

Sieker, H., S. Bandermann, M. Becker, and U. Raasch. 2006. Urban stormwater management demonstration projects in the Emscher Region. First SWITCH Scientific Meeting. Univ. of Birmingham, UK.

Spirn, A. 1980. The role of natural processes in the design of cities. Ann. Am. Acad. Pol. Soc. Sci. 451:98–105.

Spirn, A. 1984. The granite garden: Urban nature and human design. Basic Books, New York.

Steiner, F. 2002. Human ecology: Following nature's lead. Island Press, Washington.

Steiner, F. 2006. Landscape planning: A method applied to a growth management example. Springer, New York.

Tjallingii, S. 1995. Ecopolis: Strategies for ecologically sound urban development. Backhuys, Leiden.

Tjallingii, S., S. Spijker, and T.J.de Vries. 1995. Ecologisch Stadsbeheer. Rep. 163. Wageningen, the Netherlands.

Turner, T. 1992. Open space planning in London, from standards per 1000 to green strategy. Town Plan. Rev. 63:365–386.

UK Environment Agency. 2008. Sustainable drainage systems (SUDS) guidance. Available at http://www.environment-agency.gov.uk/business/sectors/39909.aspx (verified 18 May 2010).

United Nations Development Programme. 1996. Urban agriculture: Food, jobs and sustainable cities. UN Publication Series, Habitat II, Vol. 1. UNDP, New York.

UN-HABITAT. 2008. State of the world's cities 2008/2009—Harmonious cities. Earthscan Ltd., New York.

Vähä-Piikkiö, I., and O. Maijala. 2005. Helsinki. p. 163–169. *In* COST C11 Action final report, green structure and urban planning. Office for Official Publications of the European Communities, Luxembourg.

Verheij, R., J. Maas, and P. Groenewegen. 2008. Urban–rural health differences and the availability of green space. Eur. Urban Reg. Stud. 4:307–316.

Walker, B., A. Kinzig, and J. Langridge. 1999. Plants attribute diversity, resilience and ecosystem function: The nature and significance of dominant and minor species. Ecosystems 2:95–113.

Walmsley, A. 2006. Greenways: Multiplying and diversifying in the 21st century. Landsc. Urban Plan. 76:252–290.

Werquin, A., B. Duhem, G. Lindholm, B. Opperman, S. Pauleit, and S. Tjallingii. 2005. COST C11 Action final report, green structure and urban planning. Office for Official Publications of the European Communities, Luxembourg.

Woods Ballard, B., and R. Kellagher (ed.) 2007. The SUDS manual. C697. CIRIA Construction Industry Research, London.

Index